1990

ALLE ZEIT WACH
1842

NICOLAS BOURBAKI

ELEMENTS OF MATHEMATICS

General Topology

Chapters 1–4

Springer-Verlag
Berlin Heidelberg New York
London Paris Tokyo

Originally published as
ÉLÉMENTS DE MATHÉMATIQUE,
TOPOLOGIE GÉNÉRALE

Mathematics Subject Classification (1980): 54XX

Distribution rights worldwide:
Springer-Verlag Berlin Heidelberg New York London Paris Tokyo

ISBN 3-540-19374-X Springer-Verlag Berlin Heidelberg New York
ISBN 0-387-19374-X Springer-Verlag New York Berlin Heidelberg
2nd printing 1989

Library of Congress Cataloging-in-Publication Data
Bourbaki, Nicolas. [Topologie générale. English] General topology / Nicolas Bourbaki. p. cm.–(Elements of mathematics) Translation of: Topologie générale. Includes bibliographies and indexes.
ISBN 0-387-19374-X (U.S. : v. 1)
1. Topology. I. Title. II. Series: Bourbaki, Nicolas. Eléments de mathématique. English.
QA611.B65913 1988 514–dc 19 88-31206

Printed in France

Bookbinding: J. Schäffer, Grünstadt
2141/3140-543210

ADVICE TO THE READER

1. This series of volumes, a list of which is given on pages 9 and 10, takes up mathematics at the beginning, and gives complete proofs. In principle, it requires no particular knowledge of mathematics on the reader's part, but only a certain familiarity with mathematical reasoning and a certain capacity for abstract thought. Nevertheless, it is directed especially to those who have a good knowledge of at least the content of the first year or two of a university mathematics course.

2. The method of exposition we have chosen is axiomatic and abstract, and normally proceeds from the general to the particular. This choice has been dictated by the main purpose of the treatise, which is to provide a solid foundation for the whole body of modern mathematics. For this it is indispensable to become familiar with a rather large number of very general ideas and principles. Moreover, the demands of proof impose a rigorously fixed order on the subject matter. It follows that the utility of certain considerations will not be immediately apparent to the reader unless he has already a fairly extended knowledge of mathematics; otherwise he must have the patience to suspend judgment until the occasion arises.

3. In order to mitigate this disadvantage we have frequently inserted examples in the text which refer to facts the reader may already know but which have not yet been discussed in the series. Such examples are always placed between two asterisks : * ... *. Most readers will undoubtedly find that these examples will help them to understand the text, and will prefer not to leave them out, even at a first reading. Their omission would of course have no disadvantage, from a purely logical point of view.

4. This series is divided into volumes (here called " Books "). The first six Books are numbered and, in general, every statement in the text

assumes as known only those results which have already been discussed in the preceding volumes. This rule holds good within each Book, but for convenience of exposition these Books are no longer arranged in a consecutive order. At the beginning of each of these Books (or of these chapters), the reader will find a precise indication of its logical relationship to the other Books and he will thus be able to satisfy himself of the absence of any vicious circle.

5. The logical framework of each chapter consists of the *definitions*, the *axioms*, and the *theorems* of the chapter. These are the parts that have mainly to be borne in mind for subsequent use. Less important results and those which can easily be deduced from the theorems are labelled as "propositions," "lemmas", "corollaries", "remarks", etc. Those which may be omitted at a first reading are printed in small type. A commentary on a particularly important theorem appears occasionally under the name of "scholium".

To avoid tedious repetitions it is sometimes convenient to introduce notations or abbreviations which are in force only within a certain chapter or a certain section of a chapter (for example, in a chapter which is concerned only with commutative rings, the word "ring" would always signify "commutative ring"). Such conventions are always explicitly mentioned, generally at the beginning of the chapter in which they occur.

6. Some passages in the text are designed to forewarn the reader against serious errors. These passages are signposted in the margin with the sign

☡ ("dangerous bend").

7. The Exercises are designed both to enable the reader to satisfy himself that he has digested the text and to bring to his notice results which have no place in the text but which are nonetheless of interest. The most difficult exercises bear the sign ¶.

8. In general, we have adhered to the commonly accepted terminology, except where there appeared to be good reasons for deviating from it.

9. We have made a particular effort to always use rigorously correct language, without sacrificing simplicity. As far as possible we have drawn attention in the text to *abuses of language*, without which any mathematical text runs the risk of pedantry, not to say unreadability.

10. Since in principle the text consists of the dogmatic exposition of a theory, it contains in general no references to the literature. Bibliographical references are gathered together in *Historical Notes*, usually

at the end of each chapter. These notes also contain indications, where appropriate, of the unsolved problems of the theory.

The bibliography which follows each historical note contains in general only those books and original memoirs which have been of the greatest importance in the evolution of the theory under discussion. It makes no sort of pretence to completeness; in particular, references which serve only to determine questions of priority are almost always omitted.

As to the exercises, we have not thought it worthwhile in general to indicate their origins, since they have been taken from many different sources (original papers, textbooks, collections of exercises).

11. References to a part of this series are given as follows:

a) If reference is made to theorems, axioms, or definitions presented *in the same section*, they are quoted by their number.

b) If they occur *in another section of the same chapter*, this section is also quoted in the reference.

c) If they occur *in another chapter in the same Book*, the chapter and section are quoted.

d) If they occur *in another Book*, this Book is first quoted by its title.

The *Summaries of Results* are quoted by the letter R: thus *Set Theory*, R signifies "*Summary of Results of the Theory of Sets*".

CONTENTS

CONTENTS OF THE ELEMENTS OF MATHEMATICS SERIES

I. Theory of sets

1. Description of formal mathematics. 2. Theory of sets. 3. Ordered sets; cardinals; natural numbers. 4. Structures.

II. Algebra

1. Algebraic structures. 2. Linear algebra. 3. Tensor algebras, exterior algebras, symmetric algebras. 4. Polynomials and rational fractions. 5. Fields. 6. Ordered groups and fields. 7. Modules over principal ideal rings. 8. Semi-simple modules and rings. 9. Sesquilinear and quadratic forms.

III. General topology

1. Topological structures. 2. Uniform structures. 3. Topological groups. 4. Real numbers. 5. One-parameter groups. 6. Real number spaces, affine and projective spaces. 7. The additive groups $\mathbf{R}^n$. 8. Complex numbers. 9. Use of real numbers in general topology. 10. Function spaces.

IV. Functions of a real variable

1. Derivatives. 2. Primitives and integrals. 3. Elementary functions. 4. Differential equations. 5. Local study of functions. 6. Generalized Taylor expansions. The Euler-Maclaurin summation formula. 7. The gamma function. Dictionary.

V. Topological vector spaces

1. Topological vector spaces over a valued field. 2. Convex sets and locally convex spaces. 3. Spaces of continuous linear mappings.

VI. Integration

Lie groups and Lie algebras

Commutative algebra

Spectral theories

INTRODUCTION

Most branches of mathematics involve structures of a type different from the *algebraic* structures (groups, rings, fields, etc.) which are the subject of the book Algebra of this series: namely structures which give a mathematical content to the intuitive notions of *limit*, *continuity* and *neighbourhood*. These structures are the subject matter of the present book.

Historically, the ideas of limit and continuity appeared very early in mathematics, notably in geometry, and their role has steadily increased with the development of analysis and its applications to the experimental sciences, since these ideas are closely related to those of *experimental determination* and *approximation*. But since most experimental determinations are *measurements*, that is to say determinations of one or more *numbers*, it is hardly surprising that the notions of limit and continuity in mathematics were featured at first only in the theory of real numbers and its outgrowths and fields of application (complex numbers, real or complex functions of real or complex variables, Euclidean geometry and related geometries).

In recent times it has been realized that the domain of applicability of these ideas far exceeds the real and complex numbers of classical analysis (see the Historical Note to Chapter I). Their essential content has been extracted by an effort of analysis and abstraction, and the result is a tool whose usefulness has become apparent in many branches of mathematics.

In order to bring out what is essential in the ideas of limit, continuity and neighbourhood, we shall begin by analysing the notion of *neighbourhood* (although historically it appeared later than the other two). If we start from the physical concept of approximation, it is natural to say that a subset A of a set E is a neighbourhood of an element *a* of A if, whenever we replace *a* by an element that "approximates" *a*, this new element will also belong to A, provided of course that the "error" involved is small enough; or, in other words, if all the points of E which are "sufficiently near" *a* belong to A. This definition is meaningful whenever precision can be given to the concept of sufficiently small error or of an element sufficiently near another. In this direction, the first idea was to suppose that the "distance" between two elements can be measured by a (positive) real number. Once the "distance" between any two elements of a set has been defined, it is clear how the "neighbourhoods" of an element *a* should be defined : a subset will be a neighbourhood of *a* if it contains all elements whose distance from *a* is less than some preassigned strictly positive number. Of course, we cannot expect to develop an interesting theory from this definition unless we impose certain conditions or axioms on the "distance" (for example, the inequalities relating the distances between the three vertices of a triangle which hold in Euclidean geometry should continue to hold for our generalized distance). In this way we arrive at a vast generalization of Euclidean geometry. It is convenient to continue to use the language of geometry : thus the elements or a set on which a "distance" has been defined are called *points*, and the set itself is called a *space*. We shall study such spaces in Chapter IX.

So far we have not succeeded in freeing ourselves from the real numbers. Nevertheless, the spaces so defined have a great many properties which can be stated without reference to the "distance" which gave rise to them. For example, every subset which contains a neighbourhood of *a* is again a neighbourhood of *a*, and the intersection of two neighbourhoods of *a* is a neighbourhood of *a*. These properties and others have a multitude of consequences which can be deduced without any further recourse to the "distance" which originally enabled us to define neighbourhoods. We obtain statements in which there is no mention of magnitude or distance.

We are thus led at last to the general concept of a topological space, which does not depend on any preliminary theory of the real numbers. We shall say that a set E carries a *topological structure* whenever we have associated with each element of E, by some means or other, a family of subsets of E which are called *neighbourhoods* of this element — provided of course that these neighbourhoods satisfy certain conditions (the *axioms* of topological structures). Evidently the choice of axioms to be imposed is to some extent arbitrary, and historically has been the subject of a great deal of experiment (see the Historical Note to Chapter I). The system of axioms finally arrived at is broad enough for the present needs of mathematics, without falling into excessive and pointless generality.

A set carrying a topological structure is called a *topological space* and its elements are called *points*. The branch of mathematics which studies topological structures bears the name of *Topology* (etymologically, "science of place", not a particularly expressive name), which is preferred nowadays to the earlier (and synonymous) name of *Analysis situs*.

To formulate the idea of neighbourhood we started from the vague concept of an element "sufficiently near" another element. Conversely, a topological structure now enables us to give precise meaning to the phrase "such and such a property holds for all points *sufficiently near* a": by definition this means that the set of points which have this property is a neighbourhood of a for the topological structure in question.

From the notion of neighbourhood there flows a series of other notions whose study is proper to topology: the interior of a set, the closure of a set, the frontier of a set, open sets, closed sets, and so on (see Chapter I, § 1). For example, a subset A is an *open* set if, whenever a point a belongs to A, all the points sufficiently near a belong to A; in other words, if A is a neighbourhood of each of its points. The axioms for neighbourhoods have certain consequences for all these notions; for example, the intersection of two open sets is an open set (because we have supposed that the intersection of two neighbourhoods of a is a neighbourhood of a). Conversely, we can start from one of these derived notions instead of starting from the notion of a neighbourhood; for example, we may suppose that the open sets are known, and take as axioms the properties of the family of open sets (one of these properties has just been stated, by way of example). We can then verify that, from knowledge of the open sets, the neighbourhoods can be reconstructed; the axioms for neighbourhoods are now consequences of the new axioms for open sets that we took as a starting point. Thus a topological structure can be defined in various different ways which are basically equivalent. In this book we shall start from the notion of *open set*, because the corresponding axioms are the simplest.

Once topological structures have been defined, it is easy to make precise the idea of *continuity*. Intuitively, a function is continuous at a point if its value varies as little as we please whenever the argument remains sufficiently near the point in question. Thus continuity will have an exact meaning whenever the space of arguments and the space of values of the function are topological spaces. The precise definition is given in Chapter I, § 2.

As with continuity, the idea of a *limit* involves two sets, each endowed with suitable structures, and a mapping of one set into the other. For example, the limit of a sequence of real numbers a_n involves the set $\mathbf{N}$ of natural numbers, the set $\mathbf{R}$ of real numbers, and a mapping of the former set into the latter. A real number a is then said to be a limit of the sequence if, whatever neighbourhood V of a we take, this neighbourhood contains all the a_n except for a finite number of values of n; that is, if the set of natural numbers n for which a_n belongs to V is a subset of $\mathbf{N}$ whose complement is finite. Note that $\mathbf{R}$ is assumed to carry a topological structure, since we are speaking of neigbourhoods; as to the set $\mathbf{N}$, we have made a certain family of subsets play a particular part, namely those subsets whose complement is finite. This is a general fact: whenever we speak of limit, we are considering a mapping f of a set E into a topological space F, and we say that f has a point a of F as a limit if the set of elements x of E whose image $f(x)$ belongs to a neighbourhood V of a [this set is just the "inverse image" $\overset{-1}{f}(V)$] belongs, whatever the neighbourhood V, to a certain family $\mathfrak{F}$ of subsets of E, given beforehand. For the notion of limit to have the essential properties ordinarily attributed to it, the family $\mathfrak{F}$ must satisfy certain axioms, which are stated in Chapter I, § 6. Such a family $\mathfrak{F}$ of subsets of E is called a *filter* on E. The notion of a filter, which is thus inseparable from that of a limit, appears also in other contexts in topology; for example, the neighbourhoods of a point in a topological space form a filter.

The general study of all these notions is the essential purpose of Chapter I. In addition, particular classes of topological spaces are considered there, spaces which satisfy more restrictive axioms, or spaces obtained by particular procedures from other given spaces.

As we have already said, a topological structure on a set enables one to give an exact meaning to the phrase "whenever x is sufficiently near a, x has the property $P\{x\}$". But, apart from the situation in which a "distance" has been defined, it is not clear what meaning ought to be given to the phrase "every pair of points x, y which are sufficiently near each other has the property $P\{x, y\}$", since *a priori* we have no means of comparing the neighbourhoods of two different points. Now the notion of a pair of points near to each other arises frequently in classical analysis (for example, in propositions which involve uniform continuity). It is therefore important that we should be able to give a precise meaning to this notion in full generality, and we are thus led to define structures which are richer than topological structures, namely *uniform structures*. They are the subject of Chapter II.

The other chapters of this Book are devoted to questions in which, in addition to a topological or uniform structure, there is some other structure present. For example a *group* which carries a suitable topology (compatible in a certain sense with the group structure) is called a *topological group*. Topological groups are studied in Chapter III, and we shall see there in particular how every topological group can be endowed with certain uniform structures.

In Chapter IV we apply the preceding principles to the field of rational numbers. This enables us to define the field of real numbers; because of its importance, we study it in considerable detail. In the succeeding chapters, starting from the real numbers, we define certain topological spaces which are of particular interest in applications of topology to classical geometry : finite-dimensional vector spaces, spheres, projective spaces, etc. We consider also certain topological groups closely related to the additive group of real numbers, which we characterize axiomatically, and this leads us to the definition and elementary properties of the most important functions of classical analysis: the exponential, logarithmic and trigonometric functions.

In Chapter IX we revert to general topological spaces, but now with a new instrument, namely the real numbers, at our disposal. In particular we study spaces whose topology is defined by means of a "distance"; these spaces have properties, some of which cannot be extended to more general spaces. In Chapter X we study sets of mappings of a topological space into a uniform space (function spaces); these sets, suitably topologized, have interesting properties which already play an important part in classical analysis.

CHAPTER 1

Topological Structures

1. OPEN SETS, NEIGHBOURHOODS, CLOSED SETS

1. OPEN SETS

DEFINITION 1. *A topological structure* (or, more briefly, a *topology*) *on a set* X *is a structure given by a set* $\mathfrak{O}$ *of subsets of* X, *having the following properties* (called *axioms of topological structures*) :

(O_I) *Every union of sets of* $\mathfrak{O}$ *is a set of* $\mathfrak{O}$.

(O_{II}) *Every finite intersection of sets of* $\mathfrak{O}$ *is a set of* $\mathfrak{O}$.

The sets of $\mathfrak{O}$ *are called open sets of the topological structure defined by* $\mathfrak{O}$ *on* X.

DEFINITION 2. *A topological space is a set endowed with a topological structure.*

The elements of a topological space are often called *points.* When a topology has been defined on a set X, this set is said to be the set *underlying* the topological space X.

Axiom (O_I) implies in particular that the union of the empty subset of $\mathfrak{O}$, i.e. the *empty set* belongs to $\mathfrak{O}$. Axiom (O_{II}) implies that the intersection of the empty subset of $\mathfrak{O}$, i.e. *the set* X, belongs to $\mathfrak{O}$.

To show that a set $\mathfrak{O}$ of subsets of X satisfies (O_{II}), it is often convenient to prove separately that it satisfies the following two axioms, whose conjunction is equivalent to (O_{II}) :

$(O_{II\,a})$ *The intersection of two sets of* $\mathfrak{O}$ *belongs to* $\mathfrak{O}$.

$(O_{II\,b})$ X *belongs to* $\mathfrak{O}$.

Examples of topologies. Given any set X, the set of subsets of X consisting of X and $\emptyset$ satisfies axioms (O_I) and (O_{II}) and therefore defines a topology on X. So does the set $\mathfrak{P}(X)$ of all subsets of X: the topology it defines is the *discrete topology* on X, and the set X with this topology is called a *discrete space.*

A *covering* $(U_\iota)_{\iota \in I}$ of a subset A of a topological space X is said to be *open* if all the U_ι are open in X.

Definition 3. *A homeomorphism of a topological space* X *onto a topological space* X′ *is an isomorphism of the topological structure of* X *onto that of* X′; that is to say, in accordance with the general definitions *a bijection of* X *onto* X′ *which transforms the set of open subsets of* X *into the set of open subsets of* X′.

X and X′ are said to be *homeomorphic* if there is a homeomorphism of X onto X′.

Example. If X and X′ are two discrete spaces, any bijection of X onto X′ is a homeomorphism.

The following criterion follows immediately from the definition of a homeomorphism: *for a bijection* f *of a topological space* X *onto a topological space* X′ *to be a homeomorphism, it is necessary and sufficient that the image under* f *of each open set in* X *is an open set in* X′, *and that the inverse image under* f *of each open set in* X′ *is an open set in* X.

2. NEIGHBOURHOODS

Definition 4. *Let* X *be a topological space and* A *any subset of* X. *A neighbourhood of* A *is any subset of* X *which contains an open set containing* A. *The neighbourhoods of a subset* $\{x\}$ *consisting of a single point are also called neighbourhoods of the point* x.

It is clear that every neighbourhood of a subset A of X is also a neighbourhood of each subset $B \subset A$; in particular, it is a neighbourhood of each point of A. Conversely, suppose A is a neighbourhood of each of the points of a set B, and let U be the union of the open sets contained in A; then $U \subset A$, and since each point of B belongs to an open set contained in A, we have $B \subset U$; but U is open by virtue of (O_I), hence A is a neighbourhood of B. In particular:

Proposition 1. *A set is a neighbourhood of each of its points if and only if it is open.*

The everyday sense of the word "neighbourhood" is such that many of the properties which involve the mathematical idea of neighbourhood appear as the mathematical expression of intuitive properties; the choice of this term thus has the advantage of making the language more expressive. For this purpose it is also permissible to use the expressions "sufficiently near" and "as near as we please" in some statements. For example, Proposition 1 can be stated in the following form: a set A is open if and only if, for each $x \in A$, all the points *sufficiently near* x belong to A. More generally, we shall say that a property holds for all points *sufficiently near* a point x, if it holds at all points of some neighbourhood of x.

Let us denote by $\mathfrak{B}(x)$ the set of all neighbourhoods of x. The sets $\mathfrak{B}(x)$ have the following properties:

(V_I) *Every subset of* X *which contains a set belonging to* $\mathfrak{B}(x)$ *itself belongs to* $\mathfrak{B}(x)$.

(V_{II}) *Every finite intersection of sets of* $\mathfrak{B}(x)$ *belongs to* $\mathfrak{B}(x)$.

(V_{III}) *The element* x *is in every set of* $\mathfrak{B}(x)$.

Indeed, these three properties are immediate consequences of Definition 4 and axiom (O_{II}).

(V_{IV}) *If* V *belongs to* $\mathfrak{B}(x)$, *then there is a set* W *belonging to* $\mathfrak{B}(x)$ *such that, for each* $y \in W$, V *belongs to* $\mathfrak{B}(y)$.

By Proposition 1, we may take W to be any open set which contains x and is contained in V.

This property may be expressed in the form that *a neighbourhood of* x *is also a neighbourhood of all points sufficiently near to* x.

These four properties of the sets $\mathfrak{B}(x)$ are *characteristic*. To be precise, we have:

PROPOSITION 2. *If to each element* x *of a set* X *there corresponds a set* $\mathfrak{B}(x)$ *of subsets of* X *such that the properties* (V_I), (V_{II}), (V_{III}) *and* (V_{IV}) *are satisfied, then there is a unique topological structure on* X *such that, for each* $x \in X$, $\mathfrak{B}(x)$ *is the set of neighbourhoods of* x *in this topology.*

By Proposition 1, if there is a topology on X satisfying these conditions, the set of open sets for this topology is necessarily the set $\mathfrak{D}$ of subsets A of X such that *for each* $x \in A$ *we have* $A \in \mathfrak{B}(x)$; hence the *uniqueness* of this topology if it exists.

The set $\mathfrak{D}$ certainly satisfies axioms (O_I) and (O_{II}): for (O_I), this follows immediately from (V_I), and for (O_{II}), from (V_{II}). It

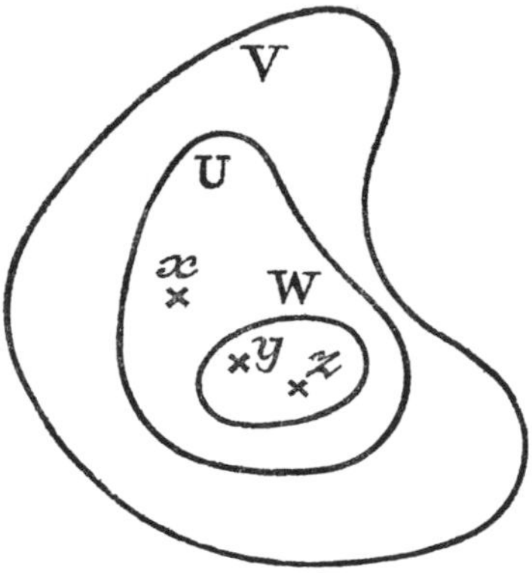

Figure 1.

remains to show that, in the topology defined by $\mathfrak{O}$, $\mathfrak{B}(x)$ is the set of neighbourhoods of x for each $x \in X$. It follows from (V_I) that every neighbourhood of x belongs to $\mathfrak{B}(x)$. Conversely, let V be a set belonging to $\mathfrak{B}(x)$, and let U be the set of points $y \in X$ such that $V \in \mathfrak{B}(y)$; if we can show that $x \in U$, $U \subset V$ and $U \in \mathfrak{O}$, then the proof will be complete. We have $x \in U$ since $V \in \mathfrak{B}(x)$; also $U \subset V$, for every point $y \in U$ belongs to V by reason of (V_{III}) and the hypothesis $V \in \mathfrak{B}(y)$. It remains to show that $U \in \mathfrak{O}$, i.e. that $U \in \mathfrak{B}(y)$ for each $y \in U$; now (Fig. 1) if $y \in U$ then by (V_{IV}) there is a set W such that for each $z \in W$ we have $V \in \mathfrak{B}(z)$; since $V \in \mathfrak{B}(z)$ means that $z \in U$, it follows that $W \subset U$, and therefore, by (V_I), that $U \in \mathfrak{B}(y)$.
Q.E.D.

Proposition 2 shows that a topology on X can be defined by means of the sets $\mathfrak{B}(x)$ of neighbourhoods of points of X, subject only to the axioms (V_I), (V_{II}), (V_{III}) and (V_{IV}).

Example. We may define a topology on the set $\mathbf{Q}$ of rational numbers by taking for open sets all *unions of bounded open intervals*; the set of these subsets certainly satisfies (O_I), and to see that it satisfies (O_{II}) it is enough to remark that if the intersection of two open intervals $]a, b[$ and $]c, d[$ is not empty, then it is the interval $]\alpha, \beta[$, where $\alpha = \sup(a, c)$ and $\beta = \inf(b, d)$. We get the same topology by defining for each $x \in \mathbf{Q}$ the set $\mathfrak{B}(x)$ of neighbourhoods of x to be the *set of subsets containing an open interval to which* x *belongs*. The topological space obtained by assigning this topology to $\mathbf{Q}$ is called the *rational line* (cf. Chapter IV, § 1, no. 2). Notice that in this space every open interval is an open set. * We can define a topology on the set $\mathbf{R}$ of real numbers in the same way; $\mathbf{R}$ with this topology is called the *real line* (cf. § 2, Exercise 5 and Chapter IV, § 1, no. 3). *

3. FUNDAMENTAL SYSTEMS OF NEIGHBOURHOODS; BASES OF A TOPOLOGY

DEFINITION 5. *In a topological space* X, *a fundamental system of neighbourhoods of a point* x (resp. *of a subset* A *of* X) *is any set* $\mathfrak{S}$ *of neighbourhoods of* x (resp. A) *such that for each neighbourhood* V *of* x (resp. A) *there is a neighbourhood* $W \in \mathfrak{S}$ *such that* $W \subset V$.

If $\mathfrak{S}$ is a fundamental system of neighbourhoods of a subset A of X, then every finite intersection of sets of $\mathfrak{S}$ contains a set of $\mathfrak{S}$.

Examples. 1) In a *discrete* space (no. 1) the set $\{x\}$ alone constitutes a fundamental system of neighbourhoods of the point x.

2) On the rational line **Q** the set of all open intervals containing a point x is a fundamental system of neighbourhoods of this point. So is the set of open intervals $]x - 1/n, x + 1/n[$, and the set of closed intervals $[x - 1/n, x + 1/n]$, where n runs through all the integers > 0, or through any infinite strictly increasing sequence of integers > 0. * There are analogous results for the real line. *

DEFINITION 6. *A base of the topology of a topological space* X *is any set* $\mathfrak{B}$ *of open subsets of* X *such that every open subset of* X *is the union of sets belonging to* $\mathfrak{B}$.

PROPOSITION 3. *If* X *is a topological space, then for a set* $\mathfrak{B}$ *of open subsets of* X *to be a base of the topology of* X *it is necessary and sufficient that for each* $x \in X$ *the set of* $V \in \mathfrak{B}$ *such that* $x \in V$ *is a fundamental system of neighbourhoods of* x.

It is clear that the condition is necessary. Conversely, if it is satisfied, then given any open set U and any $x \in U$ there is an open set $V_x \in \mathfrak{B}$ such that $x \in V_x \subset U$. The union of the sets V_x for $x \in U$ is therefore equal to U. This completes the proof.

Examples. 1) The discrete topology has as a base the set of subsets of X which consist of a single point.

2) The set of bounded open intervals is by definition a base of the topology of the rational line (no. 2). * Likewise, the set of bounded open intervals is a base of the topology of the real line. *

4. CLOSED SETS

DEFINITION 7. *In a topological space* X, *the complements of the open sets of* X *are called closed sets.*

Taking complements, we find that the axioms (O_I) and (O_{II}) take the following form:

$(\mathbf{O}'_I)$ *Every intersection of closed sets is a closed set.*

$(\mathbf{O}'_{II})$ *Every finite union of closed sets is a closed set.*

The empty set and the whole space X are closed (and therefore *both open and closed*; cf. § 11).

On the rational line, every interval of the form $[a, \rightarrow[$ is a closed set, for its complement $]\leftarrow, a[$ is open; likewise, every interval of the form $]\leftarrow, a]$ is a closed set; hence so is every bounded closed interval $[a, b]$, since it is the intersection of the intervals $[a, \rightarrow[$ and $]\leftarrow, b]$.

The set $\mathbf{Z}$ of rational integers is closed in the rational line, since its complement $\bigcup_{n \in \mathbf{Z}}]n, n+1[$ is open.

A *covering* $(F_\iota)_{\iota \in I}$ of a subset A of a topological space X is said to be *closed* if each of the F_ι is closed in X.

A *homeomorphism* f of a topological space X onto a topological space X′ (no. 1) can be characterized as a bijection of X onto X′ *such that the image under* f *of every closed subset of* X *is a closed subset of* X′ *and the inverse image under* f *of every closed subset of* X′ *is a closed subset of* X.

5. LOCALLY FINITE FAMILIES

DEFINITION 8. *A family* $(A_\iota)_{\iota \in I}$ *of subsets of a topological space* X *is said to be locally finite if for each* $x \in X$ *there is a neighbourhood* V *of* x *such that* $V \cap A_\iota = \varnothing$ *for all but a finite number of indices* $\iota \in I$. *A set* $\mathfrak{S}$ *of subsets of* X *is said to be locally finite if the family of subsets defined by the identity map of* $\mathfrak{S}$ *onto itself is locally finite.*

It is clear that if $(A_\iota)_{\iota \in I}$ is a locally finite family of subsets and if $B_\iota \subset A_\iota$ for each $\iota \in I$, then the family $(B_\iota)_{\iota \in I}$ is locally finite.

Every *finite* family of subsets of a topological space X is obviously locally finite; the converse is not true in general.

* For example, in $\mathbf{R}$, the open covering formed by the interval $]\leftarrow, 1[$ and the intervals $]n, \rightarrow[$ for each integer $n \geqslant 0$ is locally finite; and each interval $]n, \rightarrow[$ meets an infinite number of sets of this covering. *

PROPOSITION 4. *The union of a locally finite family of closed subsets of a topological space* X *is closed in* X.

Let $(F_\iota)_{\iota \in I}$ be a locally finite family of closed subsets of X, and suppose that $x \in X$ does not belong to $F = \bigcup_{\iota \in I} F_\iota$; then x has a neighbourhood V which meets only those F_ι whose indices belong to a *finite* subset J of I. For each $\iota \in J$ let U_ι be the complement of F_ι; then $\complement F$ contains the set $V \cap \bigcap_{\iota \in J} U_\iota$, which is a neighbourhood of x since each U_ι is open and contains x. Hence, by Proposition 1 of no. 2, $\complement F$ is open and therefore F is closed in X.

We note that the union of an *arbitrary* family of closed subsets of X is not necessarily closed; for example, on the rational line **Q**, the set]2, 1[is the union of the closed sets

$$\left[\frac{1}{n}, 1 - \frac{1}{n}\right] \quad \text{for } n > 2,$$

but is not closed.

6. INTERIOR, CLOSURE, FRONTIER OF A SET; DENSE SETS

DEFINITION 9. *In a topological space* X, *a point* x *is said to be an interior point of a subset* A *of* X *if* A *is a neighbourhood of* x. *The set of interior points of* A *is called the interior of* A *and is denoted by* $\mathring{A}$.

According to Definitions 9 and 4, a point x is an interior point of A if there is an open set contained in A which contains x; it follows that $\mathring{A}$ is the union of all the open sets contained in A, and hence is *the largest open set contained in* A: in other words, if B is an *open* set contained in A, then $B \subset \mathring{A}$. Consequently, if A and B are two subsets of X such that $B \subset A$, then $\mathring{B} \subset \mathring{A}$; and A is a neighbourhood of B if and only if $B \subset \mathring{A}$.

Remark. The interior of a non-empty set can be empty; this is the case for a set consisting of a single point which is not open, for example on the rational line * (or the real line) *.

Proposition 1 of no. 2 can be restated as follows:

A set is open if and only if it coincides with its interior.

The property (V_{II}) of no. 2 implies that every point which is an interior point of each of two subsets A and B is an interior point of $A \cap B$; consequently

(1) $$\overset{\circ}{\widehat{A \cap B}} = \mathring{A} \cap \mathring{B}.$$

Every point which is interior to the complement of a set A is said to be an *exterior* point of A, and the set of these points is called the *exterior* of A in X; a point $x \in X$ which is an exterior point of A is therefore characterized by the property that x *has a neighbourhood which does not meet* A.

DEFINITION 10. *The closure of a subset* A *of a topological space* X *is the set of all points* $x \in X$ *such that every neighbourhood of* x *meets* A, *and is denoted by* $\overline{A}$.

This definition can be reformulated by saying that a point x lies in the closure of a set A if there are points of A *as near* x *as we please to.*

Every point which is not in the closure of A is exterior to A, and conversely; thus we have the formulae (which are duals of each other)

$$(2) \qquad \complement\overline{A} = \overset{\circ}{\widehat{\complement A}}, \qquad \complement\mathring{A} = \overline{\complement A}.$$

Hence, to any proposition on interiors of sets, there corresponds by *duality* a proposition on closures, and vice versa. In particular, the closure of a set A is *the smallest closed set which contains* A; in other words, if B is a closed set such that $A \subset B$, then $\overline{A} \subset B$. If A and B are two subsets of X such that $A \subset B$, then $\overline{A} \subset \overline{B}$.

A set is closed if and only if it coincides with its closure.

The dual of formula (1) is

$$(3) \qquad \overline{A \cup B} = \overline{A} \cup \overline{B}.$$

PROPOSITION 5. *Let* A *be an* open *set in* X; *then for every subset* B *of* X *we have*

$$(4) \qquad A \cap \overline{B} \subset \overline{A \cap B}.$$

For suppose $x \in A \cap \overline{B}$; then if V is any neighbourhood of x, $V \cap A$ is a neighbourhood of x, since A is open; hence $V \cap A \cap B$ is not empty and therefore x lies in the closure of $A \cap B$.

If x lies in the closure of A but not in A, then every neighbourhood of x contains a point of A *other than* x; but if $x \in A$ it can happen that x has a neighbourhood which contains no point of A except x. We say then that x is an *isolated point* of A. In particular, x is isolated in the whole space X if and only if $\{x\}$ is an open set.

A closed set which has no isolated points is called a *perfect* set.

DEFINITION 11. *In a topological space* X, *a point* x *is said to be a frontier point of a set* A *if* x *lies in the closure of* A *and in the closure of* $\complement$A. *The set of frontier points of* A *is called the frontier of* A.

The frontier of A is therefore the set $\overline{A} \cap \overline{\complement A}$, which is *closed*. A frontier point x of A is characterized by the property that every neighbourhood of x contains at least one point of A and at least one point of $\complement$A; x may or may not belong to A. The frontier of A is the same as the frontier of $\complement$A. The interior of A, the exterior of A and the frontier of A are mutually disjoint and their union is the whole space X.

DEFINITION 12. *A subset* A *of a topological space* X *is said to be dense in* X (or simply *dense*, if there is no ambiguity about X) *if* $\overline{A} = X$, *i.e. if every non-empty open set* U *of* X *meets* A.

Examples. * We shall see in Chapter IV, § 1 that the set of rational numbers and its complement are dense on the real line. *

In a discrete space X the only dense subset of X is X itself. On the other hand, every non-empty subset of X is dense in the topology on X for which the only open sets are ∅ and X.

PROPOSITION 6. *If* $\mathfrak{B}$ *is a base of the topology of a topological space* X, *there is a dense set* D *in* X *such that* Card (D) ⩽ Card ($\mathfrak{B}$).

We may restrict ourselves to the case in which none of the sets of $\mathfrak{B}$ is empty (the non-empty sets of $\mathfrak{B}$ already form a base of the topology of X). For each $U \in \mathfrak{B}$, let x_V be a point of U; it follows from Proposition 3 of no. 3 that the set D of the points x_U is dense in X, and we have Card (D) ⩽ Card ($\mathfrak{B}$) (*Set Theory*, Chapter III, § 3, no. 2, Proposition 3).

2. CONTINUOUS FUNCTIONS

1. CONTINUOUS FUNCTIONS

DEFINITION 1. *A mapping* f *of a topological space* X *into a topological space* X' *is said to be continuous at a point* $x_0 \in X$ *if, given any neighbourhood* V' *of* $f(x_0)$ *in* X', *there is a neighbourhood* V *of* x_0 *in* X *such that the relation* $x \in V$ *implies* $f(x) \in V'$.

Definition 1 may be restated in the following more intuitive form: to say that f is continuous at the point x_0 means that $f(x)$ *is as near as we please to* $f(x_0)$ *whenever* x *is sufficiently near* x_0.

The relation "for each $x \in \mathrm{V}$, $f(x) \in \mathrm{V}'$" is equivalent to $f(\mathrm{V}) \subset \mathrm{V}'$ or again to $\mathrm{V} \subset \overset{-1}{f}(\mathrm{V}')$; in view of the neighbourhood axiom $(\mathrm{V_I})$, we see that Definition 1 is equivalent to the following: $f: \mathrm{X} \to \mathrm{X}'$ *is said to be continuous at the point* x_0 *if, for each neighbourhood* V' *of* $f(x_0)$ *in* X', $\overset{-1}{f}(\mathrm{V}')$ *is a neighbourhood of* x_0 *in* X. Moreover, it is sufficient that $\overset{-1}{f}(\mathrm{V}')$ is a neighbourhood of x_0 for each neighbourhood V' belonging to a *fundamental system of neighbourhoods* of $f(x_0)$ in X' (§ 1, no. 3).

PROPOSITION 1. *Let* f *be a mapping of a topological space* X *into a topological space* X'. *If* f *is continuous at* x, *and if* x *lies in the closure of a subset* A *of* X, *then* $f(x)$ *lies in the closure of* $f(\mathrm{A})$.

Let V' be a neighbourhood of $f(x)$ in X'. Since f is continuous at x, $\overset{-1}{f}(\mathrm{V}')$ is a neighbourhood of x in X. Hence $\overset{-1}{f}(\mathrm{V}')$ meets A, from which it follows that V' meets $f(\mathrm{A})$, and therefore $f(x)$ is in the closure of $f(\mathrm{A})$.

PROPOSITION 2. *Let* $\mathrm{X}, \mathrm{X}', \mathrm{X}''$ *be three topological spaces; let* f *be a mapping of* X *into* X', *continuous at* $x \in \mathrm{X}$; *let* g *be a mapping of* X' *into* X'', *continuous at* $f(x)$. *Then the composition* $h = g \circ f: \mathrm{X} \to \mathrm{X}''$ *is continuous at* x.

Let V'' be a neighbourhood of $h(x) = g(f(x))$ in X''. Since g is continuous at $f(x)$ it follows that $\overset{-1}{g}(\mathrm{V}'')$ is a neighbourhood of $f(x)$ in X'. But f is continuous at x, hence $\overset{-1}{f}(\overset{-1}{g}(\mathrm{V}'')) = \overset{-1}{h}(\mathrm{V}'')$ is a neighbourhood of x in X, and therefore h is continuous at x.

DEFINITION 2. *A mapping of a topological space* X *into a topological space* X' *is said to be continuous on* X (or just *continuous*) *if it is continuous at each point of* X.

Examples. 1) The identity mapping of a topological space X onto itself is continuous.

2) A constant map of a topological space into a topological space is continuous.

3) Every mapping of a discrete space into a topological space is continuous.

THEOREM 1. *Let f be a mapping of a topological space X into a topological space X'. Then the following statements are equivalent:*

a) *f is continuous in X.*

b) *For every subset A of X, $f(\overline{A}) \subset \overline{f(A)}$.*

c) *The inverse image under f of every closed subset of X' is a closed subset of X.*

d) *The inverse image under f of every open subset of X' is an open subset of X.*

We have already seen that a) implies b) (Proposition 1). To show that b) implies c), let F' be a closed subset of X' and let $F = \overset{-1}{f}(F')$; then by hypothesis $f(\overline{F}) \subset \overline{f(F)} \subset \overline{F'} = F'$, hence $\overline{F} \subset \overset{-1}{f}(F') = F \subset \overline{F}$, so that $F = \overline{F}$ and F is closed. By virtue of the relation $\complement \overset{-1}{f}(A') = \overset{-1}{f}(\complement A')$ for every subset A' of X', c) implies d). Finally, suppose that d) is satisfied. Let x be any point of X and let V' be any neighbourhood of $f(x)$ in X'; then there is an open set A' in X' for which

$$f(x) \in A' \subset V'$$

and hence $x \in \overset{-1}{f}(A') \subset \overset{-1}{f}(V')$. By hypothesis, $\overset{-1}{f}(A')$ is open in X, so that $\overset{-1}{f}(V')$ is a neighbourhood of x in X. Thus d) implies a).

Remarks. 1) Let $\mathfrak{B}$ be a base (§ 1, no. 3) of the topology of X'; then for $f: X \to X'$ to be continuous, it is necessary and sufficient that $\overset{-1}{f}(U')$ is open in X for every $U' \in \mathfrak{B}$.

Examples. Let a be any rational number. The mapping $x \to a + x$ of the rational line $\mathbf{Q}$ into itself is continuous on $\mathbf{Q}$, for the inverse image under this mapping of an open interval $]b, c[$ is the open interval

$$]b - a, c - a[.$$

Likewise, the mapping $x \to ax$ is continuous on $\mathbf{Q}$; this is clear if $a = 0$, for then $ax = 0$ for all x; if $a \neq 0$ then the inverse image under this mapping of the open interval $]b, c[$ is the open interval with end-points b/a and c/a.

2) The *direct* image of an open (resp. closed) set of X under a continuous mapping $f: X \to X'$ is not necessarily open (resp. closed) in X' (cf. § 5).

Example. * The mapping $f: x \to 1/(1 + x^2)$ of $\mathbf{R}$ into itself is continuous, but $f(\mathbf{R})$ is the half-open interval $]0, 1]$, which is neither open nor closed in $\mathbf{R}$. *

THEOREM 2. 1) *If $f: X \to X'$ and $g: X' \to X''$ are two continuous mappings, then $g \circ f: X \to X''$ is continuous.*

2) *For a bijection f of a topological space* X *onto a topological space* X′ *to be a homeomorphism, it is necessary and sufficient that f and the inverse of f are continuous* (or, as is also said, that f is *bicontinuous*).

The first assertion is an immediate consequence of Proposition 2; the second follows from Theorem 1, d) and the definition of a homeomorphism (§ 1, no. 1).

Remarks. 1) It is possible to have a *continuous bijection* of a topological space X onto a topological space X′ *which is not bicontinuous:* for example, take X′ to be the rational line **Q**, and X to be the set **Q** with the discrete topology; then the identity map $X \to X'$ is continuous but is not a homeomorphism.

2) To verify that a continuous bijection $f: X \to X'$ is a homeomorphism, it is enough to show that for each $x \in X$ and each neighbourhood V of x, $f(V)$ is a neighbourhood of $f(x)$ in X′.

3) Let X be a topological space, and for each $x \in X$ let $\mathfrak{B}(x)$ be the set of all neighbourhoods of x. Let x_0 be a point of X; for each $x \in X$, define a set $\mathfrak{B}_0(x)$ of subsets of X as follows: $\mathfrak{B}_0(x_0) = \mathfrak{B}(x_0)$, and if $x \neq x_0$ then $\mathfrak{B}_0(x)$ is to be the set of all subsets of X which contain x. It is immediately verified (§ 1, no. 2, Proposition 2) that the sets $\mathfrak{B}_0(x)$ are sets of neighbourhoods of points of X for a topology on X; let X_0 denote the topological space thus obtained, and let $j: X_0 \to X$ denote the identity map, which is continuous but not in general bicontinuous. A mapping f of X into a topological space X′ is continuous *at the point* x_0 if and only if the composition $X_0 \xrightarrow{j} X \xrightarrow{f} X'$ is *continuous* on X_0; this follows immediately from the definitions.

2. COMPARISON OF TOPOLOGIES

Theorem 2 of no. 1 shows that we may take the continuous mappings as *morphisms* of topological structures (*Set Theory*, Chapter IV, § 2, no. 1); from now on, we shall assume that we have made this choice of morphisms. In accordance with the general definitions (*Set Theory*, Chapter IV, § 2, no. 2), this allows us to define an *ordering* on the set of topologies on a given set X:

DEFINITION 3. *Given two topologies $\mathfrak{T}_1$, $\mathfrak{T}_2$ on the same set* X, *we say that $\mathfrak{T}_1$ is finer than $\mathfrak{T}_2$ (and that $\mathfrak{T}_2$ is coarser than $\mathfrak{T}_1$) if, denoting by X_i the set* X *with the topology $\mathfrak{T}_i$ ($i = 1, 2$), the identity mapping $X_1 \to X_2$ is continuous. If in addition $\mathfrak{T}_1 \neq \mathfrak{T}_2$, we say that $\mathfrak{T}_1$ is strictly finer than $\mathfrak{T}_2$ (and that $\mathfrak{T}_2$ is strictly coarser than $\mathfrak{T}_1$).*

Two topologies, one of which is finer than the other, are said to be *comparable.*

The criteria for a mapping to be continuous (no. 1, Definition 1 and Theorem 1) give the following proposition:

PROPOSITION 3. *Given two topologies* $\mathcal{T}_1$, $\mathcal{T}_2$ *on a set* X, *the following statements are equivalent* :

a) $\mathcal{T}_1$ *is finer than* $\mathcal{T}_2$.

b) *For each* $x \in X$, *each neighbourhood of* x *for* $\mathcal{T}_2$ *is a neighbourhood of* x *for* $\mathcal{T}_1$.

c) *For each subset* A *of* X, *the closure of* A *in the topology* $\mathcal{T}_1$ *is contained in the closure of* A *in the topology* $\mathcal{T}_2$.

d) *Every subset of* X *which is closed in* $\mathcal{T}_2$ *is closed in* $\mathcal{T}_1$.

e) *Every subset of* X *which is open in* $\mathcal{T}_2$ *is open in* $\mathcal{T}_1$.

Example. * In Hilbert space H consisting of sequences $x = (x_n)$ of real numbers such that

$$\|x\|^2 = \sum_{n=0}^{\infty} x_n^2 < +\infty,$$

the neighbourhoods of a point x_0 in the *strong* topology on H are the sets which contain an open ball $\|x - x_0\| < \alpha$ centred at x_0; the neighbourhoods of x_0 in the *weak* topology on H are the sets containing a set defined by arelation of the form $\sup_{1 \leqslant i \leqslant n} |(x - x_0|a_i)| \leqslant 1$, where the a_i are points of H and

$$(x|y) = \sum_{n=0}^{\infty} x_n y_n$$

if $x = (x_n)$ and $y = (y_n)$. Now if $\beta = \sup_{1 \leqslant i \leqslant n} \|a_i\|$, the relation $\|x - x_0\| \leqslant 1/\beta$ implies that $|(x - x_0|a_i)| \leqslant \|x - x_0\| \cdot \|a_i\| \leqslant 1$ for $1 \leqslant i \leqslant n$; hence the strong topology on H is *finer* than the weak topology. On the other hand, given any finite family $(a_i)_{1 \leqslant i \leqslant n}$ of points of H, there are points x in H such that $(x - x_0|a_i) = 0$ for $1 \leqslant i \leqslant n$ and such that $\|x - x_0\|$ is *arbitrarily large*; this shows that the strong topology is *strictly finer* than the weak topology. *

Remarks. 1) In the ordered set of all topologies on a set X, the topology in which the only open sets are ∅ and X is the *coarsest* and the discrete topology is the *finest*.

2) The *finer* the topology, the *more* open sets, closed sets and neighbourhoods; the *finer* the topology, the *smaller* (resp. the *larger*) the closure (resp. the *interior*) of a set; the *finer* the topology, the *fewer* dense sets.

3) If f: $X \to X'$ is a continuous mapping, it remains continuous if the topology of X is replaced by a *finer* topology and the topology of X′ is replaced by a *coarser* topology (no. 1, Theorem 2). In other words, the *finer* the topology of X and the *coarser* the topology of X′, the *more* continuous mappings there are of X into X′.

3. INITIAL TOPOLOGIES

PROPOSITION 4. *Let* X *be a set, let* $(Y_\iota)_{\iota \in I}$ *be a family of topological spaces, and for each* $\iota \in I$ *let* f_ι *be a mapping of* X *into* Y_ι. *Let* $\mathfrak{S}$ *be the set of subsets of* X *of the form* $\overset{-1}{f}_\iota(U_\iota)$ $(\iota \in I,\ U_\iota$ *open in* $Y_\iota)$, *and let* $\mathfrak{B}$ *be the set of finite intersections of sets of* $\mathfrak{S}$. *Then* $\mathfrak{B}$ *is a base of a topology* $\mathfrak{T}$ *on* X *which is the initial topological structure on* X *for the family* (f_ι) (*Set Theory*, Chapter IV, § 2, no. 3) *and in particular is the coarsest topology on* X *for which the mappings* f_ι *are continuous. More precisely, if g is a mapping of a topological space* Z *into* X, *then g is continuous at a point* $z \in Z$ (X *carrying the topology* $\mathfrak{T}$) *if and only if each of the functions* $f_\iota \circ g$ *is continuous at* z.

Let $\mathfrak{O}$ be the set of *all* unions of sets belonging to $\mathfrak{B}$; clearly $\mathfrak{O}$ satisfies axiom (O_I) since formation of unions is associative; and $\mathfrak{O}$ satisfies axiom (O_{II}) by reason of the definition of $\mathfrak{B}$ and the fact that finite intersection is distributive over arbitrary union [*Set Theory*, R § 4, formula (37)]. The set $\mathfrak{O}$ is therefore the set of open subsets of X for a topology $\mathfrak{T}$ of which $\mathfrak{B}$ is a base. We shall prove the last assertion of the proposition, which implies the others by reason of the general properties of initial structures (*Set Theory*, Chapter IV, § 2, no. 3, criterion CST 9). In the first place, the definition of $\mathfrak{S}$ shows that the f_ι are continuous on X (no. 1, Theorem 1); hence, if g is continuous at z, so are the mappings $f_\iota \circ g$ (no. 1, Proposition 2). Conversely, suppose that all the mappings $f_\iota \circ g$ are continuous at z, and let V be a neighbourhood of $g(z)$ in X; by definition, there is a finite subset J of I, and for each $\iota \in J$ an open subset U_ι of Y_ι such that V contains the set $\bigcap_{\iota \in J} \overset{-1}{f}_\iota(U_\iota)$ and $g(z)$ belongs to this set. It follows that

$$\overset{-1}{g}(V) \supset \bigcap_{\iota \in J} \overset{-1}{g}(\overset{-1}{f}_\iota(U_\iota)),$$

and the hypothesis implies that each of the sets $\overset{-1}{g}(\overset{-1}{f}_\iota(U_\iota))$ is a neighbourhood of z in Z; hence $\overset{-1}{g}(V)$ is also a neighbourhood of z in Z. This completes the proof.

Let $\mathfrak{B}_\iota$ be a *base* of the topology of $Y_\iota (\iota \in I)$; let $\mathfrak{S}'$ denote the set of subsets of X of the form $\overset{-1}{f}_\iota(U_\iota)$ for $\iota \in I$ and $U_\iota \in \mathfrak{B}_\iota$ for each $\iota \in I$; if $\mathfrak{B}'$ is the set of finite intersections of sets of $\mathfrak{S}'$, it is evident that $\mathfrak{B}'$ is a *base* of the topology $\mathfrak{T}$.

The general properties of initial structures (*Set Theory*, Chapter IV, § 2, no. 3, criterion CST 10) imply in particular the following *transitivity* property (the direct proof of which is quite straightforward) :

PROPOSITION 5. *Let* X *be a set,* $(Z_\iota)_{\iota\in I}$ *a family of topological spaces,* $(J_\lambda)_{\lambda\in L}$ *a partition of* I *and* $(Y_\lambda)_{\lambda\in L}$ *a family of sets indexed by* L. *Also for each* $\lambda \in L$ *let* h_λ *be a mapping of* X *into* Y_λ; *for each* $\lambda \in L$ *and each* $\iota \in J_\lambda$ *let* $g_{\iota\lambda}$ *be a mapping of* Y_λ *into* Z_ι, *and put* $f_\iota = g_{\iota\lambda} \circ h_\lambda$. *If each* Y_λ *carries the coarsest topology for which the mappings* $g_{\iota\lambda}(\iota \in J_\lambda)$ *are continuous, then the coarsest topology on* X *for which the* f_ι *are continuous is the same as the coarsest topology for which the* h_λ *are continuous.*

Examples. 1) *Inverse image of a topology.* Let X be a set, Y a topological space, f a mapping of X into Y; the coarsest topology $\mathfrak{T}$ on X for which f is continuous is called the *inverse image* under f of the topology of Y. It follows from Proposition 4 and the formulae for the inverse image of a union and an intersection [*Set Theory*, R, § 4, formulae (34) and (46)] that the open (resp. closed) sets in the topology $\mathfrak{T}$ are the *inverse images* under f of the open (resp. closed) sets of Y; consequently, for each $x \in X$, the sets $\overset{-1}{f}(W)$, where W runs through a fundamental system of neighbourhoods of $f(x)$ in Y, form a fundamental system of neighbourhoods of x in the topology $\mathfrak{T}$. In § 3 we shall study, under the name of *induced topology*, the particular case in which X is a subset of Y and f is the canonical injection $X \to Y$; X, with the induced topology, is then called a *subspace* of Y.

For a mapping f of a topological space X into a topological space X′ to be *continuous* it is necessary and sufficient that the topology of X is *finer* than the inverse image under f of the topology of X′.

2) *Least upper bound of a set of topologies.* Every family $(\mathfrak{T}_\iota)_{\iota\in I}$ of topologies on a set X has a *least upper bound* $\mathfrak{T}$ in the ordered set of all topologies on X, i.e. there exists a topology on X which is *coarsest* among all the topologies on X which are *finer* than each of the $\mathfrak{T}_\iota$. To see this we may apply Proposition 4, taking Y_ι to be the set X with the topology $\mathfrak{T}_\iota$, and f_ι to be the identity map $X \to Y_\iota$; $\mathfrak{T}$ is the coarsest topology for which all the mappings f_ι are continuous.

Let $\mathfrak{S}$ be an arbitrary set of subsets of a set X; amongst the topologies $\mathfrak{T}$ on X for which the sets of $\mathfrak{S}$ are *open*, there is a topology $\mathfrak{T}_0$ which is coarser than all the others and which is called the topology *generated* by $\mathfrak{S}$. For each set $U \in \mathfrak{S}$ let $\mathfrak{T}_U$ be the topology whose open sets are $\varnothing$, U and X [it is clear that this set of subsets of X satisfies (O_I) and (O_{II})]; then $\mathfrak{T}_0$ is just the *least upper bound* of the topologies $\mathfrak{T}_U$. By Proposition 4, if $\mathfrak{B}$ is the set of finite intersections of sets belonging to $\mathfrak{S}$, then $\mathfrak{B}$ is a *base* of the topology $\mathfrak{T}_0$. We say that $\mathfrak{S}$ is a *subbase* of $\mathfrak{T}_0$.

3) *Product topology.* Let $(X_\iota)_{\iota\in I}$ be a family of topological spaces. The coarsest topology on the product set $X = \prod_{\iota\in I} X_\iota$ for which the *projec-*

tions $\mathrm{pr}_\iota : X \to X_\iota$ are continuous mappings is called the *product* of the topologies of the X_ι; we shall study it in more detail in § 4.

4. FINAL TOPOLOGIES

PROPOSITION 6. *Let* X *be a set, let* $(Y_\iota)_{\iota \in I}$ *be a family of topological spaces, and for each* $\iota \in I$ *let* f_ι *be a mapping of* Y_ι *into* X. *Let* $\mathfrak{O}$ *be the set of subsets* U *of* X *such that* $\overset{-1}{f}_\iota(U)$ *is open in* Y_ι *for each* $\iota \in I$; *then* $\mathfrak{O}$ *is the set of open subsets of* X *in a topology* $\mathcal{T}$ *on* X *which is the final structure on* X *for the family* (f_ι) (*Set Theory*, Chapter IV, § 2, no. 5), *and in particular* $\mathcal{T}$ *is the finest topology on* X *for which the mappings* f_ι *are continuous. In other words, if* g *is a mapping of* X *into a topological space* Z, *then* g *is continuous* (X *carrying the topology* $\mathcal{T}$) *if and only if each of the mappings* $g \circ f_\iota$ *is continuous.*

It is immediately verified that $\mathfrak{O}$ satisfies the axioms (O_I) and (O_{II}). [*Set Theory*, R, § 4, formulae (34) and (46)]. We shall prove the last assertion of the proposition, which implies the other assertions by reason of the general properties of final structures (*Set Theory*, Chapter IV, § 2, no. 5, criterion CST 18). It is clear that the f_ι are continuous in the topology $\mathcal{T}$, by the definition of $\mathfrak{O}$ (no. 1, Theorem 1); hence if g is continuous, so is each mapping $g \circ f_\iota$ (no. 1, Theorem 2). Conversely, suppose that each $g \circ f_\iota$ is continuous, and let V be an open set in Z; by hypothesis, $\overset{-1}{f}_\iota(\overset{-1}{g}(V))$ is open in Y_ι for each $\iota \in I$; hence $\overset{-1}{g}(V) \in \mathfrak{O}$, and the proof is complete.

COROLLARY. *Under the hypotheses of Proposition* 6, *a subset* F *of* X *is closed in the topology* $\mathcal{T}$ *if and only if* $\overset{-1}{f}_\iota(F)$ *is closed in* Y_ι *for each* $\iota \in I$.

This follows from the definition of the open sets in the topology $\mathcal{T}$ by taking complements.

The general properties of final structures (*Set Theory*, Chapter IV, § 2, no. 5, criterion CST 19) imply the following *transitivity* property (the direct proof of which is also straightforward):

PROPOSITION 7. *Let* X *be a set,* $(Z_\iota)_{\iota \in I}$ *a family of topological spaces,* $(J_\lambda)_{\lambda \in L}$ *a partition of* I, *and* $(Y_\lambda)_{\lambda \in L}$ *a family of sets indexed by* L. *Also, for each* $\lambda \in L$ *let* h_λ *be a mapping of* Y_λ *into* X; *for each* $\lambda \in L$ *and each* $\iota \in J_\lambda$ *let* $g_{\lambda\iota}$ *be a mapping of* Z_ι *into* Y_λ, *and put* $f_\iota = h_\lambda \circ g_{\lambda\iota}$. *If each* Y_λ *carries the finest topology for which the mappings* $g_{\lambda\iota} (\iota \in J_\lambda)$ *are continuous, then the finest topology on* X *for which the* f_ι *are continuous is the same as the finest topology for which the* h_λ *are continuous.*

Examples

1) *Quotient topology.* Let X be a topological space, R an equivalence relation on X, $Y = X/R$ the quotient set of X with respect to the relation R, $\varphi : X \to Y$ the canonical mapping. The finest topology on Y for which φ is continuous is called the *quotient* of the topology of X by the relation R; we shall study it in more detail in § 3.

2) *Greatest lower bound of a set of topologies.* Every family $(\mathfrak{T}_\iota)_{\iota \in I}$ of topologies on a set X has a *greatest lower bound* $\mathfrak{T}$ in the set of all topologies on X, i.e. $\mathfrak{T}$ is the *finest* of all the topologies on X which are *coarser* than each of the $\mathfrak{T}_\iota$. To see this we may apply Proposition 6, taking Y_ι to be the set X with the topology $\mathfrak{T}_\iota$, and f_ι to be the identity mapping $Y_\iota \to X$. If $\mathfrak{O}_\iota$ is the set of subsets of X which are open in the topology $\mathfrak{T}_\iota$, then the set $\bigcap_{\iota \in I} \mathfrak{O}_\iota$ is the set of subsets of X which are open in $\mathfrak{T}$. $\mathfrak{T}$ is also called the *intersection* of the topologies $\mathfrak{T}_\iota$.

3) *Sum of topological spaces.* Let $(X_\iota)_{\iota \in I}$ be a family of topological spaces, X the set which is the *sum* of the X_ι (*Set Theory*, Chapter II, § 4, no. 8, Definition 8); for each $\iota \in I$, let j_ι be the canonical (injective) mapping of X_ι into X. The finest topology $\mathfrak{T}$ on X for which the mappings j_ι are continuous is called the *sum of the topologies of the* X_ι, and X with this topology is said to be the *sum* of the topological spaces X_ι. Let us identify each of the X_ι with a subset of X by means of j_ι; then a set $A \subset X$ is open (resp. closed) in the topology $\mathfrak{T}$ if and only if each of the sets $A \cap X_\iota$ is open (resp. closed) in X_ι. In particular, each of the X_ι is *both open and closed.*

The following proposition generalizes the situation of Example 3:

PROPOSITION 8. *Let* X *be a set,* $(X_\lambda)_{\lambda \in L}$ *a family of subsets of* X. *Suppose each* X_λ *carries a topology* $\mathfrak{T}_\lambda$ *such that, for each pair of indices* (λ, μ):

1) $X_\lambda \cap X_\mu$ *is open* (resp. *closed*) *in each of the topologies* $\mathfrak{T}_\lambda$, $\mathfrak{T}_\mu$.

2) *The topologies induced on* $X_\lambda \cap X_\mu$ *by* $\mathfrak{T}_\lambda$ *and* $\mathfrak{T}_\mu$ *coincide. Let* $\mathfrak{T}$ *be the finest topology on* X *for which the injections* $j_\lambda : X_\lambda \to X$ *are continuous. Then, for each* $\lambda \in L$, X_λ *is open* (resp. *closed*) *in* X *in the topology* $\mathfrak{T}$, *and the topology induced by* $\mathfrak{T}$ *on* X_λ *coincides with* $\mathfrak{T}_\lambda$.

In view of Proposition 6 and its corollary, it is enough to show that for each λ and each subset A_λ of X_λ the following statements are equivalent:

(i) A_λ is open (resp. closed) in the topology $\mathfrak{T}_\lambda$.

(ii) For each $\mu \in L$, $A_\lambda \cap X_\mu$ is open (resp. closed) in the topology $\mathfrak{T}_\mu$.

It is clear that (ii) implies (i) by taking $\mu = \lambda$. Conversely, if (i) is satisfied, then $A_\lambda \cap X_\mu$ is open (resp. closed) in $X_\lambda \cap X_\mu$ for the topology

$\mathfrak{T}_{\lambda\mu}$ induced on $X_\lambda \cap X_\mu$ by $\mathfrak{T}_\lambda$; but $\mathfrak{T}_{\lambda\mu}$ is also the topology induced on $X_\lambda \cap X_\mu$ by $\mathfrak{T}_\mu$; hence $A_\lambda \cap X_\mu$ is also the intersection of $X_\lambda \cap X_\mu$ with a subset B_μ of X_μ which is open (resp. closed) in the topology $\mathfrak{T}_\mu$; since $X_\lambda \cap X_\mu$ is open (resp. closed) in $\mathfrak{T}_\mu$, so is $A_\lambda \cap X_\mu$. This completes the proof.

We remark that if the union of the X_λ is different from X, then the topology induced by $\mathfrak{T}$ on $X - \left(\bigcup_{\lambda \in L} X_\lambda\right)$ is *discrete*. For if $x \in X$ belongs to none of the X_λ, then $\{x\} \cap X_\lambda = \emptyset$ is open in each topology $\mathfrak{T}_\lambda$ and therefore $\{x\}$ is open in the topology $\mathfrak{T}$.

5. PASTING TOGETHER OF TOPOLOGICAL SPACES

Let $(X_\lambda)_{\lambda \in L}$ be a family of sets, and let X be the set which is the *sum* of the X_λ (*Set Theory*, Chapter II, § 4, no. 8, Definition 8); we shall identify each X_λ with a subset of X by means of the canonical injection $j_\lambda : X_\lambda \to X$.

Let R be an equivalence relation on X such that *each equivalence class of* R *has at most one element in each* X_λ; for each pair of indices (λ, μ) let $A_{\lambda\mu}$ be the subset of X_λ consisting of the elements x for which there is an element $y \in X_\mu$ which belongs to the equivalence class of X. Clearly to each $x \in A_{\lambda\mu}$ corresponds a unique $y \in A_{\mu\lambda}$ which is congruent to x mod R; the mappings $h_{\mu\lambda} : A_{\lambda\mu} \to A_{\mu\lambda}$ so defined satisfy the following conditions:

(i) For each $\lambda \in L$, $h_{\lambda\lambda}$ is the identity mapping of $A_{\lambda\lambda} = X_\lambda$.

(ii) For each triple of indices (λ, μ, ν) of L and each $x \in A_{\lambda\mu} \cap A_{\lambda\nu}$, we have $h_{\mu\lambda}(x) \in A_{\mu\nu}$ and

$$h_{\nu\lambda}(x) = h_{\nu\mu}(h_{\mu\lambda}(x)). \tag{1}$$

Conversely, suppose that for each pair of indices (λ, μ) we are given a subset $A_{\lambda\mu}$ of X_λ and a mapping $h_{\mu\lambda} : A_{\lambda\mu} \to A_{\mu\lambda}$ satisfying the conditions (i) and (ii) above. It follows first of all from (ii) applied to the triples (λ, μ, λ) and (μ, λ, μ) that $h_{\lambda\mu} \circ h_{\mu\lambda}$ (resp. $h_{\mu\lambda} \circ h_{\lambda\mu}$) is the restriction of $h_{\lambda\lambda}$ (resp. $h_{\mu\mu}$) to $A_{\lambda\mu}$ (resp. $A_{\mu\lambda}$); hence we deduce from (i) that $h_{\lambda\mu}$ and $h_{\mu\lambda}$ are *bijections* which are inverses of each other. Now let $R\{x, y\}$ be the relation " there exist λ, μ such that $x \in A_{\lambda\mu}$, $y \in A_{\mu\lambda}$ and $y = h_{\mu\lambda}(x)$ ". It follows from (i) and from what precedes that R is *reflexive* and *symmetric*; on the other hand, if $x \in A_{\lambda\mu}$,

$$y = h_{\mu\lambda}(x) \in A_{\mu\lambda} \cap A_{\mu\nu} \qquad \text{and} \qquad z = h_{\nu\mu}(y),$$

then also $x = h_{\lambda\mu}(y)$ and therefore, by (ii), $x \in A_{\lambda\mu} \cap A_{\lambda\nu}$; the relation (1) thus shows that R is *transitive*, and therefore R is an equivalence relation on X. It follows also from (i) and from the definition of R that each equivalence class mod R has at most one element in each of the sets X_λ, and that $A_{\lambda\mu}$ is the set of all $x \in X_\lambda$ for which there is an element $y \in X_\mu$ congruent to x mod R. We say that the quotient set X/R is obtained by *pasting together the* X_λ *along the* $A_{\lambda\mu}$ *by means of the bijections* $h_{\mu\lambda}$. If $\varphi : X \to X/R$ is the canonical mapping, the restriction of φ to each X_λ is a *bijection* of X_λ onto $\varphi(X_\lambda)$.

Now suppose that each X_λ is a *topological space*, and let $\mathcal{T}_\lambda$ be its topology. Let $\mathcal{T}$ be the finest topology on the set X/R for which the mappings $\varphi \circ j_\lambda$ are continuous; $\mathcal{T}$ is the quotient by R of the topology on X which is the *sum* of the topologies $\mathcal{T}_\lambda$. We say that the topological space X/R (with the topology $\mathcal{T}$) is obtained by *pasting together the topological spaces* X_λ *along the* $A_{\lambda\mu}$ *by means of the bijections* $h_{\mu\lambda}$. The *open* (resp. *closed*) subsets of X/R are thus the canonical images of the subsets B of X which are *saturated* with respect to R and are such that $B \cap X_\lambda$ is *open* (resp. *closed*) in X_λ for each $\lambda \in L$.

Since the restriction of φ to each X_λ is a bijection onto the subset $X'_\lambda = \varphi(X_\lambda)$ of X/R, we can transport the topology $\mathcal{T}_\lambda$ to X'_λ by means of this bijection, so that X'_λ carries a topology $\mathcal{T}'_\lambda$; and the topology $\mathcal{T}$ on X/R is the *finest* for which the canonical injections $X'_\lambda \to X/R$ are continuous. In general, the topology induced by $\mathcal{T}$ on X'_λ is *coarser* than $\mathcal{T}'_\lambda$, but not identical with the latter; even if the $h_{\mu\lambda}$ are homeomorphisms (§ 3, Exercise 15). However, it follows from no. 4, Proposition 8 that, with the preceding notation:

PROPOSITION 9. *Suppose that the* $h_{\mu\lambda}$ *are homeomorphisms and that each* $A_{\lambda\mu}$ *is open* (resp. *closed*) *in* X_λ; *then each* $\varphi(X_\lambda)$ *is open* (resp. *closed*) *in* X/R *and the restriction of* φ *to* X_λ *is a homeomorphism of* X_λ *onto the subspace* $\varphi(X_\lambda)$ *of* X/R.

3. SUBSPACES; QUOTIENT SPACES

1. SUBSPACES OF A TOPOLOGICAL SPACE

Let A be a subset of a topological space X. We have defined the *topology induced* on A by the topology of X as the inverse image of the latter under the canonical injection $A \to X$ (§ 2, no. 3, Example 1). An equivalent definition is as follows:

DEFINITION 1. *Let* A *be a subset of a topological space* X. *The topology induced on* A *by the topology of* X *is that in which the open sets are the intersections with* A *of open sets of* X. *The set* A *with this topology is called a subspace of* X.

Example. The topology induced on the set **Z** of rational integers by the topology of the rational line is the *discrete* topology, for the intersection of **Z** and the open interval $]n - 1/2, n + 1/2[$ is the set $\{n\}$.

By Proposition 5 of § 2, no. 3 (or directly from Definition 1) we see that, if $B \subset A \subset X$, the *subspace* B *of* X is identical with the *subspace* B of the *subspace* A *of* X (*transitivity* of induced topologies). If $\mathfrak{S}$ is a subbase (resp. a base) of the topology of X (§ 2, no. 3, *Example* 3) its trace $\mathfrak{S}_A$ on A is a subbase (resp. a base) of the topology induced on A.

In all questions which involve the elements or subsets of A, it is essential to distinguish carefully between their properties as points (resp. subsets) of X, and their properties as points (resp. subsets) of the subspace A. We shall make this distinction by using the phrases "in A", "with respect to A", or "relative to A" to refer to properties in the latter category (possibly contrasting them with the phrases "in X", "with respect to X", "relative to X").

An *open set of the subspace* A need not be *open in* X; *in order that every set which is open in* A *should be open in* X *it is necessary and sufficient that* A *is open in* X. The condition is necessary, since A is open in A, and it is sufficient by virtue of (O_{II}) and Definition 1.

The sets which are *closed in* A are the *intersections with* A *of the closed sets in* X (§ 2, no. 3, *Example* 1); as above we see that *every set which is closed in* A *is closed in* X *if and only if* A *is closed in* X.

The neighbourhoods of a point $x \in A$ *relative to* A are the *intersections with* A *of neighbourhoods of* x *relative to* X. Every neighbourhood of x relative to A is a neighbourhood of x relative to X if and only if A is a *neighbourhood of* x *in* X.

PROPOSITION 1. *If* A *and* B *are two subsets of a topological space* X, *and* $B \subset A$, *then the closure of* B *in the subspace* A *is the intersection with* A *of the closure* $\overline{B}$ *of* B *in* X.

If $x \in A$, every neighbourhood of x in A is of the form $V \cap A$, where V is a neighbourhood of x in X. Since $V \cap B = (V \cap A) \cap B$, it follows that x lies in the closure of B with respect to A if and only if x lies in the closure of B with respect to X.

COROLLARY. *A subset* B *of* A *is dense in* A *if and only if* $\overline{B} = \overline{A}$ *in* X (i.e. *if and only if* $A \subset \overline{B}$).

It follows that if A, B, C are three subsets of X such that $A \supset B \supset C$, and if B is dense in A, and C is dense in B, then C is dense in A (*transitivity* of density), for we have $\overline{A} = \overline{B} = \overline{C}$ in X.

PROPOSITION 2. *Let* A *be a dense subset of a topological space* X; *then for each* $x \in A$ *and each neighbourhood* V *of* x *relative to* A, *the closure* $\overline{V}$ *of* V *in* X *is a neighbourhood of* x *in* X.

For V contains $U \cap A$, where U is an open subset of X which contains x, hence $\overline{V}$ contains $U \cap \overline{A} = U$ (§ 1, no. 6, Proposition 5).

PROPOSITION 3. *Let* $(A_\iota)_{\iota \in I}$ *be a family of subsets of a topological space* X, *such that one of the following properties holds:*

a) *The interiors of the* A_ι *cover* X.

b) $(A_\iota)_{\iota \in I}$ *is a locally finite closed covering of* X (§ 1, no. 5).

Under these conditions, a subset B *of* X *is open* (resp. *closed*) *in* X *if and only if each of the sets* $B \cap A_\iota$ *is open* (resp. *closed*) *in* A_ι.

Clearly if B is open (resp. closed) in X, then $B \cap A_\iota$ is open (resp. closed) in A_ι. Conversely, suppose first that condition a) is satisfied; since $(\complement B) \cap A_\iota = A_\iota - (B \cap A_\iota)$, it is enough, by duality, to consider the case in which each of the $B \cap A_\iota$ is *open* with respect to A_ι. In this case $B \cap \mathring{A}_\iota$ is open in $\mathring{A}_\iota$ for each $\iota \in I$, and therefore open in X; and since $B = \bigcup_\iota (B \cap \mathring{A}_\iota)$ by hypothesis, it follows that B is open in X.

Now suppose that b) is satisfied; by duality again, we need only consider the case in which each of the $B \cap A_\iota$ is *closed* in A_ι, and therefore closed in X. Since the family $(B \cap A_\iota)$ is locally finite and $B = \bigcup_\iota (B \cap A_\iota)$, it follows from § 1, no. 5, Proposition 4 that B is closed in X.

Remark. Let $(U_\iota)_{\iota \in I}$ be an *open* covering of a topological space X, and for each $\iota \in I$ let $\mathfrak{B}_\iota$ be a *base* of the topology of the subspace U_ι of X; then it is clear that $\mathfrak{B} = \bigcup_{\iota \in I} \mathfrak{B}_\iota$ is a *base* of the topology of X.

2. CONTINUITY WITH RESPECT TO A SUBSPACE

Let X and Y be two topological spaces, f a mapping of X into Y, B a subset of Y which contains $f(X)$. The definition of the induced topology as an initial topology (§ 2, no. 3, Proposition 4) shows that f is continuous at $x \in X$ if and only if the mapping of X into the *subspace* B of Y, having the same graph as f, is continuous at x.

Now let A be a subset of X; if f is continuous at $x \in A$ (resp. continuous on X), its *restriction* $f|A$ is a mapping of the subspace A into Y, which is continuous at x (resp. continuous on A) by Proposition 2 of § 2, no. 1. We shall sometimes say that a mapping $f: X \to Y$ is *continuous relative to* A *at* $x \in A$ (resp. *continuous relative to* A) if its restriction $f|A$ is continuous at x (resp. continuous on A).

It should be noted that $f|A$ can be continuous without f being continuous at any point of X; an example of this phenomenon is provided by the characteristic function φ_A of a subset A of X which is such that both A and its complement are dense in X (§ 2, Exercise 11), φ_A being regarded as a mapping of X into the discrete space $\{0, 1\}$. φ_A is not continuous at any point of X, but its restriction to A is constant and therefore continuous.

If A is a neighbourhood in X of a point $x \in A$, and if $f: X \to Y$ is such that $f|A$ is continuous at x, then f is continuous at x; for each neighbourhood of x in A is a neighbourhood of x in X (*local* character of continuity).

PROPOSITION 4. *Let* $(A_\iota)_{\iota \in I}$ *be a family of subsets of a topological space* X *whose interiors cover* X, *or which is a locally finite closed covering of* X. *Let* f *be a mapping of* X *into a topological space* X'. *If the restriction of* f *to each of the subspaces* A_ι *is continuous, then* f *is continuous.*

For if F' is a closed subset of X' and if $F = \overset{-1}{f}(F')$, then $F \cap A_\iota$ is closed in A_ι for each $\iota \in I$ (§ 2, no. 1, Theorem 1) and therefore F is closed in X by Proposition 3 of no. 1; the result now follows from Theorem 1 of § 2, no. 1.

3. LOCALLY CLOSED SUBSPACES

DEFINITION 2. *A subset* L *of a topological space* X *is said to be locally closed at a point* $x \in L$ *if there is a neighbourhood* V *of* x *in* X *such that* $L \cap V$ *is a closed subset of the subspace* V. L *is said to be locally closed in* X *if it is locally closed at each* $x \in L$.

Remark. Let F be a subset of X such that *for each point* x *of* X there is a neighbourhood V of x such that $V \cap F$ is closed in the subspace V; then it follows from Proposition 3 of no. 1 that F is *closed* in X. On the other hand, Proposition 5 below shows that in general there are locally closed sets which are not closed in X.

PROPOSITION 5. *For a subset* L *of a topological space* X, *the following properties are equivalent:*

a) L *is locally closed in* X.

b) L *is the intersection of an open subset and a closed subset of* X.

c) L *is open in its closure* $\bar{L}$ *in* X.

If L is locally closed, then, for each $x \in L$, there is an open neighbourhood V_x of x in X such that $L \cap V_x$ is closed in V_x; $U = \bigcup_{x \in L} V_x$ is open in X, and Proposition 3 of no. 1 shows that L is closed in U; therefore a) implies b). If $L = U \cap F$, where U is open and F closed in X, we have $\bar{L} \subset F$; hence $L \subset U \cap \bar{L} \subset U \cap F = L$, which shows that $L = U \cap \bar{L}$ is open in $\bar{L}$, so that b) implies c). Finally, if $L = U \cap \bar{L}$, where U is open in X, L is closed in U, hence locally closed, and thus c) implies a).

Corollary. *Let* $f: X \to X'$ *be a continuous map and* L′ *a locally closed subset of* X′; *then* $\overset{-1}{f}(L')$ *is locally closed in* X.

This follows immediately from Proposition 5 above and Theorem 1 of § 2, no. 1.

4. QUOTIENT SPACES

Definition 3. *Let* X *be a topological space,* R *an equivalence relation on* X. *The quotient space of* X *by* R *is the quotient set* X/R *with the topology which is the quotient of the topology of* X *by the relation* R (§ 2, no. 4, Example 1).

Unless the contrary is expressly stated, whenever we speak of X/R as a topological space, it is to be understood that we mean the quotient space of X by R. We shall often say that this topological space is the space obtained by *identifying* the points of X which belong to the same equivalence class mod R.

Let φ be the canonical mapping $X \to X/R$. By definition (§ 2, no. 4, Proposition 6 and its corollary) the *open* (resp. *closed*) sets in X/R are the sets A such that $\overset{-1}{\varphi}(A)$ is *open* (resp. *closed*) in X; in other words, the open (resp. closed) sets in X/R are in one-to-one correspondence with the open (resp. closed) subsets of X which are *saturated* with respect to R and are the canonical images of these subsets.

Proposition 6. *Let* X *be a topological space,* R *an equivalence relation on* X, φ *the canonical mapping of* X *onto* X/R; *then a mapping* f *of* X/R *into a topological space* Y *is continuous if and only if* $f \circ \varphi$ *is continuous on* X.

This is a particular case of § 2, no. 4, Proposition 6; it expresses the fact that the quotient topology is the *final* topology for the mapping φ.

Proposition 6 shows that there is a one-to-one correspondence between the continuous mappings of X/R into Y and the continuous maps of X into Y *which are constant on each equivalence class mod* R.

> *Example.* * Consider the equivalence relation $x \equiv y \pmod 1$ on the real line **R**; the quotient space of **R** by this relation is called the *one-dimensional torus* and is denoted by **T**. The equivalence class of a point $x \in \mathbf{R}$ consists of all the points $x + n$, where n runs through the set **Z** of rational integers. By Proposition 6 there is a one-to-one correspondence between the continuous functions on **T** and the continuous functions on **R** which are *periodic* with period 1. We shall return to this important example in Chapter V, § 1. *

COROLLARY. *Let* X, Y *be two topological spaces,* R (resp. S) *an equivalence relation on* X (resp. Y), *and let* $f: X \to Y$ *be a continuous mapping which is compatible with the equivalence relations* R *and* S (*Set Theory* R, § 5, no. 8); *then the mapping* $g: X/R \to Y/S$ *induced by* f (*Set Theory* R, § 5, no. 8) *is continuous.*

This is a particular case of a general property of quotient structures (*Set Theory*, Chapter IV, § 2, no. 6, criterion CST 20).

PROPOSITION 7 (Transitivity of quotient spaces). *Let* R *and* S *be two equivalence relations on a topological space* X *such that* R *implies* S, *and let* S/R *be the quotient equivalence relation on the quotient space* X/R (*Set Theory* R, § 5, no. 9). *Then the canonical bijection* $(X/R)/(S/R) \to X/S$ *is a homeomorphism.*

This is a particular case of the transitivity of final topologies (§ 2, no. 4, Proposition 7. Cf. *Set Theory*, Chapter IV, § 2, no. 3, criterion CST 21).

5. CANONICAL DECOMPOSITION OF A CONTINUOUS MAPPING

Let X and Y be two topological spaces, $f: X \to Y$ a continuous map, R the equivalence relation $f(x) = f(y)$ on X. Consider the *canonical decomposition*

$$f: X \xrightarrow{\varphi} X/R \xrightarrow{g} f(X) \xrightarrow{\psi} Y$$

where φ is the canonical (surjective) mapping of X onto the quotient space X/R, ψ is the canonical injection of the subspace $f(X)$ into Y, and g is the bijection associated with f (*Set Theory* R, § 5, no. 3). It is immediately seen that g is continuous (by Proposition 6 of no. 4); this is also a particular case of a general result on quotient structures. (Cf. *Set Theory*, Chapter IV, § 2, no. 6). But the bijection g *is not necessarily a homeomorphism.*

PROPOSITION 8. *Let* $f = \psi \circ g \circ \varphi$ *be the canonical decomposition of a continuous mapping* $f : X \to Y$, *and let* R *denote the equivalence relation*

$$f(x) = f(y).$$

Then the following three conditions are equivalent:

a) g *is a homeomorphism of* X/R *onto* $f(X)$.

b) *The image under* f *of every open set which is saturated with respect to* R *is an open set in the subspace* $f(X)$.

c) *The image under* f *of every closed set which is saturated with respect to* R *is a closed set in the subspace* $f(X)$.

For the condition b) [resp. c)] expresses that the image under g of every open (resp. closed) set in X/R is an open (resp. closed) set in $f(X)$.

Example. Let X be a topological space, $(X_\iota)_{\iota\in I}$ a *covering* of X, Y the *sum* of the subspaces X_ι of X; then there is a partition $(Y_\iota)_{\iota\in I}$ of Y into subspaces which are both open and closed, and for each $\iota \in I$ there is a homeomorphism $f_\iota : Y_\iota \to X_\iota$. Let $f : Y \to X$ be the continuous mapping which agrees with f_ι on Y_ι for each $\iota \in I$, and let R be the equivalence relation $f(x) = f(y)$; the quotient space Y/R is thus obtained by "pasting together" the Y_ι (§ 2, no. 5). Consider the bijection $g : Y/R \to X$ associated with f; in general g is not a homeomorphism, as is shown by the example in which each X_ι consists of a single point and X is not discrete. However, if the interiors of the X_ι cover X, or if (X_ι) is a locally finite closed covering of X, then g is a homeomorphism: for if U is any open subset of Y which is saturated with respect to R, then for each $\iota \in I$ the set

$$f(U) \cap X_\iota = f_\iota(U \cap Y_\iota)$$

is open in X_ι, and the assertion follows from Proposition 3 of no. 1.

The following proposition gives a simple *sufficient* condition for g to be a homeomorphism:

PROPOSITION 9. *Let* $f : X \to Y$ *be a continuous surjection, and let* R *denote the equivalence relation* $f(x) = f(y)$. *If there is a continuous section* $s : Y \to X$ *associated with* f (*Set Theory*, Chapter II, § 3, no. 8, Definition 11), *then the mapping* $g : X/R \to Y$ *associated with* f *is a homeomorphism, and* s *is a homeomorphism of* Y *onto the subspace* $s(Y)$ *of* X.

For if $\varphi : X \to X/R$ is the canonical mapping, then g and $\varphi \circ s$ are bijective, continuous and inverse to each other; likewise s and the restriction of f to $s(Y)$ are bijective, continuous and inverse to each other.

If R is an equivalence relation on a topological space X and

$$\varphi : X \to X/R$$

is the canonical mapping, a *continuous section* $s : X/R \to X$ associated with φ is also called a *continuous section* of X with respect to R (cf. *Set Theory*, Chapter II, § 6, no. 2); the subspace $s(X/R)$ of X is then homeomorphic to X/R. If we are given $s(X/R)$, then s is uniquely determined; $s(X/R)$ is often called, by abuse of language, a (continuous) *section* of X with respect to R.

A continuous section of a topological space with respect to an equivalence relation need not exist (Exercise 12).

6. QUOTIENT SPACE OF A SUBSPACE

Let X be a topological space, A a subspace of X, R an equivalence relation on X, f the canonical map $X \to X/R$, g the restriction of f to A. The equivalence relation $g(x) = g(y)$ on A is just the relation R_A *induced* by R on A (*Set Theory* R, § 5, no. 5). Let $g = \psi \circ h \circ \varphi$ be the canonical decomposition of g, so that if j is the canonical injection of A into X we have a commutative diagram (*)

(1) $$\begin{array}{ccccccc} A & \xrightarrow{\varphi} & A/R_A & \xrightarrow{h} & f(A) & \xrightarrow{\psi} & X/R. \\ & {\scriptstyle j}\searrow & & X & & \nearrow{\scriptstyle f} & \end{array}$$

PROPOSITION 10. *The canonical bijection* $h: A/R_A \to f(A)$ *is continuous. Furthermore, the following three statements are equivalent*:

a) h *is a homeomorphism.*

b) *Every open subset of* A *which is saturated with respect to* R_A *is the intersection with* A *of an open subset of* X *which is saturated with respect to* R.

c) *Every closed subset of* A *which is saturated with respect to* R_A *is the intersection with* A *of a closed subset of* X *which is saturated with respect to* R.

The first part of the proposition is immediate (no. 5). The second part follows from Proposition 8 of no. 5: if B is an open (resp. closed) subset of A which is saturated with respect to R_A, and $g(B) = f(B)$ is the intersection with $f(A)$ of an open (resp. closed) subset C of X/R, then B is the intersection with A of the open (resp. closed) subset $\overset{-1}{f}(C)$

(*) This expression means that $f \circ j = \psi \circ h \circ \varphi$.

of X, which is saturated with respect to R; and conversely, if B is the intersection with A of an open (resp. closed) subset D which is saturated with respect to R, then $f(B)$ is the intersection of $f(A)$ and $f(D)$, which is open (resp. closed) in X/R.

Corollary 1. *If* A *is an open (resp. closed) subset of* X *which is saturated with respect to* R, *then the canonical mapping* $h : A/R_A \to f(A)$ *is a homeomorphism.*

For if A is open (resp. closed) in X and saturated with respect to R, and if $B \subset A$ is open (resp. closed) in A and saturated with respect to R_A, then B is open (resp. closed) in X and saturated with respect to R.

Corollary 2. *If there is a continuous mapping* $u : X \to A$ *such that* $u(x)$ *is congruent to* x mod R *for each* $x \in X$, *then* $f(A) = X/R$ *and the canonical mapping* $h : A/R_A \to X/R$ *is a homeomorphism.*

Since each equivalence class mod R meets A, the canonical image of A/R_A in X/R is the whole of X/R; on the other hand, if U is open in A and is saturated with respect to R_A, it follows from the hypothesis that $\overset{-1}{u}(U)$ is the set obtained by saturating U with respect to R; since u is continuous, $\overset{-1}{u}(U)$ is open in X (§ 2, no. 1, Theorem 1). The corollary follows from this fact by virtue of Proposition 10.

Example. * Let R denote the equivalence relation $x \equiv y \pmod 1$ on the real line **R** (no. 4, *Example*) and let A denote the closed interval [0, 1]; A contains at least one point of each equivalence class mod R. The canonical mapping of A/R_A onto the torus **T** is a homeomorphism; for if F is closed in A (and hence in **R**), in order to saturate F with respect to the relation R we have to take the union of the closed sets $F + n$ (for all $n \in \mathbf{Z}$), which evidently form a locally finite family, so that their union is closed (§ 1, no. 5, Proposition 4); the assertion follows from this. We remark that A/R_A is obtained by identifying the points 0 and 1 in A. *

4. PRODUCT OF TOPOLOGICAL SPACES

1. PRODUCT SPACES

Definition 1. *Given a family* $(X_\iota)_{\iota \in I}$ *of topological spaces, the product space of this family is the product set* $X = \prod_{\iota \in I} X_\iota$ *with the topology which is the product of the topologies of the* X_ι (§ 2, no. 3, *Example* 3). *The spaces* $X_\iota (\iota \in I)$ *are called the factors of* X.

By virtue of § 2, no. 3, Proposition 4, the product topology on X has as a *base* the set $\mathfrak{B}$ of *finite* intersections of sets of the form $\overset{-1}{\mathrm{pr}}_\iota(\mathrm{U}_\iota)$, where U_ι is open in X_ι; these sets are products $\prod_{\iota \in \mathrm{I}} \mathrm{A}_\iota$, where A_ι is *open* in X_ι for each $\iota \in \mathrm{I}$ and $\mathrm{A}_\iota = \mathrm{X}_\iota$ *for all but a finite number of indices* These sets will be called *elementary sets.*

If $\mathfrak{B}_\iota$ is a base of the topology of X_ι (for each $\iota \in \mathrm{I}$), it is clear that the elementary sets $\prod_{\iota \in \mathrm{I}} \mathrm{A}_\iota$ such that $\mathrm{A}_\iota \in \mathfrak{B}_\iota$ for each index ι such that $\mathrm{A}_\iota \neq \mathrm{X}_\iota$ form another base of the product topology. The elementary sets of this type which contain a given point $x \in \mathrm{X}$ thus form a *fundamental system of neighbourhoods* of x (§ 1, no. 3, Proposition 3).

If I is a *finite* set, the construction of the product topology from the topologies of the factors X_ι is simpler: the elementary sets are just products $\prod_{\iota \in \mathrm{I}} \mathrm{A}_\iota$, where A_ι is any open subset of X_ι, for *each* $\iota \in \mathrm{I}$ (cf. Exercise 9).

Examples. * The product $\mathbf{R}^n$ of n spaces identical with the real line $\mathbf{R}$ is called *real number space of n dimensions*; $\mathbf{R}^2$ is also called the *real plane* (cf. Chapter VI, § 1). Likewise, starting from the rational line $\mathbf{Q}$, we define the *rational number space of n dimensions* $\mathbf{Q}^n$ (*rational plane* for $n = 2$).

The topology of the spacc $\mathbf{R}^n$ has as a base the set of all products of n open intervals in $\mathbf{R}$, which are called *open boxes of n dimensions.* The open boxes which contain a point $x \in \mathbf{R}^n$ form a fundamental system of neighbourhoods of this point. Likewise the products of n closed intervals in $\mathbf{R}$ are called *closed boxes of n dimensions.* The closed boxes to which x is interior also form a fundamental system of neighbourhoods of x. There are analogous results for $\mathbf{Q}^n$. *

PROPOSITION 1. *Let $f = (f_\iota)$ be a mapping of a topological space* Y *into a product space* $\mathrm{X} = \prod_{\iota \in \mathrm{I}} \mathrm{X}_\iota$. *Then f is continuous at a point $a \in \mathrm{Y}$ if and only if f_ι is continuous at a for each* $\iota \in \mathrm{I}$.

Since $f_\iota = \mathrm{pr}_\iota \circ f$, this is just a particular case of Proposition 4 of § 2, no. 3.

COROLLARY 1. *Let* $(\mathrm{X}_\iota)_{\iota \in \mathrm{I}}$, $(\mathrm{Y}_\iota)_{\iota \in \mathrm{I}}$ *be two families of topological spaces with the same set of indices. For each* $\iota \in \mathrm{I}$, *let f_ι be a mapping of* X_ι *into* Y_ι. *In order that the product mapping* $f : (x_\iota) \rightarrow (f_\iota(x_\iota))$ *of*

$$\prod_{\iota \in \mathrm{I}} \mathrm{X}_\iota \quad \textit{into} \quad \prod_{\iota \in \mathrm{I}} \mathrm{Y}_\iota$$

should be continuous at a point $a = (a_\iota)$ it is necessary and sufficient that f_ι is continuous at a_ι for each $\iota \in \mathrm{I}$.

f can be written as $x \to (f_\iota(\mathrm{pr}_\iota(x)))$, so that by Proposition 1 the condition is sufficient. Conversely, for each $\varkappa \in I$ let $g_\varkappa$ be the mapping of $X_\varkappa$ into $\prod_{\iota \in I} X_\iota$ such that $\mathrm{pr}_\varkappa(g_\varkappa(x_\varkappa)) = x_\varkappa$ and $\mathrm{pr}_\iota(g_\varkappa(x_\varkappa)) = a_\iota$ whenever $\iota \neq \varkappa$; then $g_\varkappa$ is continuous at the point $a_\varkappa$, by Proposition 1. Since $f_\varkappa = \mathrm{pr}_\varkappa \circ f \circ g_\varkappa$ it follows that if f is continuous at a, then $f_\varkappa$ is continuous at $a_\varkappa$.

COROLLARY 2. *Let* X, Y *be two topological spaces. In order that a mapping* $f: X \to Y$ *should be continuous it is necessary and sufficient that the mapping* $g: x \to (x, f(x))$ *is a homeomorphism of* X *onto the graph* G *of* f *(considered as a subspace of the product space* $X \times Y$).

Since $f = \mathrm{pr}_2 \circ g$, the condition is sufficient. It is also necessary, for if f is continuous, then g is bijective and continuous (Proposition 1) and the inverse of g is the restriction of pr_1 to G, which is continuous (cf. *Set Theory*, Chapter IV, § 2, no. 4, criterion CST 17).

PROPOSITION 2 (Associativity of topological products). *Let* $(X_\iota)_{\iota \in I}$ *be a family of topological spaces,* $(J_\varkappa)_{\varkappa \in K}$ *a partition of the set* I, *and for each* $\varkappa \in K$ *let* $X'_\varkappa = \prod_{\iota \in J_\varkappa} X_\iota$ *be the product of the spaces* X_ι *for* $\iota \in J_K$. *Then the canonical mapping of the product space*

$$\prod_{\iota \in I} X_\iota \text{ onto the product space } \prod_{\varkappa \in K} X'_\varkappa$$

is a homeomorphism.

This is a particular case of transitivity of initial topologies (§ 2, no. 3, Proposition 5; cf. *Set Theory*, Chapter IV, § 2, no. 4, criterion CST 13).

Generally we *identify* the product spaces $\prod_{\iota \in I} X_\iota$ and $\prod_{\varkappa \in K} X'_\varkappa$ by means of the canonical mapping.

COROLLARY. *Let* σ *be a permutation of the set* I. *Then the mapping* $(x_\iota) \to (x_{(\sigma)\iota})$ *is a homeomorphism of*

$$\prod_{\iota \in I} X_\iota \text{ onto } \prod_{\iota \in I} X_{\sigma(\iota)}.$$

Take $K = I$ and $J_\iota = \{\sigma(\iota)\}$ for each $\iota \in I$ in Proposition 2.

PROPOSITION 3. *Let* X *be a set,* $(Y_\iota)_{\iota \in I}$ *a family of topological spaces, and for each* $\iota \in I$ *let* f_ι *be a mapping of* X *into* Y_ι. *Let* f *be the mapping* $x \to (f_\iota(x))$ *of* X *into* $Y = \prod_{\iota \in I} Y_\iota$, *and let* $\mathfrak{T}$ *be the coarsest topology on* X *for which the mappings* f_ι *are continuous. Then* $\mathfrak{T}$ *is the inverse image under* f *of the topology induced on* $f(X)$ *by the product topology on* Y.

This is another particular case of transitivity of initial topologies (§ 2, no. 3, Proposition 5; cf. *Set Theory*, Chapter IV, § 2, no. 4, criterion CST 15).

COROLLARY. *For each* $\iota \in I$ *let* A_ι *be a subspace of* Y_ι. *Then the topology induced on* $A = \prod_{\iota \in I} A_\iota$ *by the product topology on* $\prod_{\iota \in I} Y_\iota$ *is the product of the topologies of the subspaces* A_ι.

Let j_ι be the canonical injection $A_\iota \to Y_\iota$, and apply Proposition 3 to the mappings $f_\iota = j_\iota \circ \mathrm{pr}_\iota$ (cf. *Set Theory*, Chapter IV, § 2, no. 4, criterion CST 14).

2. SECTION OF AN OPEN SET; SECTION OF A CLOSED SET; PROJECTION OF AN OPEN SET. PARTIAL CONTINUITY

PROPOSITION 4. *Let* X_1, X_2 *be two topological spaces; then for each* $a_1 \in X_1$, *the mapping* $x_2 \to (a_1, x_2)$ *is a homeomorphism of* X_2 *onto the subspace* $\{a_1\} \times X_2$ *of* $X_1 \times X_2$.

This is a particular case of Corollary 1 of Proposition 1 of no. 1 applied to the constant function $x_2 \to a_1$.

The mapping $x_2 \to (a_1, x_2)$ is a *continuous section* (§ 3, no. 5) with respect to the equivalence relation $\mathrm{pr}_2 z = \mathrm{pr}_2 z'$ in $X_1 \times X_2$; the quotient space $X_1 \times X_2$ by this equivalence relation is therefore homeomorphic to X_2.

COROLLARY. *The section* $A(x_1)$ *of an open* (resp. *closed*) *set* A *of the product* $X_1 \times X_2$ *at an arbitrary point* $x_1 \in X_1$ (*) *is open* (resp. *closed*) *in* X_2.

PROPOSITION 5. *The projection of an open set* U *of the product* $X_1 \times X_2$ *onto either factor is an open set.*

For example, we have $\mathrm{pr}_2 U = \bigcup_{x_1 \in X_1} U(x_1)$, and the proposition follows from the Corollary to Proposition 4 and axiom (O_I).

Remark 1. The projection of a *closed* subset of $X_1 \times X_2$ onto a factor *need not be closed.* For example, in the rational plane $\mathbf{Q}^2$, the hyperbola whose equation is $x_1 x_2 = 1$ is a closed set, but both its projections are equal to the complement of the point 0 in $\mathbf{Q}$, and this is not a closed set.

(*) By the *section* $A(x_1)$ *of* A *at* x_1 is meant the set of all $x_2 \in X_2$ such that $(x_1, x_2) \in A$ (cf. *Set Theory* R, § 3, no. 7).

PROPOSITION 6. *Let* X_1, X_2, Y *be three topological spaces,* f *a mapping of the product space* $X_1 \times X_2$ *into* Y. *If* f *is continuous at the point* (a_1, a_2) *then the partial mapping* $x_2 \to f(a_1, x_2)$ *of* X_2 *into* Y *is continuous at the point* a_2.

For this mapping is the composition of f and the mapping $x_2 \to (a_1, x_2)$; hence the result follows from Proposition 4.

Proposition 6 is often expressed by saying that a continuous function of two variables is continuous with respect to each of them separately.

Remark 2. It is possible for *all* the partial mappings determined by a map $f: X_1 \times X_2 \to Y$ to be continuous *without* f *being continuous on* $X_1 \times X_2$ (cf. Chapter IX, § 5, Exercise 23). * For example if f is the mapping of the real plane $\mathbf{R}^2$ into $\mathbf{R}$ defined by

$$f(x, y) = xy/(x^2 + y^2) \qquad \text{if} \qquad (x, y) \neq (0, 0)$$

and $f(0, 0) = 0$, then all the partial mappings are continuous; but f is not continuous at $(0, 0)$, since $f(x, x) = 1/2$ if $x \neq 0$.

If g is a mapping of X_1 into Y, continuous at a point a_1, then the mapping $(x_1, x_2) \to g(x_1)$ of $X_1 \times X_1 \to Y$ is continuous at all points (a_1, x_2), for it is the composition of g and the projection onto X_1.

The results of this subsection are easily extended to an arbitrary product $\prod_{\iota \in I} X_\iota$ of topological spaces by remarking that this product is homeomorphic to the product $\left(\prod_{\iota \in J} X_\iota\right) \times \left(\prod_{\iota \in K} X_\iota\right)$ for any partition (J, K) of I (no. 1, Proposition 2).

3. CLOSURE IN A PRODUCT

PROPOSITION 7. *In a product space* $\prod_{\iota \in I} X_\iota$, *the closure of a product of sets* $\prod_{\iota \in I} A_\iota$ *is the same as the product* $\prod_{\iota \in I} \overline{A}_\iota$ *of their closures.*

Suppose that $a = (a_\iota)$ lies in the closure of $\prod_\iota A_\iota$; then for each $\varkappa \in I$, $a_\varkappa = \mathrm{pr}_\varkappa a$ is in the closure of $A_\varkappa$ because of the continuity of $\mathrm{pr}_\varkappa$ (§ 2, no. 1, Theorem 1) and therefore $a \in \prod_\iota \overline{A}_\iota$. Conversely, let $b = (b_\iota) \in \prod_\iota \overline{A}_\iota$, and let $\prod_\iota V_\iota$ be any elementary set containing b; for each $\iota \in I$, V_ι contains a point $x_\iota \in A_\iota$; hence $\prod_\iota V_\iota$ contains the point $(x_\iota) \in \prod_\iota A_\iota$ and therefore b lies in the closure of $\prod_\iota A_\iota$.

COROLLARY. *A product* $\prod_\iota A_\iota$ *of non-empty sets is closed in the product space* $\prod_\iota X_\iota$ *if and only if* A_ι *is closed in* X_ι *for each* $\iota \in I$.

If I is *finite*, a product $\prod_{\iota \in I} A_\iota$ is open provided that A_ι is open in X_ι for each $\iota \in I$; but this is no longer so if I is infinite.

PROPOSITION 8. *Let* $a = (a_\iota)$ *be any point of a product space* $X = \prod_{\iota \in I} X_\iota$; *then the set* D *of points* $x \in X$ *such that* $\mathrm{pr}_\iota x = a_\iota$ *except for a finite number of indices* ι *is dense in* X.

For each $x \in X$ and each elementary set $V = \prod_{\iota \in I} U_\iota$ which contains x, we have $U_\iota = X_\iota$ except for indices ι belonging to a finite subset J of I; if we take $y_\iota = x_\iota$ for $\iota \in J$ and $y_\iota = a_\iota$ for $\iota \notin J$, it is clear that

$$y = (y_\iota) \in D \quad \text{and} \quad y \in V;$$

hence the result.

4. INVERSE LIMITS OF TOPOLOGICAL SPACES

Let I be a partially ordered (but not necessarily directed) set (*), in which the order relation is written $\alpha \leqslant \beta$. For each $\alpha \in I$, let X_α be a topological space, and for each pair (α, β) such that $\alpha \leqslant \beta$ let $f_{\alpha\beta}$ be a mapping of X_β into X_α. We say that $(X_\alpha, f_{\alpha\beta})$ is an *inverse system of topological spaces* if: 1) $(X_\alpha, f_{\alpha\beta})$ is an *inverse system of sets*; 2) $f_{\alpha\beta}$ is a *continuous* mapping whenever $\alpha \leqslant \beta$. Let X denote the set $\varprojlim X_\alpha$, and for each $\alpha \in I$ let f_α be the canonical mapping $X \to X_\alpha$; then the *coarsest* topology on X for which the f_α are continuous is said to be the *inverse limit* (with respect to the $f_{\alpha\beta}$) of the topologies of the X_α, and the set X with this topology is called the *inverse limit of the inverse system of topological spaces* $(X_\alpha, f_{\alpha\beta})$. Whenever we speak of $\varprojlim X_\alpha$ as a topological space, it is always to be understood that the topology of this space is the inverse limit of the topologies of the X_α, unless the contrary is expressly stated.

The set X is the subset of the product $\prod_{\alpha \in I} X_\alpha$ consisting of those points x such that

$$\mathrm{pr}_\alpha(x) = f_{\alpha\beta}(\mathrm{pr}_\beta(x)) \tag{1}$$

whenever $\alpha \leqslant \beta$. It follows from Proposition 3 of no. 1 that the inverse limit of the topologies of the X_α is the same as the topology *induced* on X by the topology of the product space $\prod_{\alpha \in I} X_\alpha$. If, for each $\alpha \in I$, Y_α is

(*) That is, a set endowed with a reflexive and transitive relation (*Set Theory* R, § 6, no. 1).

a subspace of X_α such that the Y_α form an *inverse system of subsets* of the X_α (*Set Theory* chapter III, § 7, nº 2), then it is clear that the topological space $\varprojlim Y_\alpha$ is a *subspace* of $\varprojlim X_\alpha$.

Let $(X'_\alpha, f'_{\alpha\beta})$ be another inverse system of topological spaces indexed by the same set I, and for each $\alpha \in I$ let $u_\alpha : X_\alpha \to X'_\alpha$ be a *continuous* mapping such that (u_α) is an *inverse system of mappings*; then $u = \varprojlim u_\alpha$ is a continuous mapping of $X = \varprojlim X_\alpha$ into $X' = \varprojlim X'_\alpha$. For if f'_α is the canonical mapping $X' \to X'_\alpha$, we have $f'_\alpha \circ u = u_\alpha \circ f_\alpha$, so that $f'_\alpha \circ u$ is continuous for each $\alpha \in I$, and the assertion follows from Proposition 4 of § 2, no. 3.

Finally, suppose I is a *directed* set, and let J be a *cofinal* subset of I; let Z be the inverse limit of the inverse system of topological spaces $(X_\alpha, f_{\alpha\beta})_{\alpha\in J, \beta\in J}$. Then the canonical bijection $g : X \to Z$ (*Set Theory*, Chapter III, § 7, no. 2, proposition 3) is a *homeomorphism*. For we have $\mathrm{pr}_\lambda(g(x)) = \mathrm{pr}_\lambda(x)$ for each $\lambda \in J$; hence g is continuous (no. 1, Proposition 1); and if h is the inverse of g, then for each $\alpha \in I$ there exists $\lambda \in J$ such that $\alpha \leqslant \lambda$, and therefore $\mathrm{pr}_\alpha(h(z)) = f_{\alpha\lambda}(\mathrm{pr}_\lambda(z))$, which shows that h is continuous (no. 1, Proposition 1), since the $f_{\alpha\lambda}$ are continuous.

PROPOSITION 9. *Let* I *be a directed set and* J *a cofinal subset of* I. *Let* $(X_\alpha, f_{\alpha\beta})$ *be an inverse system of topological spaces indexed by* I; *let* $X = \varprojlim X_\alpha$ *and let* $f_\alpha : X \to X_\alpha$ *be the canonical mapping. Then the family of sets* $\overset{-1}{f}_\alpha(U_\alpha)$, *where* α *runs through* J *and* U_α *runs through a base* $\mathfrak{B}_\alpha$ *of the topology of* X_α *for each* $\alpha \in J$, *is a base of the topology of* X.

From § 2, no. 3 we know that the *finite intersections* of sets of the form $\overset{-1}{f}_\alpha(U_\alpha)$ ($\alpha \in I$, U_α open in X_α) form a base of the topology of X. If $(\alpha_i)_{1\leqslant i\leqslant n}$ is a finite family of indices of I, then there exists $\gamma \in J$ such that $\alpha_i \leqslant \gamma$ for $1 \leqslant i \leqslant n$; hence $f_{\alpha_i} = f_{\alpha_i\gamma} \circ f_\gamma$; if we put

$$V_\gamma = \bigcap_i \overset{-1}{f}_{\alpha_i\gamma}(U_{\alpha_i}),$$

then we have

$$\overset{-1}{f}_\gamma(V_\gamma) = \bigcap_i \overset{-1}{f}_{\alpha_i}(U_{\alpha_i});$$

but V_γ is open and is therefore a union of sets belonging to $\mathfrak{B}_\gamma$. Hence the result.

COROLLARY. *Let* A *be a subset of* X *and let* A_α *denote* $f_\alpha(A)$ *for each* $\alpha \in I$. *Then*:

(i) *The* A_α (resp. *the* $\overline{A}_\alpha$) *form an inverse system of subsets of the* X_α, *and* $\overline{A} = \bigcap_\alpha \overset{-1}{f}_\alpha(\overline{A}_\alpha) = \varprojlim \overline{A}_\alpha$.

(ii) *If* A *is closed in* X, $A = \varprojlim A_\alpha = \varprojlim \overline{A}_\alpha$.

The first assertion of (i) follows from the relations $f_\alpha = f_{\alpha\beta} \circ f_\beta$ for $\alpha \leqslant \beta$ and from the continuity of the $f_{\alpha\beta}$ (§ 2, no. 1, Theorem 1). Let A′ denote

$$\bigcap_\alpha \overset{-1}{f}_\alpha(\overline{A}_\alpha);$$

then it is clear that A′ is closed and contains A, so that $\overline{A} \subset A'$. Conversely, let $x \in A'$; we have to show that x lies in the closure of A. By virtue of Proposition 9 it is enough to prove that every neighbourhood of x which is of the form $\overset{-1}{f}_\alpha(U_\alpha)$, with $\alpha \in I$ and U_α open in X_α, meets A. Now, by hypothesis, $f_\alpha(x) \in U_\alpha$, and since $f_\alpha(x) \in \overline{A}$ we have $U_\alpha \cap A_\alpha \neq \varnothing$, which means that $A \cap \overset{-1}{f}_\alpha(U_\alpha)$ is not empty.

To establish (ii) it is enough to remark that, without any restriction on A, we have $A \subset \varprojlim A_\alpha \subset \varprojlim \overline{A}_\alpha$; now if A is closed, then from (i)

$$A = \varprojlim \overline{A}_\alpha$$

and (ii) follows.

Example. Let I be a directed set and $(X_\alpha)_{\alpha \in I}$ a family of subsets of a set Y, such that $X_\alpha \supset X_\beta$ whenever $\alpha \leqslant \beta$. For each $\alpha \in I$ let $\mathcal{T}_\alpha$ be a topology on X_α such that $\mathcal{T}_\beta$ is *finer* than the topology induced on X_β by $\mathcal{T}_\alpha$ whenever $\alpha \leqslant \beta$. If we take $f_{\alpha\beta}$ to be the canonical injection $X_\beta \to X_\alpha$ for $\alpha \leqslant \beta$, then $\varprojlim X_\alpha$ may be identified canonically with the *intersection* X of the X_α, with the topology which is the *least upper bound* (§ 2, no. 3, Example 2) of the topologies induced on X by the $\mathcal{T}_\alpha$.

5. OPEN MAPPINGS AND CLOSED MAPPINGS

1. OPEN MAPPINGS AND CLOSED MAPPINGS

DEFINITION 1. *Let* X, X′ *be two topological spaces. A mapping* $f: X \to X'$ *is open* (resp. *closed*) *if the image under* f *of each open* (resp. *closed*) *set of* X *is open* (resp. *closed*) *in* X′.

In particular, $f(X)$ is then an open (resp. closed) subset of X′.

Examples. 1) Let A be a subspace of a topological space X. Then the canonical injection $j: A \to X$ is open (resp. closed) ifand only if A is open (resp. closed) in X (§ 3, no. 1).

2) For a bijection f of a topological space X onto a topological space X′ to be a *homeomorphism* it is necessary and sufficient that f is *continuous and open*, or *continuous and closed*.

3) Let f be a *surjection* of a set X onto a topological space X′; if we give X the topology which is the *inverse image* under f of the topology of X′ (§ 2, no. 3, Example 1), then f is continuous, open and closed.

4) In a product space

$$X = \prod_{\iota \in I} X_\iota,$$

each projection $\mathrm{pr}_\iota : X \to X_\iota$ is a continuous open mapping, but is not necessarily closed (§ 4, no. 2, Proposition 5).

* 5) A holomorphic function on an open subset A of **C** is an open mapping of A into **C**. *

6) Let X, X′ be two topological spaces and f a continuous, but not bicontinuous, bijection of X onto X′. Then the inverse bijection $g : X' \to X$ is an open and closed mapping of X′ onto X, but is not continuous.

PROPOSITION 1. *Let* X, X′, X″ *be three topological spaces, and let* $f : X \to X'$, $g : X' \to X''$ *be two mappings. Then:*

a) *If* f *and* g *are open* (resp. *closed*), *so is* $g \circ f$.

b) *If* $g \circ f$ *is open* (resp. *closed*) *and if* f *is continuous and surjective, then* g *is open* (resp. *closed*).

c) *If* $g \circ f$ *is open* (resp. *closed*) *and if* g *is continuous and injective, then* f *is open* (resp. *closed*).

From Definition 1. a) follows immediately. To prove b) it is enough to remark that every open (resp. closed) subset A′ of X′ can be written as $f(A)$, where $A = \overset{-1}{f}(A')$ is open (resp. closed) in X (§ 2, no. 1, Theorem 1); hence $g(A') = g(f(A))$ is open (resp. closed) in X″. Finally, to prove c), we remark that $f(A) = \overset{-1}{g}(g(f(A))$ for every subset A of X; by hypothesis, if A is open (resp. closed) in X, then $g(f(A))$ is open (resp. closed) in X″, hence $f(A)$ is open (resp. closed) in X′ by § 2, no. 1, Theorem 1.

PROPOSITION 2. *Let* X, Y *be two topological spaces,* f *a mapping of* X *into* Y. *For each subset* T *of* Y *let* f_T *denote the mapping of* $\overset{-1}{f}(T)$ *into* T *which agrees with* f *on* $\overset{-1}{f}(T)$.

a) *If* f *is open* (resp. *closed*), f_T *is open* (resp. *closed*).

b) *Let* $(T(\iota))_{\iota \in I}$ *be a family of subsets of* Y *whose interiors cover* Y, *or which is a locally finite closed covering of* Y; *if all the* $f_{T(\iota)}$ *are open* (resp. *closed*), *then* f *is open* (resp. *closed*).

a) If A is an open (resp. closed) subset of $\overset{-1}{f}(T)$, then there is an open (resp. closed) subset B of X such that $A = B \cap \overset{-1}{f}(T)$, and therefore

$f_T(A) = f(B) \cap T$; by hypothesis, $f(B)$ is open (resp. closed), so that $f_T(A)$ is open (resp. closed) in T.

b) Let B be an open (resp. closed) subset of X, and let B_ι denote $B \cap \overset{-1}{f}(T(\iota))$; then $f(B) \cap T(\iota) = f_{T(\iota)}(B_\iota)$. Since $f_{T(\iota)}(B_\iota)$ is open (resp. closed) in $T(\iota)$ by hypothesis, it follows that $f(B)$ is open (resp. closed) in Y, by Proposition 3 of § 3, no. 1.

COROLLARY. *Let* $(T(\iota))_{\iota \in I}$ *be a family of subsets of* Y *whose interiors cover* Y, *or which is a locally finite closed covering of* Y. *If* $f : X \to Y$ *is continuous and if each of the* $f_{T(\iota)}$ *is a homeomorphism of* $\overset{-1}{f}(T(\iota))$ *onto* $T(\iota)$, *then* f *is a homeomorphism of* X *onto* Y.

For f is clearly bijective, and is open by virtue of Proposition 2.

2. OPEN EQUIVALENCE RELATIONS AND CLOSED EQUIVALENCE RELATIONS

DEFINITION 2. *An equivalence relation* R *on a topological space* X *is said to be open* (resp. *closed*) *if the canonical mapping of* X *onto* X/R *is open* (resp. *closed*).

It comes to the same thing to say that the *saturation* of each open (resp. closed) subset of X with respect to R is open (resp. closed) in X (§ 3, no. 4).

Examples. 1) Let X be a topological space, Γ a *group of homeomorphisms* of X onto itself, and let R be the equivalence relation

$$\text{"there exists } \sigma \in \Gamma \text{ such that } y = \sigma(x)\text{"}$$

between x and y (thus R is the equivalence relation whose classes are the *orbits* of Γ in X. The relation R is *open*, for the saturation of a subset A of X with respect to R is open so is each $\sigma(A)$ and hence so is their union.

* On the real line **R** the equivalence relation $x \equiv y \pmod 1$ is open, since it is derived as above from the group of translations

$$x \to x + n \qquad (n \in \mathbf{Z})$$

(see Chapter III, § 2, no. 4). *

2) Let X be the sum of a family (X_ι) of subspaces of X, and let X/R be the space obtained by *pasting together* the X_ι along *open* subsets $A_{\iota\varkappa}$ by means of bijections $h_{\varkappa\iota}$ (§ 2, no. 5); and suppose that $h_{\varkappa\iota}$ is a *homeomorphism* of $A_{\iota\varkappa}$ onto $A_{\varkappa\iota}$ for each pair of indices $(\iota, \varkappa)$.

Then the relation R is *open*. For if U is open in X, the saturation of U is the union of the $h_{\varkappa\iota}(U \cap A_{\iota\varkappa})$; since $U \cap A_{\iota\varkappa}$ is open in $A_{\iota\varkappa}$, $h_{\varkappa\iota}(U \cap A_{\iota\varkappa})$ is open in $A_{\varkappa\iota}$ and therefore in X.

3) With the notation of Example 2, suppose now that the $A_{\iota\varkappa}$ are *closed* and the $h_{\varkappa\iota}$ are *homeomorphisms*; further suppose that for each index ι there is only a *finite* number of indices $\varkappa$ such that $A_{\iota\varkappa} \neq \varnothing$ (i.e. each X_ι is "stuck" to only a finite number of $X_\varkappa$). Then the relation R is *closed*. For if F is any closed subset of X, the saturation of F is the union of the sets $h_{\varkappa\iota}(F \cap A_{\iota\varkappa}) \subset A_{\varkappa\iota}$; the assumptions made imply that this family is locally finite, and $h_{\varkappa\iota}(F \cap A_{\iota\varkappa})$ is closed in $A_{\varkappa}$, and therefore in X. The result therefore follows from Proposition 4 of § 1, no. 5.

Proposition 3. *Let* X, Y *be two topological spaces, let* $f: X \to Y$ *be a continuous mapping, let* R *be the equivalence relation* $f(x) = f(y)$ *on* X, *and let* $X \xrightarrow{p} X/R \xrightarrow{h} f(X) \xrightarrow{i} Y$ *be the canonical decomposition of* f. *Then the following three statements are equivalent:*

a) f *is an open mapping.*

b) *The three mappings* p, h, i *are open.*

c) *The equivalence relation* R *is open,* h *is a homeomorphism, and* $f(X)$ *is an open subset of* Y.

Also the preceding proposition remains true if "open" is replaced by "closed" throughout.

Since the injection i is continuous, it follows from Proposition 1 c) of no. 1 that if f is open then so is $h \circ p$; since p is surjective and continuous, Proposition 1 b) shows that h is open; h is in any case a continuous bijection; thus h is a homeomorphism, and therefore, from Proposition 1 a) of no. 1, $p = \overset{-1}{h} \circ (h \circ p)$ is an open mapping. Also [no. 1, Proposition 1 b)] $i \circ h$ is open; hence [no. 1, Proposition 1 a)] so is $i = (i \circ h) \circ \overset{-1}{h}$. This proves that a) implies b). Conversely, Proposition 1 a) of no. 1 shows that b) implies a). Finally, the equivalence of b) and c) follows immediately from the definitions.

The proof in the case of closed mappings is analogous, *mutatis mutandis*.

Proposition 4. *Let* R *be an open* (resp. *closed*) *equivalence relation on a topological space* X, *and* f *the canonical mapping* $X \to X/R$. *Let* A *be a subset of* X *and suppose that one of the following two conditions is satisfied:*

a) A *is open* (resp. *closed*) *in* X.

b) A *is saturated with respect to* R.

Then the relation R_A *induced on* A *is open* (resp. *closed*) *and the canonical mapping of* A/R_A *onto* $f(A)$ *is a homeomorphism.*

Consider the commutative diagram (1) of § 3, no. 6, which gives the canonical decomposition of $f \circ j$. Under condition a), j is open (resp. closed) and so is f by hypothesis; hence $f \circ j$ is open (resp. closed) [no. 1, Proposition 1 a)], and the result follows from Proposition 3. Under condition b) we have

$$A = \overset{-1}{f}(f(A)),$$

and $h \circ \varphi$ is the mapping of A into $f(A)$ which agrees with f on A; by virtue of Proposition 2 a) of no. 1, $h \circ \varphi$ is open (resp. closed), and the result again follows from Proposition 3, applied to $h \circ \varphi$.

3. PROPERTIES PECULIAR TO OPEN MAPPINGS

PROPOSITION 5. *Let* X, Y *be two topological spaces,* f *a mapping of* X *into* Y, $\mathfrak{B}$ *a base of the topology of* X. *Then the following statements are equivalent:*

a) f *is an open mapping.*

b) *For each* $U \in \mathfrak{B}$, $f(U)$ *is open in* Y.

c) *For each* $x \in X$ *and each neighbourhood* V *of* x *in* X, $f(V)$ *is a neighbourhood of* $f(x)$ *in* Y.

The equivalence of a) and b) follows immediately from the definitions and from (O_1); the equivalence of a) and c) is a consequence of Proposition 1 of § 1, no. 2.

PROPOSITION 6. *Let* R *be an equivalence relation on a topological space* X; *then the following three conditions are equivalent:*

a) *The relation* R *is open.*

b) *The interior of each subset which is saturated with respect to* R *is saturated with respect to* R.

c) *The closure of each subset which is saturated with respect to* R *is saturated with respect to* R.

By taking complements (§ 1, no. 6, formula (2)) we see that b) and c) are equivalent. Let us show that b) implies a): suppose condition b) is satisfied and let U be an open subset of X, V its saturation with respect to R; then $\mathring{V} \supset U$, and since by hypothesis $\mathring{V}$ is saturated, it follows that $\mathring{V} = V$, and therefore the saturation of U is open. Conversely, suppose condition a) is satisfied, and let A be a saturated set; if B is the saturation of $\mathring{A}$, then $\mathring{A} \subset B \subset A$, and since B is open by hypothesis it follows that $B = \mathring{A}$.

PROPOSITION 7. *Let* R *be an open equivalence relation on a topological space* X, *and let* $\varphi : X \to X/R$ *be the canonical mapping. If* A *is any subset of* X *which is saturated with respect to* R, *then the closure* (resp. *the interior*) *of* $\varphi(A)$ *in* X/R *is* $\varphi(\overline{A})$ (resp. $\varphi(\mathring{A})$).

Each of the two assertions of the proposition can be deduced from the other by taking complements and using formula (2) of § 1, no. 6 and the fact that if B is a saturated subset of X then $\varphi(\complement B) = \complement\varphi(B)$. By virtue of Proposition 6, $\overline{A}$ is saturated; hence $\varphi(\overline{A})$ is closed in X/R, and since $A \subset \overline{A}$ we have $\overline{\varphi(A)} \subset \varphi(\overline{A})$, so that $\varphi(A) \subset \varphi(\overline{A})$. But since φ is continuous, $\varphi(\overline{A}) \subset \overline{\varphi(A)}$ (§ 2, no. 1, Theorem 1), and the result follows.

PROPOSITION 8. *Let* $(X_\iota)_{\iota\in I}$, $(Y_\iota)_{\iota\in I}$ *be two families of topological spaces indexed by the same set* I. *For each* $\iota \in I$ *let* f_ι *be an open mapping of* X_ι *into* Y_ι, *and suppose that* f_ι *is surjective for all but a finite number of indices. Then the product mapping* $f : (x_\iota) \to (f_\iota(x_\iota))$ *of* $\prod_{\iota\in I} X_\iota$ *into* $\prod_{\iota\in I} Y_\iota$ *is open.*

By virtue of Proposition 5 we need only prove that the image under f of any elementary set $\prod_{\iota\in I} A_\iota$ in $\prod_{\iota\in I} X_\iota$ is open in $\prod_{\iota\in I} Y_\iota$. But this image is $\prod_{\iota\in I} f_\iota(A_\iota)$, and the hypotheses imply that $f_\iota(A_\iota)$ is open in Y_ι for each $\iota \in I$, and that $f_\iota(A_\iota) = Y_\iota$ for all but a finite number of indices; whence the result.

COROLLARY. *Let* $(X_\iota)_{\iota\in I}$ *be a family of topological spaces, and for each* $\iota \in I$ *let* R_ι *be an equivalence relation on* X_ι, *and let* f_ι *be the canonical mapping* $X_\iota \to X_\iota/R_\iota$. *Let* R *be the equivalence relation in* $X = \prod_{\iota\in I} X_\iota$

$$\text{"for each } \iota \in I,\ \mathrm{pr}_\iota(x) = \mathrm{pr}_\iota(y) \pmod{R_\iota}\text{"}$$

between x *and* y, *and let* f *be the product mapping* $(x_\iota) \to (f_\iota(x_\iota))$ *of* X *into* $\prod_{\iota\in I} (X_\iota/R_\iota)$. *If each of the relations* R_ι *is open, then the relation* R *is open, and the bijection associated with* f *is a homeomorphism of* X/R *onto* $\prod_{\iota\in I} (X_\iota/R_\iota)$.

R is the relation $f(x) = f(y)$. Since f is continuous and open by Proposition 8 above and § 4, no. 1, Corollary 1 of Proposition 1, the result follows from Proposition 3 of no. 2.

In particular, if R (resp. S) is an *open* equivalence relation on a topological space X (resp. Y), then the canonical bijection of

$$(X \times Y)/(R \times S) \qquad \text{onto} \qquad (X/R) \times (Y/S)$$

is a homeomorphism. If R and S are not assumed to be open, this bijection is continuous but is not necessarily a homeomorphism, even if one of the relations R, S is the relation of equality (Exercise 6).

4. PROPERTIES PECULIAR TO CLOSED MAPPINGS

PROPOSITION 9. *Let* X, X′ *be two topological spaces. A necessary and sufficient condition for a mapping* $f: X \to X'$ *to be continuous and closed is that* $f(\overline{A}) = \overline{f(A)}$ *for every subset* A *of* X.

The condition is sufficient, for it obviously implies that f is closed, and it also implies that f is continuous by reason of § 2, no. 1, Theorem 1. Conversely, if f is continuous and closed, we have $f(A) \subset f(\overline{A}) \subset \overline{f(A)}$ by § 2, no. 1, Theorem 1; also $f(\overline{A})$ is closed in X′ by hypothesis; hence $f(\overline{A}) = \overline{f(A)}$.

PROPOSITION 10. *Let* R *be an equivalence relation on a topological space* X. *Then* R *is closed if and only if every equivalence class* M *mod* R *possesses a fundamental system of neighbourhoods which are saturated with respect to* R.

Suppose R is closed, and let U be an arbitrary open neighbourhood of M; since $F = \complement U$ is closed in X, the saturation S of F with respect to R is closed in X. Since M is saturated with respect to R, we have $M \cap S = \emptyset$, and thus $V = \complement S$ is an open neighbourhood of M, saturated with respect to R and contained in U.

To prove the converse, let F be any closed subset of X. Let T be the saturation of F with respect to R, let x be a point of $\complement T$, and let M be the equivalence class of x; then $M \cap T = \emptyset$ and *a fortiori* $M \cap F = \emptyset$, so that $U = \complement F$ is a neighbourhood of M. Hence there is a neighbourhood $V \subset U$ of M such that V is saturated with respect to R; V does not meet F, hence does not meet T, so that $\complement T$ is a neighbourhood of M and therefore of x. This shows that $\complement T$ is open (§ 1, no. 2, Proposition 1), i.e. T is closed.

Remark. Proposition 10 implies the following: if R is closed and if φ denotes the canonical mapping $X \to X/R$, then for each $x \in X$ and each *neighbourhood* U *of the equivalence class of* x *in* X, $\varphi(U)$ is a neighbourhood of $\varphi(x)$ in X/R. It should be carefully noticed that this statement by no means implies that for every *neighbourhood* V *of* x, $\varphi(V)$ is a neighbourhood of $\varphi(x)$; in other words (no. 3, Proposition 5) a closed equivalence relation is not necessarily open (Exercise 2). Conversely, an open equivalence relation is not necessarily closed (no. 1, Example 4);

for if U is a neighbourhood in X of an equivalence class M, then for each $x \in M$ and each neighbourhood $V \subset U$ of x, the saturation of V is certainly a neighbourhood of M in X, but this neighbourhood is *not necessarily contained in* U.

Finally, there are equivalence relations other than equality which are both open and closed (Exercise 3) and equivalence relations which are neither open nor closed (§ 8, Exercise 10).

6. FILTERS

1. DEFINITION OF A FILTER

DEFINITION 1. *A filter on a set* X *is a set* $\mathfrak{F}$ *of subsets of* X *which has the following properties:*

(F_I) *Every subset of* X *which contains a set of* $\mathfrak{F}$ *belongs to* $\mathfrak{F}$.

(F_{II}) *Every finite intersection of sets of* $\mathfrak{F}$ *belongs to* $\mathfrak{F}$.

(F_{III}) *The empty set is not in* $\mathfrak{F}$.

It follows from (F_{II}) and (F_{III}) that every finite intersection of sets of $\mathfrak{F}$ is *non-empty.*

A filter $\mathfrak{F}$ on X defines a structure on X, the axioms of which are (F_I), (F_{II}) and (F_{III}); this structure is called a *structure of a filtered set*, and the set X endowed with this structure is called a *set filtered by* $\mathfrak{F}$.

Axiom (F_{II}) is equivalent to the conjunction of the following two axioms:

($F_{II\,a}$) The intersection of two sets of $\mathfrak{F}$ belongs to $\mathfrak{F}$.

($F_{II\,b}$) X belongs to $\mathfrak{F}$.

Axioms ($F_{II\,b}$) and (F_{III}) show that *there is no filter on the empty set.*

In order for a set of subsets which satisfies (F_I) also to satisfy ($F_{II\,b}$) it is necessary and sufficient that it is *not empty*. A set of subsets which satisfies (F_I) also satisfies (F_{III}) if and only if it is different from $\mathfrak{P}(X)$.

Examples of filters. 1) If $X \neq \emptyset$, the set of subsets consisting of X alone is a filter on X. More generally, the set of all subsets of X which contain a given non-empty subset A of X is a filter on X.

2) In a topological space X, the *set of all neighbourhoods* of an arbitrary non-empty subset A of X (and in particular the set of all neighbourhoods of a point of X) is a filter, called the *neighbourhood filter* of A.

3) If X is an *infinite* set, the *complements of the finite subsets* of X are the elements of a filter. The filter of complements of finite subsets of the set **N** of integers $\geqslant 0$ is called the *Fréchet filter.*

2. COMPARISON OF FILTERS

DEFINITION 2. *Given two filters* $\mathfrak{F}$, $\mathfrak{F}'$ *on the same set* X, $\mathfrak{F}'$ *is said to be finer than* $\mathfrak{F}$, *or* $\mathfrak{F}$ *is coarser than* $\mathfrak{F}'$, *if* $\mathfrak{F} \subset \mathfrak{F}'$. *If also* $\mathfrak{F} \neq \mathfrak{F}'$, *then* $\mathfrak{F}'$ *is said to be strictly finer than* $\mathfrak{F}$, *or* $\mathfrak{F}$ *strictly coarser than* $\mathfrak{F}'$.

Two filters are said to be *comparable* if one is finer than the other. The set of all filters on X is *ordered* by the relation "$\mathfrak{F}$ is coarser than $\mathfrak{F}'$"; this relation is induced by the inclusion relation in $\mathfrak{P}(\mathfrak{P}(X))$.

Let $(\mathfrak{F}_\iota)_{\iota\in I}$ be any *non-empty* family of filters on a set X (which must therefore be non-empty); then the set

$$\mathfrak{F} = \bigcap_{\iota\in I} \mathfrak{F}_\iota$$

satisfies axioms (F_I), (F_{II}) and (F_{III}) and is therefore a filter; $\mathfrak{F}$ is called the *intersection* of the family of filters $(\mathfrak{F}_\iota)_{\iota\in I}$ and is obviously the greatest lower bound of the set of the $\mathfrak{F}_\iota$ in the ordered set of all filters on X.

The filter formed by the single set X is the *smallest* element of the ordered set of all filters on X. We shall see in no. 4 that, if X has more than one element, the set of all filters on X has no greatest element.

Given a set $\mathfrak{G}$ of subsets of a set X, let us consider whether there are any filters on X which *contain* $\mathfrak{G}$. If such a filter exists then by (F_{II}) it contains also the set $\mathfrak{G}'$ of *finite intersections* of sets of $\mathfrak{G}$ (including X, which is the intersection of the empty subset of $\mathfrak{G}$); hence a *necessary* condition for such a filter to exist is that the empty subset of X is not in $\mathfrak{G}'$. This condition is also *sufficient*, for by (F_I) any filter which contains $\mathfrak{G}'$ also contains the set $\mathfrak{G}''$ of subsets of X which contain a set of $\mathfrak{G}'$. Now $\mathfrak{G}''$ clearly satisfies (F_I); it satisfies (F_{II}) by reason of the definition of $\mathfrak{G}'$; and finally it satisfies (F_{III}) because the empty subset of X does not belong to $\mathfrak{G}'$. Hence $\mathfrak{G}''$ is the *coarsest filter which contains* $\mathfrak{G}$, and we have proved:

PROPOSITION 1. *A necessary and sufficient condition that there should exist a filter on* X *containing a set* $\mathfrak{G}$ *of subsets of* X *is that no finite subset of* $\mathfrak{G}$ *has an empty intersection.*

The filter $\mathfrak{G}''$ defined above is said to be *generated* by $\mathfrak{G}$, and $\mathfrak{G}$ is said to be a *subbase* of $\mathfrak{G}''$.

Example. Let $\mathfrak{S}$ be any set of subsets of a set X, and let $\mathfrak{T}$ be the topology on X *generated* by $\mathfrak{S}$ (§ 2, no. 3, Example II). Since the set of finite intersections of sets of $\mathfrak{S}$ is a base of $\mathfrak{T}$, it follows from the proof of Proposition 1 above and from Proposition 3 of § 1, no. 3 that for each

$x \in X$ the *neighbourhood filter* of x for $\mathfrak{T}$ is *generated* by the set $\mathfrak{S}(x)$ of sets of $\mathfrak{S}$ which contain x.

Corollary 1. *Let $\mathfrak{F}$ be a filter on a set X, and A a subset of X. Then there is a filter $\mathfrak{F}'$ which is finer than $\mathfrak{F}$ and such that $A \in \mathfrak{F}'$, if and only if A meets all the sets of $\mathfrak{F}$.*

Corollary 2. *A set Φ of filters on a non-empty set X has a least upper bound in the set of all filters on X if and only if, for all finite sequences $(\mathfrak{F}_i)_{1 \leqslant i \leqslant n}$ of elements of Φ and all $A_i \in \mathfrak{F}_i$ $(1 \leqslant i \leqslant n)$, the intersection $A_1 \cap \cdots \cap A_n$ is not empty.*

For this condition expresses that the union $\mathfrak{G}$ of the filters $\mathfrak{F} \in \Phi$ satisfies the condition of Proposition 1.

Corollary 3. *The ordered set of all filters on a non-empty set X is inductive.*

For every linearly ordered set Φ of filters on X satisfies the condition of Corollary 2 of Proposition 1, since the sets A_ι all belong to the same $\mathfrak{F}_j$ by hypothesis, and we can apply (F_{II}).

3. BASES OF A FILTER

If $\mathfrak{G}$ is a subbase of a filter $\mathfrak{F}$ on X (no. 2), then $\mathfrak{F}$ is not in general the set of subsets of X which contain a set of $\mathfrak{G}$; for $\mathfrak{G}$ to have this property it is necessary and sufficient that every finite intersection of sets of $\mathfrak{G}$ should contain a set of $\mathfrak{G}$. Hence the following proposition:

Proposition 2. *Let $\mathfrak{B}$ be a set of subsets of a set X. Then the set of subsets of X which contain a set of $\mathfrak{B}$ is a filter if and only if $\mathfrak{B}$ has the following two properties:*

(B_I) *The intersection of two sets of $\mathfrak{B}$ contains a set of* B.

(B_{II}) *$\mathfrak{B}$ is not empty, and the empty subset of X is not in $\mathfrak{B}$.*

Definition 3. *A set $\mathfrak{B}$ of subsets of a set X which satisfies axioms (B_I) and (B_{II}) is said to be a base of the filter it generates. Two filter bases are said to be equivalent if they generate the same filter.*

If $\mathfrak{G}$ is a subbase of a filter $\mathfrak{F}$, then the set $\mathfrak{G}'$ of *finite intersections* of sets of $\mathfrak{G}$ is a *base* of $\mathfrak{F}$ (no. 2).

Proposition 3. *A subset $\mathfrak{B}$ of a filter $\mathfrak{F}$ on X is a base of $\mathfrak{F}$ if and only if every set of $\mathfrak{F}$ contains a set of $\mathfrak{B}$.*

If $\mathfrak{B}$ is a base of $\mathfrak{F}$, then clearly every set of $\mathfrak{F}$ contains a set of $\mathfrak{B}$; conversely, if every set of $\mathfrak{F}$ contains a set of $\mathfrak{B}$, then the set of subsets of X containing a set of $\mathfrak{B}$ coincides with $\mathfrak{F}$ by reason of (F_I).

PROPOSITION 4. *On a set* X, *a filter* $\mathfrak{F}'$ *with base* $\mathfrak{B}'$ *is finer than a filter* $\mathfrak{F}$ *with base* $\mathfrak{B}$ *if and only if every set of* $\mathfrak{B}$ *contains a set of* $\mathfrak{B}'$.

This is an immediate consequence of Definitions 2 and 3.

COROLLARY. *Two filter bases* $\mathfrak{B}$, $\mathfrak{B}'$ *on a set* X *are equivalent if and only if every set of* $\mathfrak{B}$ *contains a set of* $\mathfrak{B}'$ *and every set of* $\mathfrak{B}'$ *contains a set of* $\mathfrak{B}$.

Examples of filter bases. 1) Let X be a topological space. Proposition 3 shows that the bases of the neighbourhood filter of a point $x \in X$ are precisely the *fundamental systems of neighbourhoods* of x (§ 1, no. 3, Definition 5).

2) Let X be a non-empty directed set with respect to a relation (σ) (*Set Theory*, Chapter III, § 1, no. 10). For each $a \in X$, the set $S(a)$ of all $x \in X$ such that $a(\sigma)x$ will be called the *section* of X relative to the element a. Then *the set* $\mathfrak{S}$ *of sections of* X *is a filter base*, for it clearly satisfies (B_{II}), and if a, b are any two elements of X, then there is by hypothesis an element $c \in X$ such that $a(\sigma)c$ and $b(\sigma)c$, and therefore

$$S(c) \subset S(a) \cap S(b),$$

so that (B_I) is satisfied. The filter generated by $\mathfrak{S}$ is called the *section filter* of the directed set X.

For example, the *Fréchet filter* (no. 1) is the section filter of the ordered set **N**, considered as directed by the relation $\leqslant$.
Let $\mathfrak{F}$ be a filter on a set Z. Since $\mathfrak{F}$ is directed with respect to the relation $\supset$ [by reason of axiom (F_{II})] we can define a *section filter* on $\mathfrak{F}$; here a section of $\mathfrak{F}$ relative to a set $A \in \mathfrak{F}$ is the set $S(A)$ of all $M \in \mathfrak{F}$ such that $M \subset A$. This filter is called the *section filter of the filter* $\mathfrak{F}$.

4. ULTRAFILTERS

DEFINITION 4. *An ultrafilter on a set* X *is a filter* $\mathfrak{F}$ *such that there is no filter on* X *which is strictly finer than* $\mathfrak{F}$ (in other words, a *maximal* element in the ordered set of all filters on X).

Since the ordered set of all filters on X is inductive (no. 2, Proposition 1, Corollary 3), Zorn's lemma (*Set Theory*, R, § 6, no. 10) shows that:

THEOREM 1. *If* $\mathfrak{F}$ *is any filter on a set* X, *there is an ultrafilter finer than* $\mathfrak{F}$.

PROPOSITION 5. *Let* $\mathfrak{F}$ *be an ultrafilter on a set* X. *If* A *and* B *are two subsets of* X *such that* $A \cup B \in \mathfrak{F}$, *then either* $A \in \mathfrak{F}$ *or* $B \in \mathfrak{F}$.

If the proposition is false, there exist subsets A and B of X such that $A \notin \mathfrak{F}$ and $B \notin \mathfrak{F}$ and $A \cup B \in \mathfrak{F}$. Let $\mathfrak{G}$ be the set of subsets M of X such that $A \cup M \in \mathfrak{F}$. It is straightforward to check that $\mathfrak{G}$ is a filter on X, and $\mathfrak{G}$ is strictly finer than $\mathfrak{F}$, since $B \in \mathfrak{G}$; but this contradicts the hypothesis that $\mathfrak{F}$ is an ultrafilter.

COROLLARY. *If the union of a finite sequence* $(A_i)_{1 \leqslant i \leqslant n}$ *of subsets of* X *belongs to an ultrafilter* $\mathfrak{F}$, *then at least one of the* A_i *belongs to* $\mathfrak{F}$.

Proof is by induction on n.

In particular, if $(A_i)_{1 \leqslant i \leqslant n}$ is a *covering* of X, then at least one of the A_i belongs to $\mathfrak{F}$.

Proposition 5 *characterizes* the ultrafilters; more generally, we have :

PROPOSITION 6. *Let* $\mathfrak{G}$ *be a subbase of a filter on a set* X. *If for each subset* Y *of* X *we have either* $Y \in \mathfrak{G}$ *or* $\complement Y \in \mathfrak{G}$, *then* $\mathfrak{G}$ *is an ultrafilter on* X.

Let $\mathfrak{F}$ be a filter containing $\mathfrak{G}$ (there is one, by hypothesis); then $\mathfrak{F}$ coincides with $\mathfrak{G}$; for if $Y \in \mathfrak{F}$ then $\complement Y \notin \mathfrak{F}$; hence $\complement Y \notin \mathfrak{G}$ and therefore $Y \in \mathfrak{G}$.

Example of an ultrafilter. The set of all subsets of a non-empty set X which contain a given element $a \in X$ is an ultrafilter; for it is a filter, and if Y is any subset of X then either $a \in Y$ or $a \in \complement Y$. Such ultrafilters are called *trivial.*

Apart from this example, we shall never prove the existence of an ultrafilter (even on a countably infinite set) except by using Theorem 1 (and therefore the axiom of choice).

Remark. If X contains at least two elements, there are at least two distinct ultrafilters on X, and therefore the ordered set of filters on X has no greatest element.

PROPOSITION 7. *Every filter* $\mathfrak{F}$ *on a set* X *is the intersection of the ultrafilters finer than* $\mathfrak{F}$.

Clearly this intersection contains $\mathfrak{F}$. Conversely, let A be a subset of X which does not belong to $\mathfrak{F}$, and let A′ denote $\complement A$; A contains no set of $\mathfrak{F}$; hence every $M \in \mathfrak{F}$ meets A′ and therefore (no. 2, Proposition 1, Corollary 1) there is a filter $\mathfrak{F}'$ which is finer than $\mathfrak{F}$ and contains A′. If $\mathfrak{U}$ is an ultrafilter finer than $\mathfrak{F}'$ (Theorem 1) it follows that $A \notin \mathfrak{U}$. This completes the proof.

5. INDUCED FILTER

PROPOSITION 8. *Let* $\mathfrak{F}$ *be a filter on a set* X *and* A *a subset of* X. *Then the trace* $\mathfrak{F}_A$ *of* $\mathfrak{F}$ *on* A *is a filter if and only if each set of* $\mathfrak{F}$ *meets* A.

Since $(M \cap N) \cap A = (M \cap A) \cap (N \cap A)$ we see that $\mathfrak{F}_A$ satisfies (F_{II}); again, if $M \cap A \subset P \subset A$ then $P = (M \cup P) \cap A$, whence $\mathfrak{F}_A$ satisfies (F_I). Hence $\mathfrak{F}_A$ is a filter if and only if it satisfies (F_{III}), i.e. if and only if each set of $\mathfrak{F}$ meets A.

In particular, if $A \in \mathfrak{F}$ then $\mathfrak{F}_A$ is a filter on A, by (F_{II}) and (F_{III}).

DEFINITION 5. *Let* A *be a subset of a set* X *and* $\mathfrak{F}$ *a filter on* X. *If the trace of* $\mathfrak{F}$ *on* A *is a filter on* A, *this filter is said to be induced by* $\mathfrak{F}$ *on* A.

If a filter $\mathfrak{F}$ on X induces a filter on $A \subset X$, then the trace on A of a base of $\mathfrak{F}$ is a base of $\mathfrak{F}_A$, by reason of Proposition 3 of no. 3.

Example. Let X be a topological space, A a subset of X, x a point of X. In order that the trace on A of the *neighbourhood filter* $\mathfrak{V}$ of x should be a filter on A, it is necessary and sufficient that every neighbourhood of x meets A, i.e. that x lies in the *closure* of A (§ 1, no. 6, Definition 10).

This example of an induced filter is of interest for two reasons: first because it plays an important role in the theory of limits (§ 7, no. 5) and secondly because *every filter can be defined in this way.* Indeed, let $\mathfrak{F}$ be a filter on a set X and let X′ be the set obtained by *adjoining* a new element ω to X, X being identified with the complement of $\{\omega\}$ in X′ (*Set Theory,* R, § 4, no. 5); let $\mathfrak{F}'$ be the filter on X′ consisting of the sets $M \cup \{\omega\}$ where M runs through $\mathfrak{F}$. For each point $x \neq \omega$ of X′, let $\mathfrak{V}(x)$ be the set of all subsets of X′ which contain x, and let $\mathfrak{V}(\omega)$ be $\mathfrak{F}'$; then the $\mathfrak{V}(x)$ for $x \in X'$ obviously satisfy axioms (V_I), (V_{II}), (V_{III}) and (V_{IV}) and therefore define a topology on X′ for which they are the neighbourhood filters of points. Finally ω lies in the *closure* of X in this topology, and $\mathfrak{F}$ is induced by $\mathfrak{F}' = \mathfrak{V}(\omega)$ on X. The topology thus defined on X′ (resp. the set X′ with this topology) is called the *topology* (resp. the *topological space*) *associated with* $\mathfrak{F}$.

PROPOSITION 9. *An ultrafilter* $\mathfrak{U}$ *on a set* X *induces a filter on a subset* A *of* X *if and only if* $A \in \mathfrak{U}$; *and if this condition is satisfied then* $\mathfrak{U}_A$ *is an ultrafilter on* A.

This is an immediate consequence of Propositions 5 and 6 of no. 4.

6. DIRECT IMAGE AND INVERSE IMAGE OF A FILTER BASE

Let $\mathfrak{B}$ be a filter base on a set X, and let f be a mapping of X into a set X′; then $f(\mathfrak{B})$ is a *filter base* on X′, for the relation $M \neq \emptyset$ implies $f(M) \neq \emptyset$, and we have $f(M \cap N) \subset f(M) \cap f(N)$. If $\mathfrak{B}_1$ is a base of a filter which is *finer* than the filter of base $\mathfrak{B}$, then $f(\mathfrak{B}_1)$ is a base of a filter *finer* than the filter of base $f(\mathfrak{B})$ (no. 3, Proposition 4).

PROPOSITION 10. *If* $\mathfrak{B}$ *is an ultrafilter base on a set* X *and if* f *is a mapping of* X *into a set* X′, *than* $f(\mathfrak{B})$ *is an ultrafilter base on* X′.

Let M′ be a subset of X′. If $\overset{-1}{f}(M')$ contains a set M of $\mathfrak{B}$, then M′ contains $f(M)$; if not, then $\complement\overset{-1}{f}(M') = \overset{-1}{f}\complement(M')$ contains a set N of $\mathfrak{B}$ (no. 4, Proposition 5) and therefore $\complement M'$ contains $f(N)$. Hence the result follows from Proposition 6 of no. 4.

Consider in particular the case where f is the canonical injection $A \to X$ of a subset A of a set X. If $\mathfrak{B}$ is a filter base on A then $f(\mathfrak{B})$ is a filter base on X. The filter $\mathfrak{F}$ on X generated by $f(\mathfrak{B})$ is called the *filter generated by* $\mathfrak{B}$ *when* $\mathfrak{B}$ *is considered as a filter base on* X. If $\mathfrak{B}$ is an *ultrafilter base* on A it is also an *ultrafilter base* on X by reason of Proposition 10.

Let us next examine whether the *inverse image* of a filter base is a filter base. Let $\mathfrak{B}'$ be a filter base on a set X′, and let f be a mapping of a set X into X′; then $\overset{-1}{f}(\mathfrak{B}')$ is a filter base on X *if and only if* $\overset{-1}{f}(M') \neq \varnothing$ *for each* $M' \in \mathfrak{B}'$. This is an immediate consequence of the relation $\overset{-1}{f}(M' \cap N') = \overset{-1}{f}(M') \cap \overset{-1}{f}(N')$ and of Definition 3 of no. 3. This condition can also be expressed by saying that *every set of* $\mathfrak{B}'$ *meets* $f(X)$ [or that the trace of $\mathfrak{B}'$ on $f(X)$ is a filter base]. If this condition is satisfied, then $f(\overset{-1}{f}(\mathfrak{B}'))$ is a base of a filter *finer* than the filter of base $\mathfrak{B}$.

If $\mathfrak{B}$ is a filter base on X it is clear that the above condition is satisfied by $\mathfrak{B}' = f(\mathfrak{B})$; $\overset{-1}{f}(f(\mathfrak{B}))$ is then a base of a filter *coarser* than the filter of base $\mathfrak{B}$.

Let A be a subset of a set X, φ the canonical injection $A \to X$; if $\mathfrak{B}$ is a filter base on X then $\overset{-1}{\varphi}(\mathfrak{B})$ is the same as $\mathfrak{B}_A$. If we express this as a filter base of A by means of the above condition, we recover part of Proposition 8 of no. 5.

7. PRODUCT OF FILTERS

Let $(X_\iota)_{\iota \in I}$ be a family of sets, and for each $\iota \in I$ let $\mathfrak{B}_\iota$ be a *filter base* on X_ι. Let $\mathfrak{B}$ be the set of subsets of the product set $X = \prod_{\iota \in I} X_\iota$ which are of the form $\prod_{\iota \in I} M_\iota$, where $M_\iota = X_\iota$ except for a *finite* number of indices and where $M_\iota \in \mathfrak{B}_\iota$ for each ι such that $M_\iota \neq X_\iota$. The formula $\left(\prod_{\iota \in I} M_\iota\right) \cap \left(\prod_{\iota \in I} N_\iota\right) = \prod_{\iota \in I} (M_\iota \cap N_\iota)$ shows that $\mathfrak{B}$ is a *filter base* on X. Note that the filter of base $\mathfrak{B}$ is also generated by the sets $\overset{-1}{\mathrm{pr}}_{\varkappa}(\mathrm{M}_{\varkappa})$, where $\mathrm{M}_{\varkappa} \in \mathfrak{B}_{\varkappa}$ and $\varkappa$ runs through I, since

$$\overset{-1}{\mathrm{pr}}_{\varkappa}(\mathrm{M}_{\varkappa}) = \mathrm{M}_{\varkappa} \times \prod_{\iota \neq \varkappa} \mathrm{X}_{\iota}.$$

DEFINITION 6. *Given a filter* $\mathfrak{F}_{\iota}$ *on each set* X_{ι} *of a family of sets* $(\mathrm{X}_{\iota})_{\iota \in \mathrm{I}}$, *the product of the filters* $\mathfrak{F}_{\iota}$ *is the filter on* $\mathrm{X} = \prod_{\iota \in \mathrm{I}} \mathrm{X}_{\iota}$ *which has as a base the set of subsets of* X *of the form* $\prod_{\iota \in \mathrm{I}} \mathrm{M}_{\iota}$, *where* $\mathrm{M}_{\iota} \in \mathfrak{F}_{\iota}$ *for each* $\iota \in \mathrm{I}$ *and* $\mathrm{M}_{\iota} = \mathrm{X}_{\iota}$ *for all but a finite number of indices. The product filter is denoted by* $\prod_{\iota \in \mathrm{I}} \mathfrak{F}_{\iota}$.

The reader may easily verify that the product of the filters $\mathfrak{F}_{\iota}$ can also be defined as the *coarsest* filter $\mathfrak{G}$ on X such that $\mathrm{pr}_{\iota}(\mathfrak{G}) = \mathfrak{F}_{\iota}$ for each $\iota \in \mathrm{I}$.

The preceding remarks show that if $\mathfrak{B}_{\iota}$ is a base of $\mathfrak{F}_{\iota}$ for each $\iota \in \mathrm{I}$, then $\mathfrak{B}$ is a *base* of the product filter $\prod_{\iota \in \mathrm{I}} \mathfrak{F}_{\iota}$ (no. 3, Proposition 3).

On a product $\mathrm{X} = \prod_{\iota \in \mathrm{I}} \mathrm{X}_{\iota}$ of topological spaces, the neighbourhood filter of any point $x = (x_{\iota})$ is the *product* of the neighbourhood filters of the x_{ι} (§ 4, no. 1).

The construction of a product filter $\mathfrak{F} = \prod_{\iota \in \mathrm{I}} \mathfrak{F}_{\iota}$ is simpler when the index set I is *finite*: a base of $\mathfrak{F}$ is then formed by *all* products $\prod_{\iota \in \mathrm{I}} \mathrm{M}_{\iota}$, where $\mathrm{M}_{\iota} \in \mathfrak{F}_{\iota}$ for each $\iota \in \mathrm{I}$. If $\mathrm{I} = \{1, 2, \ldots, n\}$ we write

$$\mathfrak{F}_1 \times \mathfrak{F}_2 \times \cdots \times \mathfrak{F}_n$$

instead of $\prod_{\iota \in \mathrm{I}} \mathfrak{F}_{\iota}$.

8. ELEMENTARY FILTERS

DEFINITION 7. *Let* $(x_n)_{n \in \mathrm{N}}$ *be an infinite sequence of elements of a set* X. *The elementary filter associated with the sequence* (x_n) *is the filter generated by the image of the Fréchet filter* (no. 1) *by the mapping* $n \to x_n$ *of* N *into* X.

It comes to the same thing to say that the elementary filter associated with (x_n) is the set of subsets M of X such that $x_n \in \mathrm{M}$ except for a *finite* number of values of n. If S_n denotes the set of all x_p such that $p \geqslant n$, then the sets S_n form a *base* of the elementary filter associated with the sequence (x_n).

The elementary filter associated with an infinite *subsequence* of a sequence (x_n) is *finer* than the elementary filter associated with (x_n) (cf. Exercise 15).

By definition, every elementary filter has a countable base. Conversely:

PROPOSITION 11. *If a filter* $\mathfrak{F}$ *has a countable base, it is the intersection of the elementary filters which are finer than* $\mathfrak{F}$.

Let us arrange the countable base of $\mathfrak{F}$ as a sequence $(A_n)_{n \in \mathbf{N}}$; if we put

$$B_n = \bigcap_{p=0}^{n} A_p,$$

then the B_n again form a base of $\mathfrak{F}$ (no. 3, Proposition 3) and we have $B_{n+1} \subset B_n$ for each n. Let a_n be any element of B_n for each $n \in \mathbf{N}$; then it is clear that $\mathfrak{F}$ is coarser than the elementary filter associated with (a_n). Hence the intersection $\mathfrak{I}$ of the elementary filters which are finer than $\mathfrak{F}$ exists and is finer than $\mathfrak{F}$; if $\mathfrak{I}$ is *strictly finer* than $\mathfrak{F}$ there exists a set $M \in \mathfrak{I}$ such that $B_n \cap \complement M \neq \emptyset$ for each n; if $b_n \in B_n \cap \complement M$, the elementary filter associated with the sequence (b_n) is finer than $\mathfrak{F}$ and does not contain M. This contradicts the definition of $\mathfrak{I}$.

Remark. A filter which is *coarser* than a filter with a countable base need not possess a countable base; for example, if X is an uncountable infinite set, then the filter consisting of the complements of finite subsets of X has no countable base (otherwise the set of finite subsets of X would be countable, contrary to assumption); nevertheless this filter is coarser than every elementary filter associated with an infinite sequence of distinct elements of X.

9. GERMS WITH RESPECT TO A FILTER

Let $\mathfrak{F}$ be a filter on a set X. On the set $\mathfrak{P}(X)$ of all subsets of X, the relation

"there exists $V \in \mathfrak{F}$ such that $M \cap V = N \cap V$"

between M and N is an *equivalence relation* R, for R is obviously reflexive and symmetric, and if M, N, P are three subsets of X such that $M \cap V = N \cap V$ and $N \cap W = P \cap W$, where V and W belong to $\mathfrak{F}$, it follows that $M \cap (V \cap W) = N \cap (V \cap W) = P \cap (V \cap W)$ and $V \cap W \in \mathfrak{F}$, so that R is transitive. The equivalence class mod R of a subset M of X is called the *germ of* M *with respect to* $\mathfrak{F}$; the quotient set $\mathfrak{P}(X)/R$ is called the *set of germs of subsets of* X (*with respect to* $\mathfrak{F}$).

The mappings $(M, N) \to M \cap N$ and $(M, N) \to M \cup N$ of

$$\mathfrak{P}(X) \times \mathfrak{P}(X)$$

into $\mathfrak{P}(X)$ are compatible with the equivalence relations $R \times R$ and R

(*Set Theory*, R, § 5, no. 8). For if $M \equiv M'$ (mod R) and $N \equiv N'$ (mod R) then there exist V and W in $\mathfrak{F}$ such that

$$M \cap V = M' \cap V \quad \text{and} \quad N \cap W = N' \cap W,$$

so that
$$(M \cap N) \cap (V \cap W) = (M' \cap N') \cap (V \cap W)$$
and
$$\begin{aligned}(M \cup N) \cap (V \cap W) &= (M \cap (V \cap W)) \cup (N \cap (V \cap W)) \\ &= (M' \cap (V \cap W)) \cup (N' \cap (V \cap W)) \\ &= (M' \cup N') \cap (V \cap W).\end{aligned}$$

Passing to the quotients, these mappings induce two mappings of $(\mathfrak{P}(X)/R) \times (\mathfrak{P}(X)/R)$ into $\mathfrak{P}(X)/R$, which we denote (by abuse of language) by $(\xi, \eta) \to \xi \cap \eta$ and $(\xi, \eta) \to \xi \cup \eta$ respectively. It is a straightforward exercise to verify that with respect to these laws of composition [defined throughout $\mathfrak{P}(X)/R$] every element is idempotent, and that each law is commutative and associative and distributive with respect to the other. Further, the relations $\xi = \xi \cap \eta$ and $\eta = \xi \cup \eta$ are equivalent; if we denote them (by abuse of language) by $\xi \subset \eta$, it is easily verified that this relation is an *ordering* on $\mathfrak{P}(X)/R$, with respect to which $\mathfrak{P}(X)/R$ is a *lattice* which has the germ of $\varnothing$ as least element and the germ of X as greatest element. Note that the relation $\xi \subset \eta$ means that there exist $M \in \xi$, $N \in \eta$ and $V \in \mathfrak{F}$ such that $M \cap V \subset N \cap V$.

Now let X′ be another set, and let Φ be the set of all mappings of a set of $\mathfrak{F}$ into X′. The relation on Φ

"there exists $V \in \mathfrak{F}$ such that f and g are defined and agree on V"

between f and g is an *equivalence relation* S; it is clear that S is reflexive and symmetric, and if f, g, h are three elements of Φ such that f and g are defined on $V \in \mathfrak{F}$, and g and h are defined and agree on $W \in \mathfrak{F}$, then f and h are defined and agree on $V \cap W \in \mathfrak{F}$, so that S is transitive. The equivalence class mod S of a mapping f of a set $V \in \mathfrak{F}$ into X′ is called the *germ of f (with respect to* $\mathfrak{F}$), and the quotient set $\tilde{\Phi} = \Phi/S$ is called the *set of germs of mappings of* X *into* X′ (*with respect to* $\mathfrak{F}$).

Remarks. 1) Every mapping f of a subset M of X into X′, where M belongs to $\mathfrak{F}$, is equivalent mod S to a mapping f_1 of X into X′ (which justifies the above terminology) : it is sufficient to extend f to X, e.g. by giving it a constant value on X — M.

2) The *characteristic functions* φ_M and φ_N of two subsets M and N of X have the same germ with respect to $\mathfrak{F}$ if and only if M and N have the same germ with respect to $\mathfrak{F}$.

Let X″ be a third set, φ a mapping of X′ into X″, Φ' the set of all mappings of a set of $\mathfrak{F}$ into X″. For each $f \in \Phi$, $\varphi \circ f$ belongs to Φ'; further, it is immediately seen that if $g \in \Phi$ has the same germ as f with respect to $\mathfrak{F}$, then $\varphi \circ f$ and $\varphi \circ g$ have the same germ with respect to $\mathfrak{F}$.

This germ therefore depends only on the germ $\tilde{f}$ of f with respect to $\mathfrak{F}$ and is denoted by $\varphi(\tilde{f})$. We thus define a mapping (denoted by φ, by abuse of language) of the set $\tilde{\Phi}$ of germs of mappings of X into X′, into the set $\tilde{\Phi}'$ of germs of mappings of X into X″.

Now let X'_i $(1 \leqslant i \leqslant n)$ be sets and

$$Y = \prod_{i=1}^{n} X'_i$$

their product; let Φ_i (resp. Φ) denote the set of all mappings of a set of $\mathfrak{F}$ into X'_i (resp. Y). If $f_i \in \Phi_i$ for $1 \leqslant i \leqslant n$ and if $M_i \in \mathfrak{F}$ is the domain and f_i, then the mapping $t \to (f_1(t), \ldots, f_n(t))$ is defined on

$$\bigcap_{i=1}^{n} M_i$$

and hence belongs to Φ; we denote this mapping (by abuse of language) by $(f_1, \ldots, f_n)$. Furthermore, if f_i and g_i belong to Φ_i and have the same germ with respect to $\mathfrak{F}$ (for $1 \leqslant i \leqslant n$), it is immediately seen that $(f_1, \ldots, f_n)$ and $(g_1, \ldots, g_n)$ have the same germ with respect to $\mathfrak{F}$; this germ therefore depends only on the germs $\tilde{f}_i$ of the f_i. If we denote it by $\Gamma(\tilde{f}_1, \ldots, \tilde{f}_n)$ then Γ is clearly a *bijection* of the product set

$$\prod_{i=1}^{n} \tilde{\Phi}_i$$

onto the set $\tilde{\Phi}$, where $\tilde{\Phi}_i$ (resp. $\tilde{\Phi}$) denotes the set of germs of mappings of X into X'_i (resp. Y) with respect to $\mathfrak{F}$; hence, by abuse of language, we shall generally write $(\tilde{f}_1, \ldots, \tilde{f}_n)$ instead of $\Gamma(\tilde{f}_1, \ldots, \tilde{f}_n)$ whenever there is no risk of confusion.

From what has been said, every mapping ψ of Y into a set X″ defines a mapping $(\tilde{f}_1, \ldots, \tilde{f}_n) \to \psi(\tilde{f}_1, \ldots, \tilde{f}_n)$ of

$$\prod_{i=1}^{n} \tilde{\Phi}_i$$

into the set $\tilde{\Phi}'$ of germs of mappings of X into X″.

In particular, if $n = 2$ and if X'_1, X'_2 and X″ are all equal to the same set X′ (so that ψ is a *law of composition* defined throughout X′), then ψ induces a law of composition defined throughout the set $\tilde{\Phi}$ of germs of mappings of X into X′. It is easily verified that if the law given on X′ is associative (resp. commutative) then so is the corresponding law on $\tilde{\Phi}$; if the law ψ on X′ has an identity element e', then the germ with respect to $\mathfrak{F}$ of the constant mapping $x \to e'$ is an identity element for the corresponding law on $\tilde{\Phi}$. Finally, if the law on X′ has an identity element e', then the germ $\tilde{f}$ of $f \in \Phi$ has an inverse in $\tilde{\Phi}$ if and only if there exists $V \in \mathfrak{F}$, contained in the domain of f, such that $f(t)$ is invertible in X' for each $t \in V$; if, for each $t \in V$, $g(t)$ denotes the inverse of $f(t)$ then the germ $\tilde{g}$ of g is the inverse of $\tilde{f}$ in $\tilde{\Phi}$. In particular if X' is a *group* with respect to the law ψ, then $\tilde{\Phi}$ is a group with respect to the corresponding law; likewise, if X' is a *ring* (resp. an *algebra* over a ring A) then $\tilde{\Phi}$ is a ring (resp. an algebra over A) with respect to the corresponding laws of composition.

10. GERMS AT A POINT

One of the commonest situations to which the definitions and results of no. 9 apply is that in which $\mathfrak{F}$ is the *neighbourhood filter* of a point a of a topological space X; instead of "*germs with respect to* $\mathfrak{F}$" we then speak of "*germs at the point* a". Notice that there is only one germ of neighbourhoods of a, namely the germ of the whole space X. The germs of *closed* sets are identical with the germs of sets which are *locally closed at the point* a, for if L is locally closed at a, then the germs of L and $\overline{L}$ at a are equal (§ 3, no. 1, Proposition 1). It follows that if ξ, η are two germs of locally closed sets at a, then so are $\xi \cup \eta$ and $\xi \cap \eta$.

Since a is in each $V \in \mathfrak{F}$, $f(a)$ is defined for each mapping f whose domain belongs to $\mathfrak{F}$; furthermore, if f and g have the same germ at a we must have $f(a) = g(a)$, so that $f(a)$ depends only on the germ $\tilde{f}$ of f at a, and is called the *value* of $\tilde{f}$ at a and is denoted by $\tilde{f}(a)$. It should be emphasized that the relation $\tilde{f}(a) = \tilde{g}(a)$ does not in general imply that $\tilde{f} = \tilde{g}$.

Let X', X'' be two topological spaces; b a point of X''; g, g' two mappings of X' into X'' having the same germ at b. If f, f' are two mappings of X into X' which are *continuous* at a and have the same germ at a and are such that $f(a) = b$, then $g \circ f$ and $g' \circ f'$ have *the same germ at the point* a; for if V' is a neighbourhood of b in X' such that $g(x') = g'(x')$ for all $x' \in V'$, then there is a neighbourhood V of a such that $f(V) \subset V'$, $f'(V) \subset V'$ and $f(x) = f'(x')$ for all $x \in V$, and the assertion follows. The germ of $g \circ f$ at a is then called the *composition* of the germs $\tilde{g}$ and $\tilde{f}$ of g and f respectively and is denoted by $\tilde{g} \circ \tilde{f}$.

7. LIMITS

1. LIMIT OF A FILTER

DEFINITION 1. *Let* X *be a topological space and* $\mathfrak{F}$ *a filter on* X. *A point* $x \in X$ *is said to be a limit point* (or simply a *limit*) *of* $\mathfrak{F}$, *if* $\mathfrak{F}$ *is finer than the neighbourhood filter* $\mathfrak{B}(x)$ *of* x; $\mathfrak{F}$ *is also said to converge* (or to be *convergent*) *to* x. *The point* x *is said to be a limit of a filter base* $\mathfrak{B}$ *on* X, *and* $\mathfrak{B}$ *is said to converge to* x, *if the filter whose base is* $\mathfrak{B}$ *converges to* x.

This definition, together with Proposition 4 of § 6, no. 3, gives the following criterion:

PROPOSITION 1. *A filter base* $\mathfrak{B}$ *on a topological space* X *converges to* x *if and only if every set of a fundamental system of neighbourhoods of* x *contains a set of* $\mathfrak{B}$.

In accordance with the terminology introduced in § 1, no. 2 we can state Proposition 1 in the following way: $\mathfrak{B}$ converges to x if and only if there are sets of $\mathfrak{B}$ *as near as we please to* x.

If a filter $\mathfrak{F}$ converges to x, then every filter *finer* than $\mathfrak{F}$ also converges to x, by reason of Definition 1. Likewise, if the topology of X is replaced by a *coarser* topology, the neighbourhood filter of x is replaced by a *coarser* filter (§ 2, no. 2, Proposition 3), and therefore $\mathfrak{F}$ still converges to x in this new topology.

We can therefore say that *the finer the topology, the fewer convergent filters there are in this topology*. In particular, in the discrete topology, the only convergent filters are the neighbourhood filters, for these are the trivial ultrafilters on X (§ 6, no. 4).

Let Φ be a set of filters on X, all of which converge to the same point x; the neighbourhood filter $\mathfrak{B}(x)$ is coarser than all the filters of Φ, hence also coarser than the *intersection* $\mathfrak{J}$ of these filters; in other words, $\mathfrak{J}$ also converges to x.

PROPOSITION 2. *A filter* $\mathfrak{F}$ *on a topological space* X *converges to a point* x *if and only if every ultrafilter which is finer than* $\mathfrak{F}$ *converges to* x.

This is an immediate consequence of the preceding remarks and Proposition 7 of § 6, no. 4.

In general a filter can have *several distinct limit points*; we shall revert to this question in § 8, no. 1.

2. CLUSTER POINT OF A FILTER BASE

DEFINITION 2. *In a topological space* X, *a point* x *is a cluster point of a filter base* $\mathfrak{B}$ *on* X *if it lies in the closure of all the sets of* $\mathfrak{B}$.

If x is a cluster point of a filter base $\mathfrak{B}$, it is also a cluster point of every *equivalent* filter base by reason of § 6, no. 3, corollary to Proposition 4; in particular, x is a cluster point of the *filter* whose base is $\mathfrak{B}$.

PROPOSITION 3. *A point* x *is a cluster point of a filter base* $\mathfrak{B}$ *if and only if every set of a fundamental system of neighbourhoods of* x *meets every set of* $\mathfrak{B}$.

This follows immediately from the definitions.

This proposition and Corollary 2 to Proposition 1 of § 6, no. 2 show that the property "x is a cluster point of the filter $\mathfrak{F}$" is equivalent to the property "there is a filter which is finer than both $\mathfrak{F}$ and the neighbourhood filter of x". In other words :

PROPOSITION 4. *A point* x *is a cluster point of a filter* $\mathfrak{F}$ *if and only if there is a filter finer than* $\mathfrak{F}$ *which converges to* x.

In particular, every *limit point* of a filter $\mathfrak{F}$ is a *cluster point* of $\mathfrak{F}$.

COROLLARY. *An ultrafilter* $\mathfrak{U}$ *converges to a point* x *if and only if* x *is a cluster point of* $\mathfrak{U}$.

If x is a cluster point of a filter $\mathfrak{F}$, it is also a cluster point of every filter *coarser* than $\mathfrak{F}$; likewise, if we replace the topology of X by a *coarser* topology, x remains a cluster point of $\mathfrak{F}$ in the new topology.

The set of cluster points of a filter base $\mathfrak{B}$ on X is by definition the set $\bigcap_{M \in \mathfrak{B}} \overline{M}$, whence

PROPOSITION 5. *The set of cluster points of a filter base on a topological space* X *is closed in* X.

PROPOSITION 6. *Let* $\mathfrak{B}$ *be a filter base on a subset* A *of a topological space* X. *Then every cluster point of* $\mathfrak{B}$ *in* X *belongs to* $\overline{A}$; *and conversely every point of* $\overline{A}$ *is a limit point of a filter on* A.

The first assertion is trivial; on the other hand, if $x \in \overline{A}$, the trace on A of the neighbourhood filter of x in X is a filter on A which evidently converges to x.

Remark. A filter on a topological space need have no cluster points (and *a fortiori* no limit points); for example, in an *infinite discrete space* the filter of complements of finite subsets has no cluster points. Spaces in which every filter has a cluster point play an important role in mathematics, and we shall study them in § 9.

3. LIMIT POINT AND CLUSTER POINT OF A FUNCTION

DEFINITION 3. *Let* f *be a mapping of a set* X *into a topological space* Y, *and let* $\mathfrak{F}$ *be a filter on* X. *A point* $y \in Y$ *is said to be a limit point* (or simply a *limit*) (resp. *cluster point*) *of* f *with respect to the filter* $\mathfrak{F}$ *if* y *is a limit point* (resp. *cluster point*) *of the filter base* $f(\mathfrak{F})$.

The relation "y is a limit of f with respect to the filter $\mathfrak{F}$" is written $\lim_{\mathfrak{F}} f = y$, or $\lim_{x, \mathfrak{F}} f(x) = y$, or $\lim_{x} f(x) = y$ if there is no risk of confusion.

From Definition 3 and Propositions 1 (no. 1) and 3 (no. 2) we deduce the following criteria :

PROPOSITION 7. *A point $y \in Y$ is a limit of f with respect to the filter $\mathfrak{F}$ if and only if, for each neighbourhood V of y in Y, there is a set $M \in \mathfrak{F}$ such that $f(M) \subset V$ (i.e. $\overset{-1}{f}(V) \in \mathfrak{F}$ for each neighbourhood V of y).*
A point $y \in Y$ is a cluster point of f with respect to $\mathfrak{F}$ if and only if for each neighbourhood V of y and each $M \in \mathfrak{F}$ there is a point $x \in M$ such that $f(x) \in V$.

Examples. 1) A sequence of points $(x_n)_{n \in \mathbf{N}}$ of a topological space is a mapping $n \to x_n$ of $\mathbf{N}$ into X. In analysis one frequently uses the notions of limit point and cluster point of such a mapping *with respect to the Fréchet filter* (§ 6, no. 1) on $\mathbf{N}$; if y is a limit of $n \to x_n$ with respect to the Fréchet filter, y is said to be a *limit of the sequence* (x_n) *as* n *tends to infinity*, and we write $\lim_{n \to \infty} x_n = y$. A cluster point of the mapping $n \to x_n$ with respect to the Fréchet filter is called a *cluster point of the sequence* (x_n).
Thus a point $y \in X$ is a limit (resp. cluster point) of a sequence (x_n) of points of X if it is a limit point (resp. cluster point) of the *elementary filter associated with* (x_n) (§ 6, no. 8).

The point y is a limit of a sequence (x_n) in X if and only if, for *every* neighbourhood V of y in X, *all but a finite number of the terms of the sequence* (x_n) *are in* V, i.e. there is an integer n_0 such that $x_n \in V$ for all $n \geqslant n_0$. Likewise y is a cluster point of the sequence (x_n) if and only if, for *every* neighbourhood V of y in X and *every* integer n_0, there is an integer $n \geqslant n_0$ such that $x_n \in V$.

2) More generally, let f be a mapping of a *directed* set A into a topological space X. If $x \in X$ is a limit (resp. cluster point) of f with respect to the *section filter* of A then x is said to be a *limit* (resp. *cluster point*) *of* f *with respect to the directed set* A, and we write $x = \lim_{z \in A} f(z)$.

If y is a limit (resp. cluster point) of a mapping $f: X \to Y$ with respect to a filter $\mathfrak{F}$ on X, then y remains a limit (resp. cluster point) of f with respect to $\mathfrak{F}$ if we replace the topology of Y by a *coarser* topology, or if we replace the filter $\mathfrak{F}$ by a *finer* (resp. *coarser*) filter.

PROPOSITION 8. *Let f be a mapping of a set X into a topological space Y; then $y \in Y$ is a cluster point of f with respect to $\mathfrak{F}$ if and only if there is a filter $\mathfrak{G}$ on X which is finer than $\mathfrak{F}$ and such that y is a limit of f with respect to $\mathfrak{G}$.*

For if y is a cluster point of f with respect to $\mathfrak{F}$, and if $\mathfrak{V}$ is the neighbourhood filter of y, then $\overset{-1}{f}(\mathfrak{V})$ is a filter base on X since every set of $\overset{-1}{f}(\mathfrak{V})$ meets every set of $\mathfrak{F}$ (§ 6, no. 6). This remark shows also that there is a filter $\mathfrak{G}$ on X which is finer than both $\mathfrak{F}$ and the filter with base $\overset{-1}{f}(\mathfrak{V})$ (§ 6, no. 2, Proposition 1, Corollary 2), hence that y is a limit point of f with respect to $\mathfrak{G}$.

Notice finally that if f is a mapping of a set X into a topological space Y, the set of cluster points of f with respect to a filter $\mathfrak{F}$ on X is *closed* in Y (no. 2, Proposition 5) and possibly empty.

Remark. If $y \in \mathrm{Y}$ is a limit (resp. cluster point) of a mapping $f : \mathrm{X} \to \mathrm{Y}$ with respect to a filter $\mathfrak{F}$ on X, then y is also a limit (resp. cluster point) of every function $g : \mathrm{X} \to \mathrm{Y}$ which has *the same germ* as f with respect to $\mathfrak{F}$ (§ 6, no. 9); y is said to be a *limit* (resp. *cluster point*) *of the germ* $\tilde{f}$ *of* f with respect to $\mathfrak{F}$.

4. LIMITS AND CONTINUITY

Let X, Y be two topological spaces, f a mapping of X into Y, $\mathfrak{V}$ the neighbourhood filter in X of a point $a \in \mathrm{X}$. Instead of saying that $y \in \mathrm{Y}$ is a limit of f with respect to the filter $\mathfrak{V}$ and writing $y = \lim_{\mathfrak{V}} f$, we use the special notation

$$y = \lim_{\iota \in \mathrm{I}} f(x),$$

and we say that y is a *limit of* f *at the point* a, or that $f(x)$ *tends to* y *as* x *tends to* a. Similarly, instead of saying that y is a cluster point of f with respect to $\mathfrak{V}$, we say that y *is a cluster point of* f *at the point* a.

A consideration of the definition of continuity (§ 2, no. 1, Definition 1) Proposition 7 of no. 3 shows that:

Proposition 9. *A mapping* f *of a topological space* X *into a topological space* Y *is continuous at a point* $a \in \mathrm{X}$ *if and only if* $\lim_{\iota \in \mathrm{I}} f(x) = f(a)$.

Corollary 1. *Let* X, Y *be two topological spaces,* f *a mapping of* X *into* Y *which is continuous at a point* $a \in \mathrm{X}$; *then, for every filter base* $\mathfrak{B}$ *on* X *which converges to* a, *the filter base* $f(\mathfrak{B})$ *converges to* $f(a)$. *Conversely if, for every ultrafilter* $\mathfrak{U}$ *on* X *which converges to* a, *the ultrafilter base* $f(\mathfrak{U})$ *converges to* $f(a)$, *then* f *is continuous at* a.

The first assertion is an immediate consequence of Proposition 9. To prove the second, suppose that f is not continuous at a; then there is a neighbourhood W of $f(a)$ in Y such that $\overset{-1}{f}(W)$ does not belong to the filter $\mathfrak{B}$ of neighbourhoods of a in X. Hence (§ 6, no. 4, Proposition 7) there is an ultrafilter $\mathfrak{U}$, finer than $\mathfrak{B}$, which does not contain $\overset{-1}{f}(W)$ and therefore contains its complement $A = X - \overset{-1}{f}(W)$ (§ 6, no. 4, Proposition 5); since $f(A) \cap W = \emptyset$, $f(\mathfrak{U})$ does not converge to $f(a)$.

Corollary 2. *Let* g *be a mapping of a set* Z *into a topological space* X, *which has a limit* a *with respect to a filter* $\mathfrak{F}$ *on* Z; *then if the map* $f: X \to Y$ *is continuous at* a, *the composition* $f \circ g$ *has* $f(a)$ *as a limit point with respect to* $\mathfrak{F}$.

5. LIMITS RELATIVE TO A SUBSPACE

Let X, Y be two topological spaces, let A be a subset of X, and let $a \in X$ be a point of the *closure* of A (but not necessarily in A). Let $\mathfrak{F}$ be the *trace* on A of the neighbourhood filter of a in X. If f is a mapping of A into Y, then instead of saying that $y \in Y$ is a limit of f with respect to $\mathfrak{F}$ and writing $y = \lim_{\mathfrak{F}} f$, we write

$$y = \lim_{x \to a,\, x \in A} f(x)$$

and we say that y is a *limit of* f *at* a, *relative to the subspace* A, or that $f(x)$ *tends to* y *as* x *tends to* a *while remaining in* A. We have then $y \in \overline{f(A)}$.

If $A = \complement\{a\}$ where a is *not* an isolated point of X, then we write $y = \lim_{x \to a,\, x \neq a} f(x)$ instead of $y = \lim_{x \to a,\, x \in A} f(x)$.

We make analogous definitions for cluster points.

If f is the *restriction* to A of a mapping $g: X \to Y$, we say that g has a limit (resp. cluster point) y relative to A at a point $a \in \overline{A}$, if y is a limit (resp. cluster point) of f at a, relative to A.

Let B be a subset of A and let $a \in X$ be a point of the closure of B; if y is a limit at a, *relative to* A, of a map f: $A \to Y$, then y is also a limit of f at a, *relative to* B; the converse is not necessarily true. But if V is a *neighbourhood* in X of a point $a \in \overline{A}$ and if f has a limit y at a, *relative to* $V \cap A$, then y is still a limit of f at a, *relative to* A.

Let a be a *non-isolated* point of X, so that a is in the closure of $\complement\{a\}$. Then a mapping f: $X \to Y$ is *continuous at* a if and only if we have $f(a) = \lim_{x \to a,\, x \neq a} f(x)$; this follows immediately from the definitions.

6. LIMITS IN PRODUCT SPACES AND QUOTIENT SPACES

PROPOSITION 10. *Let* X *be a set, let* $(Y_\iota)_{\iota \in I}$ *be a family of topological spaces, and for each* $\iota \in I$ *let* f_ι *be a mapping of* X *into* Y_ι. *Let* X *be given the coarsest topology* $\mathcal{T}$ *for which the* f_ι *are continuous. Then a necessary and sufficient condition for a filter* $\mathfrak{F}$ *on* X *to converge to* $a \in X$ *is that for each* $\iota \in I$ *the filter base* $f_\iota(\mathfrak{F})$ *should converge to* $f_\iota(a)$ *in* Y_ι.

The condition is necessary since the f_ι are continuous (no. 4, Proposition 9, Corollary 1). Conversely, suppose that the condition is satisfied, and let V be an open neighbourhood of a in X. By the definition of $\mathcal{T}$ (§ 2, no. 3, Proposition 4) there is a finite subset J of I, and for each $\iota \in J$ an open subset U_ι of Y_ι, such that $f_\iota(a) \in U_\iota$ for $\iota \in J$ and such that V contains the set

$$\bigcap_{\iota \in J} \overset{-1}{f}_\iota(U_\iota).$$

The hypothesis implies that $\overset{-1}{f}_\iota(U_\iota) \in \mathfrak{F}$ (no. 3, Proposition 7); since J is finite, it follows that

$$M = \bigcap_{\iota \in J} \overset{-1}{f}_\iota(U_\iota)$$

belongs to $\mathfrak{F}$, and $M \subset V$. This completes the proof.

COROLLARY 1. *A filter* $\mathfrak{F}$ *on a product space* $X = \prod_{\iota \in I} X_\iota$ *converges to a point* x *if and only if for each* $\iota \in I$ *the filter base* $\mathrm{pr}_\iota(\mathfrak{F})$ *converges to* $\mathrm{pr}_\iota(x)$.

COROLLARY 2. *Let* $f = (f_\iota)$ *be a mapping of a set* X *into a product space* $Y = \prod_{\iota \in I} Y_\iota$. *Then* f *has a limit* $y = (y_\iota)$ *with respect to a filter* $\mathfrak{F}$ *on* X *if and only if for each* $\iota \in I$ f_ι *has limit* y_ι *with respect to* $\mathfrak{F}$.

PROPOSITION 11. *Let* R *be an* open *equivalence relation on a topological space* X *and let* φ *be the canonical mapping* $X \to X/R$. *Then for each* $x \in X$ *and each filter base* $\mathfrak{B}'$ *on* X/R *which converges to* $\varphi(x)$, *there is a filter base* $\mathfrak{B}$ *on* X *which converges to* x *and is such that* $\varphi(\mathfrak{B})$ *is equivalent to* $\mathfrak{B}'$.

If U is any neighbourhood of x in X, then $\varphi(U)$ is a neighbourhood of $\varphi(x)$ in X/R (§ 5, no. 3, Proposition 5), hence there is a set $M' \in \mathfrak{B}'$ such that $M' \subset \varphi(U)$; if we put $M = U \cap \overset{-1}{\varphi}(M')$, then $M' = \varphi(M)$. This shows that as M' runs through $\mathfrak{B}'$ and U runs through the neighbourhood filter of x, the sets $U \cap \overset{-1}{\varphi}(M')$ form a filter base $\mathfrak{B}$ on X; clearly $\mathfrak{B}$ converges to x and $\varphi(\mathfrak{B})$ is equivalent to $\mathfrak{B}'$.

8. HAUSDORFF SPACES AND REGULAR SPACES

1. HAUSDORFF SPACES

PROPOSITION 1. *Let* X *be a topological space. Then the following statements are equivalent:*

(H) *Any two distinct points of* X *have disjoint neighbourhoods.*

(H^{i}) *The intersection of the closed neighbourhoods of any point of* X *consists of that point alone.*

(H^{ii}) *The diagonal of the product space* $X \times X$ *is a closed set.*

(H^{iii}) *For every set* I, *the diagonal of the product space* $Y = X^I$ *is closed in* Y.

(H^{iv}) *No filter on* X *has more than one limit point.*

(H^{v}) *If a filter* $\mathfrak{F}$ *on* X *converges to* x, *then* x *is the only cluster point of* $\mathfrak{F}$.

We shall prove the implications

$$(H) \Longrightarrow (H^{i}) \Longrightarrow (H^{v}) \Longrightarrow (H^{iv}) \Longrightarrow (H)$$

and

$$(H) \Longrightarrow (H^{iii}) \Longrightarrow (H^{ii}) \Longrightarrow (H).$$

(H) $\Longrightarrow$ (H^{i}): If $x \neq y$ there is an open neighbourhood U of x and an open neighbourhood V of y such that $U \cap V = \emptyset$; hence $y \notin \overline{U}$.

(H^{i}) $\Longrightarrow$ (H^{v}): Let $y \neq x$; then there is a closed neighbourhood V of x such that $y \notin V$, and by hypothesis there exists $M \in \mathfrak{F}$ such that $M \subset V$; thus $M \cap \complement V = \emptyset$. But $\complement V$ is a neighbourhood of y; hence y is not a cluster point of $\mathfrak{F}$.

(H^{v}) $\Longrightarrow$ (H^{iv}): Clear, since every limit point of a filter is also a cluster point.

(H^{iv}) $\Longrightarrow$ (H): Suppose $x \neq y$ and that every neighbourhood V of x meets every neighbourhood W of y. Then the sets $V \cap W$ form a base of a filter which has both x and y as limit points, which is contrary to hypothesis.

(H) $\Longrightarrow$ (H^{iii}): Let $(x) = (x_\iota)$ be a point of X^I which does not belong to the diagonal Δ. Then there are at least two indices λ, μ such that $x_\lambda \neq x_\mu$. Let V_λ (resp. V_μ) be a neighbourhood of x_λ (resp. x_μ) in X, such that $V_\lambda \cap V_\mu = \emptyset$; then the set $W = V_\lambda \times V_\mu \times \prod_{\iota \neq \lambda, \mu} X_\iota$ (where $X_\iota = X$ if $\iota \neq \lambda, \mu$) is a neighbourhood of x in X^I (§ 4, no. 1) which does not meet Δ. Hence Δ is closed in X^I.

(H^{iii}) $\Longrightarrow$ (H^{ii}): Obvious.

$(\mathrm{H^{ii}}) \Longrightarrow (\mathrm{H})$: If $x \neq y$ then $(x, y) \in \mathrm{X} \times \mathrm{X}$ is not in the diagonal Δ, hence (§ 4, no. 1) there is a neighbourhood V of x and a neighbourhood W of y in X such that $(\mathrm{V} \times \mathrm{W}) \cap \Delta = \varnothing$, which means that

$$\mathrm{V} \cap \mathrm{W} = \varnothing.$$

DEFINITION 1. *A topological space satisfying the conditions of Proposition 1 is called a Hausdorff or separated space; the topology of such a space is said to be a Hausdorff topology.*

Axiom (H) is *Hausdorff's axiom.*

Examples. Any discrete space is Hausdorff. The rational line **Q** is Hausdorff, for if x, y are two rational numbers such that $x < y$ then there is a rational number z such that $x < z < y$, and the neighbourhoods $]\leftarrow, z[$ of x and $]z, \rightarrow[$ of y do not intersect.

A set X having at least two points and carrying the coarsest topology (§ 2, no. 2) is not a Hausdorff space.

Let f: $\mathrm{X} \rightarrow \mathrm{Y}$ be a mapping of a set X into a *Hausdorff* space Y; then it follows immediately from Proposition 1 that f can have *at most one limit* with respect to a filter $\mathfrak{F}$ on X, and that if f has y as a limit with respect to $\mathfrak{F}$, then y is *the only cluster point* of f with respect to $\mathfrak{F}$.

PROPOSITION 2. *Let f, g be two continuous mappings of a topological space* X *into a* Hausdorff *space* Y; *then the set of all* $x \in \mathrm{X}$ *such that* $f(x) = g(x)$ *is closed in* X.

For this set is the inverse image of the diagonal of $\mathrm{Y} \times \mathrm{Y}$ under the mapping $x \rightarrow (f(x), g(x))$, which is continuous (§ 4, no. 1, Proposition 1). The result therefore follows from $(\mathrm{H^{ii}})$ and § 2, no. 1, Theorem 1.

COROLLARY 1 (Principle of extension of identities). *Let f, g be two continuous mappings of a topological space* X *into a Hausdorff space* Y. *If* $f(x) = g(x)$ *at all points of a dense subset of* X, *then* $f = g$.

In other words, a continuous map of X into Y (Hausdorff) is uniquely determined by its values at all points of a dense subset of X.

COROLLARY 2. *If f is a continuous mapping of a topological space* X *into a* Hausdorff *space* Y, *then the graph of f is closed in* $\mathrm{X} \times \mathrm{Y}$.

For this graph is the set of all $(x, y) \in \mathrm{X} \times \mathrm{Y}$ such that $f(x) = y$, and the two mappings $(x, y) \rightarrow y$ and $(x, y) \rightarrow f(x)$ are continuous.

PROPOSITION 3. *Let* $(x_i)_{1 \leqslant i \leqslant n}$ *be a finite family of distinct points of a Hausdorff space* X; *then each* x_i *has a neighbourhood* V_i *in* X *such that the* V_i $(1 \leqslant i \leqslant n)$ *are mutually disjoint.*

The proof is by induction on n: the case $n = 2$ is just the axiom (H). Let then W_i $(1 \leqslant i \leqslant n-1)$ be a neighbourhood of x_i such that the W_i are mutually disjoint. On the other hand, for $1 \leqslant i \leqslant n-1$ there is a neighbourhood T_i of x_i and a neighbourhood U_i of x_n which do not intersect. If we take V_i to be $W_i \cap T_i$ for $1 \leqslant i \leqslant n-1$ and

$$V_n = \bigcap_{i=1}^{n-1} U_i,$$

the conditions of the proposition are satisfied.

Corollary. *Every finite Hausdorff space is discrete.*

Proposition 4. *Every finite subset of a Hausdorff space is closed.*

For every subset consisting of a single point is closed by reason of axiom (HI).

Proposition 5. *Let X be a topological space and suppose that for each pair of distinct points x, y of X there is a continuous mapping f of X into a Hausdorff space X′ such that $f(x) \neq f(y)$. Then X is Hausdorff.*

Let V′ and W′ be disjoint neighbourhoods of $f(x)$ and $f(y)$ respectively in X′; then $\overset{-1}{f}(V')$ and $\overset{-1}{f}(W')$ are disjoint neighbourhoods of x and y respectively in X.

Corollary. *Every topology which is finer than a Hausdorff topology is Hausdorff.*

2. SUBSPACES AND PRODUCTS OF HAUSDORFF SPACES

A subspace A *of a Hausdorff space* X *is Hausdorff*, as we see by applying Proposition 5 of no. 1 to the canonical injection $A \to X$. Conversely, we have

Proposition 6. *If every point of a topological space* X *has a* closed *neighbourhood which is a Hausdorff subspace of* X, *then* X *is Hausdorff.*

Let $x \in X$ and let V be a closed neighbourhood of x in X such that the subspace V is Hausdorff. Then the closed neighbourhoods of x in V have $\{x\}$ as their intersection (axiom (H′)); but they are also closed neighbourhoods of x in X (§ 3, no. 1) and therefore X satisfies (HI).

There exist *non-Hausdorff* spaces in which every point has a Hausdorff neighbourhood (Exercise 7).

Proposition 7. *Every product of Hausdorff spaces is Hausdorff. Conversely, if a product of non-empty spaces is Hausdorff, then each factor is a Hausdorff space.*

Let $X = \prod_{\iota \in I} X_\iota$ be a product of topological spaces. Then if x, y are two distinct points of X, we have $pr_\iota x \neq pr_\iota y$ for some index ι, and Proposition 5 of no. 1 shows that X is Hausdorff if the X_ι are. Conversely, if X is Hausdorff and the X_ι are non-empty, then each X_ι is homeomorphic to a subspace of X (§ 4, no. 2, Proposition 4) and is therefore Hausdorff.

COROLLARY 1. *Let* X *be a set, let* $(Y_\iota)_{\iota \in I}$ *be a family of Hausdorff topological spaces, and for each* $\iota \in I$ *let* f_ι *be a mapping of* X *into* Y_ι. *Let* X *carry the coarsest topology* $\mathcal{T}$ *for which the* f_ι *are continuous. Then a necessary and sufficient condition for* X *to be Hausdorff is that for each pair of distinct points* x, y *of* X *we have* $f_\iota(x) \neq f_\iota(y)$ *for some index* $\iota \in I$.

The condition is sufficient by reason of Proposition 5 of no. 1. Conversely, suppose X is Hausdorff; let $Y = \prod_{\iota \in I} Y_\iota$ and let $f = (f_\iota)_{\iota \in I}$ be the mapping $x \to (f_\iota(x))$. By Proposition 7 above Y is Hausdorff, and by Proposition 3 of § 4, no. 1, $\mathcal{T}$ is the inverse image under f of the topology of Y. If $f(x) = f(y)$ for two distinct points x, y of X it is clear that every open set (in the topology $\mathcal{T}$) which contains x also contains y, contrary to the hypothesis that X is Hausdorff.

COROLLARY 2. *Let* $(X_\alpha, f_{\alpha\beta})$ *be an inverse system of topological spaces. If the* X_α *are Hausdorff, then* $X = \varprojlim X_\alpha$ *is Hausdorff and is a closed subspace of* $\prod_\alpha X_\alpha$.

The first assertion follows from the fact that X is a subspace of the Hausdorff space $\prod_\alpha X_\alpha$ (Proposition 7). To show that X is closed in the product space, let $F_{\alpha\beta}$ $(\alpha \leqslant \beta)$ be the subset of $\prod_\alpha X_\alpha$ consisting of points x for which $pr_\alpha x = f_{\alpha\beta}(pr_\beta x)$; the $F_{\alpha\beta}$ are closed in $\prod_\alpha X_\alpha$ (no. 1, Proposition 2), hence so is their intersection X.

Evidently, any *sum* of Hausdorff spaces (§ 2, no. 4, Example 3) is a Hausdorff space.

3. HAUSDORFF QUOTIENT SPACES

Let us look for conditions under which a quotient space X/R is *Hausdorff* (in which case the equivalence relation R is said to be *Hausdorff*). In the first place, if X/R is Hausdorff, then the subsets of X/R consisting of a single point are *closed* (no. 1, Proposition 4) and hence *each equivalence class* mod R *is closed in* X. But this necessary condition is not sufficient.

The definition of open sets in X/R gives rise to the following necessary and sufficient condition: X/R *is Hausdorff if and only if any two distinct equivalence classes in* X *are contained in disjoint saturated open subsets of* X. We shall give other more usable conditions.

PROPOSITION 8. *A necessary condition for the quotient space* X/R *to be Hausdorff is that the graph* C *of* R *is closed in* X × X. *If the equivalence relation* R *is open, this condition is also sufficient.*

Let φ : X → X/R be the canonical mapping; then C is the inverse image under $\varphi \times \varphi$: X × X → (X/R) × (X/R) of the diagonal Δ of (X/R) × (X/R). The first part of the proposition therefore follows from the continuity of $\varphi \times \varphi$ [Axiom (H^{II}) and Theorem 1 of § 2, no. 1]. If R is open then (X/R) × (X/R) can be identified with the quotient space (X × X)/(R × R) (§ 5, no. 3, corollary to Proposition 8), and Δ is then identified with the canonical image in (X × X)/(R × R) of the set C, which is saturated with respect to R × R. Hence Δ is closed in X × X and therefore X is Hausdorff.

If R is not open, there are examples where C is closed but R is not Hausdorff (Exercises 10 and 28).

To show that X/R is Hausdorff we can also apply Proposition 5 of no. 1: M and N being two distinct equivalence classes of R it is sufficient that there should be a continuous mapping f of an open subset A of X, saturated with respect to R and containing M and N, into a Hausdorff space X′, such that 1) f is constant on each equivalence class mod R contained in A, 2) f takes distinct values on M and N. For since A/R_A can be identified with an open subset of X/R (§ 3, no. 6, Proposition 10, Corollary 1), we can apply Proposition 5 of no. 1 to the mapping g : $A/R_A \to X'$ induced by f, since g is continuous (§ 3, no. 4, Proposition 6).

In particular:

PROPOSITION 9. *If* f *is a continuous mapping of a topological space* X *into a Hausdorff space* X′, *and* R *is the equivalence relation* $f(x) = f(y)$, *then the quotient space* X/R *is Hausdorff.*

PROPOSITION 10. *If* X *is a Hausdorff space, and if* X *has a continuous section* s *with respect to the equivalence relation* R, *then* X/R *is Hausdorff and* s(X/R) *is closed in* X.

For (§ 3, no. 5) X/R is homeomorphic to the subspace s(X/R) of X, which is Hausdorff. Furthermore s(X/R) is the set of all $x \in X$ such that $s(\varphi(x)) = x$, where φ : X → X/R is the canonical mapping; hence the second assertion follows from no. 1, Proposition 2.

4. REGULAR SPACES

PROPOSITION 11. *The following properties of a topological space* X *are equivalent:*

(O_{III}) *The set of closed neighbourhoods of any point of* X *is a fundamental system of neighbourhoods of the point.*

(O'_{III}) *Given any closed subset* F *of* X *and any point* $x \notin F$ *there is a neighbourhood of* x *and a neighbourhood of* F *which do not intersect.*

(O_{III}) $\Longrightarrow$ (O'_{III}): If F is closed and $x \notin F$, then there is a closed neighbourhood V of x contained in the neighbourhood $\complement F$ of x; V and $\complement V$ are neighbourhoods of x and F respectively, and have no point in common.

(O'_{III}) $\Longrightarrow$ (O_{III}): If W is an open neighbourhood of $x \in X$, then there is a neighbourhood U of x and a neighbourhood V of $\complement W$ which are disjoint, and therefore $\overline{U} \subset W$.

DEFINITION 2. *A topological space is said to be regular if it is Hausdorff and satisfies axiom* (O_{III}); *its topology is then said to be regular.*

A discrete space is regular. * We shall see in § 9 that every *locally compact* space (in particular the real line **R**) is regular. *

PROPOSITION 12. *Every subspace of a regular space is regular.*

Let A be a subspace of a regular space X. Since X is Hausdorff, so is A (no. 2); on the other hand, every neighbourhood of a point $x \in A$ with respect to A is of the form $V \cap A$, where V is a neighbourhood of x in X. Since X is regular there is a neighbourhood W of x in X which is closed in X and contained in V; $W \cap A$ is then a neighbourhood of x in A, closed in A and contained in $V \cap A$. Hence the result. Conversely:

PROPOSITION 13. *If every point* x *of a topological space* X *has a* closed *neighbourhood which is a regular subspace of* X, *then* X *is regular.*

X is Hausdorff by Proposition 6 of no. 2. Let x be any point of X and let V be a closed regular neighbourhood of x. If U is any neighbourhood of x contained in V, then U is a neighbourhood of x relative to V; hence by hypothesis there is a neighbourhood W of x in V which is closed in V and contained in U. But W is a neighbourhood of x in X since V is a neighbourhood of x in X, and W is closed in X since V is closed in X.

Remarks. 1) There are examples of *non-Hausdorff* spaces in which every point has a regular neighbourhood (Exercise 7).

2) There are spaces which are *Hausdorff but not regular* (Exercise 20).

3) A topology which is *finer* than a regular topology need not be regular (Exercise 20).

5. EXTENSION BY CONTINUITY; DOUBLE LIMIT

THEOREM 1. *Let* X *be a topological space,* A *a dense subset of* X, $f : A \to Y$ *a mapping of* A *into a regular space* Y. *A necessary and sufficient condition for* f *to extend to a continuous mapping* $\bar{f} : X \to Y$ *is that, for each* $x \in X$, $f(y)$ *tends to a limit in* Y *when* y *tends to* x *while remaining in* A. *The continuous extension* $\bar{f}$ *of* f *to* X *is then unique.*

The uniqueness of $\bar{f}$ follows from the principle of extension of identities (no. 1, Proposition 2, Corollary 1). It is clear that the condition is necessary, for if $\bar{f}$ is continuous on X, then for each $x \in X$ we have

$$\bar{f}(x) = \lim_{y \to x,\, y \in A} \bar{f}(y) = \lim_{y \to x,\, y \in A} f(y)$$

(§ 7, no. 5). Conversely, suppose that the condition is satisfied and *define*

$$\bar{f}(x) = \lim_{y \to x,\, y \in A} f(y)$$

for each $x \in X$; $\bar{f}(x)$ is a well-defined element of Y, since Y is Hausdorff. We have to show that $\bar{f}$ is *continuous* at each point $x \in X$. Let then V′ be a *closed* neighbourhood of $\bar{f}(x)$ in Y; then by hypothesis there is an *open* neighbourhood V of x in X such that $f(V \cap A) \subset V'$. Since V is a neighbourhood of each of its points, we have

$$\bar{f}(z) = \lim_{y \to z,\, y \in V \cap A} f(y)$$

for each $z \in V$, and from this it follows that $\bar{f}(z) \in \overline{f(V \cap A)} \subset V'$, since V′ is closed. The result now follows from the fact that the closed neighbourhoods of $f(x)$ form a fundamental system of neighbourhoods of $f(x)$ in Y.

The mapping $\bar{f}$ is said to be obtained by *extending* f *by continuity to* X.

In the statement of Theorem 1 the hypothesis that Y is regular cannot be weakened without imposing additional restrictions on X, A or f (Exercise 19).

COROLLARY. *Let* $\mathfrak{F}_1$ *be a filter on a set* X_1, *and* $\mathfrak{F}_2$ *a filter on a set* X_2; *let* $\mathfrak{F}_1 \times \mathfrak{F}_2$ *be the product filter* (§ 6, no. 7) *on* $X = X_1 \times X_2$, *and let* f *be a mapping of* X *into a regular space* Y. *Suppose that*

a) $\lim_{\mathfrak{F}_1 \times \mathfrak{F}_2} f$ *exists.*

b) $\lim_{x_2, \mathfrak{F}_2} f(x_1, x_2) = g(x_1)$ *exists for all* $x_1 \in X_1$.

Then $\lim_{x_1, \mathfrak{F}_1} g(x_1)$ *exists and is equal to* $\lim_{\mathfrak{F}_1 \times \mathfrak{F}_2} f$.

Let $X'_1 = X_1 \cup \{\omega_1\}$ (resp. $X'_2 = X_2 \cup \{\omega_2\}$) be the topological space *associated with the filter* $\mathfrak{F}_1$ (resp. $\mathfrak{F}_2$) (§ 6, no. 5, Example). In the product space $X' = X'_1 \times X'_2$ let X'' be the union of the subspaces $X_1 \times X'_2$ and $\{(\omega_1, \omega_2)\}$. X is clearly a dense subspace of X'', and the hypotheses imply that $f(y_1, y_2)$ tends to a limit when (y_1, y_2) tend to any point (x_1, x_2) of X'' whilst remaining in X. The existence of the extension of f by continuity to X'' then follows from Theorem 1. Since also (ω_1, ω_2) lies in the closure of $X_1 \times \{\omega_2\}$ relative to X'', the result follows immediately (§ 7, no. 5).

6. EQUIVALENCE RELATIONS ON A REGULAR SPACE

Proposition 14. *Let* X *be a regular space,* R *a closed equivalence relation on* X. *Then the graph* C *of* R *in* $X \times X$ *is closed.*

Let (a, b) be a point of $X \times X$ in the closure of C, and let V (resp. W) be a *closed* neighbourhood of a (resp. a neighbourhood of b) in X; then there is a point $(x, y) \in C \cap (V \times W)$. Since $x \in V$, y belongs to the saturation S of V with respect to R; hence each neighbourhood W of b meets S. By hypothesis S is *closed*, and therefore $b \in S$. Now let B be the saturation of $\{b\}$ with respect to R, then each closed neighbourhood V of a meets B; since by hypothesis B is *closed* and X is *regular*, it follows that $a \in B$ and therefore that $(a, b) \in C$. This completes the proof.

Corollary. *On a regular space, every equivalence relation which is both open and closed is Hausdorff.*

This follows from Proposition 14 and Proposition 8 of no. 3.

Proposition 15. *Let* X *be a regular space,* F *a closed subset of* X, R *the equivalence relation on* X *obtained by identifying all the points of* F [in other words, the equivalence relation whose equivalence classes are F (if $F \neq \emptyset$) and the sets $\{x\}$ where $x \in \complement F$]. *Then the quotient space* X/R *is Hausdorff.*

Let M and N be two distinct equivalence classes in X. If each of them consists of a single point in the complement of F, then there exist two disjoint open neighbourhoods of M and N in the Hausdorff subspace $\complement F$; these are neighbourhoods of M and N in X and are saturated with respect to R. If $M = F$ (so that $F \neq \emptyset$) and $N = \{b\}$ where $b \notin F$, then since X is regular there is an open neighbourhood of b and an open neighbourhood of F which do not intersect; these neighbourhoods are saturated with respect to R, and the proposition is proved.

Note that the quotient space X/R is not necessarily regular (Chapter IX, § 4, Exercise 14).

9. COMPACT SPACES AND LOCALLY COMPACT SPACES

1. QUASI-COMPACT SPACES AND COMPACT SPACES

DEFINITION 1. *A topological space* X *is said to be quasi-compact if it satisfies the following axiom:*

(C) *Every filter on* X *has at least one cluster point.*

A topological space is said to be compact if it is quasi-compact and Hausdorff.

It follows immediately from this axiom that if f is a mapping of a set Z into a quasi-compact space X, and $\mathfrak{F}$ is any filter on Z, then f has at least one cluster point with respect to $\mathfrak{F}$. In particular, every sequence of points of a quasi-compact space has at least one cluster point; but this condition is not equivalent to (C) (Exercise 11).

We give three axioms each of which is *equivalent to axiom* (C) :

(C′) *Every ultrafilter on X is convergent.*

(C′) ⟹ (C) : If $\mathfrak{F}$ is a filter on X then there is an ultrafilter finer than $\mathfrak{F}$ (§ 6, no. 4, Theorem 1). Since this ultrafilter converges to a point x, x is a cluster point of $\mathfrak{F}$.

(C) ⟹ (C′) : For if an ultrafilter has a cluster point then it converges to this point (§ 7, no. 2, Corollary to Proposition 4).

If f is a mapping of a set Z into a quasi-compact space X, and $\mathfrak{U}$ is an ultrafilter on Z, f has at least one limit point with respect to $\mathfrak{U}$ (§ 6, no. 6, Proposition 10).

(C″) *Every family of closed subsets of* X *whose intersection is empty contains a finite subfamily whose intersection is empty.*

(C) ⟹ (C″) : Let $\mathfrak{G}$ be a family of closed subsets of X with empty intersection. If every finite subfamily of $\mathfrak{G}$ has a non-empty intersection, then $\mathfrak{G}$ generates a filter (§ 6, no. 2, Proposition 1) which has a cluster point by hypothesis. This point belongs to all the sets of $\mathfrak{G}$ (since they are closed); so we have a contradiction.

(C″) $\Longrightarrow$ (C) : For if (C) is false then there is a filter $\mathfrak{F}$ on X which has no cluster point; hence the closures of the sets of $\mathfrak{F}$ form a family of closed subsets of X contradicting axiom (C″).

(C‴) (Axiom of Borel-Lebesgue) *Every open covering of* X *contains a finite open covering of* X.

(C‴) $\Longleftrightarrow$ (C″) by taking complements.

If X is quasi-compact, then every *locally finite* covering $\mathfrak{R}$ of X is *finite*. For each point of X has an open neighbourhood which meets only a finite number of sets of $\mathfrak{R}$, and by (C‴) a finite number of these neighbourhoods covers X.

Examples. 1) Every *finite* space is *quasi-compact*, and more generally every space in which there is only a finite number of open sets is quasi-compact. A finite space is compact if and only if it is discrete, for a finite Hausdorff space is discrete (§ 8, no. 1, Corollary to Proposition 3). Conversely, *every compact discrete space is finite*, for in such a space the sets consisting of a single point are open; hence the space is finite by (C‴).

2) Let X be a set, and give X the topology in which the closed sets are X and all finite subsets of X [this set of subsets clearly satisfies axioms (O'_I) and (O'_{II}) of § 1, no. 4]. The topological space so defined is *quasi-compact*. For if $(F_\iota)_{\iota\in I}$ is a family of closed subsets of X with empty intersection, then F_α is finite for some $\alpha \in I$. Let a_k $(1 \leqslant k \leqslant n)$ be the elements of F_α; then by hypothesis for each index k there is an index $\iota_k \in I$ such that $a_k \notin F_{\iota_k}$; the intersection of the F_{ι_k} $(1 \leqslant k \leqslant n)$ and F_α is therefore empty, whence axiom (C″) is satisfied. If X is infinite it is not Hausdorff.

Remark. Quasi-compact (non-Hausdorff) spaces are of use mainly in applications of topology to algebraic geometry and are seldom featured in other mathematical theories, where on the contrary compact spaces play an important role.

THEOREM 1. *Let* $\mathfrak{F}$ *be a filter on a quasi-compact space* X *and let* A *be the set of cluster points of* $\mathfrak{F}$. *Then every neighbourhood of* A *belongs to* $\mathfrak{F}$.

Let V be a neighbourhood of A and suppose it possible that every set of $\mathfrak{F}$ meets $\complement V$. The intersections of the sets of $\mathfrak{F}$ with $\complement V$ then form a base of a filter $\mathfrak{G}$ on X; X is quasi-compact, so that $\mathfrak{G}$ has at least one cluster point y, which does not belong to A, because the neighbourhood V of A does not meet some of the sets of $\mathfrak{G}$. But since $\mathfrak{G}$ is *finer* than $\mathfrak{F}$, y is also a cluster point of $\mathfrak{F}$, which is contrary to hypothesis.

Corollary. *For a filter on a compact space to converge it is necessary and sufficient that it has a single cluster point.*

Necessity by § 8, no. 1, Proposition 1; sufficiency by Theorem 1 above.

2. REGULARITY OF A COMPACT SPACE

Proposition 1. *Let* X *be a compact space,* x *a point of* X. *In order that a filter base* $\mathfrak{B}$ *formed of closed neighbourhoods of* x *should be a fundamental system of neighbourhoods of* x *it is necessary and sufficient that the intersection of the sets of* $\mathfrak{B}$ *consists of* x *alone.*

The condition is necessary since X is Hausdorff (§ 8, no. 1, Proposition 1). It is sufficient, for it signifies that x is the only cluster point of $\mathfrak{B}$; hence $\mathfrak{B}$ converges to x by the Corollary to Theorem 1 of no. 1.

Corollary. *Every compact space is regular.*

For it follows from axiom (H^{i}) (§ 8, no. 1, Proposition 1) that the filter base formed by *all* the closed neighbourhoods of an arbitrary point of the space satisfies the condition of Proposition 1.

The following proposition amplifies the Corollary to Proposition 1:

Proposition 2. *Let* X *be a compact space and let* A, B *be two disjoint closed subsets of* X. *Then there exist two open sets* U, V, *such that* $U \cap V = \emptyset$ *and* $A \subset U$ *and* $B \subset V$.

Suppose the conclusion is false. If every neighbourhood U of A meets every neighbourhood V of B, then the sets $U \cap V$ form a filter base $\mathfrak{B}$ on X, which therefore has a cluster point $x \in X$. Now x must lie in A, since if y is any point of X not in A there is a neighbourhood of y and a neighbourhood of A which do not intersect, since X is regular, and thus y cannot be a cluster point of $\mathfrak{B}$. Similarly x must lie in B and we have a contradiction.

This proposition has important consequences which will be examined in Chapter IX, § 4.

The non-Hausdorff quasi-compact space X of Example 2 of no. 1 does not satisfy axiom (O_{III}), nor *a fortiori* the property stated in Proposition 2, since any two non-empty open sets of this space always intersect.

3. QUASI-COMPACTS SETS; COMPACT SETS; RELATIVELY COMPACT SETS

Definition 2. *A subset* A *of a topological space* X *is said to be a quasi-compact* (resp. *compact*) *set if the subspace* A *is quasi-compact* (resp. *compact*).

A subset A of a topological space X is a quasi-compact set if and only if every covering of A by *open sets of* X contains a finite covering of A; this follows from axiom (C‴). In a Hausdorff space, the notions of quasi-compact and compact sets are the same, since every subspace is Hausdorff.

Examples. 1) In a topological space X, every finite subset is quasi-compact; the empty set and every set consisting of one point are compact.
2) In a topological space X, let $(x_n)_{n \in \mathbf{N}}$ be an infinite sequence of points which converges to a point a; then the set A consisting of the points x_n $(n \in \mathbf{N})$ and a is quasi-compact. For if (U_ι) is a covering of A by open sets of X, then $a \in U_\varkappa$ for some index $\varkappa$. $U_\varkappa$ is a neighbourhood of a and therefore there is only a finite number of indices n_k such that $x_{n_k} \notin U_\varkappa$. For each index k let ι_k be an index such that $x_{n_k} \in U_{\iota_k}$; then $U_\varkappa$ and the U_{ι_k} form a finite open covering of A.

PROPOSITION 3. *Every closed subset of a quasi-compact* (resp. *compact*) *space is quasi-compact* (resp. *compact*).

This is an immediate consequence of axiom (C″) if we remark that if A is closed in X then every set which is closed in A is closed in X.

PROPOSITION 4. *Every compact subset of a Hausdorff space is closed.*

Let A be a compact subset of a Hausdorff space X, and let x be any point of $\overline{A}$; we have to show that $x \in A$. By hypothesis, every neighbourhood of x meets A, and therefore the neighbourhood filter $\mathfrak{B}$ of x in X induces a filter $\mathfrak{B}_A$ on A; A is compact whence $\mathfrak{B}_A$ has a cluster point $y \in A$. Since the filter $\mathfrak{B}$ is coarser than the filter on X generated by $\mathfrak{B}_A$ (considered as a filter base on X), y is also a cluster point of $\mathfrak{B}$; hence $y = x$, because $\mathfrak{B}$ converges to x in X and X is Hausdorff (§ 8, no. 1, Proposition 1).

COROLLARY. *In a compact space* X *a subset* A *is compact if and only if it is closed in* X.

PROPOSITION 5. *The union of a finite family of quasi-compact subsets of a topological space is quasi-compact.*

It is sufficient to show that if A and B are two quasi-compact subsets of a topological space X, then $A \cup B$ is quasi-compact. Let $\mathfrak{R}$ be covering of $A \cup B$; then $\mathfrak{R}$ is a covering of A and a covering of B; hence $\mathfrak{R}$ contains a finite covering $\mathfrak{R}_1$ of A and a finite covering $\mathfrak{R}_2$ of B; $\mathfrak{R}_1 \cup \mathfrak{R}_2$ is thus a finite covering of $A \cup B$ contained in $\mathfrak{R}$.

DEFINITION 3. *A subset* A *of a topological space* X *is said to be relatively quasi-compact* (resp. *relatively compact*) *in* X *if* A *is contained in a quasi-compact* (resp. *compact*) *subset of* X.

To abbreviate, we say also that A is a "relatively quasi-compact set" (resp. "relatively compact set") when there is no ambiguity about X. In a *Hausdorff* space the notions of relatively quasi-compact set and relatively compact set are the same.

PROPOSITION 6. *If* X *is a Hausdorff space, a subset* A *of* X *is relatively compact if and only if* $\overline{A}$ *is compact.*

If A is relatively compact, then $\overline{A}$ is compact by Proposition 4 and its corollary; the reverse implication is self-evident.

PROPOSITION 7. *If* A *is a relatively quasi-compact subset of a topological space* X, *then every filter base on* A *has a cluster point in* X.

For if $A \subset K$, where K is a quasi-compact subset of X, then every filter base on A has a cluster point in K.

The converse of this proposition is not valid without restriction on X (Exercise 22).

Remark. In a non-Hausdorff space, a compact set need not be closed, and its closure need not be quasi-compact (Exercise 5); the intersection of two compact sets need not be quasi-compact (Exercise 5); the union of two compact sets need not be compact (Exercise 5).

4. IMAGE OF A COMPACT SPACE UNDER A CONTINUOUS MAPPING

THEOREM 2. *If* f *is a continuous mapping of a quasi-compact space* X *into a topological space* X′, *then the set* $f(X)$ *is quasi-compact.*

Let $\mathfrak{R}$ be a covering of $f(X)$ by open sets in X′; then $\overset{-1}{f}(\mathfrak{R})$ is an open covering of X (§ 2, no. 1, Theorem 1); hence there is a finite subset $\mathfrak{S}$ of $\mathfrak{R}$ such that $\overset{-1}{f}(\mathfrak{S})$ is a covering of X; but then $\mathfrak{S}$ is a covering of $f(X)$ and the theorem is proved.

COROLLARY 1. *Let* f *be a continuous mapping of a topological space* X *into a* Hausdorff *space* X′. *Then the image under* f *of any quasi-compact* (resp. *relatively quasi-compact*) *set in* X *is a compact* (resp. *relatively compact*) *set in* X′.

COROLLARY 2. *Every continuous mapping* f *of a quasi-compact space* X *into a Hausdorff space* X′ *is closed. If also* f *is bijective, then* f *is a homeomorphism.*

This follows immediately from Corollary 1 and Proposition 4 of no. 3.

In particular :

COROLLARY 3. *A Hausdorff topology which is coarser than the topology of a quasi-compact space must coincide with the latter.*

COROLLARY 4. *Let* X *be a topological space and* R *a Hausdorff equivalence relation on* X.

a) *If there is a quasi-compact set* K *in* X *which meets every equivalence class mod* R, *then* X/R *is compact and the canonical mapping of* K/R_K *onto* X/R *is a homeomorphism.*

b) *If* K *also meets each equivalence class in only one point, then* K *is a continuous section of* X *with respect to the relation* R (§ 3, no. 5).

Let f be the restriction to K of the canonical mapping $X \to X/R$. Since X/R is Hausdorff it follows from Corollary 1 that X/R is compact and from Corollary 2 that f is closed; hence the bijection $K/R_K \to X/R$ associated with f is a homeomorphism (§ 5, no. 2, Proposition 3). This deals with a); b) follows immediately, since we now have $K/R_K = K$.

5. PRODUCT OF COMPACT SPACES

THEOREM 3 (Tychonoff). *Every product of quasi-compact* (resp. *compact*) *spaces is quasi-compact* (resp. *compact*). *Conversely, if a product of non-empty spaces is quasi-compact* (resp. *compact*) *then each of the factors is quasi-compact* (resp. *compact*).

In view of the characterization of Hausdorff product spaces given in § 8, no. 2, Proposition 7 it is enough to prove the assertions for quasi-compact spaces. If $X = \prod_{\iota \in I} X_\iota$ is quasi-compact and non-empty, then $X_\iota = \mathrm{pr}_\iota(X)$ is quasi-compact by reason of Theorem 2 of no. 4. Conversely, suppose the X_ι are quasi-compact and let $\mathfrak{U}$ be an ultrafilter on X; then for each $\iota \in I$, $\mathrm{pr}_\iota(\mathfrak{U})$ is an ultrafilter base on X_ι (§ 6, no. 6, Proposition 10) which therefore converges by reason of axiom (C′); hence $\mathfrak{U}$ is convergent (§ 7, no. 6, Corollary 1 of Proposition 10) and therefore X is quasi-compact.

COROLLARY. *For a subset of a product of topological spaces to be relatively quasi-compact it is necessary and sufficient that each of its projections should be relatively quasi-compact in the corresponding factor.*

Necessity follows from Theorem 2 of § 4. To prove sufficiency, let A be a subset of $\prod_\iota X_\iota$ such that, for each index ι, $\mathrm{pr}_\iota(A)$ is contained in a quasi-compact subset K_ι of X_ι; then A is contained in the quasi-compact subset $\prod_\iota K_\iota$ of $\prod_\iota X_\iota$.

6. INVERSE LIMITS OF COMPACT SPACES

PROPOSITION 8. *Let* $(X_\alpha, f_{\alpha\beta})$ *be an inverse system of compact spaces indexed by a directed set* I *such that* $f_{\alpha\alpha}$ *is the identity mapping for each* $\alpha \in I$. *Let* $X = \varprojlim X_\alpha$ *be the inverse limit and* $f_\alpha : X \to X_\alpha$ *the canonical mapping* (§ 4, no. 4). *Then*

a) X *is compact and for each* $\alpha \in I$ *we have*

$$f_\alpha(X) = \bigcap_{\beta \geqslant \alpha} f_{\alpha\beta}(X_\beta). \tag{1}$$

b) *If the* X_α *are all non-empty then* X *is non-empty.*

X is a closed subspace of $\prod_\alpha X_\alpha$ (§ 8, no. 2, Proposition 7, Corollary 2) which is compact by Theorem 3 of no. 5 and Proposition 3 of no. 3. The other assertions are consequences of *Set Theory*, chap. III, § 7, no. 4, th. 1. We apply this theorem by taking $\mathfrak{S}_\alpha$ to be the set of closed subsets of X_α. The conditions (i) and (ii) are just axioms (O_I') and (C'') respectively; condition (iii) is satisfied since $\{x_\alpha\}$ is closed and $f_{\alpha\beta}$ continuous (§ 2, no. 1, Theorem 1), and lastly condition (iv) is satisfied by reason of Corollary 2 of Theorem 2 of no. 4.

COROLLARY 1. *Let* $(X_\alpha, f_{\alpha\beta})$ *be an inverse system of topological spaces indexed by a directed set, such that for each pair of indices* α, β *for which* $\alpha \leqslant \beta$ *and for each* $x_\alpha \in X_\alpha$, $\overset{-1}{f}_{\alpha\beta}(x_\alpha)$ *is compact. Then equation* (1) *is valid and* $\overset{-1}{f}_\alpha(x_\alpha)$ *is compact for each index* α *and each* $x_\alpha \in X_\alpha$.

For each $x_\alpha \in \bigcap_{\beta \geqslant \alpha} f_{\alpha\beta}(X_\beta)$ and each $\beta \geqslant \alpha$, let L_β denote $\overset{-1}{f}_{\alpha\beta}(x_\alpha)$. If $\alpha \leqslant \beta \leqslant \gamma$, then we have $f_{\beta\gamma}(L_\gamma) \subset L_\beta$ and the set of all indices $\beta \geqslant \alpha$ is cofinal in the index set. It follows immediately that the L_β ($\beta \geqslant \alpha$) form an inverse system of topological spaces (with the restrictions of the $f_{\beta\gamma}$ as mappings), whose inverse limit L is homeomorphic to $\overset{-1}{f}_\alpha(x_\alpha)$. Since by hypothesis the L_β are compact and not empty, the corollary follows from Proposition 8.

COROLLARY 2. *Let* $(X_\alpha, f_{\alpha\beta})$ *and* $(X'_\alpha, f'_{\alpha\beta})$ *be two inverse systems of topological spaces indexed by the same directed set* I, *and let* (u_α) *be an inverse system of mappings* $u_\alpha : X_\alpha \to X'_\alpha$. *Let* $X = \varprojlim X_\alpha$, $X' = \varprojlim X'_\alpha$, $u = \varprojlim u_\alpha$. *Then:*

a) *If* $x' = (x'_\alpha) \in X'$ *is such that* $\overset{-1}{u}_\alpha(x'_\alpha)$ *is compact and non-empty for each* $\alpha \in I$, *then* $\overset{-1}{u}(x')$ *is compact and non-empty.*

b) *If the* X_α *are compact, the* X'_α *Hausdorff and the* u_α *surjective and continuous, then* u *is surjective.*

Let L_α denote $\overset{-1}{u}_\alpha(x'_\alpha)$; then clearly the L_α form an inverse system of topological spaces (with the restrictions of the $f_{\alpha\beta}$ as maps) and $\overset{-1}{u}(x' = L)$ is the inverse limit of the L_α; assertion a) therefore follows from Proposition 8. Assertion b) is an immediate consequence, in view of Proposition 3 of no. 3.

7. LOCALLY COMPACT SPACES

Definition 4. *A topological space* X *is said to be locally compact if it is Hausdorff and if every point of* X *has a compact neighbourhood.*

Clearly a compact space is locally compact, but the converse is false; for example, every *discrete* space is locally compact, but not compact if *infinite.*

* As we shall see in Chapter IV, § 2, the *real line* **R** is locally compact, but not compact. *

Proposition 9. *Every locally compact space is regular.*

Let X be a locally compact space, then every point $x \in X$ has a compact neighbourhood V; since X is Hausdorff, V is closed (no. 3, Proposition 4). On the other hand, V is a regular subspace of X (no. 2, Corollary to Proposition 1) and therefore X is regular (§ 8, no. 4, Proposition 13).

Corollary. *In a locally compact space every point has a fundamental system of compact neighbourhoods.*

For the intersection of a closed neighbourhood of x and a compact neighbourhood of x is a compact neighbourhood of x (no. 3, Proposition 3).

There exist *non-Hausdorff* topological spaces in which every point has a fundamental system of compact neighbourhoods (Exercise 5).

The Corollary to Proposition 9 may be generalized as follows:

Proposition 10. *In a locally compact space* X, *every compact set* K *has a fundamental system of compact neighbourhoods.*

Let U be any neighbourhood of K. For each $x \in K$ there is a compact neighbourhood $W(x)$ of x contained in U. The interiors of the sets

$W(x)$ form an open covering of K as x runs through K; hence there exist a finite number of points $x_i \in K$ $(1 \leqslant i \leqslant n)$ such that the interiors of the $W(x_i)$ cover K. The union V of the $W(x_i)$ is therefore a compact neighbourhood of K contained in U (no. 3, Proposition 5).

PROPOSITION 11. *Let* X *be a locally compact space and* F *a subset of* X *such that* $F \cap K$ *is compact whenever* K *is a compact subset of* X. *Then* F *is closed in* X.

In view of Proposition 4 of no. 3, this follows from Proposition 3 a) of § 3, no. 1.

PROPOSITION 12. *In a Hausdorff space* X, *every locally compact subspace* A *is locally closed.*

By hypothesis, for every $x \in A$ there is a neighbourhood V of x in X such that $V \cap A$ is compact and therefore closed in V (no. 3, Proposition 4).

PROPOSITION 13. *Every locally closed subspace of a locally compact space* X *is locally compact.*

Let A be a locally closed subspace of X; then for each $x \in A$ there is a neighbourhood U of x in X such that $U \cap A$ is closed in U. Let $V \subset U$ be a compact neighbourhood of x in X; $V \cap A = V \cap (U \cap A)$ is closed in V and therefore compact (no. 3, Proposition 3). Since $V \cap A$ is a neighbourhood of x in A, the result is proved (it is clear that A is Hausdorff).

Theorem 1 (no. 1) and Corollary 2 of Theorem 2 (no. 4) *do not extend* to locally compact spaces which are not compact.

For example, in an infinite discrete space X, the filter consisting of those sets which contain a given point $x \in X$ and have a finite complement has x as its only cluster point but does not converge to x. Since any mapping f of X into a Hausdorff space X′ is continuous, the image under f of an arbitrary subset of X (which is closed in X, since X is discrete) will not in general be a closed subset of X′.

The proposition corresponding to Theorem 3 of no. 5 is the following:

PROPOSITION 14. a) *Let* $(X_\iota)_{\iota \in I}$ *be a family of locally compact spaces such that* X_ι *is compact for all but a finite number of indices. Then the product space* $X = \prod_{\iota \in I} X_\iota$ *is locally compact.*

b) *Conversely, if the product of a family* $(X_\iota)_{\iota \in I}$ *of non-empty topological spaces is locally compact, then the factors* X_ι *are compact for all but a finite number of indices, and the factors which are not compact are locally compact.*

a) Let $x = (x_\iota)$ be a point of X. For each index $\iota \in I$ such that X_ι is locally compact but not compact, let V_ι be a compact neighbourhood of x_ι in X_ι, and for all other indices ι put $V_\iota = X_\iota$. Then $\prod_{\iota \in I} V_\iota$ is a compact neighbourhood of x in X (no. 5, Theorem 3). Also X is Hausdorff by § 8, no. 2, Proposition 7, and is therefore locally compact.

b) If $X = \prod_{\iota \in I} X_\iota$ is locally compact and the X_ι non-empty, then each of the X_ι is homeomorphic to a closed subspace of X (§ 4, no. 2, Proposition 4 and § 4, no. 3, Corollary to Proposition 7), hence locally compact by Proposition 13. Let $a = (a_\iota)$ be a point of X and let V be a compact neighbourhood of a; since we have $\mathrm{pr}_\iota V = X_\iota$ for all but a finite number of indices (§ 4, no. 1), it follows from no. 4, Corollary 1 to Theorem 2, that the X_ι are compact except for a finite number of indices.

8. EMBEDDING OF A LOCALLY COMPACT SPACE IN A COMPACT SPACE

THEOREM 4 (Alexandroff). *If X is any locally compact space, there exists a compact space X′ and a homeomorphism f of X onto the complement of a point in X′. Furthermore, if X'_1 is another compact space such that there is a homeomorphism f_1 of X onto the complement of a point in X'_1, then there is a unique homeomorphism g of X′ onto X'_1 such that $f_1 = g \circ f$.*

Let us begin by proving the second assertion of the theorem. Let

$$f(X) = X' - \{\omega\} \qquad \text{and} \qquad f_1(X) = X'_1 - \{\omega_1\}.$$

If the homeomorphism g exists it must be unique, for by definition we have $g(x') = f_1(\overset{-1}{f}(x'))$ if $x' \neq \omega$ and therefore $g(\omega) = \omega_1$. It remains to show that the bijection $g\colon X' \to X'_1$ thus defined is bicontinuous; since X′ and X'_1 are interchangeable we need only show that the image under g of a neighbourhood of a point $x' \in X'$ is a neighbourhood of $g(x')$ in X'_1. This is obvious from the definition of g if $x' \neq \omega$. If $x' = \omega$, let V′ be an open neighbourhood of ω in X′; then $X' - V' = K$ is closed in X′ and therefore compact (no. 3, Proposition 3) and is contained in $f(X)$; hence $g(K) = f_1(\overset{-1}{f}(K))$ is compact (no. 4, Theorem 2, Corollary 1). It follows that $g(V') = X'_1 - g(K)$ is an open neighbourhood of ω_1 (no. 3, Proposition 4). Hence g is a homeomorphism.

To prove the first part of the theorem, let X′ be a set which is the sum of X and a set consisting of a single point ω. We define a topology on X′ by taking the set $\mathfrak{O}$ of open subsets of X′ to consist of all open subsets of X and all subsets of the form $(X - K) \cup \{\omega\}$, where K is a compact subset of X. Since any intersection of compact subsets of X is compact (no. 3, Propositions 3 and 4) and since any closed subset of a compact set is compact (no. 3, Proposition 3), it follows that $\mathfrak{O}$ satisfies axiom (O_I); and since any finite union of compact subsets of X is compact (no. 3, Proposition 5), $\mathfrak{O}$ also satisfies (O_{II}). Every compact subset of X is closed in X (no. 3, Proposition 4) and therefore the topology induced on X by that of X′ is the original topology on X. Thus it remains to show that X′ is compact. In the first place, X′ is *Hausdorff*. For if x, y are any two distinct points of X, they have disjoint open neighbourhoods V, W respectively in X, and V, W are open in X′; on the other hand each $x \in X$ has a compact neighbourhood K in X, which is also a neighbourhood of x in X′, while

$$(X - K) \cup \{\omega\}$$

is a neighbourhood of ω in X′ and manifestly does not meet K. Finally, X′ is *quasi-compact*. Let $(U_\lambda)_{\lambda \in L}$ be an open covering of X′; then for at least one index $\mu \in L$ we have $U_\mu = (X - K_\mu) \cup \{\omega\}$ where K_μ is a compact subset of X. Hence there is a finite subset H of L such that the sets U_λ (for $\lambda \in H$) cover K_μ; if $J = H \cup \{\mu\}$, then $(U_\lambda)_{\lambda \in J}$ is a finite open covering of X′, and the proof is complete.

Notice that if X is already *compact*, then ω is an *isolated* point of the compact space X′; hence X′ is the *sum* (§ 2, no. 4, Example 3) of the space X and the space $\{\omega\}$.

When a compact space X′ has been constructed as above from a locally compact space X by adjoining an element ω, it is often said that ω is the " point at infinity " of X′, and that X′ is obtained from X by adjoining a point at infinity. X′ is also called the *Alexandroff compactification* or the *one-point compactification* of the locally compact space X.

* *Example*. If we apply Alexandroff's theorem to the real plane $\mathbf{R}^2$, we get a compact space homeomorphic to the sphere $\mathbf{S}_2$ whose equation is $x_1^2 + x_2^2 + x_3^2 = 1$ in $\mathbf{R}^3$. A homeomorphism of these two spaces may be described as follows: the point ω (the point at infinity) adjoined to $\mathbf{R}^2$ is mapped to $(0, 0, 1) \in \mathbf{S}_2$, and every point (x_1, x_2) of $\mathbf{R}^2$ is mapped to the point where the line joining the points $(0, 1, 1)$ and $(x_1, x_2, 0)$ in $\mathbf{R}^3$ meets $\mathbf{S}_2$ again. This homeomorphism is known as *stereographic projection*.

9. LOCALLY COMPACT σ-COMPACT SPACES

DEFINITION 5. *A locally compact space* X *is said to be σ-compact or countable at infinity if it is a countable union of compact subsets.*

Examples. 1) A discrete space is σ-compact if and only if it is countable. * 2) The real line $\mathbf{R}$ is locally compact and σ-compact, since it is the union of the compact intervals $[-n, +n]$ for $n \in \mathbf{N}$. *

Remark. A Hausdorff space can be a countable union of compact subspaces without being locally compact. * An example is Hilbert space with the *weak* topology, as we shall show in a later volume.

PROPOSITION 15. *If* X *is a locally compact* σ*-compact space, there is a sequence* (U_n) *of relatively compact open subsets of* X *which cover* X, *such that* $\overline{U}_n \subset U_{n+1}$ *for each* n.

Let X be the union of a sequence (K_n) of compact sets. Let U_1 be a relatively compact open neighbourhood of K_1 (no. 7, Proposition 10) and define U_n inductively for $n > 1$ to be a relatively compact open neighbourhood of $\overline{U}_{n-1} \cup K_n$ (no. 3, Proposition 5; no. 7, Proposition 10). The sequence (U_n) clearly has the required properties.

COROLLARY 1. *With the notation of Proposition* 15, *every compact subset* K *of* X *is contained in some* U_n.

For K can be covered by a finite number of the U_k, by axiom (C''').

COROLLARY 2. *Let* X *be a locally compact space and let* X' *be the compact space obtained by adjoining a point at infinity* ω *to* X (no. 8). *Then* X *is* σ*-compact if and only if the point* ω *has a countable fundamental system of neighbourhoods in* X'.

If X is σ-compact we can construct a sequence of subsets U_n of X as in Proposition 15, and the neighbourhoods $X' - \overline{U}_n$ of ω in X' form a fundamental system of neighbourhoods of ω by reason of Corollary 1. The converse follows from the fact that the complements of open neighbourhoods of ω are compact subsets of X.

Clearly every *closed* subspace of a locally compact σ-compact space is locally compact and σ-compact. Likewise any *finite* product of locally compact σ-compact spaces is locally compact and σ-compact.

Notice, however, that an *open* subspace of a compact space need not be σ-compact, as Alexandroff's theorem (no. 8, Theorem 4) shows.

10. PARACOMPACT SPACES

DEFINITION 6. *A topological space is said to be paracompact if it is Hausdorff and satisfies the following axiom:*

(PC) *Every open covering* $\mathfrak{R}$ *of* X *has a* locally finite *open refinement* $\mathfrak{R}'$. (*Set Theory*, Chapter II, § 4, no. 6, Definition 5).

Every *compact* space is clearly paracompact. Every *discrete* space X is paracompact, for the open covering formed by all sets consisting of a single point of X is locally finite and is finer than every open covering of X.

PROPOSITION 16. *Every closed subspace* F *of a paracompact space* X *is paracompact.*

Certainly F is Hausdorff. On the other hand, if (V_ι) is an open covering in the subspace F, then each V_ι is of the form $V_\iota = U_\iota \cap F$, where U_ι is open in X. Consider the open covering $\mathfrak{R}$ of X formed by $\complement F$ and the U_ι; since X is paracompact, $\mathfrak{R}$ has a locally finite refinement $\mathfrak{R}'$, and the intersections with F of the sets belonging to $\mathfrak{R}'$ form a locally finite open covering of F which is finer than the given covering (V_ι).

On the other hand an open subspace of a compact space need not be paracompact (Exercise 11).

PROPOSITION 17. *The product of a paracompact space and a compact space is paracompact.*

Let X be a paracompact space, Y a compact space, $\mathfrak{R}$ an open covering of $X \times Y$. For each $(x, y) \in X \times Y$ there is an open neighbourhood $V(x, y)$ of x in X and an open neighbourhood $W(x, y)$ of y in Y such that $V(x, y) \times W(x, y)$ is contained in some set belonging to $\mathfrak{R}$. For each $x \in X$, the sets $W(x, y)$ as y runs through Y form an open covering of Y; hence there exist a finite number of points

$$y_i \in Y \qquad (1 \leqslant i \leqslant n(x))$$

such that the $W(x, y_i)$ cover Y. Let $U(x)$ denote

$$\bigcap_{i=1}^{n(x)} V(x, y_i);$$

then each of the open sets $U(x) \times W(x, y_i)$ is contained in a set of $\mathfrak{R}$. Now let $(T_\iota)_{\iota \in I}$ be a locally finite open covering of X which refines the covering $(U(x))_{x \in X}$. For each $\iota \in I$, let x_ι be a point of X such that $T_\iota \subset U(x_\iota)$, and let us denote by $S_{\iota, k}$ the sets $W(x_\iota, y_k)$ corresponding to this point x_ι $(1 \leqslant k \leqslant n(x_\iota))$. Clearly the sets

$$T_\iota \times S_{\iota, k} \qquad (\iota \in I,\ 1 \leqslant k \leqslant n(x_\iota) \text{ for each } \iota \in I)$$

form an open covering of $X \times Y$ which refines $\mathfrak{R}$, and the proof will be complete if we show that this covering is locally finite. Let (x, y) be any point of $X \times Y$; there is a neighbourhood Q of x which meets only a finite number of sets T_ι, and therefore the neighbourhood $Q \times Y$ of (x, y) meets only a finite number of sets $T_\iota \times S_{\iota, k}$.

On the other hand the product of two paracompact spaces need not be paracompact (see Chapter IX, § 5, Exercise 16).

PROPOSITION 18. *The sum (§ 2, no. 4, Example 3) of a family* $(X_\iota)_{\iota \in I}$ *of paracompact spaces is paracompact.*

Let X be the sum of the X_ι, and let $(V_\lambda)_{\lambda \in L}$ be an open covering of X. The covering formed by the open sets $X_\iota \cap Y_\lambda$ is finer than (V_λ). For each $\iota \in I$, let $(U_{\iota, \mu})_{\mu \in M_\iota}$ be a locally finite open refinement of the covering $(V_\lambda \cap X_\iota)_{\lambda \in L}$; then the open covering of X formed by the

$$U_{\iota, \mu} \quad (\iota \in I, \ \mu \in M_\iota \text{ for each } \iota \in I)$$

is locally finite and refines the original covering (V_λ).

THEOREM 5. *A locally compact space* X *is paracompact if and only if* X *is the sum of a family of locally compact* σ*-compact spaces.*

Suppose X is paracompact, and for each $x \in X$ let V_x be a relatively compact open neighbourhood of x in X. Then by hypothesis there exists a locally finite open covering $(U(\alpha))_{\alpha \in A}$ of X which refines the covering $(V_x)_{x \in X}$. The $U(\alpha)$ are therefore relatively compact. Every compact subset K of X meets only a finite number of the sets $U(\alpha)$, for the non-empty sets $U(\alpha) \cap K$ form a locally finite open covering of the compact space X; hence they must be finite in number (no. 1). Now let R be the following relation between two points x, y of X: "there exists a finite sequence $(\alpha_i)_{1 \leqslant i \leqslant n}$ of indices in A such that $x \in U(\alpha_1)$, $y \in U(\alpha_n)$ and $U(\alpha_i)$ meets $U(\alpha_{i+1})$ for $1 \leqslant i \leqslant n - 1$". It is immediately verified that R is an *equivalence relation*, and that each equivalence class $\bmod R$ is an *open* subset of X [since the $U(\alpha)$ are open sets]. X is therefore the *sum* of the locally compact subspaces (no. 7, Proposition 13) formed by the equivalence classes $\bmod R$, and it remains to show that each of these subspaces is the union of a *countable* subfamily of the family $(U(\alpha))$.

Let x be any point of X, and define a sequence (C_n) of relatively compact open subsets of X by induction on n as follows; C_1 is the union of the sets $U(\alpha)$ which contain x, and for each $n > 1$, C_n is the union of the sets $U(\alpha)$ which meet C_{n-1}. It is immediately verified by induction on n that each of the C_n is relatively compact and is the union of a *finite* number of sets $U(\alpha)$. Furthermore, the equivalence class of x with respect to R is the *union* of the C_n: for if $(\alpha_i)_{1 \leqslant i \leqslant n}$ is a sequence of indices such that $x \in U(\alpha_1)$ and $U(\alpha_i)$ meets $U(\alpha_{i+1})$ for $1 \leqslant i \leqslant n - 1$, then one sees by induction on i that $U(\alpha_i) \subset C_i$ for $1 \leqslant i \leqslant n$. It follows that the equivalence classes $\bmod R$ are σ-compact, and this completes the proof of the first part of the theorem.

To prove the converse, we may assume (by Proposition 18) that X is *σ-compact.* Let $\mathfrak{R} = (G_\lambda)_{\lambda \in L}$ be any open covering of X, and let (U_n) be a sequence of relatively compact open sets in X which have the properties stated in Proposition 15 of no. 9. Let K_n denote the compact set $\overline{U}_n - U_{n-1}$ ($U_n = \varnothing$ if $n \leqslant 0$). The open set $U_{n+1} - \overline{U}_{n-2}$ is a neighbourhood of K_n by construction; hence for each $x \in K_n$ there is a neighbourhood W_x of x contained in one of the sets G_λ and contained also in $U_{n+1} - \overline{U}_{n-2}$. Since K_n is compact, a finite number of the sets W_x cover K_n; let H_{ni} $(1 \leqslant i \leqslant p_n)$ be these sets. Then the family $\mathfrak{R}'$ of sets H_{ni} ($n \geqslant 1$, $1 \leqslant i \leqslant p_n$ for each n) is an open covering of X which refines $\mathfrak{R}$, and hence to complete the proof we have to show that $\mathfrak{R}'$ is *locally finite.* Let z be any point of X, n the smallest integer such that $z \in U_n$; then since $z \notin U_{n-1}$, there is a neighbourhood T of z which is contained in U_n and does not meet U_{n-2}. It follows that T meets only those sets H_{mi} for which $n - 2 \leqslant m \leqslant n + 1$, i.e. T meets only a finite number of sets of $\mathfrak{R}'$.

In the course of the proof we have also established the following result:

COROLLARY. *Let* X *be a locally compact paracompact space. Then every open covering* $\mathfrak{R}$ *of* X *has a locally finite open refinement* $\mathfrak{R}'$ *formed of relatively compact sets. If* X *is σ-compact then* $\mathfrak{R}'$ *can be taken to be countable.*

10. PROPER MAPPINGS

In this section we denote by ι_X *the identity mapping of a set* X *onto itself.*

1. PROPER MAPPINGS

If $f\colon X \to Y$ and $f'\colon X' \to Y'$ are two *continuous closed* mappings, the product $f \times f'\colon X \times X' \to Y \times Y'$ is not necessarily a closed map, even if f is of the form ι_X.

Example. Every constant mapping into a Hausdorff space is closed. But it f is the constant mapping $\mathbf{Q} \to 0$, then $f \times \iota_{\mathbf{Q}}$ is the mapping $(x, y) \to (0, y)$ of $\mathbf{Q}^2$ into $\mathbf{Q}^2$, so it is the second projection and is not closed (§ 4, no. 2, Remark 1).

DEFINITION 1. *Let* f *be a mapping of a topological space* X *into a topological space* Y. f *is said to be proper if* f *is continuous and if the mapping* $f \times \iota_Z$: $X \times Z \to Y \times Z$ *is closed, for every topological space* Z.

We shall give other characterizations of proper mappings in no. 2 and 3.

If in Definition 1 we take the space Z to consist of a single point, we see that:

Proposition 1. *Every proper mapping is closed.*

Proposition 2. *Let* $f: X \to Y$ *be a continuous* injection. *Then the following three statements are equivalent:*

a) *f is proper.*

b) *f is closed.*

c) *f is a homeomorphism of X onto a closed subset of Y.*

We have just seen that a) implies b). Since the equivalence relation $f(x) = f(x')$ is the equality relation, the quotient space of X with respect to this relation can be identified with X; hence b) implies c) by reason of § 5, no. 2, Proposition 3. Finally, if c) is satisfied then $f \times \iota_Z$ is a homeomorphism of $X \times Z$ onto a closed subspace of $Y \times Z$ and is therefore a closed mapping; hence c) implies a).

Proposition 3. *Let* $f: X \to Y$ *be a continuous mapping. If T is any subset of Y, let* f_T *denote the mapping* $\overset{-1}{f}(T) \to T$ *which agrees with* f *on* $\overset{-1}{f}(T)$.

a) *If f is proper, so is f_T.*

b) *Let* $(T(\iota))_{\iota \in I}$ *be a family of subsets of Y whose interiors cover Y, or which is a locally finite closed covering of Y; then if each of the mappings* $f_{T(\iota)}$ *is proper, so is* f.

Let Z be a topological space. If T is any subset of Y, we have

$$f_T \times \iota_Z = (f \times \iota_Z)_{T \times Z};$$

if f is proper, then $f \times \iota_Z$ is closed, hence so is $(f \times \iota_Z)_{T \times Z}$ [§ 5, no. 1, Proposition 2 a)], whence a) is proved. If now $(T(\iota))_{\iota \in I}$ satisfies one of the two conditions stated in b), then the covering $(T(\iota) \times Z)_{\iota \in I}$ of $Y \times Z$ has the same property; if the $f_{T(\iota)}$ are proper then the mappings

$$(f \times \iota_Z)_{T(\iota) \times Z}$$

are closed, whence $f \times \iota_Z$ is closed [§ 5, no. 1, Proposition 2 b)]. This completes the proof.

Proposition 4. *Let I be a finite set and for each* $i \in I$ *let* $f_i : X_i \to Y_i$ *be a continuous mapping. Let* $X = \prod_{i \in I} X_i$, $Y = \prod_{i \in I} Y_i$, *and let* $f: X \to Y$

be the product mapping $(x_i) \to (f_i(x_i))$. *Then:*

a) *If each of the* f_i *is proper, then* f *is proper.*

b) *If* f *is proper and if the* X_i *are non-empty, then each of the* f_i *is proper.*

(We shall see in no. 2, Theorem 1, Corollary 3, that this proposition extends to infinite products.)

By induction it is enough to consider the case where $I = \{1, 2\}$.

a) Suppose that f_1, f_2 are proper, and let Z be a topological space; $f_1 \times f_2 \times \iota_Z$ is the composition of $\iota_{Y_1} \times f_2 \times \iota_Z$ and

$$f_1 \times \iota_{X_2} \times \iota_Z;$$

these two mappings are closed by hypothesis, hence so is $f_1 \times f_2 \times \iota_Z$ [§ 5, no. 1, Proposition 1 a)], whence $f_1 \times f_2$ is proper.

b) Now suppose f is proper. Let F be a closed subset of $X_2 \times Z$ and let G be the image of F in $Y_2 \times Z$ under the mapping $f_2 \times \iota_Z$. Then the image of $X_1 \times F$ in $Y_1 \times Y_2 \times Z$ under $f_1 \times f_2 \times \iota_Z$ is $f_1(X_1) \times G$. By hypothesis, this is closed in $Y_1 \times Y_2 \times Z$; if $X_1 \neq \emptyset$, then $f_1(X_1)$ is not empty, which implies that G is closed in $Y_2 \times Z$ (§ 4, no. 3, Corollary to Proposition 7); hence f_2 is proper. Similarly f_1 is proper if $X_2 \neq \emptyset$.

Proposition 5. *Let* $f: X \to X'$ *and* $g: X' \to X''$ *be two continuous mappings.*

a) *If* f *and* g *are proper, then* $g \circ f$ *is proper.*

b) *If* $g \circ f$ *is proper and* f *is surjective, then* g *is proper.*

c) *If* $g \circ f$ *is proper and* g *is injective, then* f *is proper.*

d) *If* $g \circ f$ *is proper and* X' *is Hausdorff, then* f *is proper.*

Let Z be a topological space. We have

$$(g \circ f) \times \iota_Z = (g \times \iota_Z) \circ (f \times \iota_Z);$$

if f and g are proper, then $f \times \iota_Z$ and $g \times \iota_Z$ are closed; hence [§ 5, no. 1, Proposition 1 a)] $(g \circ f) \times \iota_Z$ is closed; this proves a). The proof of b) [resp. c)] runs along the same lines, using part b) [resp. c)] of Proposition 1 of § 5, no. 1, and remarking that if f is surjective (resp. if g is injective) then $f \times \iota_Z$ is surjective (resp. $g \times \iota_Z$ is injective). Finally, to prove d), consider the commutative diagram

$$\text{(1)} \qquad \begin{array}{ccc} X & \xrightarrow{\varphi} & X \times X' \\ f \downarrow & & \downarrow (g \circ f) \times \iota_{X'} \\ X' & \xrightarrow[\psi]{} & X'' \times X' \end{array}$$

where $\varphi(x) = (x, f(x))$ and $\psi(x') = (g(x'), x')$. The mapping φ (resp. ψ) is a homeomorphism of X (resp. X′) onto the graph of f (resp. the reflection of the graph of g) (§ 4, no. 1, Proposition 1, Corollary 2). Further, since X′ is Hausdorff, the graph $\varphi(X)$ of f is closed in $X \times X'$ (§ 8, no. 1, Proposition 2, Corollary 2). Hence (Proposition 2) φ is proper; on the other hand Proposition 4 shows that $(g \circ f) \times \iota_{X'}$ is proper. By a) above and the commutativity of the diagram (1), $\psi \circ f$ is proper; but ψ is injective and therefore f is proper by c) above.

Remark. If X′ is not Hausdorff it can happen that $g \circ f$ is proper and f not; for example, take X and X″ to consist of one point and X′ to consist of two points, with the coarsest topology.

COROLLARY 1. *If $f : X \to Y$ is a proper mapping, then the restriction of f to a closed subset F of X is a proper mapping of F into Y.*

For this restriction is the composition $f \circ j$, where $j : F \to X$ is the canonical injection, which is proper by Proposition 2.

COROLLARY 2. *Let $f : X \to Y$ be a proper mapping, where X is Hausdorff. Then the subspace $f(X)$ of Y is Hausdorff.*

By reason of Proposition 5 c) we need only consider the case where $f(X) = Y$. Then the diagonal of $Y \times Y$ is the image under $f \times f$ of the diagonal of X, which is closed (§ 8, no. 1, Proposition 1); $f \times f$ is proper (Proposition 4); hence the diagonal of $Y \times Y$ is closed (Proposition 1) and therefore Y is Hausdorff (§ 8, no. 1, Proposition 1).

COROLLARY 3. *Let I be a finite set and for each $i \in I$, let $f_i : X \to Y_i$ be a proper mapping. If X is Hausdorff, then the mapping $x \to (f_i(x))$ of X into $\prod_{\iota \in I} Y_i$ is proper.*

This mapping is the composition of the product mapping $(x_i) \to (f_i(x_i))$ of X^I into $\prod_i Y_i$ and the diagonal mapping of X into X^I; since the latter is proper (by Proposition 2 and § 8, no. 1, Proposition 1) the conclusion follows from Proposition 4 and Proposition 5 a).

COROLLARY 4. *Let X and Y be two topological spaces, $f : X \to Y$ a continuous mapping, R the equivalence relation $f(x) = f(y)$ on X, and*

$$X \xrightarrow{p} X/R \xrightarrow{h} f(X) \xrightarrow{i} Y$$

the canonical decomposition of f. Then for f to be proper it is necessary and sufficient that p is proper, h a homeomorphism and $f(X)$ a closed subset of Y.

The conditions are sufficient by virtue of Proposition 5 a) and Proposition 2. Conversely, if f is proper, then f is closed; hence $f(\mathrm{X})$ is closed in Y and h is a homeomorphism (§ 5, no. 2, Proposition 3); also $h \circ p$ is proper by Proposition 5 c); hence $p = \overset{-1}{h} \circ (h \circ p)$ is proper by Proposition 5 a).

2. CHARACTERIZATION OF PROPER MAPPINGS BY COMPACTNESS PROPERTIES

In this subsection we shall denote by P a space consisting of a single point, with its unique topology.

Lemma 1. *Let* X *be a topological space such that the constant mapping* $\mathrm{X} \to \mathrm{P}$ *is proper. Then* X *is quasi-compact.*

(We shall see a little later on (Theorem 1, Corollary 1) that this property *characterizes* quasi-compact spaces.)

We may restrict ourselves to the case where X is not empty. Let $\mathfrak{F}$ be a filter on X, and $\mathrm{X}' = \mathrm{X} \cup \{\omega\}$ the topological space associated with $\mathfrak{F}$ (§ 6, no. 5, Example). Let Δ be the subset of $\mathrm{X} \times \mathrm{X}'$ consisting of all (x, x) where $x \in \mathrm{X}$, and let $\mathrm{F} = \overline{\Delta}$ be the closure of Δ in $\mathrm{X} \times \mathrm{X}'$. In view of the hypothesis on X, the image of F under the projection $\mathrm{X} \times \mathrm{X}' \to \mathrm{X}'$ is closed in X'; this image contains X and therefore contains ω, which lies in the closure of X; in other words, there is a point $x \in \mathrm{X}$ such that $(x, \omega) \in \mathrm{F}$. By the definition of the topology of $\mathrm{X} \times \mathrm{X}'$, this means that, for each neighbourhood V of x in X and each $\mathrm{M} \in \mathfrak{F}$, we have $(\mathrm{V} \times \mathrm{M}) \cap \Delta \neq \emptyset$, i.e. $\mathrm{V} \cap \mathrm{M} \neq \emptyset$, so that x is a cluster point of the filter $\mathfrak{F}$, and therefore X is quasi-compact.

Q.E.D.

THEOREM 1. *Let* $f : \mathrm{X} \to \mathrm{Y}$ *be a continuous mapping. Then the following four statements are equivalent:*

a) f *is proper.*

b) f *is closed and* $\overset{-1}{f}(y)$ *is quasi-compact for each* $y \in \mathrm{Y}$.

c) *If* $\mathfrak{F}$ *is a filter on* X *and if* $y \in \mathrm{Y}$ *is a cluster point of* $f(\mathfrak{F})$ *then there is a cluster point* x *of* $\mathfrak{F}$ *such that* $f(x) = y$.

d) *If* $\mathfrak{U}$ *is an ultrafilter on* X *and if* $y \in \mathrm{Y}$ *is a limit point of the ultrafilter base* $f(\mathfrak{U})$, *then there is a limit point* x *of* $\mathfrak{U}$ *such that* $f(x) = y$.

a) $\Longrightarrow$ b) : If f is proper then f is closed (no. 1, Proposition 1) and for each $y \in \mathrm{Y}$ the mapping $f_{\{y\}} : \overset{-1}{f}(y) \to \{y\}$ is proper [no. 1, Proposition 3a)]. By Lemma 1, this implies that $\overset{-1}{f}(y)$ is quasi-compact.

b) $\Longrightarrow$ c) : Suppose $\mathfrak{F}$ and y satisfy the hypotheses of c). Let $\mathfrak{B}$ be the filter base on X formed by the closures of the sets of $\mathfrak{F}$. Since f is closed, we have $f(\overline{\mathrm{M}}) = \overline{f(\mathrm{M})}$ for each $\mathrm{M} \in \mathfrak{F}$ (§5, no. 4, Proposition 9). This shows that the sets $\overline{\mathrm{M}} \cap \overset{-1}{f}(y)$ are non-empty for all $\mathrm{M} \in \mathfrak{F}$, and hence form a filter base on $\overset{-1}{f}(y)$ whose elements are closed subsets of $\overset{-1}{f}(y)$. Since $\overset{-1}{f}(y)$ is quasi-compact, there is a point $x \in \overset{-1}{f}(y)$ which belongs to all the sets $\overline{\mathrm{M}}$ as M runs through $\mathfrak{F}$. Hence $f(x) = y$ and x is a cluster point of $\mathfrak{F}$.

c) $\Longrightarrow$ d) : Trivial.

d) $\Longrightarrow$ a) : We show first that if d) is satisfied, then f *is a closed mapping.* Let A be a non-empty closed subset of X and let $\mathfrak{F}$ be the filter of subsets of X which contain A. Then A is the set of cluster points of $\mathfrak{F}$. Let B be the set of cluster points of the filter base $f(\mathfrak{F})$ on Y; B is closed and clearly contains $f(\mathrm{A})$; we shall show that $\mathrm{B} = f(\mathrm{A})$. Let $y \in \mathrm{B}$ and let $\mathfrak{V}$ be the neighbourhood filter of y in Y; then by hypothesis every set of $\mathfrak{W} = \overset{-1}{f}(\mathfrak{V})$ meets every set of $\mathfrak{F}$; hence $\mathfrak{W}$ is a filter base on X and there is an ultrafilter $\mathfrak{U}$ on X which is finer than both $\mathfrak{F}$ and the filter whose base is $\mathfrak{W}$ (§ 6, no. 2, Proposition 1, Corollary 1 and no. 4, Theorem 1). The ultrafilter whose base is $f(\mathfrak{U})$ is finer than $\mathfrak{V}$ and therefore converges to y. By virtue of d) there is a point $x \in \mathrm{X}$ such that $f(x) = y$ and $\mathfrak{U}$ converges to x; since $\mathfrak{U}$ is finer than $\mathfrak{F}$, x is a cluster point of $\mathfrak{F}$; hence $x \in \mathrm{A}$. This shows that $\mathrm{B} = f(\mathrm{A})$ and therefore that f is closed.

To complete the proof we have to show that $f \times \iota_{\mathrm{Z}}$ is closed for every topological space Z. From what has been proved it is enough to show that if f satisfies condition d), then so does $f \times \iota_{\mathrm{Z}}$. This is a consequence of the following general lemma:

Lemma 2. *If* $(f_\iota)_{\iota \in \mathrm{I}}$ *is a family of continuous mappings* $f_\iota : \mathrm{X}_\iota \to \mathrm{Y}_\iota$ *each of which satisfies condition* d) *of Theorem* 1, *then the product mapping*

$$f : (x_\iota) \to (f_\iota(x_\iota))$$

also satisfies d).

Let $\mathfrak{U}$ be an ultrafilter on $\mathrm{X} = \prod_\iota \mathrm{X}_\iota$, and let $y = (y_\iota)$ be a point of $\mathrm{Y} = \prod_\iota \mathrm{Y}_\iota$ such that $f(\mathfrak{U})$ converges to y. This means that each of the ultrafilter bases $\mathrm{pr}_\iota(f(\mathfrak{U})) = f_\iota(\mathrm{pr}_\iota(\mathfrak{U}))$ converges to y_ι (§ 7, no. 6, Proposition 10, Corollary 1). By virtue of condition d), for each $\iota \in \mathrm{I}$ there exists $x_\iota \in \mathrm{X}_\iota$ such that $f_\iota(x_\iota) = y_\iota$ and $\mathrm{pr}_\iota(\mathfrak{U})$ converges to x_ι; but then $\mathfrak{U}$ converges to $x = (x_\iota)$ (*loc. cit.*) and we have $f(x) = y$. This completes the proof of Lemma 2 and hence of Theorem 1.

COROLLARY 1. *A topological space* X *is quasi-compact if and only if the mapping* $X \to P$ *is proper.*

Apply a) $\Leftrightarrow$ b) to $X \to P$.

COROLLARY 2. *Every continuous mapping* f *of a quasi-compact space* X *into a* Hausdorff *space* Y *is proper.*

The composition $X \xrightarrow{f} Y \to P$ is proper by Corollary 1; hence f is proper by no. 1, Proposition 5 d). Alternatively we may apply the criterion b) of Theorem 1, using § 9, no. 4, Theorem 2, Corollary 2.

COROLLARY 3. *If* (f_ι) *is a family of proper mappings, then the product mapping* $(x_\iota) \to (f_\iota(x_\iota))$ *is proper.*

In view of Theorem 1, this is just Lemma 2 above.

If we apply this corollary to the family of mappings $X_\iota \to P$ and use Corollary 1, we have Tychonoff's theorem (§ 9, no 5, Theorem 3).

COROLLARY 4. *Let* X *be a Hausdorff space, and let* $f_\iota : X \to Y_\iota$ *be a family of proper mappings. Then the mapping* $f: x \to (f_\iota(x))$ *of* X *into* $\prod_\iota Y_\iota$ *is proper.*

The proof is the same as in the case of a finite family (no. 1, Proposition 5, Corollary 3), using Corollary 3 above and the fact that the diagonal of X^I is closed (§ 8, no. 1, Proposition 1).

COROLLARY 5. *If* X *is any quasi-compact space and* Y *is any topological space, then the projection* $pr_2 : X \times Y \to Y$ *is proper.*

For we may identify Y with $P \times Y$ and pr_2 with the product of $X \to P$ and ι_Y, both of which are proper mappings.

Example. Let X be a set, and let $f: X \to X'$ be a mapping of X *onto* a topological space X'; topologize X with the inverse image under f of the topology of X'. Then f is *proper*, for f is closed (§ 5, no. 1, Example 3) and the inverse image of a point of X' is a subspace of X whose topology is the coarsest topology and is therefore quasi-compact.

Remark. When Y is *Hausdorff*, condition d) of Theorem 1 is equivalent to the following:

d') *If* $\mathfrak{U}$ *is an ultrafilter on* X *such that* $f(\mathfrak{U})$ *is a convergent filter base, then* $\mathfrak{U}$ *is convergent.*

For if $\mathfrak{U}$ converges to x and $f(\mathfrak{U})$ converges to y, then the uniqueness of the limit in Y and the continuity of f show that we must have $y = f(x)$. Likewise, Y being Hausdorff, condition c) of Theorem 1 is equivalent to:

c′) *If* $\mathfrak{F}$ *is a filter on* X *such that* $f(\mathfrak{F})$ *has a cluster point, then* $\mathfrak{F}$ *has a cluster point.*

For c) $\Longrightarrow$ c′) $\Longrightarrow$ d′) $\Longrightarrow$ d) $\Longrightarrow$ c).

On the other hand, if Y is not Hausdorff, then d′) no longer implies d); for example, take X to consist of one point and Y to consist of two points, with the coarsest topology.

PROPOSITION 6. *Let* $f : X \to Y$ *be a proper mapping, and let* K *be a quasi-compact subset of* Y. *Then* $\overset{-1}{f}(K)$ *is quasi-compact.*

By Proposition 3 of no. 1, the mapping $f_K : \overset{-1}{f}(K) \to K$ is proper. Since $K \to P$ is a proper mapping (Theorem 1, Corollary 1) it follows from no. 1, Proposition 5 a) that the composition $\overset{-1}{f}(K) \xrightarrow{f_K} K \to P$ is proper, whence $\overset{-1}{f}(K)$ is quasi-compact by Theorem 1, Corollary 1.

3. PROPER MAPPINGS INTO LOCALLY COMPACT SPACES

PROPOSITION 7. *Let* f *be a continuous mapping of a Hausdorff space* X *into a locally compact space* Y. *Then* f *is proper if and only if the inverse image under* f *of every compact subset of* Y *is compact. Further, if* f *is proper then* X *is locally compact.*

If f is proper and K is a compact subset of Y, then $\overset{-1}{f}(K)$ is compact by Proposition 6 of no. 2. Conversely, if this condition is satisfied, let (U_α) be a covering of Y by relatively compact open sets. Then the sets $\overset{-1}{f}(\overline{U}_\alpha)$ are compact in X and their interiors cover X; since X is Hausdorff this shows that X is locally compact. Furthermore, each of the mappings $f_{\overline{U}_\alpha}: \overset{-1}{f}({}_\alpha\overline{U}) \to \overline{U}_\alpha$ is proper (no. 2, Theorem 1, Corollary 2) and therefore f is proper by Proposition 3 b) of no. 1.

COROLLARY. *Let* X, X′ *be two locally compact spaces, and let* Y (resp. Y′) *be the compact space obtained by adjoining a point at infinity* ω (resp. ω') *to* X (resp. X′) (§ 9, no. 8). *Then a continuous mapping* $f : X \to X'$ *is proper if and only if its extension* $\overline{f} : Y \to Y'$, *such that* $\overline{f}(\omega) = \omega'$, *is continuous.*

By Proposition 7, f is proper if and only if, for each compact subset K′ of X′, $\overset{-1}{f}(X' - K') = X - \overset{-1}{f}(K')$ is the complement of a compact subset of X; by the definition of the neighbourhoods of ω (resp. ω') in Y (resp. Y′) this is so if and only if $\overline{f}$ is continuous at ω.

4. QUOTIENT SPACES OF COMPACT SPACES AND LOCALLY COMPACT SPACES

PROPOSITION 8. *Let* X *be a compact space,* R *an equivalence relation on* X, C *the graph of* R *in* $X \times X$, f *the canonical mapping* $X \to X/R$. *Then the following conditions are equivalent:*

a) C *is closed in* $X \times X$.

b) R *is closed.*

c) f *is proper.*

d) X/R *is Hausdorff.*

Furthermore, if these conditions are satisfied then X/R *is compact.*

R is closed if and only if f is closed; hence b) implies c) by reason of Theorem 1 b) of no. 2. That c) implies d) is a particular case of no. 1, Proposition 5, Corollary 2. d) implies a) for any topological space X (§ 8, no. 3, Proposition 8). It remains to show that a) implies b). If F is a closed subset of X its saturation (with respect to R) is $pr_2(C \cap (F \times X))$; by hypothesis, $C \cap (F \times X)$ is a closed subset of the compact space $X \times X$, and is therefore compact (§ 9, no. 3, Proposition 3); the result now follows from the continuity of pr_2 (§ 9, no. 4, Theorem 2, Corollary 2).

Finally it is clear that if X/R is Hausdorff then it is compact (§ 9, no. 4, Theorem 2).

PROPOSITION 9. *Let* X *be a locally compact space,* R *an equivalence relation on* X, C *the graph of* R *in* $X \times X$, f *the canonical mapping* $X \to X/R$; *let* X′ *be the compact space obtained by adjoining a point at infinity* ω *to* X, *and let* R′ *be the equivalence relation on* X′ *whose graph is* $C' = C \cup \{(\omega, \omega)\}$. *Then the following conditions are equivalent:*

a) f *is proper.*

b) *The saturation of each compact subset of* X *with respect to* R *is compact.*

c) R′ *is closed.*

d) *The restriction of* pr_2 *to* C *is proper.*

e) R *is closed and the equivalence classes with respect to* R *are compact.*

Furthermore, if these conditions are satisfied, X/R *is locally compact.*

a) $\Longrightarrow$ b): Since $X/R = f(X)$ and f is proper, X/R is Hausdorff (no. 1, Proposition 5, Corollary 2); hence the image under f of every compact subset K of X is compact (§ 9, no. 4, Theorem 2, Corollary 1). The saturation of K with respect to R is $\overset{-1}{f}(f(K))$ and is therefore compact by no. 2, Proposition 6.

b) $\Longrightarrow$ c) : If F' is closed in X' and does not contain ω, then F' is a compact subset of X; hence its saturation with respect to R', which is the same as its saturation with respect to R, is compact and *a fortiori* closed in X'. If $\omega \in F'$ and if $F = F' \cap X = F' - \{\omega\}$, then the saturation of F' with respect to R' is the union of $\{\omega\}$ and the saturation H of F with respect to R; hence it is enough to show that H is *closed* in X (i.e. that R is a *closed* relation). For this, it is enough to show that if K is any compact subset of X then $H \cap K$ is compact (§ 9, no. 7, Proposition 11). Now the saturation L of K with respect to R is compact by hypothesis, and $H \cap L$ is the saturation of $F \cap L$, which is also compact; *a fortiori* $H \cap K = (H \cap L) \cap K$ is compact.

c) $\Longrightarrow$ d) : Since X' is regular (§ 9, no. 2, Corollary to Proposition 1), C' is *closed* in $X' \times X'$ (§ 8, no. 6, Proposition 14) and therefore compact. It follows that C' is the one-point compactification of C (§ 9, no. 8, Theorem 4). Since the restriction to C' of $pr_2 : X' \times X' \to X'$ is continuous at ω, the result follows from no. 3, Corollary to Proposition 7.

d) $\Longrightarrow$ e) : If F is any closed subset of X, then $C \cap (F \times X)$ is closed in C, whence the saturation of F with respect to R, which is equal to $pr_2(C \cap (F \times X))$, is closed in X (no. 1, Proposition 1). Also the equivalence class of $x \in X$ mod R is homeomorphic to the inverse image of $\{x\}$ under the restriction of pr_2 to C and is therefore compact [no. 2, Theorem 1 b)].

e) $\longrightarrow$ a) : If R is closed, then by definition f is closed, and for each $z \in X/R$, $\overset{-1}{f}(z)$ is an equivalence class mod R and is therefore compact; hence f is proper by Theorem 1 b) of no. 2.

Finally we have to prove that X/R is locally compact. X'/R' is compact by c) and Proposition 8; the relation R is that induced on X by R'; X is open in X' and is saturated with respect to R'; hence X/R is homeomorphic to the image $f'(X)$ of X under the canonical mapping $f' : X' \to X'/R'$ (§ 3, no. 6, Proposition 10, Corollary 1). Now $f'(X)$ is open in X'/R', and hence is a locally compact subspace of X'/R'.

Q.E.D.

Corollary. *Let* X *be a Hausdorff space,* Y *a topological space,* $f : X \to Y$ *a proper mapping. Then for* X *to be compact* (resp. *locally compact*) *it is necessary and sufficient that* $f(X)$ *is compact* (resp. *locally compact*), *and it is sufficient that* Y *is compact* (resp. *locally compact*).

If X is compact (resp. locally compact) the fact that $f(X)$ is compact (resp. locally compact) is a consequence of no. 1, Proposition 5, Corollary 4 and of Propositions 8 and 9 (the case where X is compact is also a consequence of no. 1, Proposition 5, Corollary 2 and § 9, no. 4, Theorem 2). Conversely, if $Z = f(X)$ is compact (resp. locally compact) then since

$f_Z : X \to f(X)$ is proper [no. 1, Proposition 3 a)] it follows that X is compact (resp. locally compact) by reason of Proposition 6 of no. 2 and Proposition 7 of no. 3. Finally, if Y is compact (resp. locally compact) then so is $f(X)$ which is closed in Y (no. 1, Proposition 1 and § 9, no. 7, Proposition 13).

Remark. If X is locally compact but not compact, then a *closed* equivalence relation R on X *need not be Hausdorff* (Chapter IX, § 4, Exercise 14); and even if R is Hausdorff, X/R need not be locally compact (Exercise 17). However, we have the following criterion:

PROPOSITION 10. *Let* X *be a locally compact space,* R *an* open Hausdorff *equivalence relation on* X, *and let* $f : X \to X/R$ *be the canonical mapping. Then* X/R *is locally compact, and if* K′ *is any compact subset of* X/R *there is a compact subset* K *of* X *such that* $f(K) = K'$.

The first assertion is a consequence of the facts that each $x \in X$ has a compact neighbourhood V and that $f(V)$ is a compact neighbourhood of $f(x)$ (§ 5, no. 3, Proposition 5 and § 9, no. 4, Theorem 2, Corollary 1). For each $y \in K'$ let $V(y)$ be a compact neighbourhood of some point of $\overset{-1}{f}(y)$ in X, so that $f(V(y))$ is a compact neighbourhood of y. There are a finite number of points $y_i \in K'$ such that the $f(V(y_i))$ cover K′. Let K_1 be the compact set $\bigcup_i V(y_i)$ in X; we have $K' \subset f(K_1)$; hence $K = K_1 \cap \overset{-1}{f}(K')$ is compact (because it is closed in K_1) and $f(K) = K'$.

11. CONNECTEDNESS

1. CONNECTED SPACES AND CONNECTED SETS

DEFINITION 1. *A topological space* X *is said to be connected if it is not the union of two disjoint non-empty open sets.*

An equivalent definition is obtained by replacing the words "open sets" by "closed sets". X is connected if and only if the only subsets of X which are both open and closed are the empty set and the whole space X.

If X is connected and if A, B are two non-empty open (resp. closed) subsets such that $A \cup B = X$, then $A \cap B \neq \emptyset$.

Examples. * 1) We shall see in Chapter IV, § 2, no. 5 that the real line is connected, and that the rational line is not. *

2) A discrete space which has more than one point is not connected.

Observe that if $(U_\iota)_{\iota \in I}$ is a *partition* of a topological space X consisting of *open* sets [necessarily *non-empty*, from the definition of a partition (*Set Theory*, R, § 4, no. 4)] then each of the U_ι is *both open and closed* in X, for the complement of U_ι is the union of the $U_\varkappa$ for which $\varkappa \neq \iota$. The open subsets of X are the subsets A such that $A \cap U_\iota$ is open in U_ι for each $\iota \in I$, so that X can be identified with the *sum* of the U_ι (§ 2, no. 4, Example 3) and is not connected if I has more than one element.

Definition 2. *A subset* A *of a topological space* X *is said to be a connected set if the subspace* A *of* X *is connected.*

For A to be a connected subset of X it is necessary and sufficient that, for each covering of A by two open (or closed) subsets B, C of X such that $A \cap B$ and $A \cap C$ are non-empty, we have $A \cap B \cap C \neq \emptyset$.

Examples. In any topological space, the empty set and every set consisting of a single point are connected. In a *Hausdorff* space X, every *finite* set consisting of more than one point is not connected, and more generally every subset of X which has more than one point and which has at least one *isolated* point is not connected.

If a *dense* subset A is connected, then the whole space X is connected; otherwise there would exist two non-empty disjoint open subsets M, N of X such that $M \cup N = X$, and $M \cap A$, $N \cap A$ would be two disjoint non-empty open subsets of A whose union is A. Hence we have

Proposition 1. *If* A *is a connected set, then every set* B *such that* $A \subset B \subset \overline{A}$ *is connected.*

Proposition 2. *The union of a family of connected sets whose intersection is non-empty is connected.*

Let $(A_\iota)_{\iota \in I}$ be a family of connected subsets of X, all of which contain the same point x; we have to show that

$$A = \bigcup_\iota A_\iota$$

is connected. If not, there are two open sets B and C such that $B \cap A$ and $C \cap A$ are non-empty, and $A \subset B \cup C$ and $A \cap B \cap C = \emptyset$. x belongs to one of the sets B, C, say $x \in B$; on the other hand one of the sets A_ι, say $A_\varkappa$, meets C; we have therefore $A_\varkappa \subset B \cup C$, $A_\varkappa \cap B \cap C = \emptyset$ and $B \cap A_\varkappa$ and $C \cap A_\varkappa$ are non-empty. Hence $A_\varkappa$ is not connected, which is a contradiction.

Corollary. *Let* $(A_n)_{n \geqslant 0}$ *be an infinite sequence of connected sets such that* $A_{n+1} \cap A_n \neq \emptyset$ *for all* $n \geqslant 0$. *Then the union* $\bigcup_{n=0}^{\infty} A_n$ *is connected.*

By induction on n we see immediately that the set $B_n = \bigcup_{i=0}^{n} A_i$ is connected for all n, by Proposition 2. The sets B_n have a non-empty intersection; hence their union, equal to $\bigcup_{n=0}^{\infty} A_n$, is connected by Proposition 2.

PROPOSITION 3. *Let* A *be a subset of a topological space* X. *If* B *is a connected subset of* X *which meets both* A *and* $\complement A$, *then* B *meets the frontier of* A.

For otherwise the intersections of B with the interior and exterior of A would be two open subsets of B which form a partition of B, and B would not be connected.

COROLLARY. *In a connected space* X, *every non-empty set other than* X *itself has at least one frontier point.*

2. IMAGE OF A CONNECTED SET UNDER A CONTINUOUS MAPPING

PROPOSITION 4. *Let* A *be a connected subset of a topological space* X, *and let* f *be a continuous mapping of* X *into a topological space* X′. *Then* $f(A)$ *is connected.*

Suppose $f(A)$ is not connected. Then there exist two sets M′, N′ which are open in $f(A)$ and which form a partition of $f(A)$; hence $A \cap \overset{-1}{f}(M')$ and $A \cap \overset{-1}{f}(N')$ are open in A and form a partition of A; this contradicts the hypothesis that A is connected.

The *inverse image* of a connected set under a continuous mapping need not be connected; consider for example a mapping of a discrete space into a space consisting of one point.

From Proposition 4 we derive another characterization of *non-connected* spaces:

PROPOSITION 5. *For a topological space* X *to be not connected it is necessary and sufficient that there exists a surjective continuous mapping of* X *onto a discrete space containing more than one point.*

The condition is sufficient by Proposition 4. Conversely, if X is not connected, there exist two non-empty disjoint open subsets A, B whose union is X, and the mapping f of X onto a discrete space of two elements $\{a, b\}$, defined by $f(A) = \{a\}$ and $f(B) = \{b\}$, is continuous.

3. QUOTIENT SPACES OF A CONNECTED SPACE

PROPOSITION 6. *Every quotient space of a connected space is connected.*

This is an immediate consequence of Proposition 4 of no. 2.

PROPOSITION 7. *Let* X *be a topological space and* R *an equivalence relation on* X. *If the quotient space* X/R *is connected, and if each equivalence class mod* R *is connected, then* X *is connected.*

Suppose X is not connected. Then there is a partition of X into two open sets A, B. The sets A, B are *saturated* with respect to R; for if $x \in A$ then the equivalence class M of x cannot meet B, otherwise the sets $A \cap M$, $B \cap M$ would form a partition of M into two sets open in M, which is impossible since M is connected. The canonical images of A and B are therefore open sets in X/R and form a partition of X/R; this contradicts the assumption that X/R is connected.

4. PRODUCT OF CONNECTED SPACES

PROPOSITION 8. *Every product of connected spaces is connected. Conversely, if a product of non-empty spaces is connected, then each of the factors is connected.*

Let $X = \prod_{\iota \in I} X_\iota$ be a product of topological spaces. If the X_ι are non-empty, we have $X_\iota = \mathrm{pr}_\iota X$ for each $\iota \in I$; hence if X is connected so are the X_ι (no. 2, Proposition 4). Conversely, suppose that each of the X_ι is connected and X is not. By Proposition 5 of no. 2, there exists a continuous surjective mapping $f: X \to X'$, where X' is a discrete space which contains more than one point. Let $a = (a_\iota)$ be any point of X, and $\varkappa$ any index; the partial mapping $f_\varkappa : X_\varkappa \to X'$, defined by $f_\varkappa(x) = f((y_\iota))$ where $y_\varkappa = x$ and $y_\iota = a_\iota$ if $\iota \neq \varkappa$, is continuous on $X_\varkappa$; since $X_\varkappa$ is connected, $f_\varkappa$ must be *constant* on $X_\varkappa$. It follows immediately by induction that $f(x) = f(a)$ for all points $x = (x_\iota)$ such that $x_\iota = a_\iota$ for all but a finite number of indices $\iota \in I$. But these points x form a *dense* subset of X (§ 4, no. 3, Proposition 8). Hence f is continuous on X and constant on a dense subset of X, and therefore constant on X (§ 8, no. 1, Proposition 2, Corollary 1). But this contradicts the definition of f.

5. COMPONENTS

Given a point x of a topological space X, the union of the connected subsets of X which contain x is connected (no. 1, Proposition 2); it is therefore *the largest connected subset of* X *which contains* x.

DEFINITION 3. *The component* (or *connected component*) *of a point of a topological space* X *is the largest connected subset of* X *which contains this point. The components of a subset* A *of* X *are the components of the points of* A, *relative to the subspace* A *of* X.

If a space is connected, the component of each point is the whole space. If a space X is such that for each pair (x, y) of points of X there is a connected set containing x and y, then X is connected.

A space X is said to be *totally disconnected* if the component of each point of X consists of the point alone. A subset A of X is a *totally disconnected set* if the subspace A of X is totally disconnected.

A *discrete* space is totally disconnected, but one should beware of confusing these two notions; for example, we shall see in Chapter IV, § 2, no. 5, that the rational line, which is not a discrete space, is totally disconnected.

A set which is *both open and closed* contains the component of each of its points, whence *the component of a point is contained in the intersection of the sets which are both open and closed and which contain this point.* However, the component of a point is not necessarily equal to this intersection (cf. Exercise 9 and Chapter II, § 4, no. 4, Proposition 6).

PROPOSITION 9. *The component of any point in a topological space* X *is a closed set. The relation "y belongs to the component of x" is an equivalence relation* $R\{x, y\}$ *on* X, *and the equivalence classes are the components of* X. *The quotient space* X/R *is totally disconnected.*

The first part of the proposition is an immediate consequence of Definition 3 and the fact that the closure of a connected set is connected (no. 1, Proposition 1). Since the union of connected sets which have a point in common is connected (no. 1, Proposition 2), the relation R is transitive, hence is an equivalence relation (since it is obviously reflexive and symmetric) and the equivalence class of x with respect to R is the component of x. It remains to show that X/R is totally disconnected. Let $f: X \rightarrow X/R$ be the canonical mapping, and let F be a closed set in X/R *containing at least two distinct points*; the inverse image $\overset{-1}{f}(F)$ of F is closed in X, saturated with respect to R, and contains at least two distinct components of X and hence is *not connected.* Hence there exist two non-empty closed sets B, C in X such that $B \cap C = \varnothing$ and $B \cup C = \overset{-1}{f}(F)$. The component of any point x of $\overset{-1}{f}(F)$ *in* $\overset{-1}{f}(F)$ is the same as the component of x *in* X (by the definition of R) and therefore B and C, which are both open and closed *in* $\overset{-1}{f}(F)$, are saturated with respect to R. Hence $f(B)$ and $f(C)$ are closed in X/R, and $f(B) \cup f(C) = F$ and $f(B) \cap f(C) = \varnothing$; this shows that F is *not connected* and consequently that X/R is totally disconnected.

PROPOSITION 10. *In a product space* $X = \prod_{\iota \in J} X_\iota$, *the component of* $x = (x_\iota)$ *in* X *is the product of the components of* x_ι *in the factors* X_ι.

This product set is connected (no. 4, Proposition 8). Conversely, if A is a connected subset of X which contains x, then $\mathrm{pr}_\iota(A)$ is a connected set (no. 2, Proposition 4) which contains x; since $A \subset \prod_\iota \mathrm{pr}_\iota(A)$, it follows that A is contained in the product of the components of the x_ι.

6. LOCALLY CONNECTED SPACES

DEFINITION 4. *A topological space* X *is said to be locally connected if each point of* X *has a* fundamental system *of connected neighbourhoods.*

* We shall see in Chapter IV, § 2, no. 5, that the real line is a locally connected space. *

The existence, at each point x of a space X, of *one* connected neighbourhood of x by no means implies that X is locally connected. In particular, X can be *connected* but *not locally connected* (Exercises 2 and 13). Conversely, a space can be locally connected but not connected (e.g. a discrete space which contains more than one point).

PROPOSITION 11. *A necessary and sufficient condition for a space* X *to be locally connected is that every component of an open set in* X *is open in* X.

The condition is sufficient, since the component of x relative to an open neighbourhood of x is then a neighbourhood of x in X.

Conversely, let A be an open subset of a locally connected space X, let B be a component of A, and let $x \in B$. Let V be a connected neighbourhood of x contained in A; by the definition of components, V is contained in B; hence B is open in X (§ 1, no. 2, Proposition 1).

The components of a locally connected space X therefore form a partition of X into *open* sets, and hence X is the *sum* (§ 2, no. 4, Example 3) of its components.

COROLLARY. *Let* U *be an open subset of a locally connected space* X, *and let* V *be a component of* U. *Then the frontier of* V *(relative to* X*) is contained in the frontier of* U.

For V is open and closed in U, hence a frontier point of V (relative to X) cannot belong to U, for it would also be a frontier point of V relative to U, and there is none.

PROPOSITION 12. *Every quotient space of a locally connected space is locally connected.*

Let X be a locally connected space, R an equivalence relation on X, $\varphi : X \to X/R$ the canonical mapping. Let A be an open subset of

X/R and C a component of A. Then $\overset{-1}{\varphi}(C)$ is a union of components of $\overset{-1}{\varphi}(A)$: for if $x \in \overset{-1}{\varphi}(C)$ and if K is the component of x in $\overset{-1}{\varphi}(A)$ then $\varphi(K)$ is connected (no. 2, Proposition 4), is contained in A, and contains $\varphi(x)$; hence $\varphi(K) \subset C$ by the definition of C, and therefore $K \subset \overset{-1}{\varphi}(C)$. Since X is locally connected and $\overset{-1}{\varphi}(A)$ is open in X, it follows from Proposition 11 that $\overset{-1}{\varphi}(C)$ is open in X; consequently C is open in X/R and hence, by Proposition 11 again, X/R is locally connected.

PROPOSITION 13. a) *Let* $(X_\iota)_{\iota \in I}$ *be a family of locally connected spaces such that* X_ι *is connected for all but a finite number of indices* $\iota \in I$. *Then the product space* $X = \prod_{\iota \in I} X_\iota$ *is locally connected.*

b) *Conversely, if the product of a family* (X_ι) *of non-empty topological spaces is locally connected, then each* X_ι *is locally connected, and* X_ι *is connected for all but a finite number of indices.*

a) Let J be the finite subset of I such that X_ι is not connected if and only if $\iota \in J$. Let

$$U = \prod_{\iota \in I} U_\iota$$

be an elementary set containing a point $x = (x_\iota)$ of X and let K be the finite subset of I such that $U_\iota \neq X_\iota$ if and only if $\iota \in K$. Let V_ι be X_ι for $\iota \notin J \cup K$, and let V_ι be a connected neighbourhood of x_ι contained in U_ι for $\iota \in J \cup K$; then

$$V = \prod_{\iota \in I} V_\iota$$

is connected (by Proposition 8 of no. 4) and is a neighbourhood of x contained in U. Hence X is locally connected.

b) Let $a = (a_\iota)$ be a point of X and let V be a connected neighbourhood of a in X. Since we have $\mathrm{pr}_\iota V = X_\iota$ except for a finite number of indices (§ 4, no. 1) it follows from no. 2, Proposition 4 that the X_ι are connected, for all but a finite number of indices. On the other hand, for each $\varkappa \in I$, each $a_\varkappa \in X_\varkappa$ and each neighbourhood $V_\varkappa$ of $a_\varkappa$ in $X_\varkappa$, there is a point x of X such that $\mathrm{pr}_\varkappa x = a_\varkappa$, and

$$V = V_\varkappa \times \prod_{\iota \neq \varkappa} X_\iota$$

is a neighbourhood of x in X; V therefore contains a connected neighbourhood W of x, whose projection $\mathrm{pr}_\varkappa W$ is a connected neighbourhood of $a_\varkappa$ contained in $V_\varkappa$ (no. 2, Proposition 4 and § 4, no. 2, Proposition 5). Hence each $X_\varkappa$ is locally connected.

7. APPLICATION: THE POINCARÉ-VOLTERRA THEOREM

THEOREM 1. *Let* X *be a topological space satisfying axiom* (O_{III}) (*but not necessarily Hausdorff*), *and suppose* X *is connected and locally connected. Let* Y *be a topological space whose topology has a* countable *base, and let* p: $X \to Y$ *be a continuous mapping such that, for each* $y \in Y$, $\overset{-1}{p}(y)$ *is a discrete subspace of* X. *Finally let* $\mathfrak{V}$ *be a set of subsets of* X *whose interiors cover* X *and such that*:

(i) *The restriction of* p *to each* $V \in \mathfrak{V}$ *is a closed mapping of* V *into* Y.
(ii) *Every* $V \in \mathfrak{V}$ *has a countable subset which is dense in* V.

Then the space X *is the union of a countable family of open sets each of which is contained in a set of* $\mathfrak{V}$.

Let $\mathfrak{B}$ be a countable base of the topology of Y. We shall say that a pair (W, U) is *distinguished* if (i) $U \in \mathfrak{B}$ and (ii) W is a component of $\overset{-1}{p}(U)$ contained in a set of $\mathfrak{V}$.

LEMMA 1. *If* x *is any point of* X, *there is a distinguished pair* (W, U) *such that* $x \in W$.

The inverse image $\overset{-1}{p}(p(x))$ is discrete and therefore there is a neighbourhood of x in X all of whose points x' other than x have an image $p(x') \neq p(x)$; since X satisfies (O_{III}), there is a *closed* neighbourhood V of x with this property, and we may assume also that V is contained in a set of $\mathfrak{V}$. Let F be the frontier of V in X. By condition (i) of the theorem, $p(F)$ is closed in Y; and since $p(F)$ does not contain $p(x)$, there is a set $U \in \mathfrak{B}$ which contains $p(x)$ and does not meet $p(F)$. Let W be the component of x in $\overset{-1}{p}(U)$; then it is enough to show that $W \subset \mathring{A}$. If this were not so, then W would meet F (no. 1, Proposition 3) and therefore $p(F)$ would meet U, contrary to the definition of U.

LEMMA 2. *If* (W, U) *is a distinguished pair then the set of all distinguished pairs* (W′, U′) *such that* W′ *meets* W *is countable.*

Since $\mathfrak{B}$ is countable it is enough to show that, given $U' \in \mathfrak{B}$, the set of distinguished pairs (W′, U′) such that W′ meets W is countable. Now these sets W′ are open, since X is locally connected (no. 6, Proposition 11) and mutually disjoint since they are components of $\overset{-1}{p}(U')$; hence the sets $W' \cap W$ are open and mutually disjoint. But W contains a countable subset which is dense in W; hence the set of W′ such that $W' \cap W$ is not empty is also countable.

To prove Theorem 1, consider the following relation R between two points x, x' of X: "There exists a finite sequence of distinguished pairs (W_i, U_i) $(1 \leqslant i \leqslant n)$ such that $x \in W_1$ and $x' \in W_n$ and that $W_i \cap W_{i+1} \neq \emptyset$ for $\leqslant 1$ $i \leqslant n - 1$."

Lemma 1 states that R is reflexive, and it is readily verified that R is symmetric and transitive, so that R is an equivalence relation; also, since the W_i are open, every equivalence class mod R is open in X. But X is connected; hence there can be only one equivalence class, i.e. any two points of X are congruent mod R. We shall deduce from this that X is the union of a *countable* family of first elements of distinguished pairs, and this will prove Theorem 1. For this, let x be any point of X, and define by induction on n a sequence (C_n) of open subsets of X as follows: by Lemma 1 there is a distinguished pair (W_1, U_1) such that $x \in W_1$, and we take $C_1 = W_1$; if $n > 1$ then C_n is to be the union of all first elements W of distinguished pairs (W, U) such that W meets C_{n-1}. By induction on n one shows immediately, by virtue of Lemma 2, that C_n is a *countable* union of first elements of distinguished pairs. Finally, every $x' \in X$ belongs to some C_n; for there is a finite sequence

$$(W'_i, U'_i)_{1 \leqslant i \leqslant m}$$

of distinguished pairs such that $x \in W'_1$, $x' \in W'_m$ and $W'_i \cap W'_{i+1} \neq \emptyset$ for $1 \leqslant i \leqslant m - 1$, and by induction on i we see that $W'_i \subset C_{i+1}$ for all i, so that $x' \in C_{m+1}$.

Q.E.D.

COROLLARY 1. *Let Y be a regular space whose topology has a countable base* (*). *Let X be a connected and locally connected space, and let $p : X \to Y$ be a continuous mapping with the following property: for each $x \in X$ there is a closed neighbourhood V of x in X such that the restriction of p to V is a homeomorphism of V onto a closed subspace of Y. Then X is regular and the topology of X has a countable base.*

First, the hypotheses imply that X is regular (§ 8, no. 4, Proposition 13). Let us show that the conditions of Theorem 1 are satisfied if we take $\mathfrak{V}$ to be the set of all closed subsets V of X such that the restriction of p to V is a homeomorphism of V onto a closed subspace of Y. By hypothesis, the interiors of the sets of $\mathfrak{V}$ cover Y and, by virtue of the assumption on Y, each $V \in \mathfrak{V}$ has a countable base and therefore contains a countable dense subset (§ 1, no. 6, Proposition 6). Furthermore, if $x \in \overset{-1}{p}(y)$, there is a neighbourhood $V \in \mathfrak{V}$ of x in X such that V con-

(*) * It can be shown that these conditions imply that the topology of Y is *metrizable* (Chapter IX, § 4, Exercise 22). *

tains no point of $\overset{-1}{p}(y)$ other than x, and hence $\overset{-1}{p}(y)$ is a discrete space. We may therefore apply Theorem 1, which shows that X is the union of a countable family $(T_n)_{n \geqslant 0}$ of open sets, such that each subspace T_n has a countable base $(U_{mn})_{m \geqslant 0}$. Then the U_{mn} $(m \geqslant 0, n \geqslant 0)$ form a base of the topology of X (§ 3, no. 1, Remark).

COROLLARY 2. *Let* X *be locally compact, connected and locally connected, and suppose each point of* X *has a neighbourhood which has a countable base. Let* Y *be a Hausdorff space whose topology has a countable base, and let* p : X → Y *be a continuous mapping such that, for each* $y \in$ Y, $\overset{-1}{p}(y)$ *is a discrete subspace of* X. *Then the topology of* X *has a countable base.*

For each $x \in$ X, let V_x be a compact neighbourhood of x in X which has a countable base. It follows from § 9, no. 4, Theorem 2, Corollary 2, that the set $\mathfrak{V}$ of the V_x satisfies the conditions of Theorem 1, and we complete the proof as in Corollary 1.

> Notice that, in this corollary, it can happen that the restriction of p to an arbitrarily small neighbourhood V of a point of X *is not a homeomorphism of* V *onto* $p(\text{V})$.

COROLLARY 3 (Poincaré-Volterra Theorem). *Let* Y *be a locally compact, locally connected space whose topology has a countable base. Let* X *be a connected Hausdorff space, and let* p : X → Y *be a continuous mapping which has the following property : for each* $x \in$ X *there is an open neighbourhood* U *of* x *in* X *such that the restriction of* p *to* U *is a homeomorphism of* U *onto an open subspace of* Y. *Then* X *is locally compact and locally connected, and the topology of* X *has a countable base.*

Clearly X is locally connected. Also each $x \in$ X has an open neighbourhood U in X such that the restriction of p to U maps U homeomorphically onto an open subspace $p(\text{U})$ of Y. Since $p(\text{U})$ is a locally compact subspace of Y (§ 9, no. 7, Proposition 13), there is a compact neighbourhood W of $p(x)$ contained in $p(\text{U})$, whence $\text{U} \cap \overset{-1}{p}(\text{W})$ is a compact neighbourhood of x contained in U; thus X is locally compact, since by hypothesis X is Hausdorff. $\text{U} \cap \overset{-1}{p}(\text{W})$, being compact, is closed in X (§ 9, no. 3, Proposition 4) and the conditions of Corollary 1 are therefore satisfied; hence the topology of X has a countable base.

EXERCISES

§ 1

1) Find all possible topologies on a set of two elements or three elements.

2) *a*) Let X be an *ordered* set. Show that the set of intervals

$$[x, \rightarrow[\qquad (\text{resp.}\ ]\leftarrow, x])$$

is a base of a topology on X; this topology is called the *right* (resp. *left*) topology of X. In the right topology, any intersection of open sets is an open set, and the closure of $\{x\}$ is the interval $]\leftarrow, x]$.

b) A topological space is said to be a *Kolmogoroff space* if it satisfies the following condition: given any two distinct points x, x' of X, there is a neighbourhood of one of these points which does not contain the other. Show that an ordered set with the right topology is a Kolmogoroff space.

c) Let X be a Kolmogoroff space in which every intersection of open sets is an open set. Show that $x \in \overline{\{x'\}}$ is an order relation between x and x' in X, and that if this relation is written as $x \leqslant x'$ then the given topology on X is identical with the right topology determined by this ordering.

d) Deduce from *c*) that if X is a Kolmogoroff space then every finite non-empty subset of X has at least one isolated point. Hence, if X has no isolated point, every non-empty open subset of X is infinite.

3) If A is any subset of a topological space X, let $\alpha(A)$ denote $\overset{\circ}{\overline{A}}$ and $\beta(A)$ denote $\overline{\overset{\circ}{A}}$. Then $A \subset B$ implies $\alpha(A) \subset \alpha(B)$ and $\beta(A) \subset \beta(B)$.

a) Show that if A is open then $A \subset \alpha(A)$, and that if A is closed then $\beta(A) \subset A$.

b) Deduce from *a*) that if A is any subset of X, then $\alpha(\alpha(A)) = \alpha(A)$ and $\beta(\beta(A)) = \beta(A)$.

c) Give an example of a set A *(on the real line)$_*$ such that the seven sets A, $\mathring{A}$, $\overline{A}$, $\alpha(A)$, $\beta(A)$, $\beta(\overline{A})$, $\alpha(\mathring{A})$ are all distinct and satisfy no relations of inclusion except the following: $\mathring{A} \subset A \subset \overline{A}$, $\mathring{A} \subset \alpha(\mathring{A}) \subset \beta(A) \subset \overline{A}$, $\mathring{A} \subset \alpha(A) \subset \beta(\overline{A}) \subset \overline{A}$, $\alpha(\mathring{A}) \subset \alpha(A)$, $\beta(A) \subset \beta(\overline{A})$.

d) Show that if U, V are two open sets such that $U \cap V = \emptyset$, then also $\alpha(U) \cap \alpha(V) = \emptyset$ [use *b*)].

4) *a*) Give an example * on the real line $_*$ of two open sets A, B such that the four sets $A \cap \overline{B}$, $B \cap \overline{A}$, $\overline{A \cap B}$ and $\overline{A} \cap \overline{B}$ are all distinct.

b) * Give an example of two intervals A, B on the real line, such that $\overline{A} \cap \overline{B}$ is not contained in $\overline{A \cap B}$. $_*$

5) If A is any subset of a topological space X, the frontier of A is denoted by $\mathrm{Fr}(A)$.

a) Show that $\mathrm{Fr}(\overline{A}) \subset \mathrm{Fr}(A)$, $\mathrm{Fr}(\mathring{A}) \subset \mathrm{Fr}(A)$, and give an example * on the real line $_*$ where these three sets are distinct.

b) Let A, B be two subsets of X; show that

$$\mathrm{Fr}(A \cup B) \subset \mathrm{Fr}(A) \cup \mathrm{Fr}(B)$$

and give an example * on the real line $_*$ where these sets are distinct. If $\overline{A} \cap \overline{B} = \emptyset$, show that $\mathrm{Fr}(A \cup B) = \mathrm{Fr}(A) \cup \mathrm{Fr}(B)$.

c) If A and B are open in X, show that

$$\begin{aligned}(A \cap \mathrm{Fr}(B)) \cup (B \cap \mathrm{Fr}(A)) &\subset \mathrm{Fr}(A \cap B)\\ &\subset (A \cap \mathrm{Fr}(B)) \cup (B \cap \mathrm{Fr}(A)) \cup (\mathrm{Fr}(A) \cap \mathrm{Fr}(B))\end{aligned}$$

and give an example * on the real line $_*$ where these three sets are distinct.

6) Show that a subset A of a topological space X meets each dense subset of X if and only if the interior of A is not empty.

7) Consider the following four properties of a given topological space X:

(D_{I}) The topology of X has a countable base.

(D_{II}) X has a countable dense subset.

(D_{III}) Every subset of X, all of whose points are isolated, is countable.

(D_{IV}) Every set of mutually disjoint non-empty open subsets of X is countable.

Show that (D_I) implies (D_{II}) and (D_{III}), and that each of (D_{II}) and (D_{III}) imply (D_{IV}) (*).

8) If a subset A of a topological space X has no isolated points, then the same is true of its closure $\overline{A}$ in X.

9) Let X be a set and let $M \to \overline{M}$ be a mapping of $\mathfrak{P}(X)$ onto itself such that: 1) $\overline{\varnothing} = \varnothing$; 2) for every $M \subset X$ we have $M \subset \overline{M}$; 3) for every $M \subset X$ we have $\overline{\overline{M}} = \overline{M}$; 4) for all $M \subset X$ and $N \subset X$ we have

$$\overline{M \cup N} = \overline{M} \cup \overline{N}.$$

Show that there is a unique topology on X such that $\overline{M}$ is the closure of M with respect to this topology, for all $M \subset X$ (define the topology by means of its closed sets).

§ 2

1) Let f be a mapping of a topological space X into a topological space X'. Prove that the following statements are equivalent:

a) f is continuous on X.

b) For every subset A' of X', $\overset{-1}{f}(\mathring{A}') \subset (\overset{-1}{f}(A'))^{\circ}$.

c) For every subset A' of X', $\overline{\overset{-1}{f}(A')} \subset \overset{-1}{f}(\overline{A'})$.

Show by an example that, if f is continuous, the sets $\overline{\overset{-1}{f}(A')}$ and $\overset{-1}{f}(\overline{A'})$ can be distinct.

2) Let X, X' be two ordered sets, each topologized with the right topology (§ 1, Exercise 2). Show that a mapping $f: X \to X'$ is continuous if and only if it is order-preserving.

3) Let X, X' be two topological spaces, f a bijection of X onto X'; then a necessary and sufficient condition for f to be a homeomorphism

(*) For an example of a space which satisfies (D_{IV}) but neither (D_{II}) nor (D_{III}), see § 8, Exercise 6 *b*). For an example of a space satisfying (D_{II}) and (D_{III}), but not (D_I), see Chapter IX, § 5, Exercise 16. For examples of spaces where one of the two conditions (D_{II}), (D_{III}) is satisfied but not the other, see § 9, Exercise 23.

* In a *metrizable* space, (D_I) and (D_{IV}) are equivalent: Proposition 12 of Chapter IX, § 2, no. 8 shows that (D_I) and (D_{II}) are equivalent; on the other hand, if (D_{IV}) is satisfied, then for each integer $n > 0$ there is a *maximal* countable family $\mathfrak{B}_n = (B_{nm})_{m \geqslant 0}$ of mutually disjoint balls of radius $1/n$; whose union is dense. The countable set of centres of the B_{nm} is then dense. *

of X onto X′ is that the topology of X′ is the finest for which f is continuous.

4) On an ordered set X, the least upper bound of the left topology and the right topology (§ 1, Exercise 2) is the discrete topology; if X is a directed set (on the right or the left), the greatest lower bound of these two topologies is the coarsest topology on X.

5) On an ordered set X let $\mathfrak{T}_0(X)$ (resp. $\mathfrak{T}_+(X)$, $\mathfrak{T}_-(X)$) denote the topology generated by the set of all open intervals (resp. the set of right half-open intervals, the set of left half-open intervals), bounded or not.

a) Show that if X is linearly ordered, then the open (resp. right half-open, left half-open) intervals form a base of $\mathfrak{T}_0(X)$ (resp. $\mathfrak{T}_+(X)$, $\mathfrak{T}_-(X)$); and that the closed intervals are closed sets in each of these three topologies.

b) On a linearly ordered set X, show that the topology $\mathfrak{T}_+(X)$ is finer than the right topology (§ 1, Exercise 2), that the topology $\mathfrak{T}_-(X)$ is finer than the left topology, and that the topology $\mathfrak{T}_0(X)$ is the intersection of $\mathfrak{T}_+(X)$ and $\mathfrak{T}_-(X)$.

c) Let X be a linearly ordered set. Show that a necessary and sufficient condition for $\mathfrak{T}_0(X)$ to have a countable base is that there should be is a countable subset D of X such that, for each $x \in X$ and each interval $]a, b[$ containing x, there exist $\alpha \in D$ and $\beta \in D$ such that

$$a \leqslant \alpha < x < \beta \leqslant b$$

[cf. Chapter IV, § 2, Exercise 11 *c*)].

d) Show that if X is well-ordered (*Set Theory*, Chapter III, § 2) then the topology $\mathfrak{T}_+(X)$ is the discrete topology, and the topologies $\mathfrak{T}_0(X)$ and $\mathfrak{T}_-(X)$ are identical.

¶ 6) A topology on a set X is said to be *quasi-maximal* if it is maximal in the set of topologies in which X has no isolated points.

a) Show that the following properties of a topology $\mathfrak{T}$ on X are equivalent: α) $\mathfrak{T}$ is quasi-maximal; β) if X carries the topology $\mathfrak{T}$, then X has no isolated point and every subset of X which has no isolated points is open.

b) Let X be a Kolmogoroff space (§ 1, Exercise 2) in which every non-empty open set is infinite. Show that there exists a quasi-maximal topology on X which is finer than the given topology (show that the set of topologies on X in which all the non-empty open sets are infinite is inductive, and use Exercise 2 *d*) of § 1).

¶ 7) If X is any set, let $\mathfrak{P}_0(X)$ denote the set of non-empty subsets of X.

a) Let X be a non-empty topological space. Let $\mathfrak{T}_\Omega$ (resp. $\mathfrak{T}_\Phi$) denote the coarsest topology on $\mathfrak{P}_0(X)$ such that for every non-empty open (resp. closed) subset A of X, $\mathfrak{P}_0(A)$ is open (resp. closed) in $\mathfrak{P}_0(X)$. Show that in general these two topologies are not comparable, and that the mapping $x \to \{x\}$ is a homeomorphism of X onto a subspace of $\mathfrak{P}_0(X)$ if this latter set carries either of the topologies $\mathfrak{T}_\Omega$, $\mathfrak{T}_\Phi$.

b) Let D be a dense subset of X. Show that the set $\mathfrak{F}(D)$ of finite non-empty subsets of D is dense in $\mathfrak{P}_0(X)$ for either of the topologies $\mathfrak{T}_\Omega$, $\mathfrak{T}_\Phi$.

c) If $\mathfrak{P}_0(X)$ is ordered by inclusion, show that the topology $\mathfrak{T}_\Omega$ is coarser than the left topology on $\mathfrak{P}_0(X)$ (§ 1, Exercise 2); that $A \in \mathfrak{P}_0(X)$ is isolated in the topology $\mathfrak{T}_\Omega$ if and only if $A = \{x\}$, where x is an isolated point of X; and that $A \in \mathfrak{P}_0(X)$ is isolated in the topology $\mathfrak{T}_\Phi$ if and only if X is a finite discrete space and A = X.

d) Prove that if $\mathfrak{P}_0(X)$ carries the topology $\mathfrak{T}_\Omega$ (resp. $\mathfrak{T}_\Phi$) and

$$\mathfrak{P}_0(X) \times \mathfrak{P}_0(X)$$

carries the topology which is the product of $\mathfrak{T}_\Omega$ (resp. $\mathfrak{T}_\Phi$) with itself, then the mapping $(M, N) \to M \cup N$ of $\mathfrak{P}_0(X) \times \mathfrak{P}_0(X)$ into $\mathfrak{P}_0(X)$ is continuous.

e) Show that if $\mathfrak{P}_0(X)$ carries the topology $\mathfrak{T}_\Phi$, then the mapping $M \to \overline{M}$ is continuous.

f) Let X, Y be two topological spaces, $f: X \to Y$ a continuous map. Show that if both $\mathfrak{P}_0(X)$ and $\mathfrak{P}_0(Y)$ carry the topology $\mathfrak{T}_\Omega$, and if both carry the topology $\mathfrak{T}_\Phi$, then the mapping $M \to f(M)$ of $\mathfrak{P}_0(X)$ into $\mathfrak{P}_0(Y)$ is continuous.

g) Let X, Z be two topological spaces. Show that a mapping $f: Z \to \mathfrak{P}_0(X)$ is continuous when $\mathfrak{P}_0(X)$ carries the topology $\mathfrak{T}_\Omega$ (resp. $\mathfrak{T}_\Phi$) if and only if the set of all $z \in Z$ such that $f(z) \cap A \neq \emptyset$ is closed (resp. open) in Z for each closed (resp. open) subset A of X.

¶ 8) Let X be a set, and $\mathfrak{c}$ a cardinal number which is either infinite or equal to 1. The set of subsets of X consisting of X and those $M \subset X$ such that $\text{card}(M) < \mathfrak{c}$ is the set of closed subsets of X for a topology $\mathfrak{T}_\mathfrak{c}$; $\mathfrak{T}_1$ is the coarsest topology on X and $\mathfrak{T}_\mathfrak{c}$ is the discrete topology if $\mathfrak{c} > \text{card}(X)$. Every permutation of X is an automorphism of the topological space obtained by giving X a topology $\mathfrak{T}_\mathfrak{c}$.

Show that conversely any topology $\mathfrak{T}$ on X, which has the property that every permutation of X is continuous with respect to $\mathfrak{T}$, is necessarily one of the topologies $\mathfrak{T}_\mathfrak{c}$. [Let $\mathfrak{c}$ be the smallest cardinal such that $\text{card}(F) < \mathfrak{c}$ for each closed subset $F \neq X$ in the topology $\mathfrak{T}$; observe that when X is infinite and $\mathfrak{c} > \text{card}(X)$, $\mathfrak{T}$ is necessarily the discrete topology.]

9) *a*) With the hypotheses of Proposition 4 of no. 3, show that the coarsest topology on X for which the f_ι are continuous is also the *finest* topology $\mathfrak{T}$ on X which has the following property: every mapping g of a topological space Z into X such that the $f_\iota \circ g$ are continuous is continuous (with respect to $\mathfrak{T}$).

b) With the hypotheses of Proposition 6 of no. 4, show that the finest topology on X for which the f_ι are continuous is also the *coarsest* topology $\mathfrak{T}$ on X which has the following property: every mapping g of X into a topological space Z such that the $g \circ f_\iota$ are continuous is continuous (with respect to $\mathfrak{T}$).

10. On a set X, let $\mathfrak{M}$ be a directed set of topologies without isolated points. Show that the least upper bound of $\mathfrak{M}$ in the set of all topologies on X is a topology without isolated points.

¶ 11. A topological space is said to be *solvable* if it is the disjoint union of two dense subsets.

a) If a topological space X is solvable, show that X has no isolated points and that every open subspace of X is solvable. Show also conversely that if X is a topological space and $\mathfrak{B}$ is a set of non-empty solvable open subspaces of X such that every non-empty open subset of X contains a set belonging to $\mathfrak{B}$, then X is solvable (consider a *maximal* set of mutually disjoint open sets belonging to $\mathfrak{B}$).

b) If a Kolmogoroff space X is solvable, then every non-empty open subset of X is infinite [use Exercise 2 *d*) of § 1].

c) Let X be a topological space such that 1) the smallest cardinal $\mathfrak{d}$ of a non-empty open subset of X is infinite, and 2) there is a base $\mathfrak{B}$ of the topology of X such that $\text{card}(\mathfrak{B}) \leqslant \mathfrak{d}$. Show that X is solvable. (Use transfinite induction to construct two disjoint dense subsets of X). By the method of *Set Theory*, Chapter III, § 6, Exercise 24 *a*). In particular, the rational line is solvable.

d) A topological space X which has no isolated points is said to be *isodyne* if $\text{card}(U) = \text{card}(X)$ for every non-empty open subset of X. Show that in a space X with no isolated points, every non-empty open set in X contains a non-empty open subspace which is isodyne. Deduce that if every isodyne open subspace of X is soluble, then X is soluble.

§ 3

1) Let A, B be two subsets of a topological space X such that $A \supset B$. Show that:

a) The interior of B with respect to X is contained in the interior of B with respect to the subspace A of X. Give an example where these two sets are distinct.

b) The frontier of B with respect to A is contained in the intersection of A with the frontier of B with respect to X. Give an example where these two sets are distinct.

2) Let A and B be any two subsets of a topological space X.

a) Show that the intersection of A and the interior of B with respect to X is contained in the interior of $B \cap A$ with respect to A. Give an example where these two sets are distinct.

b) Show that the intersection of A and the closure of B with respect to X contains the closure of $B \cap A$ with respect to A. Give an example where the two sets are distinct.

3) A subspace A of a topological space X is discrete if and only if every point of A is isolated.

4) Let Y, Z be two subspaces of a topological space X such that $X = Y \cup Z$ and such that $\overline{A} \cap B = \overline{B} \cap A = \emptyset$, where $A = Y \cap \complement Z$ and $B = Z \cap \complement Y$.

a) Show that if M is any subset of X then the closure of M in X is the union of the closure of $M \cap Y$ in Y and the closure of $M \cap Z$ in Z. Deduce that if $M \cap Y$ is closed (resp. open) in Y and $M \cap Z$ is closed (resp. open) in Z, then M is closed (resp. open) in X.

b) Let f be a mapping of X into a topological space X′. Show that if $f|Y$ and $f|Z$ are continuous then so is f.

5) Let Y, Z be two subspaces of a topological space X such that $X = Y \cup Z$. Show that if a subset M of $Y \cap Z$ is open (resp. closed) in both Y and Z, then M is open (resp. closed) in X.

6) Let A be a locally closed subspace of a topological space X. Show that the set of open subsets U of X such that $A \subset U$ and A is closed in U has a greatest element, namely the complement in X of the frontier of A with respect to $\overline{A}$.

7) Show by examples that the union of two locally closed subsets and the complement of a locally closed subset need not be locally closed.

8) Let A be a subset of an ordered set X. Show that the topology induced on A by the right (resp. left) topology of X (§ 1, Exercise 2) is the right (resp. left) topology of A with respect to the ordering of A induced by that of X.

9) Let A be a subset of a linearly ordered set X.

a) Show that the topology induced on A by the topology $\mathfrak{T}_-(X)$ [resp. $\mathfrak{T}_+(X)$, $\mathfrak{T}_0(X)$] (§ 2, Exercise 5) is finer than the topology $\mathfrak{T}_-(A)$ [resp. $\mathfrak{T}_+(A)$, $\mathfrak{T}_0(A)$], the ordering of A being that induced by the ordering of X. If A is an interval of X, then $\mathfrak{T}_-(A)$ [resp. $\mathfrak{T}_+(A)$, $\mathfrak{T}_0(A)$] is the topology induced on A by $\mathfrak{T}_-(X)$ [resp. $\mathfrak{T}_+(X)$, $\mathfrak{T}_0(X)$].

b) Take X to be the product $\mathbf{Q} \times \mathbf{Z}$, ordered by the lexicographic ordering (*Set Theory*, Chapter III, § 2, no. 6) and take A to be the subset $\mathbf{Q} \times \{0\}$. Show that the topology $\mathfrak{T}_-(A)$ [resp. $\mathfrak{T}_+(A)$, $\mathfrak{T}_0(A)$] is strictly coarser than the topology induced on A by $\mathfrak{T}_-(X)$ [resp. $\mathfrak{T}_+(X)$, $\mathfrak{T}_0(X)$].

10) Let A be a non-empty subspace of a non-empty topological space X. Show that the topology induced on $\mathfrak{P}_0(A)$ by the topology $\mathfrak{T}_\Omega$ (resp. $\mathfrak{T}_\Phi$) on $\mathfrak{P}_0(X)$ (§ 2, Exercise 7) is the topology $\mathfrak{T}_\Omega$ (resp. $\mathfrak{T}_\Phi$) on $\mathfrak{P}_0(A)$.

11) Let X be a topological space, $\mathfrak{T}$ the topology of X, A a dense subspace of X. Show that the set of topologies on X which are finer than $\mathfrak{T}$, in which A is dense in X, and which induce the same topology as $\mathfrak{T}$ on A, has at least one maximal element; such a maximal element is called an A-*maximal* topology. Show that a topology $\mathfrak{T}_0$ on X is A-maximal if and only if every subset M of X, such that $A \cap M$ is dense in M (with respect to $\mathfrak{T}_0$) and open in A, is open in the topology $\mathfrak{T}_0$, and that the subspace $\complement A$ is then discrete in the topology induced by $\mathfrak{T}_0$, and is closed in X. Show also that if $\mathfrak{T}_0$ is A-maximal and if the topology induced by $\mathfrak{T}_0$ on A is quasi-maximal (§ 2, Exercise 6) then $\mathfrak{T}_0$ is quasi-maximal.

* 12) Let X be the subspace $[0, 1]$ of the real line $\mathbf{R}$, and let S be the equivalence relation on X whose equivalence classes are $\{0, 1\}$ and the sets $\{x\}$ for $0 < x < 1$. Show that X has no continuous section with respect to S. *

13) Let X be a topological space, and R and S two equivalence relations on X such that R implies S. Show that if there is a continuous section of X with respect to R and a continuous section of X/R with respect to S/R, then there is a continuous section of X with respect to S.

14) Let X be a topological space which is the sum of two subspaces X_1 and X_2, and let X/R be the space obtained by pasting together X_1 and X_2 along a subspace A_1 of X_1 and a subspace A_2 of X_2 by means of a homeomorphism *h* of A_1 onto A_2. Show that the canonical images of X_1 and X_2 in X/R are homeomorphic to X_1 and X_2 respectively.

* 15) Consider the following three subspaces of $\mathbf{R}$: $X_1 =]-1, 1[$, $X_2 =]-2, -1[$, $X_3 =]1, 2[$. Let X be their union; X is then the

sum of these three subspaces. Let A_2 (resp. A_3) be the set of irrational numbers contained in X_2 (resp. X_3); let B_1 (resp. B_2) be the set of rational numbers contained in $]-1, 0[$ (resp. X_2); and let C_1 (resp. C_3) be the set of rational numbers contained in $]0, 1[$ (resp. X_3). Let X/R be the space obtained by pasting together the X_i (i) along A_2 and A_3 by means of the homeomorphism $x \to x + 3$; (ii) along B_1 and B_2 by means of the homeomorphism $x \to x - 1$; (iii) along C_1 and C_3 by means of the homeomorphism $x \to x + 1$. Show that the canonical image of X_1 in X/R is not homeomorphic to X_1. If $Y = X_1 \cup B_2 \cup C_3$ is the saturation of X_1 with respect to R, show that the quotient space Y/R_Y is not homeomorphic to the canonical image of Y in X/R. $_*$

16) Let X be a non-empty topological space and R an equivalence relation on X. Show that if X/R is regarded as a subset of $\mathfrak{P}_0(X)$, then the quotient topology on X/R is coarser than the topologies induced on X/R by the topologies $\mathfrak{T}_\Omega$ and $\mathfrak{T}_\Phi$ on $\mathfrak{P}_0(X)$ (§ 2, Exercise 7).

§ 4

1) Let X, Y be two topological spaces, A a subset of X and B a subset of Y. Show that $\operatorname{Fr}(A \times B) = (\operatorname{Fr}(A) \times \overline{B}) \cup (\overline{A} \times \operatorname{Fr}(B))$.

2) Let X, Y be two topological spaces, A a *closed* subset of $X \times Y$. If $\mathfrak{B}(x)$ is the set of all neighbourhoods of a point $x \in X$, show that

$$\bigcap_{V \in \mathfrak{B}(x)} \overline{A(V)} = A(x).$$

Give an example of an open subset B of a product $X \times Y$ such that

$$\bigcap_{V \in \mathfrak{B}(x)} \overline{B(V)}$$

is different from $\overline{B_{(x)}}$.

3) Let $(X_\iota)_{\iota \in I}$ be an infinite family of topological spaces such that each X_ι contains (at least) two distinct points a_ι, b_ι such that there is a neighbourhood of b_ι which does not contain a_ι.

a) For each index $\varkappa \in I$, let $c_\varkappa$ be the point of the product space

$$X = \prod_{k+1} X_\iota$$

for which $\operatorname{pr}_\varkappa c_\varkappa = b_\varkappa$ and $\operatorname{pr}_\iota c_\varkappa = a_\iota$ whenever $\iota \neq \varkappa$. Show that every point of the set C of points $c_\varkappa$ is isolated.

b) Deduce from *a*) that the topology of X has a countable base if and only if I is countable and the topology of each X_ι has a countable base (use Exercise 7 of § 1).

c) Show that if the index set I is not countable then the point $b = (b_\iota)$ of X has no countable fundamental system of neighbourhoods.

¶ 4) Let K be the discrete space consisting of the two numbers 0, 1; let A be an infinite set, and let X be the product space K^A. Let

$$V = \prod_{\alpha \in A} V_\alpha$$

be an elementary set in X; if h is the number of indices α such that $V_\alpha \neq K$, put $\mu(V) = 2^{-h}$.

a) Show that if $U_1, \ldots, U_n$ are pairwise disjoint elementary sets, then

$$\sum_{k=1}^{n} \mu(U_k) \leqslant 1.$$

[Write each U_k in the form $W_k \times K^B$, where B (independent of k) is the complement of a finite subset of A, and W_k is a finite set; count up the number of points in W_k].

b) Deduce from *a*) that X satisfies the condition (D_{IV}) of Exercise 7 of § 1. [If there were an uncountable family (U_λ) of pairwise disjoint elementary sets, there would exist an integer $p > 0$ such that $\mu(U_\lambda) \geqslant 1/p$ for an infinite number of indices λ].

c) Show that if A is uncountable, then X does not satisfy the condition (D_{III}) of Exercise 7 of § 1 (cf. Exercise 3).

¶ 5) Let K be the discrete space consisting of the two numbers 0, 1; let A be an infinite set and let X denote the product space $K^{\mathfrak{P}(A)}$.

a) Let J be any finite subset of A; suppose J has p elements. For each subset H of J, let $\mathfrak{M}_H$ be the set of subsets L of A such that $L \cap J = H$; the $\mathfrak{M}_H$ form a partition $\tilde{\omega}_J$ of $\mathfrak{P}(A)$ into 2^p subsets. Let F_J be the subset of X consisting of those points $(x_L)_{L \subset A}$ such that $x_L = x_M$ whenever L and M belong to the same set of the partition $\tilde{\omega}_J$. F_J is a finite set, with 2^{2^p} elements. If F is the union of the F_J as J runs through all finite subsets of A, show that F is equipotent to A.

b) Show that F is dense in X. Taking $A = \mathbf{N}$, deduce that X satisfies condition (D_{II}) of § 1, Exercise 7, but does not satisfy condition (D_{III}) (Exercise 4).

6) Let X, Y, Z be three topological spaces, A an open subset of $X \times Y$ and B an open subset of $Y \times Z$. Show that $B \circ A$ is an open subset of $X \times Z$.

7) Let X be the sum of a family (X_λ) of topological spaces, Y the sum of a family (Y_μ) of topological spaces. Show that the product space $X \times Y$ is the sum of the $X_\lambda \times Y_\mu$.

8) Let $(G_\iota)_{\iota \in I}$ be a family of graphs of equivalence relations (*Set Theory*, Chapter II, § 6, no. 1) on a topological space X, and let R_ι be the relation $(x, y) \in G_\iota$; then

$$G = \bigcap_{\iota \in I} G_\iota$$

is the graph of an equivalence relation R on X. Show that there is a canonical continuous injection of the quotient space X/R into the product space $\prod_{\iota \in I} (X/R_\iota)$.

9) Let $(X_\iota)_{\iota \in I}$ be an infinite family of topological spaces. Study the topology on the product set

$$X = \prod_{\iota \in I} X_\iota \text{ generated by subsets of the form } \prod_{\iota \in I} U_\iota,$$

where U_ι is an open subset of X_ι for each $\iota \in I$ and the set of indices ι for which $U_\iota \neq X_\iota$ has cardinal less than $\mathfrak{c}$, where $\mathfrak{c}$ is a given infinite cardinal. Consider what becomes of the propositions of § 4 for this topology.

10) *a*) Let $(X_\alpha, f_{\alpha\beta})$ be an inverse system of topological spaces, and let X be the inverse limit. Let f_α be the canonical mapping $X \to X_\alpha$. For each α, let Y_α be a subspace of X_α containing $f_\alpha(X)$, such that the Y_α form an inverse system of subspaces of the X_α. Show that $\varprojlim Y_\alpha$ can be canonically identified with X.

b) Let $(X'_\alpha, f'_{\alpha\beta})$ be another inverse system of topological spaces with the same directed index set. For each α, let $u_\alpha : X_\alpha \to X'_\alpha$ be a continuous mapping such that the u_α form an inverse system of mappings. The $u_\alpha(X_\alpha)$ form an inverse system of subspaces of the X'_α; show that if $u = \varprojlim u_\alpha$ and if the f_α are surjective, then $u(X)$ is dense in the space $\varprojlim u_\alpha(X_\alpha)$. Does this result still hold good if the f_α are no longer assumed to be surjective? (Cf. *Set Theory*, Chapter III, § 1, Exercise 32.)

§ 5

1) *a*) Show that an equivalence relation R on a topological space X is open if and only if, for each subset A of X, the saturation of the interior of A is contained in the interior of the saturation of A. Give an example where R is open and these two sets can be distinct.

b) Show that an equivalence relation R on a topological space X is closed if and only if, for each subset A of X, the saturation of the closure of A contains the closure of the saturation of A. Give an example where R is closed and these two sets can be distinct.

* 2) Show that the equivalence relation S : $x \equiv y \pmod{1}$ on **R** is not closed. If $A = [0,1]$ show that the induced relation S_A on A is closed but not open. *

3) Let Γ be a *finite* group of homeomorphisms of a topological space X, and let R be the equivalence relation: "there exists $\sigma \in \Gamma$ such that $y = \sigma(x)$". Show that the relation R is both open and closed.

* 4) Let Γ be the group of homeomorphisms of the quotient space **T** of **R** consisting of the identity and the homeomorphism induced on **T** by the homeomorphism $x \to 1/2 + x$ of **R** onto itself. Let S be the equivalence relation "there exists $\sigma \in \Gamma$ such that $y = \sigma(x)$" on **T**; S is both open and closed (Exercise 3). Show that **T** has no continuous section with respect to S, although both **T** and **T**/S are compact, connected and locally connected. Let A be the canonical image in **T** of the interval $[0,1/2]$ of **R**. Show that A is locally closed in **T** but that the quotient space A/S_A is not homeomorphic to the canonical image of A in **T**/S. *

5) Let X be a topological space and let R, S be two equivalence relations on X such that R implies S.

a) Show that if S is open (resp. closed) then S/R is an open (resp. closed) equivalence relation on X/R; show also that the converse is not necessarily true. Show that S can be open (resp. closed) but not R.

b) Suppose that R is open (resp. closed). Show that S/R is open (resp. closed) if and only if S is open (resp. closed).

6) Let S be the equivalence relation on the rational line **Q** which is obtained by identifying all the points of **Z**. Show that S is closed, and that if U is the relation of equality on **Q**, then the canonical bijection of $(\mathbf{Q} \times \mathbf{Q})/(U \times S)$ onto $\mathbf{Q} \times (\mathbf{Q}/S)$ is not a homeomorphism.

7) Let X and Y be two non-empty topological spaces and $f: X \to Y$ a surjective map. Show that f is closed if and only if the mapping $y \to (\overset{-1}{f}\, y)$ of Y into $\mathfrak{P}_0(X)$ is continuous when $\mathfrak{P}_0(X)$ carries the topology $\mathfrak{T}_\Omega$ (resp. $\mathfrak{T}_\Phi$) (§ 2, Exercise 7).

Deduce that the topology induced on a quotient set X/R of a non-empty topological space X by the topology $\mathfrak{T}_\Omega$ (resp. $\mathfrak{T}_\Phi$) coincides with the quotient of the topology of X by R if and only if R is closed (resp. open) (cf. § 3, Exercise 16).

§ 6

1) Find all possible filters on a finite set.

2) If the intersection of all the sets of a filter $\mathfrak{F}$ on an infinite set X is empty, show that $\mathfrak{F}$ is finer than the filter of complements of finite subsets of X.

3) The intersection of two filters $\mathfrak{F}$ and $\mathfrak{G}$ on a set X is the set of all $M \cup N$, where $M \in \mathfrak{F}$ and $N \in \mathfrak{G}$.

4) Let X be an infinite set. Show that the filter of complements of finite subsets of X is the intersection of the elementary filters associated with infinite sequences in X all of whose terms are distinct.

5) Let $\mathfrak{F}$ and $\mathfrak{G}$ be two filters on a set X. Show that if $\mathfrak{F}$ and $\mathfrak{G}$ have a least upper bound in the set of all filters on X, then this least upper bound is the set of all $M \cap N$, where $M \in \mathfrak{F}$ and $N \in \mathfrak{G}$.

6) If to each filter on a set X we make correspond the associated topology (no. 5) on the set $X' = X \cup \{\omega\}$, then we have the following propositions:

a) If a topology $\mathfrak{T}$ on X' is *finer* than the topology associated with a filter $\mathfrak{F}$ on X, then $\mathfrak{T}$ is either the discrete topology or the topology associated with a filter finer than $\mathfrak{F}$. Converse?

b) The greatest lower bound of the topologies associated with the filters of a set Φ of filters on X is the topology assocated with the intersection of the filters belonging to Φ.

c) Let $\mathfrak{G}$ be a set of subsets of X, and let $\mathfrak{G}'$ be the set of subsets of X' of the form $M \cup \{\omega\}$ where $M \in \mathfrak{G}$. Let $\mathfrak{H}$ denote $\mathfrak{G} \cup \mathfrak{P}(X)$. If $\mathfrak{G}$ generates a filter $\mathfrak{F}$, show that $\mathfrak{H}$ is a subbase of the topology on X' associated with $\mathfrak{F}$. What topology does $\mathfrak{H}$ generate if $\mathfrak{G}$ is not a filter subbase?

d) $\mathfrak{F}$ is an ultrafilter on X if and only if the associated topology on X' is X-maximal (§ 3, Exercise 11).

7) In a topological space X, the intersection of the neighbourhood filters of all the points of a non-empty subset A of X is the neighbourhood filter of A in X.

8) *a*) Let Φ be a set of topologies on a set X. Show that the neighbourhood filter of a point $x \in X$ with respect to the intersection of the topologies of Φ is coarser than the intersection of the neighbourhood filters of x with respect to the topologies belonging to Φ.

b) Let $\mathfrak{T}_1$ denote the topology $\mathfrak{T}_0(\mathbf{Q}^2)$ (§ 2, Exercise 7) where the set $\mathbf{Q}^2$ is linearly ordered by the lexicographic ordering (*Set Theory*,

Chapter III, § 2, no. 6). Let $\mathfrak{T}_2$ denote the topology on $\mathbf{Q}^2$ obtained by transporting $\mathfrak{T}_1$ by means of the symmetry $(\xi, \eta) \to (\eta, \xi)$. Show that if $\mathfrak{T}$ is the intersection of the topologies $\mathfrak{T}_1, \mathfrak{T}_2$, then the neighbourhood filter of a point of $\mathbf{Q}^2$ with respect to $\mathfrak{T}$ is strictly coarser than the intersection of the neighbourhood filters of the point with respect to $\mathfrak{T}_1$ and $\mathfrak{T}_2$.

¶ 9) *a*) Show that every ultrafilter which is finer than the intersection of a finite number of filters is finer than at least one of them (use Proposition 5 of no. 4).

b) Give an example of an ultrafilter which is finer than the intersection of an infinite family of ultrafilters, but which is not equal to any of the ultrafilters of this family (consider the family of ultrafilters on an infinite set X each of which has a base consisting of a single point).

10) Show that the intersection of all the sets of an ultrafilter contains at most one point; if one point, then the ultrafilter consists of all the sets containing this point (use Proposition 5 of no. 4).

11) Show that if a subset A of a set X does not belong to an ultrafilter $\mathfrak{U}$ on X, then the trace of $\mathfrak{U}$ on A is the set $\mathfrak{P}(A)$.

12) On an infinite set, show that an elementary filter associated with a sequence all of whose terms are distinct is not an ultrafilter.

13) Let f be a mapping of a set X into a set X'. Then f is injective if and only if, for every filter base $\mathfrak{B}$ on X, $(f\overset{-1}{f}(\mathfrak{B}))$ is a filter base equivalent to $\mathfrak{B}$.

14) Show that if f is a mapping of a set X *onto* a set X', then the image $f(\mathfrak{F})$ of every filter $\mathfrak{F}$ on X is a filter on X'.

¶ 15) Let $n \to f(n)$ be a mapping of $\mathbf{N}$ *onto* $\mathbf{N}$ such that $\overset{-1}{f}(m)$ is *finite* for each $m \in \mathbf{N}$. If (x_n) is any sequence in a set X, put $y_n = x_{f(n)}$. Show that the elementary filters associated with the sequences (x_n) and (y_n) are the same.

Deduce that if (a_n) and (b_n) are two sequences of elements of a set X such that the filter associated with (b_n) is finer than the filter associated with (a_n), then the filter associated with (b_n) is the same as the filter associated with some subsequence of (a_n).

¶ 16) Let Φ be a countable linearly ordered set of elementary filters on a set X. Show that there is an elementary filter which is finer than all the filters of Φ (show that the union of the filters of Φ has a countable base).

17) Let X be a *lattice* (*Set Theory*, Chapter III, § 1, no. 13). A non-empty subset F of X is called a *prefilter* if it satisfies the following conditions: (i) whenever $x \in F$ and $y \geqslant x$, then $y \in F$; (ii) whenever $x \in F$

and $y \in F$, then $\inf(x, y) \in F$; (iii) $F \neq X$. Thus the prefilters on the set of subsets of a set Y, ordered by inclusion, are precisely the filters on Y. A prefilter F is said to be *prime* if the relation $\sup(x, y) \in F$ implies that either $x \in F$ or $y \in F$. A prefilter with respect to the ordering opposite to that of X is called a *coprefilter* on X.

a) The set of upper bounds of a subset A of X which does not consist only of the least element of X is a prefilter. Give examples of prefilters which are not sets of upper bounds.

b) Show that if F is a prime prefilter, then $\complement F$ is a prime coprefilter.

c) Show that if X has a least element 0, then every prefilter is contained in a maximal prefilter (use Zorn's lemma, noting that 0 belongs to no prefilter).

d) Find all prefilters on a linearly ordered set X, and show that they are all prime. Hence give examples of prime prefilters which are not maximal.

e) Let X be an ordered set of 5 elements, denoted by $0, 1, a, b, c$, the order relations being $0 < a < 1$, $0 < b < 1$, $0 < c < 1$. Show that X is a lattice and that the prefilters other than $\{1\}$ are maximal but not prime.

f) Let X be a lattice having a least element 0. Show that if $F \subset X$ is a prefilter and a is an element of X which does not belong to F, then there is a prefilter containing F and a if and only if $\inf(a, x) \neq 0$ for all $x \in F$; the smallest prefilter containing F and a is then the set of all $y \in X$ such that $y \geqslant \inf(a, x)$ for at least one element x of F.

¶ 18) Let X be a *distributive* lattice (*Set Theory*, Chapter III, § 1, Exercise 16) which has a least element 0.

a) If a is an *irreducible* element of X (*Set Theory*, Chapter III, § 4, Exercise 7), show that the set of all upper bounds of a is a prime prefilter. Give an example of a prime prefilter which is not a set of upper bounds of an irreducible element of X [cf. Exercise 17 *d*)].

b) Show that every maximal prefilter on X is prime [use Exercise 17 *f*)]. Give examples of prime non-maximal prefilters on a distributive lattice [Exercise 17 *d*)].

c) Show that every prefilter F is the intersection of the prime prefilters which contain F [if $a \notin F$, show that a maximal element U in the set of prefilters which contain F but not a is prime, by using Exercise 17 *f*)].

d) Let Ω be the set of prime prefilters on X. To each $x \in X$ let correspond the subset A_x of Ω consisting of prime prefilters F such that $x \in F$. Show that the mapping $x \to A_x$ of X into $\mathfrak{P}(\Omega)$ is injective and order-preserving, that $A_{\inf(x,y)} = A_x \cap A_y$ and that

$$A_{\sup(x,y)} = A_x \cup A_y.$$

19) Let X be a lattice which has a least element. Show that if every prefilter on X is the intersection of the prime prefilters which contain it, then X is distributive (if the mapping $x \to A_x$ is defined as in Exercise 18 *d*), this mapping still has the properties stated there).

¶ 20) A lattice X is said to be a *Boolean algebra* if it is distributive, has a least element 0, and a greatest element 1 and if, for each $x \in X$, there is an element $x' \in X$ such that $\inf(x, x') = 0$ and $\sup(x, x') = 1$; such an element is said to be a *complement* of x in X.

a) Let $x \to A_x$ be the injection of a Boolean algebra X into the set of subsets of the set Ω of prime prefilters on X, defined in Exercise 18 *d*). Show that if x' is a complement of x in X, then $A_{x'} = \complement A_x$ in Ω; deduce that an element x has only one complement x', that the complement of x' is x, and that the complement of $\inf(x, y)$ is $\sup(x', y')$; give direct proofs of these statements. The ideals of the Boolean ring A corresponding to X, other than A, are the coprefilters on X.

b) Show that, in a Boolean algebra X, every prime prefilter is maximal (remark that, for each $x \in X$, every prime prefilter must contain either x or its complement).

c) Conversely, let X be a distributive lattice which has a least element 0 and a greatest element 1, and is such that every prime prefilter on X is maximal. Show that X is a Boolean algebra. (For an element $x \in X$, consider the coprefilter C (Exercise 17) consisting of those $y \in X$ such that $\inf(x, y) = 0$; if x has no complement, there is a maximal coprefilter M containing x and C; remark that the complement U of M in X is a prime prefilter [Exercise 17 *b*) and 18 *b*)] and use Exercise 17 *f*)).

d) Show that if X is a topological space then the set $\mathfrak{B}$ of subsets of X which are both open and closed is a Boolean algebra if ordered by inclusion. If we take X to be the topological space associated with the Fréchet filter (no. 5), show that the Boolean algebra $\mathfrak{B}$ thus obtained is countably infinite and therefore cannot be isomorphic to the ordered set $\mathfrak{P}(A)$ of all subsets of a set A.

§ 7

1) Let $\mathfrak{T}_1$, $\mathfrak{T}_2$ be two topologies on the same set X. Show that $\mathfrak{T}_1$ is finer than $\mathfrak{T}_2$ if and only if every filter on X which is convergent in the topology $\mathfrak{T}_1$ converges to the same point in the topology $\mathfrak{T}_2$.

2) Let $\mathfrak{U}$ be an ultrafilter on **N** which is finer than the Fréchet filter, and let $X = \mathbf{N} \cup \{\omega\}$ be the space associated with $\mathfrak{U}$ (§ 6, no. 5). Show that a sequence (x_n) which has an infinite number of distinct terms is never convergent in X.

3) Let X, X′ be two topological spaces and $f: X \to X'$ a mapping which is continuous at a point $x_0 \in X$. Show that if $\mathfrak{B}$ is any filter base on X which has x_0 as a cluster point, then $f(x_0)$ is a cluster point of the filter base $f(\mathfrak{B})$ on X′.

4) *a*) Let X, Y be two topological spaces, $\mathfrak{F}$ a filter on X having a cluster point *a*, and $\mathfrak{G}$ a filter on Y having a cluster point *b*. Show that (a, b) is a cluster point of the product filter $\mathfrak{F} \times \mathfrak{G}$.

b) In the product space $\mathbf{Q}^2$, give an example of a sequence (x_n, y_n) which has no cluster points although each of the sequences (x_n), (y_n) has a cluster point in $\mathbf{Q}$.

5) Let X, Y be two topological spaces, G the graph in $X \times Y$ of a mapping $f: X \to Y$. Show that for each $x \in X$ the set of cluster points of f at x is the section $\overline{G}(x)$ of $\overline{G}$ (the closure of G in $X \times Y$) at x.

6) Let $(X_\iota)_{\iota \in I}$ be an *uncountable* family of discrete spaces, each of which has at least two points. On the product set consider the topology

$$X = \prod_{\iota \in I} X_\iota,$$

which is generated by all products $\prod_{\iota \in I} M_\iota$, where $M_\iota \subset X_\iota$ for all $\iota \in I$ and $M_\iota = X_\iota$ for all but a *countable* number of indices (cf. § 4, Exercise 9). Show that every countable intersection of open sets is open in X, and deduce that no point of x can have a countable fundamental system of neighbourhoods.

7) A subset A of a topological space X is said to be *primitive* if it is the set of limit points of an ultrafilter on X. Let Υ_A denote the set of ultrafilters on X whose set of limit points is A.

a) If A is a primitive subset of X, show that every open subset of X which meets A belongs to all the ultrafilters $\mathfrak{U} \in \Upsilon_A$.

b) Let A, B be two distinct primitive subsets of X. Show that there is an ultrafilter $\mathfrak{U} \in \Upsilon_A$, an ultrafilter $\mathfrak{B} \in \Upsilon_B$ and a subset M of X such that $M \in \mathfrak{U}$ and $\complement M \in \mathfrak{B}$.

c) Let $f: X \to Y$ be a continuous mapping. Show that the image under f of any primitive subset of X is contained in a primitive subset of Y.

d) If $x \in X$ and $\{x\}$ is closed in X, show that $\{x\}$ is a primitive subset of X.

§ 8

1) *a*) Let X be a topological space. Show that the following three statements are equivalent:

(Q) If x, y are any two distinct points of X, there is a neighbourhood of x which does not contain y.

(Q′) Every subset of X which consists of a single point is closed in X.

(Q″) For every $x \in X$ the intersection of the neighbourhoods of x consists of the point x alone.

A space X is said to be *accessible* if it satisfies these conditions.

b) An ordered set X with the right topology (with respect to which X is a Kolmogoroff space [§ 1, Exercise 2)] is accessible if and only if no two distinct elements of X are comparable, and then the right topology on X is the discrete topology.

2) *a*) If a topology on a set X has a finite subbase, and if X is accessible with respect to this topology, then X is finite and the topology is discrete.

b) If, in an accessible space X, *every* intersection of open sets is open, then X is discrete (cf. § 7, Exercise 6).

3) *a*) Let X be an accessible space, A a subset of X, and x a point of $\overline{A}$ which does not belong to A. Show that every neighbourhood of x contains an infinite number of points of A.

b) Let X be an accessible space and let A be a subset of X. Show that the set of points $x \in \overline{A}$ such that every neighbourhood of x contains a point of A other than x is closed in X.

c) Let X be an accessible space. Show that the intersection of all neighbourhoods of any subset A of X is equal to A.

4) Show that every subspace of a Kolmogoroff space (resp. accessible space) is a Kolmogoroff space (resp. accessible space); and that a topology which is finer than a topology of a Kolmogoroff space (resp. accessible space) is the topology of a Kolmogoroff space (resp. accessible space).

¶ 5) *a*) Let X be an infinite set. Show that all the topologies $\mathfrak{T}_{\mathfrak{c}}$ on X (§ 2, Exercise 8) such that $1 < \mathfrak{c} \leqslant \text{card}(X)$ are topologies of an accessible space but are not Hausdorff.

b) Show that the intersection of all the Hausdorff topologies on X is the non-Hausdorff topology $\mathfrak{T}_{\text{card}(\mathbf{N})}$, which is also the coarsest topology of an accessible space on X. (Use Exercise 8 of § 2, and remark that there exist Hausdorff topologies on X in which there are countably infinite subsets of X which are not closed).

6) *a*) Let X be a Hausdorff space and D a dense subset of X. Show that card (X) $\leqslant 2^{2^{\text{card(D)}}}$ (observe that every point of X is the limit of a filter base on D). Show that this upper bound for card (X) cannot be improved (§ 4, Exercise 5 *b*)) and deduce that on an infinite set A the set of ultrafilters and the set of filters are each equipotent to $\mathfrak{P}(\mathfrak{P}(A))$.

b) Let K be the discrete space consisting of the two numbers 0 and 1. Let A be an infinite set. Deduce from *a*) that if card (A) $\geqslant 2^{2^{\text{card(N)}}}$, then the product space K^A satisfies the condition (D_{IV}) of Exercise 7 of § 1, but satisfies neither (D_{II}) nor (D_{III}). (§ 4, Exercise 4).

* 7) Let X be the interval $[-1, 1]$ in **R**. Consider the following equivalence relation S on X: if $x \neq \pm 1$, the equivalence class of x consists of x and $-x$; the equivalence class of 1 (resp. -1) consists of 1 (resp. -1) alone. Show that S is open and that the quotient space X/S is accessible but not Hausdorff. Show also that every point of X/S has a neighbourhood in X/S which is a regular subspace. *

8) *a*) Let X be a Hausdorff (resp. regular) space which is the sum of a family (X_ι) of subspaces, and let X/R be the quotient space obtained by pasting together the X_ι along closed subsets $A_{\iota\varkappa}$ by means of homeomorphisms $h_{\varkappa\iota}$ (§ 3, no. 4); suppose also that for each index ι there is only a finite number of indices $\varkappa$ such that $A_{\iota\varkappa} \neq \emptyset$. Under these conditions, show that X/R is Hausdorff (resp. regular).

* *b*) Show that the space X/S of Exercise 7 can be obtained by pasting together two regular subspaces along two open sets. *

9) Let Γ be a *finite* group of homeomorphisms of a Hausdorff (resp. regular) space X, and let R be the equivalence relation "there exists $\sigma \in \Gamma$ such that $y = \sigma(x)$" between x and y in X. Show that the quotient space X/R is Hausdorff (resp. regular).

* 10) Let X be the subspace of **R** which is the complement of the set of points $1/n$, where n is a rational integer other than 0, 1 or -1. Let S be the equivalence relation on X whose equivalence classes are (i) the set $\{0\}$; (ii) the set of all non-zero integers; (iii) all sets of the form $\{x, 1/x\}$ where $x \in X$ and $0 < |x| < 1$.

a) Show that the relation S on X is neither open nor closed.

b) Show that the graph of S is closed in $X \times X$, but that the quotient space X/S is not Hausdorff. *

¶ * 11) Let X be a topological space. Let Z denote the product space $\mathbf{R}^X$, and for each $x \in X$ let Z_x be the subset of Z consisting of all $u \in \mathbf{R}^X$ such that $u(x) \in \mathbf{Q}$ and $u(y) \notin \mathbf{Q}$ for all $y \in X$ other than x. Let Y be the subspace of the product $X \times Z$ which is the union of the sets $\{x\} \times Z_x$ as x runs through X. Show that Y is Hausdorff and that X is homeomorphic to a quotient space of Y. $_*$

12) Let X be a topological space which has at least two points.

a) Show that the topologies $\mathfrak{T}_\Omega$ and $\mathfrak{T}_\Phi$ on $\mathfrak{P}_0(X)$ (§ 2, Exercise 7) are not topologies of an accessible space.

b) Let $\mathfrak{T}_\Theta$ denote the least upper bound of the topologies $\mathfrak{T}_\Omega$ and $\mathfrak{T}_\Phi$. Show that the set $\mathfrak{F}(X)$ of non-empty closed subsets of X, with the topology induced by $\mathfrak{T}_\Theta$, is a Kolmogoroff space, and that if X is accessible, so is $\mathfrak{F}(X)$.

c) Supposing that X is accessible, show that the space $\mathfrak{F}(X)$ (with the topology induced by $\mathfrak{T}_\Theta$) is Hausdorff if and only if X is regular.

¶ 13) Let X be a set and let Γ be a group of permutations of X.

a) Let $\mathfrak{F}$ be a set of subsets of X. Let $\mathfrak{S}$ be the set of all topologies on X in which all the sets of $\mathfrak{F}$ are closed and every $u \in \Gamma$ is a homeomorphism. Show that the set $\mathfrak{S}$ has a least element $\mathfrak{T}(\Gamma, \mathfrak{F})$. If X is infinite, show that there exist sets $\mathfrak{F}$ of subsets of X such that $\mathfrak{T}(\Gamma, \mathfrak{F})$ is accessible but not discrete (§ 2, Exercise 8).

b) Let $J(\Gamma)$ be the set of subsets A of X which have the following property: for each $x \notin A$ there is a mapping $f \in \Gamma$ such that $f(x) \neq x$ which leaves fixed every element of A. Show that every Hausdorff topology on X, in which every mapping $u \in \Gamma$ is a homeomorphism of X onto itself, is finer than $\mathfrak{T}_J(\Gamma) = \mathfrak{T}(\Gamma, J(\Gamma))$.
Show that the topology $\mathfrak{T}_J(\Gamma)$ is Hausdorff if and only if the group Γ satisfies the following condition: given any two distinct elements x, y of X there is a finite number of elements $u_k \in \Gamma$ such that, if $F(u_k)$ is the set of elements of X which are invariant under u_k, the $F(u_k)$ cover X and none of the $F(u_k)$ contains both x and y.

c) Let $\mathfrak{T}_0$ denote the topology of the rational line **Q** (*resp. the real line $\mathbf{R}_*$) and let Γ be the group of all homeomorphisms of **Q** (* resp. $\mathbf{R}_*$) onto itself. Show that the topology $\mathfrak{T}_J(\Gamma)$ is the same as $\mathfrak{T}_0$.

d) Let X be the subspace of the rational line consisting of 0 and the points $1/n$, where n is an integer $\geqslant 1$, and let Γ be the group of homeomorphisms of X. Show that the topology $\mathfrak{T}_J(\Gamma)$ is not Hausdorff, and that the group of homeomorphisms of X with respect to this topology is Γ.

e) Let X be a topological space such that for each pair of distinct points x, y of X there exists a homeomorphism u of X onto itself such that $u(x) = x$ and $u(y) \neq y$ (*for example, the spaces $\mathbf{R}^n{}_*$). Let G be the group of homeomorphisms of X onto itself, and for each $x \in X$ let S_x be the subgroup of G which leaves x fixed. Show that S_x is equal to its normalizer in G and that the intersection of all the subgroups

S_x consists of the identity element alone; in particular, the centre of G consists of the identity element alone.

14) Show that the axiom (O_{III}) is equivalent to each of the following axioms:

(O''_{III}) If F is any closed subset of X then the intersection of all the closed neighbourhoods of F is equal to F.

(O'''_{III}) If $\mathfrak{F}$ is any filter on X which converges to a point a, then the filter $\overline{\mathfrak{F}}$ on X which has the closures of the sets of $\mathfrak{F}$ as a base also converges to a.

15) *a*) Show that every Kolmogoroff space which satisfies axiom (O_{III}) is Hausdorff (and hence regular).

b) Construct a non-Hausdorff topology on a set of three elements which satisfies axiom (O_{III}).

16) Let (X_ι) be an infinite family of topological spaces, and give the product set $X = \prod_\iota X_\iota$ one of the topologies defined in Exercise 9 of § 4. Show that if the X_ι are Kolmogoroff (resp. accessible, Hausdorff, regular) spaces then so is X.

17) Show that the topologies $\mathfrak{T}_0(X)$, $\mathfrak{T}_+(X)$ and $\mathfrak{T}_-(X)$ (§ 2, Exercise 5) on a linearly ordered set X are regular.

18) Let X_1, X_2 be two sets, let $\mathfrak{F}_i$ be a filter on X_i $(i = 1, 2)$, and let f be a mapping of $X_1 \times X_2$ into a *regular* space Y such that, for each

$$x_1 \in X_1 \text{ there exists } \lim_{x_2, \mathfrak{F}_2} f(x_1, x_2) = g(x_1).$$

Show that a point $a \in Y$ is a cluster point of g with respect to the filter $\mathfrak{F}_1$ if and only if, given any neighbourhood V of a in Y and any set $A_1 \in \mathfrak{F}_1$, there exists $x_1 \in A_1$ and $A_2 \in \mathfrak{F}_2$ such that $f(\{x_1\} \times A_2) \subset V$.

¶ 19) Let X be a Hausdorff space which is not regular (cf. Exercise 20), and let a be a point of X such that there is a neighbourhood U of a which does not contain any closed neighbourhood of a. Let $\mathfrak{F}$ be the neighbourhood filter of a, and let $Y = \mathfrak{F} \cup \{\omega\}$ be the topological space associated with the section filter of $\mathfrak{F}$. Let Z be the set $Y \times X$ with the following topology: for any point $(y, z) \neq (\omega, a)$, the neighbourhoods of (y, z) are the same as for the product topology, and the neighbourhoods of (ω, a) are the sets containing sets of the form

$$(\{\omega\} \times V) \cup ((Y - \{\omega\}) \times X),$$

where V is a neighbourhood of a in X. Let A be the subset of Z which is the union of the sets $\{V\} \times V$ where $V \in \mathfrak{F}$. Let f be the mapping $(V, x) \to x$ of A into X. Show that f has a limit relative to A at each point of $\bar{A}$, but that there is no continuous mapping of $\bar{A}$ into X which extends f.

¶ 20) *a*) An open subset U of a topological space X is said to be *regular* if $U = \alpha(u)$ (§ 1, Exercise 3); equivalently, U is regular if U is the interior of a closed set. X is said to be *semi-regular* (and its topology $\mathfrak{T}$ is said to be *semi-regular*) if the regular open subsets of X form a base of $\mathfrak{T}$. Show that if $\mathfrak{T}$ satisfies axiom (O_{III}) then $\mathfrak{T}$ is semi-regular.

b) Prove that the intersection of two regular open sets is a regular open set. Deduce that the regular open sets with respect to $\mathfrak{T}$ form a base of a topology $\mathfrak{T}^*$ on X, which is coarser than $\mathfrak{T}$ and semi-regular; $\mathfrak{T}^*$ is said to be the semi-regular topology *associated* with $\mathfrak{T}$. Show that for each subset U of X which is open in $\mathfrak{T}$, the closure of U with respect to $\mathfrak{T}^*$ is the same as its closure in $\mathfrak{T}$, and that if X is accessible for $\mathfrak{T}$ the isolated points of X with respect to $\mathfrak{T}^*$ are the same as the isolated points of X with respect to $\mathfrak{T}$. Show that $\mathfrak{T}^*$ is Hausdorff if and only if $\mathfrak{T}$ is [use Exercise 3 *d*) of § 1], and finally that if Y is a regular space then the continuous mappings of X into Y are the same in both topologies $\mathfrak{T}$ and $\mathfrak{T}^*$.

c) Let X be a semi-regular space and let $\mathfrak{T}_0$ be its topology. Show that a topology $\mathfrak{T}$ on X is such that $\mathfrak{T}^* = \mathfrak{T}_0$ if and only if there is a set $\mathfrak{M}$ of *dense* subsets of X (in the topology $\mathfrak{T}_0$) such that every finite intersection of sets of $\mathfrak{M}$ belongs to $\mathfrak{M}$, and such that the topology $\mathfrak{T}$ is generated by the union of $\mathfrak{M}$ and the set of subsets of X which are open in the topology $\mathfrak{T}_0$. [Consider the dense open subsets (in the topology $\mathfrak{T}$) and notice that every open set in $\mathfrak{T}$ is the intersection of a dense open set (in the topology $\mathfrak{T}$) and an open set (in the topology $\mathfrak{T}_0$).] Hence construct examples of non-regular Hausdorff topologies which are finer than a regular topology (*and even finer than a topology of a compact metrizable space$_*$).

21) Let X be a Hausdorff space with no isolated points, and let $\mathfrak{T}_0$ be the topology of X. Let $\mathfrak{M}$ denote the set of complements of countable subsets A of X such that $\bar{A}$ has only a finite number of non-isolated points. Show that if $\mathfrak{T}$ is the topology on X generated by the union of $\mathfrak{M}$ and the set of subsets which are open in $\mathfrak{T}_0$, then $\mathfrak{T}^* = \mathfrak{T}_0$ (Exercise 20) and that every sequence which converges with respect to $\mathfrak{T}$ has only a finite number of distinct terms (which implies that, with respect to $\mathfrak{T}$, no point of X has a countable fundamental system of neighbourhoods, although X itself may be countable and therefore every point of X is then the intersection of a countable family of neighbourhoods of this point).

¶ 22) *a*) With the notation of Exercise 20 *c*), show that the set $E(\mathfrak{T}_0)$ of topologies $\mathfrak{T}$ such that $\mathfrak{T}^* = \mathfrak{T}_0$ is inductive with respect to the relation "$\mathfrak{T}$ is coarser than $\mathfrak{T}'$". If $\mathfrak{T}_0$ is a semi-regular topology on X, a maximal element of $E(\mathfrak{T}_0)$ is called a *submaximal* topology, and the set X with such a topology is called a *submaximal* space. Thus every topology of $E(\mathfrak{T}_0)$ is coarser than some submaximal topology.

b) A topology $\mathfrak{T}$ on X is submaximal if and only if every subset of X which is dense in the topology $\mathfrak{T}$ is open in $\mathfrak{T}$. A non-empty submaximal space is therefore not solvable (§ 2, Exercise 2).

c) Show that every subspace of a submaximal space is locally closed and submaximal (consider first the open subspaces and the closed subspaces).

d) Show that the space associated with a filter which is finer than the filter of complements of finite subsets of an infinite set is a submaximal space.

e) If M is a subset of a submaximal space X, show that the subspace $\overline{M} - \mathring{M} = \mathrm{Fr}(M)$ is discrete.

f) Conversely, let X be a topological space which has a dense open subset U such that U is submaximal and the topology of X is U-maximal (§ 3, Exercise 11). Then X is submaximal.

23) Let X be a topological space and let A be a dense subset of X such that the topology of X is A-maximal (§ 3, Exercise 11). Let f be a continuous mapping of A into a topological space Y, such that at every point of X — A the set of cluster points of f relative to A is not empty (*this will be the case, for example, if X is not empty and Y is quasi-compact$_*$). Show that f can be extended by continuity to the whole of X.

24) In the interval $]0,1[$ of the rational line, let A be the set of rational numbers of the form $k/2^n$, and B the set of rational numbers of the form $k/3^n$ (k an integer). Let X be the space obtained by giving $]0,1[$ the topology generated by the open intervals $]\alpha,\beta[$ contained in X, and the sets A and B. Show that X is Hausdorff and semi-regular (Exercise 20) but not regular.

¶ 25 *a*) On a set X, the set of regular topologies and the set of regular topologies with no isolated points are inductive sets with respect to the relation "$\mathfrak{T}$ is coarser than $\mathfrak{T}'$".

b) A topology $\mathfrak{T}$ on a set X is said to be *ultraregular* if it is maximal in the set of topologies on X which are regular and have no isolated points; hence given any regular topology on X with no isolated points, there is a finer ultraregular topology. A space is said to be *ultraregular* if its topology is ultraregular. Show that a regular topology $\mathfrak{T}$ with no isolated points is ultraregular if and only if, whenever A and B are complementary subsets of X which have no isolated points, then A and B are open sets. In particular, an ultraregular space is not solvable (§ 2, Exercise 11). (To prove that the condition is sufficient, observe that if $\mathcal{T}'$ is a regular topology with no isolated points and is finer than $\mathcal{T}$, and if U is a non-empty open set with respect to $\mathcal{T}'$ and $\overline{U}$ is its closure in the topology $\mathcal{T}'$, then $\overline{U}$ and $X - \overline{U}$ have no isolated points with respect to the topology $\mathcal{T}'$).

c) In an ultraregular space, the closure of every set which has no isolated points is open, and the interior of a dense set is dense; the intersection of two dense sets is therefore dense. Deduce that if $\mathcal{T}_0$ is an ultraregular topology, then amongst all the topologies $\mathcal{T}$ such that $\mathcal{T}^* = \mathcal{T}_0$ (Exercise 20) there is one (necessarily submaximal) which is finer than all the others.

* 26) Let (θ_n) be a sequence of irrational numbers which is dense in the interval $I = [0,1]$ of **R**. Let I_n be the quotient space of I obtained by identifying all the points $0, \theta_1, \ldots, \theta_n$, and let φ_n be the canonical mapping $I \to I_n$. Let X be the set of rational numbers contained in I; then the restriction of φ_n to X is a bijection of X onto $\varphi_n(X)$. Let $\mathcal{T}_n$ be the inverse image under this bijection of the topology induced on $\varphi_n(X)$ by the topology of I_n. Show that the greatest lower bound of the decreasing sequence $(\mathcal{T}_n)$ of Hausdorff topologies on X is not a Hausdorff topology. *

¶ 27) Let X be a topological space. Show that there exists a Kolmogoroff (resp. accessible, Hausdorff, regular) space Y and a continuous mapping $\varphi : X \to Y$ which has the following *universal* property: for every continuous mapping $f: X \to Z$ of X into a Kolmogoroff (resp. accessible, Hausdorff, regular) space Z, there is a unique continuous mapping $g : Y \to Z$ such that $f = g \circ \varphi$ (*Set Theory*, Chapter IV, § 3, no. 1). Prove that:

a) For Kolmogoroff spaces, we may take Y to be the quotient space of X by the equivalence relation $\overline{\{x\}} = \overline{\{y\}}$.

b) For accessible spaces, we may take Y to be the quotient space of X by the following equivalence relation R: the graph of R is the intersection of all graphs G of equivalences on X such that $G(x)$ is closed in x for each $x \in X$; the graph of R then contains the graph of the relation $\overline{\{x\}} \cap \overline{\{y\}} \neq \emptyset$ between x and y.

c) For Hausdorff spaces and regular spaces, show that conditions (CU_I) to (CU_{III}) of *Set Theory*, Chapter IV, § 3, no. 2, are satisfied, by using in particular Exercise 6 *a*). Give examples in which X is not regular and the mapping $\varphi : X \to Y$ of X into the corresponding universal regular space is bijective [cf. Exercise 20 *b*)].

d) If X is a topological space which satisfies axiom (O_{III}), show that the universal Kolmogoroff space corresponding to X is canonically homeomorphic to the universal regular space corresponding to X.

28) Let X be a Hausdorff space which is not regular, let F be a closed subset of X and let $a \notin F$ be a point of X such that every neighbourhood of a meets every neighbourhood of F. Let R be the equivalence relation on X obtained by identifying all the points of F. Show that R is closed and that the graph of R is closed in $X \times X$, but that R is not Hausdorff.

¶ 29) *a*) Let R be a Hausdorff equivalence relation on a non-empty topological space X, and let $\mathfrak{F}(X)$ be the space of non-empty closed subsets of X, with the topology induced by $\mathfrak{T}_\Theta$ (Exercise 12). Show that every $A \in \mathfrak{F}(X)$ which lies in the closure of X/R [considered as a subset of $\mathfrak{F}(X)$] is contained in some equivalence class of R (use *reductio ad absurdum*).

b) Suppose in addition that X is regular and R is open. Show that X/R is then a closed subset of $\mathfrak{F}(X)$ with respect to the topology $\mathfrak{T}_\Theta$ (use Proposition 11 of § 7, no. 6).

§ 9

1) Let X be a topological space, $\mathfrak{S}$ a subbase of the topology of X (§ 2, no. 3). Show that X is quasi-compact if every open covering of X formed by sets belonging to $\mathfrak{S}$ contains a finite covering of X. (Observe that if an ultrafilter $\mathfrak{U}$ on X were not convergent, then every $x \in X$ would belong to a set of $\mathfrak{S}$ not belonging to $\mathfrak{U}$; and use the corollary to Proposition 5 of § 6).

2) Show that a *well-ordered* set X with the topology $\mathfrak{T}_-(X)$ (§ 2, Exercise 5) is locally compact, and is compact if and only if X has a greatest element. Deduce that every set can be topologized in such a way that it becomes a compact space.

3) *a*) Let X be a Hausdorff space, and let A, B be two disjoint compact subsets of X. Show that there is a neighbourhood of A and a neighbourhood of B which have no common point.

b) Let X be a regular space and A a compact subset of X, and let B be a closed subset of X which does not meet A. Show that there is a neighbourhood of A and a neighbourhood of B which have no common point.

4) Show that in the space X defined in Exercise 6 of § 7, which is regular (§ 8, Exercise 16), every compact subset is finite.

5) *a) Show that the non-Hausdorff accessible space $Y = X/S$ defined in Exercise 7 of § 8 is such that every point of Y has a compact neighbourhood; that the images α, β of 1 and -1 in Y have compact neighbourhoods which are not closed; that the intersection of a compact neighbourhood of α and a compact neighbourhood of β is not quasi-compact, and that the union of these two neighbourhoods is not compact.*

b) Let X be the sum of $\mathbf{N}$ and an infinite set A. For each $x \in X$ let $\mathfrak{S}(x)$ be the filter base defined as follows : if $x \in N$, then $\mathfrak{S}(x)$ consists of the single set $\{x\}$; if $x \in A$ and if $S_{x,n}$ denotes the set whose elements are x and the integers $\geqslant n$, then $\mathfrak{S}(x)$ consists of the sets $S_{x,n}$ where n runs through $\mathbf{N}$. Show that there is a topology on X for which $\mathfrak{S}(x)$ is a fundamental system of neighbourhoods of x for every $x \in X$, and that X is accessible but not Hausdorff in this topology. Show also that X is not quasi-compact, but that X has dense compact subsets.

c) Show that two topologies on a set X, with respect to each of which X is accessible and quasi-compact, can be distinct but comparable [cf. § 8, Exercises 7 and 5 b)].

6) Let X, Y be two topological spaces, and let A (resp. B) be a quasi-compact subset of X (resp. Y). Show that if U is any neighbourhood of $A \times B$ in $X \times Y$ then there is a neighbourhood V of A in X and a neighbourhood W of B in Y such that $V \times W \subset U$.

7) a) Give an example of an inverse system $(X_\alpha, f_{\alpha\beta})$ of quasi-compact spaces whose inverse limit is empty (observe that in the space of Example 2 of no. 1, every subspace is quasi-compact).

b) Let p be a prime number. Let Z_n denote the set $\mathbf{Z}$ of rational integers, topologized with the inverse image of the discrete topology under the canonical mapping of $\mathbf{Z}$ onto $\mathbf{Z}/p^n\mathbf{Z}$. If $m \leqslant n$, let f_{mn} denote the identity mapping $Z_n \to Z_m$, which is continuous. Show that the spaces Z_n are quasi-compact, but that the inverse limit of the inverse system (Z_n, f_{mn}) is not quasi-compact (*).

8) Let X be a topological space and let $(G_\alpha)_{\alpha \in A}$ be a family of graphs of equivalence relations on X, whose index set A is directed; let R_α be the relation $(x, y) \in G_\alpha$ and let φ_α be the canonical mapping $X \to X/R_\alpha$. Suppose that the relation R_β implies R_α if $\alpha \leqslant \beta$ and let $\varphi_{\alpha\beta}$ be the canonical mapping $X/R_\beta \to X/R_\alpha$; then $(X/R_\alpha, \varphi_{\alpha\beta})$ is an inverse system of topological spaces. Let $Y = \varprojlim X/R_\alpha$.

(*) This example (unpublished) is due to D. Zelinsky.

a) Let $g: X \to Y$ be the continuous mapping $\varprojlim \varphi_\alpha$. Show that $g(x)$ is dense in Y and that, for each $x \in X$, $\overset{-1}{g}(g(x))$ is the class of x with respect to the equivalence relation whose graph is $\bigcap_\alpha G_\alpha$.

b) Suppose that all the R_α are *closed* equivalence relations, and that for each $x \in X$ and each neighbourhood V of x there is an index α such that the class of x with respect to R_α is contained in V. Show that under these conditions g is an *open* mapping of X onto $g(X)$.

c) Show that if the classes mod R_α are closed and quasi-compact for each $\alpha \in A$, then g is *surjective.*

9) Let X be an accessible space and $\mathfrak{K}$ a set of quasi-compact closed subsets of X, which contains all finite subsets of X and is such that the union of any two subsets of $\mathfrak{K}$ belongs to $\mathfrak{K}$. Let X′ be a set which is the sum of X and a set $\{\omega\}$ consisting of a single point. Show that there exists a quasi-compact accessible topology on X′ which induces the given topology on X, and which is such that the complements in X′ of the sets of $\mathfrak{K}$ form a fundamental system of open neighbourhoods of ω. Give examples of distinct topologies on X′ corresponding to distinct sets $\mathfrak{K}$.

10) Let X be a paracompact space. Show that if every open subspace of X is paracompact, then every subspace of X is paracompact. Deduce that every subspace of a locally compact space whose topology has a countable base, is paracompact [cf. Chapter IX, § 4, Exercise 20 *a*)].

¶ 11) Let X_0 be an uncountable well-ordered set which has a greatest element; let a be the smallest element of X_0 and b the smallest of the elements $x \in X_0$ such that the interval $[a, x[$ is uncountable. Let X denote the interval $[a, b[$ with the topology $\mathfrak{T}_-(X)$.

a) X is locally compact but not compact (Exercise 2) and every sequence in X has a cluster point (observe that every countable subset of X is bounded above).

b) Let f be a mapping of X into itself such that $f(x) < x$ for all sufficiently large x. Show that there exists an element $c \in X$ with the property that for each $x \in X$ there exists $y \geqslant x$ such that $f(y) \leqslant c$. [Use *reductio ad absurdum,* by constructing a sequence (z_n) of points of X such that z_{n+1} is the smallest of the elements $z' \in X$ such that $f(x) \geqslant z_n$ whenever $x \geqslant z'$].

c) Deduce from *b*) that X is not paracompact.

12) Let X be a topological space, $\mathfrak{F}(X)$ the set of non-empty closed subsets of X, $\mathfrak{K}(X)$ the set of non-empty quasi-compact subsets of X;

endow each of these sets with the topology induced by the topology $\mathfrak{T}_\Omega$ on $\mathfrak{P}_0(X)$ (§ 2, Exercise 7).

a) Show that if $\mathfrak{B} \subset \mathfrak{F}(X)$ is quasi-compact with respect to $\mathfrak{T}_\Omega$ and if X is regular, then the union of the sets $M \in \mathfrak{B}$ is closed in X.

b) Show that if $\mathfrak{B} \subset \mathfrak{K}(X)$ is quasi-compact with respect to $\mathfrak{T}_\Omega$ then the union of the sets $M \in \mathfrak{B}$ is quasi-compact.

¶ 13) With the notation of Exercise 13, endow $\mathfrak{F}(X)$ and $\mathfrak{K}(X)$ with the topology induced by the topology $\mathfrak{T}_\Theta$ on $\mathfrak{P}_0(X)$ (§ 8, Exercise 12).

a) Show that if X is quasi-compact, then so is $\mathfrak{F}(X)$. (For each open set U in X, let $c(U)$ [resp. $m(U)$] be the set of all closed subsets M of X such that $M \subset U$ (resp. $M \cap U \neq \emptyset$); the $c(U)$ and $m(U)$ form a subbase of $\mathfrak{T}_\Theta$. Then use Exercise 1).

b) Suppose that X is accessible, and let X′ be the image of X under the mapping $x \to \{x\}$ of X into $\mathfrak{F}(X)$. Let $\mathfrak{H}$ be a subspace of $\mathfrak{F}(X)$ which contains X′. Show that if $\mathfrak{H}$ is quasi-compact then so is X. [If (U_α) is an open covering of X, remark that if $\mathfrak{B}_\alpha$ denotes the set of all $M \in \mathfrak{F}(X)$ such that $M \cap U_\alpha \neq \emptyset$, then $\{\mathfrak{B}_\alpha\}$ is an open covering of $\mathfrak{F}(X)$].

c) Show that if X is locally compact, then $\mathfrak{K}(X)$ is open in $\mathfrak{F}(X)$ (use Proposition 10 of § 7).

d) Suppose that X is accessible. Show that $\mathfrak{K}(X)$ is Hausdorff (resp. regular, compact, locally compact) if and only if X is. [For the first two assertions, use Exercise 12 of § 8 and Exercise 3 of § 9; for the last two assertions, use *a*) and *b*) and Exercise 12].

¶ 14) A topological space X is said to be a *Lindelöf space* if every open covering of X contains a countable covering of X. Every space with a countable base is a Lindelöf space, and so is every quasi-compact space.

a) Every closed subspace of a Lindelöf space is a Lindelöf space. For every subspace of a Lindelöf space to be a Lindelöf space it is necessary and sufficient that every open subspace is Lindelöf (cf. Exercise 15).

b) If f is any continuous mapping of a Lindelöf space X into a topological space X′, then the subspace $f(X)$ of X′ is Lindelöf.

c) Every countable union of Lindelöf subspaces of a topological space is Lindelöf. In particular every countable space is Lindelöf.

d) A Lindelöf space X is quasi-compact if every sequence in X has a cluster point [show that axiom (C‴) is then verified].

¶ 15) *a*) Let X be a topological space in which every open subspace is Lindelöf. Show that X satisfies condition (D_{III}) of § 1, Exercise 7, and that if X is regular then every closed subset of X is a countable intersection of open sets [cf. Exercise 23 *c*)].

b) Let X be a topological space in which every open subspace is paracompact. Show that every open subspace of X is Lindelöf if and only if X satisfies condition (D_{III}) of § 1, Exercise 7.

c) Let X be a compact space. Show that every open subspace of X is Lindelöf if and only if (i) X satisfies condition (D_{III}) of § 1, Exercise 7, and (ii) every closed subset of X is a countable intersection of open sets (use *b*) and Theorem 5 of no. 10).

¶ 16) Let X be a topological space and A a subset of X. A point $a \in X$ is said to be a *condensation point* of A if every neighbourhood of a contains an *uncountable* infinity of points of A.

a) The set of condensation points of a subset A of X is closed. Give an example in which X has only one condensation point (cf. no. 8, Theorem 4).

b) Suppose X is an accessible space in which every open subspace is Lindelöf (Exercise 15). Show that for every subset A of X, the set B of condensation points of A is perfect, and that $A \cap \complement B$ is countable.

¶ 17) In a topological space X, a point a is said to be *totally adherent* to a subset A of X if for each neighbourhood U of a we have card (A) = card (A ∩ U). Show that X is quasi-compact if and only if for each infinite subset A of X there is a point of X which is totally adherent to A. (If X is not quasi-compact, show that there is an initial ordinal ω_α (*Set Theory*, Chapter III, § 6, Exercise 10) and an open covering (U_ξ) of X, where ξ runs through the set of ordinals $< \omega_\alpha$, such that for each index $\xi < \omega_\alpha$ the complement A_ξ in X of the union of the U_η with indices $\eta < \xi$ has a cardinal equal to $\aleph_\alpha$; deduce that it is possible to define a family (x_ξ) $(\xi < \omega_\alpha)$ of points of X by transfinite induction such that $x_\xi \in A_\xi$ for each ξ and $x_\xi \neq x_\eta$ whenever $< \eta\ \xi$).

¶ 18) A Hausdorff space X is said to be *absolutely closed* if it has the following property: for every homeomorphism f of X onto a subspace of a Hausdorff space X′, f(X) is closed in X′. Every compact space is absolutely closed.

a) Show that the following conditions are equivalent:

(AF) The space X is absolutely closed.

(AF′) Every filter base on X which consists of open sets has at least one cluster point.

(AF″) Every family of closed subsets of X whose intersection is empty contains a finite subfamily such that the interiors of the subsets in this subfamily have an empty intersection.

(AF‴) Every open covering of X contains a finite subfamily whose closures cover X.

b) Let X be a Hausdorff space and U a dense open subset of X. Suppose that every filter base on U which consists of open sets has at least one cluster point in X. Show that under these conditions X is absolutely closed. In particular, if A is an open subset of an absolutely closed space X, then $\bar{A}$ is absolutely closed.

c) Let X be an absolutely closed space and let f be a continuous mapping of X into a Hausdorff space X′. Show that $f(X)$ is an absolutely closed subspace of X′.

d) Show that the product of two absolutely closed spaces is absolutely closed (argue as in Proposition 17 of no. 10).

¶ 19) A Hausdorff space X is said to be *minimal* if its topology $\mathfrak{T}$ is such that every topology on X which is Hausdorff and coarser than $\mathfrak{T}$ necessarily coincides with $\mathfrak{T}$; then $\mathfrak{T}$ must be semi-regular (§ 8, Exercise 20). Show that a Hausdorff space X is minimal if and only if every filter base on X, which consists of open sets and has at most one cluster point, is convergent. It comes to the same thing to say that X is absolutely closed (Exercise 18) and that every filter base on X, which consists of open sets and has a single cluster point, converges to this point.

¶ 20) A topological space X is said to be *completely Hausdorff* (and the topology of X is said to be *completely Hausdorff*) if each pair of distinct points of X has disjoint *closed* neighbourhoods.

a) Every regular space is completely Hausdorff. Every subspace of a completely Hausdorff space is completely Hausdorff. Every product of completely Hausdorff spaces is completely Hausdorff. Every topology which is finer than a completely Hausdorff topology is completely Hausdorff.

b) Every minimal completely Hausdorff space X (Exercise 19) is compact. (By *reductio ad absurdum:* consider an ultrafilter $\mathfrak{U}$ on X and the filter base formed by the open sets which belong to $\mathfrak{U}$).

**c*) Let θ be an irrational number > 0. For every point $(x, y) \in \mathbf{Q}^2$ and every $\alpha > 0$ let $B_\alpha(x, y)$ be the set consisting of (x, y) and the points $(z, 0) \in \mathbf{Q}^2$ such that $|z - (x + \theta y)| < \alpha$ or $|z - (x - \theta y)| < \alpha$. Let $\mathfrak{B}(x, y)$ be the set of $B_\alpha(x, y)$ where α runs through the set of real numbers > 0. Show that there is a Hausdorff topology $\mathfrak{T}$ on $\mathbf{Q}^2$ such that, for each (x, y), $\mathfrak{B}(x, y)$ is a fundamental system of neighbourhoods of (x, y) for $\mathfrak{T}$. Show that, given any two points a, b of $\mathbf{Q}^2$, every closed neighbourhood of a meets every closed neighbourhood of b in the topology $\mathfrak{T}$.*

¶ 21) *a*) Let X be a Hausdorff space and let $\mathfrak{T}$ be the topology of X. Show that X is absolutely closed (Exercise 18) if and only if the semi-regular topology $\mathfrak{T}^*$ associated with $\mathfrak{T}$ (§ 8, Exercise 20) is the topology of a *minimal* space (Exercise 19).

b) If $\mathfrak{T}$ is completely Hausdorff (Exercise 20), then so is $\mathfrak{T}_*$ (use Exercise 20 of § 8). Deduce that an absolutely closed regular space is compact.

¶ 22) *a*) Let X be a Hausdorff space and let A be a dense subset of X. Show that if every filter on A has at least one cluster point in X, then X is absolutely closed [verify axiom (AF′) of Exercise 18].

b) Let X_0 be a compact space which has a non-open dense subset A, and let $\mathfrak{T}_0$ be the topology of X_0. Let $\mathfrak{T}$ be the topology on X_0 generated by A and the subsets of X_0 which are open in $\mathfrak{T}_0$, so that $\mathfrak{T}^* = \mathfrak{T}_0$ (§ 8, Exercise 20). If X denotes the non-compact absolutely closed space obtained by endowing the set X_0 with the topology $\mathfrak{T}$ (Exercise 21) show that every filter on A has a cluster point in X.

¶ 23) *a*) Let X be a *countable* absolutely closed space. Show that the set of isolated points of X is dense in X. [Use *reductio ad absurdum*, by arranging the points of X in a sequence, and constructing by induction a countable open covering (U_n) of X such that no finite union of the $\overline{U}_n$ covers X].

**b*) Give an example of a completely Hausdorff absolutely closed topological space which is not a Lindelöf space [in the product $[0, 1] \times [0, 1]$; consider the complements of the sets $A \times \{0\}$, where A runs through the set of all subsets of $[0,1]$, and apply the general method of § 8, Exercise 20 *c*)]. This space satisfies condition (D_{II}) of § 1, Exercise 1, but does not satisfy condition (D_{III}).

c) Let X_0 be a compact space with no isolated points, let $\mathfrak{M}$ be the set of complements of countable subsets of X_0 [so that the sets of $\mathfrak{M}$ are dense in X_0, cf. *a*)]. Let $\mathfrak{T}$ be the topology generated by the open sets of X_0 and the sets of $\mathfrak{M}$. Show that the space X obtained by giving X_0 the topology $\mathfrak{T}$ is absolutely closed and completely Hausdorff and is a Lindelöf space which does not satisfy condition (D_{II}) of § 1, Exercise 7; show also that no point of X has a countable fundamental system of neighbourhoods and that no sequence in X, all of whose terms are distinct, has a cluster point. If every open subspace of X_0 is Lindelöf [cf. Exercise 16 *c*)], show that the same is true of X and hence in particular that X satisfies (D_{III}) [Exercise 16 *a*)]. This is the case when $X_0 = [0,1]$. Show however that in this case there are countable closed subsets of X which are not countable intersections of open sets of X (cf. Chapter IX, § 5, no. 3). *

¶ 24) *a*) Let X be a Hausdorff space, and let Γ be the set of filters on X which have a base consisting of open sets and which have no cluster point in X. Show that the set Γ is inductive if it is not empty (i.e., if X is not absolutely closed).

b) Let Ω be the set of *maximal* filters in Γ, and let X' be the set which is the sum of X and Ω. For each $x \in X$ let $\mathfrak{B}(x)$ denote the set of neighbourhoods of x in X, and for each $\omega \in \Omega$ let $\mathfrak{B}(\omega)$ denote the set of subsets of X' of the form $\{\omega\} \cup M$, where M belongs to the filter ω. Then there is a topology on X' in which $\mathfrak{B}(x')$ is a fundamental system of neighbourhoods of x' for each $x' \in X'$. Show that, with this topology, X' is absolutely closed and has the following universal property : given any open continuous mapping $f: X \to Y$ of X into an absolutely closed space Y, there is a unique continuous mapping $g: X' \to Y$ which extends f (consider the images under f of the filters which belong to Ω, and observe that in an absolutely closed space, a filter which is maximal in the set of filters which have a base consisting of open sets must be convergent). In particular, the topology of X' is X-maximal (§ 3, Exercise 11). Deduce that there are absolutely closed spaces whose topology is quasi-maximal (§ 2, Exercise 6).

¶ 25) *a*) Let X be an accessible space, and let Φ be the set of filters on X which have a base consisting of closed sets. Show that the set Φ is inductive, and let X'' be the set of maximal filters in Φ.

b) For each open subset U of X, let U^* be the set of filters $\mathfrak{F} \in X''$ such that $U \in \mathfrak{F}$. If U, V are two open subsets of X, then

$$(U \cap V)^* = U^* \cap V^* \quad \text{and} \quad (U \cup V)^* = U^* \cup V^*$$

(remark that if $\mathfrak{F} \in X''$ is such that $U \notin \mathfrak{F}$, then $X - U \in \mathfrak{F}$). Show that the sets U^* form a base of a topology on X'' and that in this topology X'' is quasi-compact [verify axiom (C''''), using the remark above] and accessible [if $\mathfrak{F}$, $\mathfrak{G}$ are two distinct elements of X'', show that there exist two disjoint closed subsets A and B in X such that $A \in \mathfrak{F}$ and $B \in \mathfrak{G}$].

c) For each $x \in X$ let $\varphi(x)$ be the ultrafilter on X which has the single set $\{x\}$ as a base. Show that φ is a homeomorphism of X onto a dense subset of X'' and that if X is quasi-compact (and accessible), then $\varphi(X) = X''$.

d) Show that for each continuous mapping f of X into a compact space Y there is a unique continuous mapping $g: X'' \to Y$ such that $f = g \circ \varphi$.

e) Let $\tilde{X}$ be the universal Hausdorff space associated with X'' (§ 8, Exercise 27); show that the canonical mapping $\psi: X'' \to \tilde{X}$ is surjective and that $\tilde{X}$ is compact. Thus every continuous mapping of X into a compact space Y can be written uniquely in the form $f = h \circ \psi \circ \varphi$, where $h: \tilde{X} \to Y$ is continuous.

26) If X is a discrete space, show that the spaces X'' and $\tilde{X}$ defined in Exercise 25 are identical, and that if $\mathfrak{T}$ is the topology of the space X' defined in Exercise 24, then X'' can be identified with the space X' carrying the semi-regular topology $\mathfrak{T}^*$ associated with $\mathfrak{T}$ (§ 8, Exercise 20). Furthermore, the set X'' can be canonically identified with the set of ultrafilters on X. The compact space X'' is called the *ultrafilter space* of X.

¶ 27) Show that in a compact space X every isodyne [§ 2, Exercise 11 *d*)] infinite open subspace is solvable. [Using Proposition 1 of no. 2, show that there is a base $\mathfrak{B}$ of the topology of U such that card $(\mathfrak{B})$ = card (U); then use Exercise 11 *c*) of § 2.] Deduce that every locally compact space with no isolated points is solvable [use Exercise 11 *d*) of § 2]; consequently such a space is not submaximal.

28) *a*) Let K be the discrete space $\{0, 1\}$. If X is any compact space, let F be the set of mappings f of K into the set $\mathfrak{F}(X)$ of closed subsets of X such that $X = f(0) \cup f(1)$. Consider the compact space $Y = K^F$; show that for each $y = (y_f)_{f \in F}$ in Y, the intersection of the closed sets $f(y_f)$ of X consists of at most one point. The set Z of $y \in Y$ for which this intersection is not empty is a closed subset of Y. If $\varphi(y)$ is the unique point common to the $f(y_f)$ for $y \in Z$, show that φ is a continuous surjection of Z onto X.

b) Give an example of a compact space X such that there is no surjective continuous mapping of a compact space of the form K^A onto X [use Exercise 4 *b*) of § 4 and Exercise 26 of § 9].

¶ 29) A topological space X is said to be *locally quasi-compact* if each point of X has a fundamental system of quasi-compact neighbourhoods.

a) Show that in a locally quasi-compact space, every quasi-compact subset has a fundamental system of quasi-compact neighbourhoods.

b) Show that every locally closed subspace of a locally quasi-compact space is locally quasi-compact.

**c*) On the closed Euclidean disc $B : \|x\| \leqslant 1$ in the real plane $\mathbf{R}^2$, let $\mathfrak{T}_0$ be the topology induced by that of $\mathbf{R}^2$, and let $\mathfrak{T}$ be the (finer) topology defined as follows: the neighbourhood filter of a point $x \neq (1,0)$ of B is the same for $\mathfrak{T}$ as for $\mathfrak{T}_0$, but for the point $(1,0)$ a fundamental system of neighbourhoods in the topology $\mathfrak{T}$ is formed by the sets

$$\{(1,0)\} \cup (D \cap B_0),$$

where D is any open disc centred at $(1,0)$ and B_0 is the *open* disc $\|x\| < 1$. Let Y be the set B with the topology $\mathfrak{T}$, and let R be the equivalence relation on Y whose equivalence classes are the set S of all $x \neq (1,0)$

in B such that $\|x\| = 1$ and the subsets of Y which consist of a single point of $B - S$. Show that the quotient space $X = Y/R$ is quasi-compact but not locally quasi-compact. $_*$

§ 10

1) *a*) Give an example of continuous mappings $f: X \to Y$, $g: Y \to Z$ such that $g \circ f$ is proper, f is not surjective and g is not proper.
b) Give an example of two proper mappings $f: X \to Y$ and $g: X \to Z$ where X, Y, Z are quasi-compact, such that $x \to (f(x), g(x))$ is not a proper mapping of X into $Y \times Z$.

2) Let $f: X \to Y$ be a proper mapping and let A be a subspace of X. Show that if $A = \overset{-1}{f}(f(A)) \cap \overline{A}$ then the restriction $f|A$ is a proper mapping of A into $f(A)$. Is this condition necessary?

¶ 3) Let $(X_\alpha, f_{\alpha\beta})$ be an inverse system of quasi-compact Kolmogoroff spaces such that the $f_{\alpha\beta}$ are proper mappings. Show that if the X_α are non-empty then $\varprojlim X_\alpha$ is not empty (show that in a quasi-compact Kolmogoroff space, every non-empty closed subset contains a closed subset consisting of a single point).

4) Give an example of a continuous mapping $f: X \to Y$ (where X and Y are Hausdorff spaces) such that the inverse image $\overset{-1}{f}(K)$ of every compact subset K of Y is compact, but which is not proper (cf. § 9, Exercise 4).

¶ 5) Let $f: X \to Y$ be a proper mapping.
a) Show that if X is regular then $f(X)$ is a regular subspace of Y.
b) Show that if $f(X)$ is a Lindelöf subspace of Y (§ 9, Exercise 14), then X is Lindelöf (use Proposition 10 of § 5, no. 4).

¶ 6) Let X be a space which is the *sum* of subspaces X_ι $(\iota \in I)$, and let f be a continuous mapping of X into a topological space Y. For each $\iota \in I$, let $f_\iota : X_\iota \to Y$ be the restriction of f to X_ι. Show that f is proper if and only if each of the f_ι is proper and the family $(f(X_\iota))_{\iota \in I}$ is locally finite [use Theorem 1 *d*)].

¶ 7) Let $f: X \to X'$ be a proper mapping and let R (resp. R′) be an equivalence relation on X (resp. X′) with which f is compatible. Let $\overline{f}: X/R \to X'/R'$ be the mapping of the quotient spaces induced by f. Suppose that (i) the image under f of every saturated set with respect to R is saturated with respect to R′, and (ii) R′ is open. Show that in these conditions $\overline{f}$ is proper [use Theorem 1 *c*)].

8) Show that under the hypotheses of Example 3 of § 5, no. 2, the canonical mapping $X \to X/R$ is proper.

¶ 9) *a*) Let X, Y be two topological spaces and $f: X \to Y$ a surjective closed (but not necessarily continuous) mapping. Show that if $\overset{-1}{f}(y)$ is quasi-compact for each $y \in Y$, then $\overset{-1}{f}(K)$ is quasi-compact for each quasi-compact subset K of Y (either prove the result directly or else use Exercise 7 of § 5 and Exercise 12 of § 9).
b) Suppose in addition that X is regular. Show that if $\overset{-1}{f}(y)$ is closed in X for each $y \in Y$, then $\overset{-1}{f}(K)$ is closed in X for each quasi-compact subset K of Y (same methods). If also Y is locally compact then f is continuous.

10) Let X, Y be two topological spaces. A correspondence (G, X, Y) between X and Y (*Set Theory*, Chapter II, § 3, no. 1) is said to be *proper* if the restriction of the projection pr_2 to the graph G is a proper mapping $G \to Y$.
a) Let $f: X \to Y$ be a mapping. Show that f is continuous if and only if the correspondence $\overset{-1}{f}$ between Y and X is proper, and that f is a proper mapping if and only if the correspondences $\overset{-1}{f}$ (between Y and X) and f (between X and Y) are proper. Hence construct an example of a non-continuous mapping f such that the correspondence f is proper.
b) If (G, X, Y) is a proper correspondence, show that G(A) is a closed subset of Y whenever A is a closed subset of X.

¶ 11) Let (G, X, Y) be a proper correspondence (Exercise 10).
a) Show that if $\mathfrak{B}$ is a filter base on X which consists of *closed* sets, then

$$\bigcap_{M \in \mathfrak{B}} G(M) = G\Big(\bigcap_{M \in \mathfrak{B}} M\Big).$$

b) For each $y \in Y$ and each neighbourhood V of $\overset{-1}{G}(y)$ in X, show that there is a neighbourhood W of y in Y such that $\overset{-1}{G}(W) \subset V$ ("continuity of the roots as functions of the parameters") (use Exercise 6 of § 9).

¶ 12) *a*) A correspondence (G, X, Y) is proper if and only if, for every topological space Z and every correspondence (E, Z, X) whose graph E is closed in $Z \times X$, the graph $G \circ E$ is closed in $Z \times Y$. First to show that the condition is necessary, remark that $G \circ E$ is the image of $(E \times Y) \cap (Z \times G)$ under $\iota_Z \times pr_2 : Z \times G \to Z \times Y$. To show sufficiency, let δ_Y be the diagonal mapping $Y \to Y \times Y$; show that, for each subset F of $(Z \times Y) \times X$, the inverse image of $G \circ F$ under the mapping

$$\iota_Z \times \delta_Y : Z \times Y \to Z \times Y \times Y$$

is the image, under the mapping $(y, z) \to (z, y)$, of the set

$$\mathrm{pr}_{23}(\overset{-1}{F} \cap (G \times Z)),$$

where $\overset{-1}{F}$ is the image of F under the mapping $(z, y, x) \to (x, y, z)$.
b) Deduce from *a*) that the composition of two proper correspondences is a proper correspondence.
c) Let (G, X, Y) be a proper correspondence. Show that if X is Hausdorff then G is closed in $X \times Y$.

¶ 13) Let X be a locally compact space and Y any topological space.
a) Show that the following four statements are equivalent:

α) (G, X, Y) is a proper correspondence between X and Y.

β) For each Hausdorff space X′ which contains X as a subspace, G is closed in $X' \times Y$.

γ) G is closed in $X \times Y$, and for each $y \in Y$ there is a neighbourhood V of y in Y and a compact subset K of X such that $(X - K) \times V$ does not meet G.

δ) There is a compact space X′, containing X as a subspace, such that G is closed in $X' \times Y$.

[To prove that α) ⟹ β) use Exercise 12 *a*) and *c*). To show that δ) ⟹ α), use Proposition 2 of no. 1 and Corollary 5 of Theorem 1 of no. 2.]

b) Show that if (G, X, Y) is proper, then:

ζ) For every compact subset L of Y, $\overset{-1}{G}(L)$ is a compact set.

Show also that if Y is locally compact and if G is closed in $X \times Y$ and satisfies ζ), then (G, X, Y) is proper. [Use the equivalence of α) and γ).] This last assertion is no longer valid when Y is Hausdorff but not locally compact (Exercise 4).

14) Let Y be a locally compact space, X a topological space. Show that a mapping $f : X \to Y$ is continuous if and only if, for each compact space Y′ which contains Y as a subspace, the graph of f is closed in $X \times Y'$. [Use Exercises 13 and 10 *a*)].

15) Let X be a Hausdorff space, A a closed subset of X, f a proper mapping of $X - A$ into a locally compact space Y, Y′ the one-point compactification of Y, and ω the point at infinity in Y′. Let g be the mapping of X into Y′ which coincides with f on $X - A$ and maps each point of A to ω. Show that g is continuous.

¶ 16) Let X be a locally compact space, R an equivalence relation on X, C the graph of R in $X \times X$. Show that C is closed in $X \times X$ if and only if, for each compact subset K of X, the saturation of K with respect to R is closed in X. (To prove necessity, use Corollary 5 of Theorem 1 of no. 2).

*17) On the locally compact space **R**, let S be the equivalence relation obtained by identifying all the points of **Z**. Show that S is closed and **R**/S Hausdorff but not locally compact.*

*18) Let X be the locally compact subspace $[0, +\infty[$ of **R**. Let S be the equivalence relation on X whose equivalence classes are the sets $\{x, 1/x\}$ for $0 < x \leqslant 1$ and the set $\{0\}$. Show that S is open, that the graph of S is closed in $X \times X$ and that X/S is compact, but that S is not closed.*

¶ 19) Let X be a locally compact σ-compact space, R an equivalence relation on X whose graph is closed in $X \times X$. Show that X/R is Hausdorff. [Let (U_n) be a covering of X by relatively compact open sets such that $\overline{U}_n \subset U_{n+1}$. Let A, B be two disjoint closed sets which are saturated with respect to R. Define recursively two increasing sequences (V_n), (W_n) of closed sets saturated with respect to R, such that $V_n \cap W_n = \emptyset$ and such that $V_n \cap \overline{U}_n$ (resp. $W_n \cap \overline{U}_n$) is a neighbourhood of $A \cap \overline{U}_n$ (resp. $B \cap \overline{U}_n$) in the compact subspace $\overline{U}_n$. Use Proposition 8 of no. 3 and Exercise 16, as well as Proposition 2 of § 9, no. 2]. (*)

¶ 20) *a*) Let X be a non-empty compact space and R a Hausdorff equivalence relation on X. Show that X/R is closed in the space $\mathfrak{P}_0(X)$ with the topology $\mathcal{T}_\Theta$ (§ 8, Exercise 12) if and only if the relation R is open. [Use Exercise 13 of § 9 and Exercise 16 of § 3 to show that if X/R is closed in $\mathfrak{P}_0(X)$ in the topology $\mathcal{T}_\Theta$, then the quotient topology on X/R coincides with that induced by $\mathcal{T}_\Theta$, and then apply Exercise 7 of § 5. Conversely, if R is open, use Exercise 29 of § 8 to show that X/R is closed with respect to $\mathcal{T}_\Theta$].

**b*) Let X be the locally compact subspace of the real plane $\mathbf{R}^2$ consisting of the points $(1/n, y)$ where n is an integer > 0 and $-1 \leqslant y \leqslant 1$, and the points $(0, y)$ where either $-1 \leqslant y < 0$ or $0 < y \leqslant 1$ or $y = 2$. Let p be the restriction of pr_1 to X, and R the equivalence relation associated with p. Show that R is neither open nor closed, but that X/R is compact and closed in $\mathfrak{P}_0(X)$ in the topology $\mathcal{T}_\Theta$.*

(*) If the locally compact space X is not σ-compact, it can happen that the equivalence relation R on X is closed [and *a fortiori* has a closed graph (§ 8, no. 6, Proposition 14)] but that X/R is not Hausdorff [cf. Chapter IX, § 4, Exercises 9 *b*) and 14].

¶ 21) (*) *a*) Let X be an accessible space which is *locally quasi-compact* (§ 9, Exercise 29). Let X_0 be the discrete space which has the same underlying set as X, and let $\tilde{X}_0$ be the (compact) ultrafilter space of X_0 (the discrete space X_0 being identified with an open subspace of $\tilde{X}_0$) (§ 9, Exercise 26). On $\tilde{X}_0$, let R [or R(X)] be the following equivalence relation between ultrafilters $\mathfrak{U}$, $\mathfrak{V}$: " the sets of limit points of $\mathfrak{U}$ and $\mathfrak{V}$ in X are the same ". Show that R is *Hausdorff* [use the description of the topology of $\tilde{X}_0$ given in § 9, Exercise 25 *b*), Exercise 7 *b*) of § 7 and axiom (C')]. The compact space $X^c = \tilde{X}_0/R$ is in one-to-one correspondence with the set of *primitive* subsets of X (§ 7, Exercise 7). If X is Hausdorff (and therefore locally compact) then X^c can be identified with X if X is compact, and with the one-point compactification of X if X is not compact.

b) Let Y be a closed subspace of X. The ultrafilter space $\tilde{Y}_0$ can be identified with the closure of Y_0 in $\tilde{X}_0$, and Y^c with the canonical image of $\tilde{Y}_0$ in X^c.

c) Show that the restriction to X_0 of the canonical mapping $\tilde{X}_0 \to X^c$ is a bijection of X_0 onto a subspace X^c_* of X^c, that the canonical mapping $f: X^c_* \to X$ is a continuous bijection, and that the image under this mapping of any compact subset of X^c_* is a closed compact subset of X. [To show that f is continuous, use *b*); also remark that the inverse image in $\tilde{X}_0$ of a point of X^c_* consists of all the ultrafilters on X which have a single given limit point, and that if $\mathfrak{U}$ is an ultrafilter on X, then its limit point in $\tilde{X}_0$, when $\mathfrak{U}$ is regarded as an ultrafilter base on $\tilde{X}_0$, is $\mathfrak{U}$ regarded as a *point* of $\tilde{X}_0$].

**d*) Let Z be the set which is the sum of a countable number of spaces Y_n each identical with the space Y defined in Exercise 7 of § 8 and a set $\{\omega\}$ consisting of a single point. Define a topology on Z by taking as a fundamental system of neighbourhoods of a point $z \in Y_n$ its neighbourhood filter in the space Y_n, and as fundamental system of neighbourhoods of ω the set of subsets V_n of Y, where V_n is the union of $\{\omega\}$ and the Y_m where $m \geqslant n$. The space Z so defined is accessible, quasi-compact and locally quasi-compact; but the subspace Z^c_* of Z^c is not locally compact.*

e) Let Y be another locally quasi-compact accessible space, and let u be a continuous mapping of X into Y, such that the inverse image under u of each quasi-compact subset of Y is quasi-compact. Regarding u as a mapping of X^c_* into Y^c_*, show that u is continuous and can be

(*) The results in this exercise and in Exercise 7 of § 7 are due to J. Fell.

(uniquely) extended to a continuous mapping $X^c \to Y^c$. [First extend u to a continuous mapping of $\tilde{X}_0$ into $\tilde{Y}_0$ and show that the extension is compatible with the equivalence relations R(X), R(Y)].

**f*) Let X be the quasi-compact, not locally quasi-compact, accessible space defined in Exercise 29 *c*) of § 9. Show that in the corresponding space $\tilde{X}_0$ the equivalence relation R(X) is not Hausdorff.$_*$

§ 11

1) Show that the countable Hausdorff space defined in § 9, Exercise 20 *c*), is connected.

2) *a*) Let X be a topological space and $\mathfrak{T}$ the topology of X. Show that if X is connected in the semi-regular topology $\mathfrak{T}^*$ associated with $\mathfrak{T}$ (§ 8, Exercise 20), then X is connected in the original topology $\mathfrak{T}$. Deduce that there exist connected submaximal spaces (§ 8, Exercise 22).

b*) In particular construct a topology on the real line **R, with respect to which **R** is connected but no point of **R** has a fundamental system of connected neighbourhoods [cf. Chapter IV, § 2, Exercise 14 *c*)].$_*$

3) Let A, B be two subsets of a topological space X.

a) Show that if A and B are closed in X, and $A \cup B$ and $A \cap B$ are connected, then A and B are connected. Show by an example that this statement is no longer necessarily true when one of the sets A, B is not closed.

b) Suppose that A and B are connected and $\overline{A} \cap B \neq \emptyset$. Show that $A \cup B$ is connected.

¶ 4) Let X be a connected space of at least two points.

a) Let A be a connected subset of X, B a subset of $\complement A$ which is both open and closed in $\complement A$. Show that $A \cup B$ is connected. (Apply Exercise 5 of § 3 to $Y = A \cup B$ and $Z = \complement A$.)

b) Let A be a connected subset of X and B a component of the set $\complement A$. Show that $\complement B$ is connected [use *a*)].

c) Deduce from *b*) that there exist two connected subsets M, N of X which are distinct from X and such that $M \cup N = X$ and $M \cap N = \emptyset$.

*5) *a*) Give an example, in the real plane $\mathbf{R}^2$, of a decreasing sequence (A_n) of connected sets whose intersection is not connected (cf. Chapter II, § 4, Exercise 14).

b) In $\mathbf{R}^2$, let X be the set which is the union of the open half-planes $y > 0$ and $y < 0$ and the points $(n, 0)$ where n is an integer > 0.

X is connected with respect to the topology $\mathfrak{T}_0$ induced by that of $\mathbf{R}^2$. Define a sequence $(\mathfrak{T}_n)$ of topologies on X such that $\mathfrak{T}_{n+1}$ is finer than $\mathfrak{T}_n$, and such that X is connected with respect to each $\mathfrak{T}_n$ but not connected with respect to the topology which is the least upper bound of the sequence $(\mathfrak{T}_n)$ [pass from $\mathfrak{T}_n$ to $\mathfrak{T}_{n+1}$ by introducing new neighbourhoods of the point $(n+1, 0)$].$_*$

6) Let X, Y be two connected spaces and let A (resp. B) be a subset of X (resp. Y) distinct from X (resp. Y). Show that in the product space $X \times Y$ the complement of $A \times B$ is connected.

7) Let X, Y be two connected spaces and f a mapping of $X \times Y$ into a topological space Z such that all the partial mappings $f(0, y) : X \to Z$, $f(x, 0) : Y \to Z$ $(x \in Y, y \in Y)$ are continuous. Show that $f(X \times Y)$ is connected.

¶ 8) In the example of a topology on a product set $X = \prod_{\iota \in I} X_\iota$ given in § 4, Exercise 9, suppose that I is infinite and the cardinal $\mathfrak{t}$ uncountable. Show that if the X_ι are connected, regular, and have each at least two distinct points, then the connected component of a point $a = (a_\iota)$ of X is the set of $x = (x_\iota)$ such that $x_\iota = a_\iota$ except for a finite number of indices $\iota \in I$. (Reduce to the case $I = \mathbf{N}$; consider two points $b = (b_n)$, $c = (c_n)$ of X such that $b_n \neq c_n$ for all n. For each n let $(V_{nm})_{m \geqslant 0}$ be a sequence of open neighbourhoods of b_n in X_n which do not contain c_n and are such that $\overline{V}_{n,m+1} \subset V_{nm}$. Let Z be the set of points (x_n) of X with the following property: there exists an integer $k > 0$ such that $x_n \in V_{n,n-k}$ whenever $n \geqslant k$. Show that Z is both open and closed in X).

$_*$9) Show that in the locally compact space X defined in § 10, Exercise 20, there are points x whose component in X is different from the intersection of the subsets of X which are both open and closed and contain x.$_*$

10) Show that a linearly ordered set X, with the topology $\mathfrak{T}_+(X)$ or the topology $\mathfrak{T}_-(X)$ (§ 2, Exercise 5) is totally disconnected.

11) Let X be a topological space and R an equivalence relation on X such that every equivalence class mod R is contained in a component of X. Show that the components of X/R are the canonical images of the components of X (cf. Chapter III, § 2, Exercise 17).

12) Let X be a topological space. Show that the following three conditions are equivalent: α) the components of X are open sets; β) every $x \in X$ has a neighbourhood contained in every set which is both open and closed and contains x; γ) for every $x \in X$, the intersection of the sets which are both open and closed and contain x is an open set.

$_*$13) Let θ be an irrational number. Let f be the mapping of the real line $\mathbf{R}$ into the torus $\mathbf{T}^2$ for which the image $f(t)$ of each $t \in \mathbf{R}$ is the canonical image in $\mathbf{T}^2$ of $(t, \theta t) \in \mathbf{R}^2$. f is injective and continuous. Show that, in the topology on $\mathbf{R}$ which is the inverse image under f of the topology of $\mathbf{T}^2$, $\mathbf{R}$ is connected but no point of $\mathbf{R}$ has a fundamental system of connected neighbourhoods.$_*$

¶ 14) *a*) Let X be a locally compact and locally connected space, and let A be a closed subset of X. Let A^* be the union of A and the components of $\complement A$ which are relatively compact. A is said to be *full* if $A^* = A$. Show that A^* is closed and full.

b) Suppose in addition that X is connected. Show that if A is a compact subset of X, then A^* is compact (consider a compact neighbourhood V of A and the components of $\complement A$ which meet the frontier of V).

c) Suppose in addition that X is connected and σ-compact. Show that there exists an increasing sequence $(K_n)_{n \geqslant 0}$ of *full* connected compact sets whose union is X and which are such that $K_n \subset \mathring{K}_{n+1}$ for all n.

$_*$*d*) Show that the conclusion of *b*) is no longer valid if we suppose only that X is connected and locally compact, or that X is Hausdorff, connected and locally connected. [Consider the subspace of $\mathbf{R}^2$ consisting of points (x, y) such that either $y = 0$ and $0 \leqslant x \leqslant 1$, or else $y > 0$ and $x = 1/n$ where n is an integer > 0].$_*$

¶ 15) Let X be a connected, locally connected space.
a) Let M, N be two disjoint non-empty closed subsets of X. Show that $\complement(M \cup N)$ has a component whose closure meets both M and N. [If C is a component of $\complement(M \cup N)$ whose closure does not meet M, observe that C is both open and closed in $\complement N$].

b) If A is a closed subset of X distinct from X, deduce from *a*) that every component of A meets the closure of $\complement A$.

$_*$16) Let X be the subspace of the real plane $\mathbf{R}^2$ which is the union of the line $y = 0$, the closed segments with end points $(0, 1)$ and $(n, 1/(1+n))$, and the half-lines $x \leqslant n$, $y = 1/(1+n)$ (n an integer > 0). Show that X is a connected space but that there is a component C of the complement of $(0, 1)$ in X such that $(0, 1) \notin \overline{C}$ [(compare with Exercises 4 *a*) and 15 *b*)].$_*$

$_*$17) Let I_0 be the interval $[-1, 1]$ of $\mathbf{R}$ and let P_0 be the quotient space of I_0 obtained by identifying the two points $1/2$ and 1; let B_0 be the quotient space of I_0 obtained by identifying the points $1/2$ and 1 and also identifying $-1/2$ and -1. Let I be the open interval

$]-1, 1[$ in $\mathbf{R}$, and let P (resp. B) be the canonical image of I in P_0 (resp. B_0).

a) Show that no two of the three spaces I, P, B are homeomorphic (to show that I and P are not homeomorphic, notice that there are points in P whose complement is connected; similarly for the other pairs of spaces).

b) Let X be the sum of a countably infinite family of spaces homeomorphic to I and a countably infinite family of spaces homeomorphic to B. Let Y be the sum of X and P. Show that X and Y are not homeomorphic, but that (i) there is a continuous bijection of X onto Y and a continuous bijection of Y onto X; (ii) X is homeomorphic to an open subspace of Y and Y is homeomorphic to an open subspace of X.$_*$

*18) Let X be the real plane $\mathbf{R}^2$, and let R be the equivalence relation on X whose equivalence classes are the lines $y = \beta$ for $\beta > 0$ and the half-lines $x = \alpha$, $y \leqslant 0$ for all $\alpha \in \mathbf{R}$. Show that if X/R is regarded as a subset of $\mathfrak{P}_0(X)$, then the topology induced on X/R by $\mathfrak{T}_\Omega$ (§ 2, Exercise 7) is discrete, and that the topology induced on X/R by $\mathfrak{T}_\Phi$ is not discrete but makes X/R into a non-connected space. Show finally that X/R is connected in the quotient topology.$_*$

*¶ 19) Let X be a locally compact space. It can be shown that X can be identified with a dense open subspace of the universal compact space $\tilde{X}$ constructed in § 9, Exercise 25 (cf. Chapter IX, § 1, Exercise 8). Suppose also that X is connected and locally connected.

a) Show that, for every compact subset K of X, the set of components of X — K which are not relatively compact is finite [cf. Exercise 14 *b*)]. Deduce that a point of $\tilde{X} - X$ belongs to the closure of at most one of the components of X — K [if A_i ($1 \leqslant i \leqslant n$) are the non-compact components of X — K, and C_i the compact set of points of $\tilde{X} - X$ which lie in the closure of A_i, construct a compact space whose underlying set is the sum of X and the sets C_i, such that X is identified with a dense open subspace of this space].

b) Let $R'\{x, y\}$ be the following equivalence relation on $\tilde{X} - X$: "for every compact subset K of X, x and y lie in the closure of the same non-compact component of X — K". Let R be the equivalence relation on $\tilde{X}$ whose equivalence classes are the classes mod R′ and the subsets consisting of a single point of X. Show that R is Hausdorff and that X may be identified with a dense open subspace of the compact space $X^b = \tilde{X}/R$; show that in the compact space $X^b - X$ every point is the intersection of its neighbourhoods which are both open and closed. The points of $X^b - X$ are called the *ends* of the space X.

c) Let Y be a compact space such that X may be identified with a dense open subspace of Y and such that in the compact space $Y - X$ every point is the intersection of its neighbourhoods which are both open and closed (*). Show that there is a continuous mapping of X^b into Y which induces the identity mapping on X. [If $Y - X$ is the union of two disjoint non-empty closed sets A and B, show that there exist two disjoint open subsets U, Y in V such that $A \subset U$, $B \subset V$ and such that $Y - (U \cup V)$ is compact].

d) Show that if X is σ-compact, then $X^b - X$ has a countable base.

e) The real line has two ends, and $\mathbf{R}^n$ $(n \geqslant 2)$ has one. Give an example of a connected, locally connected, locally compact subspace of $\mathbf{R}^2$ for which the set of ends has the power of the continuum.*

20) Let X be a Hausdorff space in which each point has a fundamental system of neighbourhoods which are both open and closed in X. Let Y be a subspace of X, and let A be a compact subset of Y which is both open and closed in Y; show that there is a subset B of X which is both open and closed in X, such that $B \cap Y = A$.

¶ 21) *a*) Show that the following conditions on a topological space X are equivalent: (i) for each open set U in X, $\overline{U}$ is open; (ii) for each closed set F in X, $\mathring{F}$ is closed; (iii) for each disjoint pair of open sets U, V in X, we have $\overline{U} \cap \overline{V} = \emptyset$. A Hausdorff space which satisfies these conditions is said to be ***extremally disconnected***; it is then totally disconnected. *The rational line is totally disconnected but not extremally disconnected.*

b) A Hausdorff space is extremally disconnected if and only if the associated semi-regular space (§ 8, Exercise 20) is extremally disconnected.

c) Let X be a Hausdorff space and let A be a dense subset of X. Show that if A is extremally disconnected and if the topology of X is A-maximal (§ 3, Exercise 11), then X is extremally disconnected.

d) Deduce from *b*) and *c*) that, if X is any infinite set, the (compact) ultrafilter space of X is extremally disconnected (cf. § 9, Exercises 26 and 24) (**).

e) Let X be a Hausdorff space with no isolated points. Show that the topology of X is quasi-maximal (§ 2, Exercise 6) if and only if X is

(*) This condition is equivalent to saying that $Y - X$ is totally disconnected (Chapter II, § 4, no. 4, Proposition 6).

(**) For examples of extremally disconnected compact spaces with no isolated points, see Chapter II, § 4, Exercise 12 *b*).

submaximal (§ 8, Exercise 22) and extremally disconnected. (To show that the condition is sufficient, use Exercise 22 *e*) of § 8 and prove that every closed subset of X which has no isolated points is open).

f) Show that every ultraregular space (§ 8, Exercise 25) is extremally disconnected, but that an extremally disconnected compact space with no isolated points is not ultraregular (cf. § 9, Exercise 27).

g) Show that an extremally disconnected semi-regular space (§ 8, Exercise 20) is regular (cf. Chapter II, § 4, Exercise 12).

h) Show that in an extremally disconnected space there is no convergent sequence which has an infinite number of distinct terms [argue by contradiction by constructing recursively two sequences (U_n), (V_n) of mutually disjoint open sets such that if $U = \bigcup_n U_n$ and $V = \bigcup_n V_n$ then $U \cap V = \emptyset$ but $\overline{U} \cap \overline{V} \neq \emptyset$].

i) Show that in an extremally disconnected space X a non-isolated point *a* cannot have a fundamental system of simultaneously open and closed neighbourhoods which is *well-ordered* (with respect to the relation $\supset$). (Use *reductio ad absurdum*, as in *h*).

¶ 22) *a*) Show that in an extremally disconnected space every open subspace and every dense subspace is extremally disconnected.

b) Let X be a Hausdorff space in which every point has a neighbourhood which is an extremally disconnected subspace of X. Show that X is extremally disconnected.

c) Give an example of a Hausdorff space X which is not extremally disconnected but which has an extremally disconnected dense open subspace (paste together two extremally disconnected spaces).

d) Let X be a countable discrete space, $\tilde{X}$ the (compact) space of ultrafilters on X (§ 9, Exercise 26). Let $(A_n)_{n \in N}$ be a countably infinite partition of X into infinite subsets; in the closed subspace $Y = \tilde{X} - X$ of $\tilde{X}$, the sets $B_n = \tilde{X}_n \cap Y$ are open and closed and pairwise disjoint. Let $B = \bigcup_n B_n$; show that the closure $\overline{B}$ of B in Y is not open in Y, and hence that Y is not extremally disconnected (*). (Use *reductio ad absurdum*: using Exercise 20, let C be an open and closed subset of $\tilde{X}$ such that $C \cap Y = \overline{B}$; consider a point x_n in each of the sets $C \cap A_n$, let $\overline{J}$ be the closure of the set J of points x_n, and show that $\overline{J}$ does not meet B).

(*) Exercises 21 *h*) and 22 *d*) are due to R. Ricabarra.

23) Give an example of a bijective mapping f of a Hausdorff space X onto a Hausdorff space Y, such that the image under f of every compact subset of X is a compact subset of Y, and the image under f of every connected subset of X is a connected subset of Y, but such that f is not continuous (cf. § 9, Exercise 4).

24) Let X be a topological space.

a) Let the set $\mathfrak{P}_0(X)$ carry one of the topologies $\mathfrak{T}_\Omega$, $\mathfrak{T}_\Phi$ (§ 2, Exercise 7) or $\mathfrak{T}_\Theta$ (§ 8, Exercise 12). Show that if a subset $\mathfrak{B}$ of $\mathfrak{P}_0(X)$ is connected and if each $M \in \mathfrak{B}$ is connected, then $\bigcup_{M \in \mathfrak{B}} M$ is connected.

b) Let $\mathfrak{S}$ be a subset of $\mathfrak{P}_0(X)$ which contains the set of all finite non-empty subsets of X. Show that if X is connected then so is $\mathfrak{S}$ [use Exercise 7 *b*) of § 2 and Proposition 8 of § 4, considering the mappings $(x_1, \ldots, x_n) \to \{x_1, \ldots, x_n\}$].

¶ 25) Given two topological spaces X, Y, a mapping $f\colon X \to Y$ is said to be a *local homeomorphism* if for each $x \in X$ there is a neighbourhood U of x such that $f(U)$ is a neighbourhood of $f(x)$ in Y, and the mapping $U \to f(U)$ which coincides with f on U is a homeomorphism. Every local homeomorphism is a continuous open mapping.

a) Suppose that X is Hausdorff and Y locally connected. Let $f\colon X \to Y$ be a local homeomorphism with the property that there is an integer $n > 0$ such that $\overset{-1}{f}(y)$ has exactly n points for every $y \in Y$. Show that every $y \in Y$ has an open neighbourhood V such that $\overset{-1}{f}(V)$ has n components U_i $(1 \leqslant i \leqslant n)$ and such that the mapping $U_i \to V$ which coincides with f on U_i is a homeomorphism. f is then a proper mapping.

b) Suppose that X is Hausdorff and that Y is connected and locally connected. Let $f\colon X \to Y$ be a *proper* local homeomorphism. Show that f satisfies the hypotheses of *a*). [If for each integer $n > 0$, Y_n is the set of $y \in Y$ such that $\overset{-1}{f}(y)$ has at least n points, show that Y_n is both open and closed in Y].

**c*) Give an example of a surjective local homeomorphism $f\colon X \to Y$, in which Y is compact, connected and locally connected, X is locally compact, connected and locally connected, $\overset{-1}{f}(y)$ consists of at most two points for each $y \in Y$, but f is not proper (cf. § 5, Exercise 4).*

HISTORICAL NOTE

(Numbers in brackets refer to the bibliography at the end of this note.)

The ideas of limit and continuity go back to antiquity, and a complete history of them could not be written without studying systematically from this point of view not only the Greek mathematicians but also the Greek philosophers and Aristotle in particular. It would also be necessary to trace the evolution of these ideas through Renaissance mathematics and the beginnings of the differential and integral calculus. Such a study, though it would undoubtedly be interesting to undertake, would go far beyond the framework of this note.

It is Riemann who should be considered as the creator of topology, as of so many other branches of modern mathematics. He was the first to attempt to formulate the notion of a topological space; he conceived the idea of an autonomous theory of such spaces; he defined invariants (the "Betti numbers") which were to play a pre-eminent part in the later development of topology; and he was the first to apply topology to analysis (periods of abelian integrals). But the current of ideas in the first half of the nineteenth century had prepared the path for Riemann in more ways than one. In the first place, the desire to put mathematics on a firm basis, which was the cause of so many important researches throughout the nineteenth century and up to the present day, had led to a correct understanding of the notions of a convergent series and a sequence of numbers tending to a limit (Cauchy, Abel) and to the notion of a continuous function (Bolzano, Cauchy). On the other hand, the geometrical representation (by points of a plane) of the complex numbers (or, as they had hitherto been called, "imaginary" or even "impossible" numbers) which was due to Gauss and Argand, had become familiar to the majority of mathematicians; it constituted an advance of the same order as the adoption, in our century, of the language of geometry in the study of Hilbert space, and contained the germ of the possibility of a geometrical representation of every object capable of continuous variation. Gauss, who was in any case naturally led to such concepts by his researches on the foundations of geometry, on non-Euclidean geometry and on curved surfaces, seems to have had this possibility already in mind, for he uses the words "magnitude twice extended" when defining (independently of Argand and the French mathematicians) the geometrical representation of complex numbers ([1], pp. 101-103 and 175-178).

Riemann's work on algebraic functions and their integrals and his reflections on the foundations of geometry (largely inspired by his study of Gauss's work) led him to formulate a program of study, which is precisely that of modern topology, and to begin to realize this program. Here, for example, is what he says in his theory of abelian functions ([2], p. 91):

"*In the study of functions obtained by integrating exact differentials, some theorems of Analysis situs are almost indispensable. By this name, which was used by Leibnitz, although perhaps in a somewhat different sense, should be called that part of the theory of continuous magnitudes which studies these magnitudes, not independently of their position and by measuring them in terms of each other, but rather by abstracting all ideas of measurement and considering only their relations of position and inclusion. I reserve to a later occasion an investigation completely independent of all measurement...*"

And in his famous inaugural lecture "On the hypotheses which underlie Geometry" ([2], p. 272):

"*... the general concept of a magnitude many times extended* (*) *which contains as a particular case that of spatial magnitude, has remained completely unexplored...*" (p. 272)

"*... The notion of magnitude presupposes that an element is capable of different determinations. According as one can pass from one determination to another by a continuous process of transition or not, these determinations form a continuous or a discrete manifold: in the former case the determinations are called points of the manifold...*' (p. 273)

"*... Measurement consists of superposition of the magnitudes to be compared, hence in order to measure we need some means of using one magnitude as a yardstick for another. In the absence of this we can compare two magnitudes only if one is part of the other... The investigations which can be undertaken in this context form a part of the theory of magnitudes which is independent of the theory of measurement and in which the magnitudes are considered not as existing independently of their position nor as expressible in terms of a unit of measurement, but as regions in a manifold. Such investigations have become necessary in several parts of mathematics, in particular in the theory of many-valued analytic functions...* " (p. 274)

(*) As the sequel shows, Riemann means by this phrase a subset of a topological space of arbitrary dimension.

"... *The determination of position in a given manifold, whenever this is possible, can be reduced to a finite number of numerical determinations. There are however manifolds in which the determination of position requires not a finite number but an infinite sequence or even a continuous manifold of determinations of magnitudes. For example, the possible determinations of a function on a given domain, or the possible forms of a spatial figure, give manifolds of this type.*" (p. 276)

Note in this last phrase the first idea of a study of functional spaces; Riemann had already expressed the same idea in his dissertation: "*the totality of these functions*", he stated in connection with the minimal problem known as Dirichlet's principle, "*forms a connected domain which is closed in itself*" ([2], p. 30); this, though imperfectly expressed, is nevertheless the germ of the proof which Hilbert was later to give of Dirichlet's principle, and of most of the applications of function spaces to the calculus of variations.

As we have said, Riemann began the execution of this grandiose program by defining the "Betti numbers", first for a surface ([2], pp. 92-93) and later ([2], pp. 479-482; cf. also [3]) for a manifold of any dimension, and applied this definition to the theory of integrals; for this and for the considerable development of this theory since Riemann's time we refer the reader to the Historical Notes to the chapters on algebraic topology in this series of volumes.

Before a general theory of topological spaces, such as Riemann had envisaged, could be developed, it was necessary that the theory of real numbers, of sets of numbers, of sets of points on a line, in a plane and in space should be more systematically investigated than they had been in Riemann's time. Such investigations were related on the other hand to research into the nature of irrational numbers (semi-philosophical by Bolzano and essentially mathematical by Dedekind) and to progress in the theory of functions of a real variable in which Riemann himself made an important contribution by his definition of the integral and his theory of trigonometrical series, and to which du Bois-Reymond, Dini and Weierstrass, among others, contributed; they were the work of the second half of the nineteenth century, and especially the work of Cantor, who was the first to define (originally on the line, later in Euclidean space of n dimensions) the notions of point of accumulation, closed set, open set, perfect set, and obtained the essential results on the structure of these sets on the line (cf. the Historical Note to Chapter IV). In this context, not only the works of Cantor [4] should be consulted, but also his extremely interesting correspondence with Dedekind [5], where the idea of dimensionality as a topological invariant can be found clearly expressed. The later progress of the theory is traced in a semi-historical, semi-systematic form in Schœnflies' book [6]; by far the most important acquisition was the theorem of Borel-Lebesgue, namely the fact that every bounded closed subset of Euclidean n-space $\mathbf{R}^n$ (cf. Chapter VI, § 1) satisfies axiom (C''') of § 9 of this chapter (the theorem was first proved by Borel for a closed interval on the line and a countable family of open intervals covering it).

Cantor's ideas had originally met with vigorous opposition (cf. the Historical Note to Book I, Chapters I-IV). At any rate his theory of point-sets on the line and in the plane was quickly made use of and disseminated by the French and German schools of function-theory (Jordan, Poincaré, Klein, Mittag-Leffler, and later Hadamard, Bore, Baire, Lebesgue, etc.); each of the early Borel treatises, in particular, contains an elementary exposition of this theory (see for example [7]). As these ideas spread, their possible application to sets, not of points but of curves or functions, began to be considered in various quarters, as witness the title "On the limit curves of a variety of curves" of a memoir by Ascoli in 1883 [8] and a communication by Hadamard to the congress of mathematicians at Zürich in 1896 [9]; all this is closely related to the introduction of "line functions" by Volterra in 1887 and to the creation of "functional calculus" or theory of functions in which the argument is a function (cf. Voltera's book on functional analysis [10]). On the other hand, Hilbert's famous memoir [11], in which, taking up Riemann's ideas again, he proved the existence of the minimum in Dirichlet's principle and inaugurated the "direct method" in the calculus of variations, showed clearly the importance of considering sets of functions in which the Bolzano-Weierstrass principle holds, that is to say in which every sequence has a convergent subsequence. Such sets were beginning in any case to play an important part, not only in the calculus of variations but also in the theory of functions of a real variable (Ascoli, Arzelà) and a little later in the theory of functions of a complex variable (Vitali, Carathéodory, Montel). Finally the study of functional equations, and especially the solution by Fredholm of the type of equation which bears his name, made it commonplace to consider a function as an argument and a set of functions as a set of points, and as natural to use the language of geometry in this context as in Euclidean space of n dimensions (a space which equally eludes "intuition" and for this reason remained long an object of distrust to many mathematicians). In particular the memorable work of Hilbert on integral equations [12] led to the definition and geometrical study of Hilbert space by Erhard Schmidt [14], in complete analogy with Euclidean geometry.

Meanwhile the concept of an axiomatic theory had acquired more and more importance, thanks to much work on the foundations of geometry; here Hilbert's contributions [13] had a particularly decisive influence. In the course of this work, Hilbert had been led to formulate in 1902 ([13], p. 180) the first axiomatic definition of the "manifold twice extended" in the sense of Riemann, a definition which constituted, said Hilbert, "the foundation of a rigorous axiomatic treatment of *Analysis situs*". Hilbert also made use of neighbourhoods (in a sense restricted by the demands of the problem to which he limited himself).

The first attempts to abstract what is common to properties of sets of points and sets of functions are due to Fréchet [15] and F. Riesz [16]. The former started from the notion of countable limit and did not succeed in constructing a convenient and fruitful system of axioms, but at least he recognized the relationship between the principle of Bolzano-Weierstrass [which is just axiom (C) of § 9, restricted to countable sequences] and the Borel-Lebesgue theorem [axiom (C‴) of § 9]; in this connection he introduced the word "compact", although in a sense somewhat different from that in which it is used in this series of volumes. As to F. Riesz, who took as his starting point the concept of point of accumulation (or rather of "derived set", which amounts to the same thing), his theory was again incomplete and appeared only in outline form.

General topology as it is understood today began with Hausdorff ([17], Chapters 7, 8, 9), who again took up the concept of neighbourhood (by which he meant what in the terminology of this series of volumes is called an "open neighbourhood") and chose from Hilbert's axioms for neighbourhoods in the plane those which gave his theory all the precision and generality desired. The axioms he took as a starting-point were essentially (taking into account the difference between his concept of neighbourhood and ours) axioms (V_I), (V_{II}), (V_{III}), (V_{IV}) of § 1 and (H) of § 8, and the chapter in which he develops the consequences of these axioms has remained a model of axiomatic theory, abstract but adapted in advance to applications. Hausdorff's work was naturally the point of departure for later research in general topology and especially for the work of the Moscow school, which was largely directed towards the problem of metrization (cf. the Historical Note to Chapter IX); here we recall especially the definition of compact spaces (under the name of "bicompact spaces") by Alexandroff and Urysohn, and Tychonoff's proof of the compactness of products of compact spaces [19]. Finally, the introduction of filters by H. Cartan [20] has brought to topology a valuable instrument, usable in all sorts of applications (in which it replaces to advantage the notion of "Moore-Smith convergence" [18]). Furthermore, the development of the theorem on ultrafilters (Theorem 1, § 6), has clarified and simplified the theory.

BIBLIOGRAPHY

[1] C. F. GAUSS, *Werke*, Vol. 2, Göttingen, 1863.

[2] B. RIEMANN, *Gesammelte mathematische Werke*, 2nd edition, Leipzig (Teubner), 1892.

[3] B. RIEMANN, in Lettere di E. Betti a P. Tardy, *Rend. Accad. Lincei* (5) **24**[1] (1915) pp. 517-519.

[4] G. CANTOR, *Gesammelte Abhandlungen*, Berlin (Springer), 1932.

[5] G. CANTOR and R. DEDEKIND, Briefwechsel, *Actualités scientifiques et industrielles*, no. 518, Paris (Hermann), 1937.

[6] A. SCHOENFLIES, *Entwicklungen der Mengenlehre und ihrer Anwendungen*, Part I, 2nd edition, Leipzig-Berlin (Teubner), 1913.

[7] E. BOREL, *Leçons sur la théorie des fonctions*, 2nd edition, Paris (Gauthier-Villars), 1914.

[8] G. ASCOLI, Le curvi limiti de una varietà data di curve, *Mem. Accad. Lincei* (3) **18** (1883) p. 521-586.

[9] J. HADAMARD, Sur certaines applications possibles de la théorie des ensembles, *Verhandl. Intern. Math.-Kongress*, Zürich, 1898, pp. 201-202.

[10] V. VOLTERRA, *Theory of Functionals*, London-Glasgow (Blackie and Son), 1930.

[11] D. HILBERT, *Gesammelte Abhandlungen*, vol. 3, Berlin (Springer), 1935, pp. 10-37 [= *Jahresbericht des D.M.V.*, **8** (1900), p. 184, and *Math. Ann.*, **59** (1904), p. 161].

[12] D. HILBERT, *Grundzüge einer allgemeinen Theorie der Integralgleichungen*, 2nd edition, Leipzig-Berlin (Teubner), 1924.

[13] D. HILBERT, *Grundlagen der Geometrie*, 7th edition, Leipzig-Berlin (Teubner), 1930.

[14] E. SCHMIDT, Über die Auflösung linearer Gleichungen mit unendlich vielen Unbekannten, *Rend. Palermo*, **25** (1908), pp. 53-77.

[15] M. FRÉCHET, Sur quelques points du calcul fonctionnel, *Rend. Palermo*, **22** (1906), pp. 1-74.

[16] F. RIESZ, Stetigkeitsbegriff und abstrakte Mengenlehre, *Atti del IV Congresso Intern. dei Matem.*, Bologna, 1908, vol. 2, pp. 18-24.

[17] F. HAUSDORFF, *Grundzüge der Mengenlehre*, 1st edition, Leipzig (Veit), 1914.

[18] E. H. MOORE and H. L. SMITH, A general theory of limits, *Am. Journ. of Math.*, **44** (1922) pp. 102-121.

[19] A. TYCHONOFF, Über die topologische Erweiterung von Räumen, *Math. Ann.*, **102** (1930), pp. 514-561.

[20] H. CARTAN, Théorie des filtres; filtres et ultrafiltres, *C. R. Acad. Sc. Paris*, **205** (1937), pp. 595-598 and pp. 777-779.

CHAPTER II

Uniform Structures

1. UNIFORM SPACES

1. DEFINITION OF A UNIFORM STRUCTURE

DEFINITION 1. *A uniform structure* (or *uniformity*) *on a set* X *is a structure given by a set* $\mathfrak{U}$ *of subsets of* $X \times X$ *which satisfies axioms* (F_I) *and* (F_{II}) *of Chapter* I, § 6, *no.* 1 *and also satisfies the following axioms:*

(U_I) *Every set belonging to* $\mathfrak{U}$ *contains the diagonal* Δ.

(U_{II}) *If* $V \in \mathfrak{U}$ *then* $\overset{-1}{V} \in \mathfrak{U}$.

(U_{III}) *For each* $V \in \mathfrak{U}$ *there exists* $W \in \mathfrak{U}$ *such that* $W \circ W \subset V$.

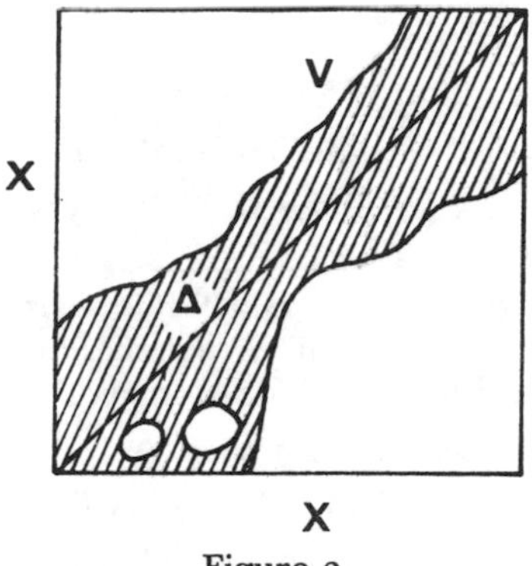

Figure 2.

The sets of $\mathfrak{U}$ *are called entourages of the uniformity defined on* X *by* $\mathfrak{U}$. *A set endowed with a uniformity is called a uniform space.*

If V is an entourage of a uniformity on X, we may express the relation $(x, x') \in V$ by saying that "x and x' are V-close".

Remarks. 1) To make the language more expressive we may use the expressions "x is close enough to y" and "x and y are as close as we please" in some statements. For example, we shall say that a relation

$R\{x, y\}$ is true *whenever* x *and* y *are close enough* if there is an entourage V such that the relation $(x, y) \in V$ *implies* $R\{x, y\}$.

2) The conjunction of axioms (U_{II}) and (U_{III}) is equivalent (assuming the other axioms of uniform structures) to the following axiom:

(U_a) *For each* $V \in \mathfrak{U}$ *there exists* $W \in \mathfrak{U}$ *such that* $W \circ \overset{-1}{W} \subset V$ (*).

Clearly (U_{II}) and (U_{III}) imply (U_a). Conversely, if (U_a) is satisfied we have $\overset{-1}{W} = \Delta \circ \overset{-1}{W} \subset V$, by (U_I); hence $W \subset \overset{-1}{V}$ and therefore [by (F_I)] $\overset{-1}{V} \in \mathfrak{U}$. Let $W' = W \cap \overset{-1}{W}$; then $W' \in \mathfrak{U}$ by what has just been proved and axiom (F_{II}), and we have $W' \circ W' \subset W \circ \overset{-1}{W} \subset V$.

Throughout this chapter we shall write $\overset{2}{V}$ instead of $V \circ V$, and in general $\overset{n}{V} = \overset{n-1}{V} \circ V = V \circ \overset{n-1}{V}$, for each integer $n > 1$ and each subset V of $X \times X$.

3) If X is not empty, then axiom (U_I) implies that no set of $\mathfrak{U}$ is empty, and therefore $\mathfrak{U}$ is a *filter* on $X \times X$. There is only one uniformity on the empty set, namely $\mathfrak{U} = \{\emptyset\}$.

DEFINITION 2. *A fundamental system of entourages of a uniformity is any set* $\mathfrak{B}$ *of entourages such that every entourage contains a set belonging to* $\mathfrak{B}$.

Axiom (U_{III}) shows that if n is any integer > 0 and V runs through a fundamental system of entourages, then the sets $\overset{n}{V}$ again form a fundamental system of entourages.

Entourages V such that $V = \overset{-1}{V}$ are called *symmetric*. If V is any entourage, then $V \cap \overset{-1}{V}$ and $V \cup \overset{-1}{V}$ are symmetric entourages, and axioms (F_{II}) and (U_{II}) show that the symmetric entourages form a *fundamental system of entourages*.

A set $\mathfrak{B}$ of subsets of $X \times X$ is a fundamental system of entourages of a uniformity on X if and only if $\mathfrak{B}$ satisfies axiom (B_I) of Chapter I, § 6, no. 3, and also satisfies the following axioms:

(U'_I) *Every set of* $\mathfrak{B}$ *contains the diagonal* Δ.

(U'_{II}) *For each* $V \in \mathfrak{B}$ *there exists* $V' \in \mathfrak{B}$ *such that* $V' \subset \overset{-1}{V}$.

(U'_{III}) *For each* $V \in \mathfrak{B}$ *there exists* $W \in \mathfrak{B}$ *such that* $\overset{2}{W} \subset V$.

If X is not empty, a fundamental system of entourages of a uniform structure on X is a *base* of the filter formed by the entourages of this structure (Chapter I, § 6, no. 3, Proposition 3).

(*) We recall (*Set Theory*, R, § 3, nos. 4 and 10) that if V and W are two subsets of $X \times X$, then the set of pairs $(x, y) \in X \times X$, such that $(x, z) \in W$ and $(z, y) \in V$ for some $z \in X$, is denoted by $V \circ W$ or VW, and that the set of pairs $(x, y) \in X \times X$ such that $(y, x) \in V$ is denoted by $\overset{-1}{V}$.

Examples of uniformities. * 1) On the set $\mathbf{R}$ of real numbers we can define a uniformity, called the *additive uniformity*, as follows: for each $\alpha > 0$ let V_α be the subset of $\mathbf{R} \times \mathbf{R}$ consisting of all pairs (x, y) such that $|x - y| < \alpha$; as α runs through the set of all real numbers > 0, the V_α form a fundamental system of entourages for the additive uniformity on $\mathbf{R}$. Similarly we can define a uniformity (again called the *additive uniformity*) on the set $\mathbf{Q}$ of rational numbers; we shall study these structures, and analogous uniform structures on *groups*, in Chapters III and IV. *
2) Let X be a set and let R be an *equivalence relation* on X; let C be the graph of R in $X \times X$. Then $\Delta \subset C$ and $\overset{2}{C} = \overset{-1}{C} = C$; (*Set Theory*, R, § 5, no. 1); the set of subsets of $X \times X$ which consists of the set C alone is therefore a fundamental system of entourages of a uniformity on X. In particular, if we take R to be the relation of equality, then $C = \Delta$ and the entourages of the corresponding uniformity are *all the subsets* of $X \times X$ which contain Δ; this uniformity is called the *discrete* uniformity on X, and the set X endowed with this uniformity is called a *discrete uniform space.*
3) On the set $\mathbf{Z}$ of rational integers we can define a uniformity, important in the theory of numbers, as follows: given a prime number p, let W_n be the set of all pairs $(x, y) \in \mathbf{Z} \times \mathbf{Z}$ such that $x \equiv y \pmod{p^n}$, for each integer $n > 0$. It is easily verified that the sets W_n (for a fixed p) form a fundamental system of entourages of a uniformity on $\mathbf{Z}$, called the *p-adic* uniformity (cf. Chapter III, § 6, Exercises 23 ff., and Chapter IX, § 3, no. 2).

In accordance with the general definitions (*Set Theory*, R, § 8, no. 5), if X and X′ are two sets each endowed with uniformities whose sets of entourages are $\mathfrak{U}$ and $\mathfrak{U}'$ respectively, then a bijection f of X onto X′ is an *isomorphism* of the uniformity of X onto that of X′ if $g(\mathfrak{U}) = \mathfrak{U}'$, where $g = f \times f$.

For example, if X and X′ are two equipotent sets, then every bijection of X onto X′ is an isomorphism of the discrete uniformity of X onto the discrete uniformity of X′.

2. TOPOLOGY OF A UNIFORM SPACE

Proposition 1. *Let* X *be a set endowed with a uniform structure* $\mathfrak{U}$, *and for each* $x \in X$ *let* $\mathfrak{B}(x)$ *be the set of subsets* $V(x)$ *of* X (*), *where* V *runs through the set of entourages of* $\mathfrak{U}$. *Then there is a unique topology on* X *such that, for each* $x \in X$, $\mathfrak{B}(x)$ *is the neighbourhood filter of* x *in this topology.*

We have to show that the $\mathfrak{B}(x)$ satisfy conditions (V_I), (V_{II}), (V_{III}) and (V_{IV}) of Chapter I, § 1, no. 2. That this is so for the first three of

(*) We recall (*Set Theory*, R, § 3, no. 7) that if V is any subset of $X \times X$ and x is any element of X, then $V(x)$ denotes the set of all $y \in X$ such that $(x, y) \in V$.

these conditions follows immediately from the fact that the entourages of $\mathcal{U}$ satisfy (F_I), (F_{II}) and (U_I). As to (V_{IV}), let V be an entourage of $\mathcal{U}$, W an entourage of $\mathcal{U}$ such that $\overset{2}{W} \subset V$; then if $(x, y) \in W$ and $(y, z) \in W$ we have $(x, z) \in V$, so that $W(y) \subset V(x)$ for all $y \in W(x)$, and therefore $V(x) \in \mathfrak{V}(y)$ for all $y \in W(x)$. This completes the proof.

DEFINITION 3. *The topology defined in Proposition* 1 *is called the topology induced by the uniform structure* $\mathcal{U}$.

> *Examples.* * 1) The topology induced by the additive uniformity on the set of real numbers is the topology of the real line (Chapter I, § 1, no. 2); similarly the topology induced by the additive uniformity on the set of rational numbers is the topology of the rational line. *
>
> 2) On any set X, the topology induced by the discrete uniformity (no. 1, Example 2) is the discrete topology.

In the future, when we speak of the topology of a uniform space X, we shall always mean the topology induced by the uniform structure of the space, unless the contrary is expressly stated. The topological space obtained by putting this topology on the set X is sometimes called the topological space *underlying* the uniform space in question. For example, when we say that a uniform space is *Hausdorff*, or *compact*, or *locally compact*, etc., we mean that the underlying topological space has this property.

If X and X′ are two uniform spaces, any isomorphism f of the uniform structure of X onto that of X′ is also a *homeomorphism* of X onto X′; we say that X is an *isomorphism* of the uniform space X onto the uniform space X′. It should be noted that a homeomorphism of X onto X′ is not necessarily an isomorphism of the uniform structure of X onto that of X′.

> In other words, *distinct* uniformities on the same set X can induce the same topology. * For example, on $]0, +\infty[$, the *additive* and the *multiplicative* uniformities (which are distinct : Chapter III, § 6, Exercise 17) induce the same topology. *
>
> For another example see § 2, no. 2, Remark 1.

PROPOSITION 2. *Let* X *be a uniform space. For every symmetric entourage* V *of* X *and every subset* M *of* X × X, VMV *is a neighbourhood of* M *in the product space* X × X, *and the closure of* M *in this space is given by the formula*

$$\overline{M} = \bigcap_{V \in \mathfrak{S}} VMV \tag{1}$$

where $\mathfrak{S}$ *denotes the set of symmetric entourages of* X.

Let V be a symmetric entourage of X. The relation $(x, y) \in VMV$ means that there is an element (p, q) of M such that $(x, p) \in V$ and $(q, y) \in V$: in other words (V being symmetric) $x \in V(p)$ and $y \in V(q)$, that is $(x, y) \in V(p) \times V(q)$. Since $V(p) \times X(q)$ is a neighbourhood of (p, q) in $X \times X$, the first part of the proposition is proved. The relations $(x, p) \in V$, $(y, q) \in V$ can also be written $p \in V(x)$, $q \in V(y)$ or $(p, q) \in V(x) \times V(y)$. As V runs through $\mathfrak{S}$, the sets $V(x) \times V(y)$ form a fundamental system of neighbourhoods of (x, y) in $X \times X$; for if U, U' are any two entourages there is always a symmetric entourage $V \subset U \cap U'$, so that $V(x) \times V(y) \subset U(x) \times U'(y)$. Hence $V(x) \times V(y)$ meets M for each $V \in \mathfrak{S}$ if and only if $(x, y) \in \overline{M}$, and formula (1) follows.

Corollary 1. *If* A *is any subset of* X *and* V *is any symmetric entourage of* X, *then* $V(A)$ *is a neighbourhood of* A *in* X, *and*

$$\overline{A} = \bigcap_{V \in \mathfrak{S}} V(A) = \bigcap_{V \in \mathfrak{U}} V(A) \tag{2}$$

where $\mathfrak{U}$ *denotes the set of all entourages in* X.

If $M = A \times A$, then $VMV = V(A) \times V(A)$ for any $V \in \mathfrak{S}$; for the relation "there exists $p \in A$ such that $(x, p) \in V$" is by definition equivalent to $x \in V(A)$. The corollary now follows from Chapter I, § 4, no. 2, Proposition 5 and no. 3, Proposition 7.

$V(A)$ is said to be the V-*neighbourhood of* A.

If V is an entourage which is *open* in $X \times X$, then $V(x)$ is open in X for each $x \in X$ (Chapter I, § 4, no. 2, Corollary to Proposition 4) and therefore $V(A)$, being the union of the $V(x)$ as x runs through A, is *open* in X. On the other hand, if V is a closed entourage in $X \times X$, $V(A)$ need not be closed in X for every subset A of X (Exercise 3).

It should also be remarked that as V runs through the set of entourages of X, the sets $V(A)$ do not necessarily form a fundamental system of neighbourhoods of A in X (Exercise 2).

Corollary 2. *The interiors* (resp. *the closures*) *of the entourages of* X *in* $X \times X$ *form a fundamental system of entourages of* X.

If V is any entourage of X, there is a symmetric entourage W such that $\overset{3}{W} \subset V$; since $\overset{3}{W}$ is a neighbourhood of W (Proposition 2), the interior of V in $X \times X$ contains W and is therefore an entourage. Furthermore, we have $W \subset \overset{3}{W} \subset \overset{3}{W} \subset V$ by Proposition 2, and hence V contains the closure of an entourage.

Corollary 3. *Every uniform space satisfies axiom* (O_{III}).

If x is any point of X and V runs through the entourages of X which are closed in $X \times X$, then the sets $V(x)$ form a fundamental system of neighbourhoods of x in X by Corollary 2, and they are closed in X (Chapter I, § 4, no. 2, Corollary to Proposition 4).

Proposition 3. *A uniform space* X *is Hausdorff if and only if the intersection of all the entourages of its uniform structure is the diagonal* Δ *of* $X \times X$. *Every Hausdorff uniform space is regular.*

The latter statement follows immediately from Corollary 3 to Proposition 2. We have seen that the closed entourages form a fundamental system of entourages (Proposition 2, Corollary 2); if their intersection is Δ, then Δ is closed in $X \times X$ and consequently X is Hausdorff (Chapter I, § 8, no. 1, Proposition 1). Conversely, if X is Hausdorff then for every point $(x, y) \notin \Delta$ there is an entourage V of X such that $y \notin V(x)$, or equivalently $(x, y) \notin V$; hence Δ is the intersection of all the entourages.

If a uniform space X is Hausdorff, we say that the uniform structure of X is *Hausdorff*. If $\mathfrak{B}$ is a fundamental system of entourages for this structure; then X is Hausdorff if and only if the intersection of all the sets of $\mathfrak{B}$ is Δ.

2. UNIFORMLY CONTINUOUS FUNCTIONS

1. UNIFORMLY CONTINUOUS FUNCTIONS

Definition 1. *A mapping* f *of a uniform space* X *into a uniform space* X′ *is said to be uniformly continuous if, for each entourage* V′ *of* X′, *there is an entourage* V *of* X *such that the relation* $(x, y) \in V$ *implies* $(f(x), f(y)) \in V'$.

In more expressive terms we may say that f is uniformly continuous if $f(x)$ and $f(y)$ are as close to each other as we please whenever x and y are close enough.

If we put $g = f \times f$, then Definition 1 means that *whenever* V′ *is an entourage of* X′, $\overset{-1}{g}(V')$ *is an entourage of* X.

Examples. 1) The identity mapping of a uniform space onto itself is uniformly continuous.

2) A constant mapping of a uniform space into a uniform space is uniformly continuous.

3) Every mapping of a discrete uniform space into a uniform space is uniformly continuous.

PROPOSITION 1. *Every uniformly continuous mapping is continuous.*

This is an immediate consequence of the definitions.

On the other hand, a continuous mapping of a uniform space X into a uniform space X′ need not be uniformly continuous, * as is shown by the example $x \rightarrow x^3$, a *homeomorphism* of **R** onto itself, which is not uniformly continuous with respect to the additive uniformity. * (See § 4, no. 1, Theorem 2.)

PROPOSITION 2. (*a*) *If* $f : X \rightarrow X'$ *and* $g : X' \rightarrow X''$ *are two uniformly continuous mappings, then* $g \circ f : X \rightarrow X''$ *is uniformly continuous.*

(*b*) *A bijection* f *of a uniform space* X *onto a uniform space* X′ *is an isomorphism if and only if* f *and the inverse of* f *are uniformly continuous.*

This follows immediately from the interpretation of Definition 1 in terms of the product mapping $f \times f$.

2. COMPARISON OF UNIFORMITIES

Proposition 2 of no. 1 shows that we can take as *morphisms* of uniform structures the uniformly continuous mappings (*Set Theory*, Chapter IV, § 2, no. 1); we shall always assume in the future that the morphisms have been so chosen. In accordance with the general definitions (*Set Theory*, Chapter IV, § 2, no. 2), this allows us to define an *order relation* on the set of uniformities on a given set X:

DEFINITION 2. *If* $\mathfrak{U}_1$ *and* $\mathfrak{U}_2$ *are two uniform structures on the same set* X, $\mathfrak{U}_1$ *is said to be finer than* $\mathfrak{U}_2$ (*and* $\mathfrak{U}_2$ *coarser than* $\mathfrak{U}_1$) *if, denoting by* X_i *the set* X *with the uniform structure* $\mathfrak{U}_i$ $(i = 1,2)$, *the identity mapping* $X_1 \rightarrow X_2$ *is uniformly continuous.*

If $\mathfrak{U}_1$ is finer than $\mathfrak{U}_2$ and distinct from $\mathfrak{U}_2$, we say that $\mathfrak{U}_1$ is *strictly finer* than $\mathfrak{U}_2$ (and that $\mathfrak{U}_2$ is *strictly coarser than* $\mathfrak{U}_1$).

Two uniformities are said to be *comparable* if one is finer than the other.

Example. In the ordered set of uniformities on a set X, the discrete uniformity is the *finest*, and the *coarsest* uniformity is that in which the set of entourages consists of the single element $X \times X$.

The following proposition is an immediate consequence of Definition 1 of no. 1:

PROPOSITION 3. *If* $\mathfrak{U}_1$ *and* $\mathfrak{U}_2$ *are two uniformities on a set* X, *then* $\mathfrak{U}_1$ *is finer than* $\mathfrak{U}_2$ *if and only if every entourage of* $\mathfrak{U}_2$ *is an entourage of* $\mathfrak{U}_1$.

COROLLARY. *Let* $\mathfrak{U}_1$ *and* $\mathfrak{U}_2$ *be two uniformities on a set* X, *and suppose that* $\mathfrak{U}_1$ *is finer than* $\mathfrak{U}_2$; *then the topology induced by* $\mathfrak{U}_1$ *is finer than the topology induced by* $\mathfrak{U}_2$.

This follows immediately from the comparison of topologies in terms of neighbourhoods (Chapter I, § 2, no. 2, Proposition 3).

Remarks. 1) It can happen that a uniformity $\mathfrak{U}_1$ is *strictly finer* than a uniformity $\mathfrak{U}_2$ but that the two induced topologies are *identical.* The following example shows this:

Let X be a non-empty set. For each *finite partition* $\varpi = (A_i)_{1 \leqslant i \leqslant n}$ of X, let V_ϖ denote

$$\bigcup_i A_i \times A_i.$$

The sets V_ϖ then form a *fundamental system of entourages* of a uniformity $\mathfrak{U}$ on X. For if ϖ is any finite partition of X we have $\Delta \subset V_\varpi$ and $V_\varpi \circ V_\varpi = \overset{-1}{V}_\varpi = V_\varpi$ (§ 1, no. 1, Example 2); and if $\varpi' = (B_j)$ and $\varpi'' = (C_k)$ are two finite partitions of X, then those of the sets $B_j \cap C_k$ which are not empty form a finite partition ϖ of X, and we have $V_\varpi \subset V_{\varpi'} \cap V_{\varpi''}$. $\mathfrak{U}$ is called the *uniformity of finite partitions* on X. The topology induced by $\mathfrak{U}$ is the *discrete* topology, since for each $x \in X$ the sets $\{x\}$ and $\complement\{x\}$ form a finite partition of X. Nevertheless, if X is infinite, it is clear that $\mathfrak{U}$ is *strictly coarser* than the discrete uniformity.

2) If $f: X \to X'$ is a uniformly continuous mapping, then f remains uniformly continuous if we replace the uniformity of X by a *finer* uniformity and that of X′ by a *coarser* uniformity (no. 1, Proposition 2). In other words, the *finer* the uniformity of X and the *coarser* the uniformity of X′, the *more* uniformly continuous mappings there are of X into X′.

3. INITIAL UNIFORMITIES

PROPOSITION 4. *Let* X *be a set, let* $(Y_\iota)_{\iota \in I}$ *be a family of uniform spaces, and for each* $\iota \in I$ *let* f_ι *be a mapping of* X *into* Y_ι. *For each* $\iota \in I$ *let* g_ι *denote* $f_\iota \times f_\iota$. *Let* $\mathfrak{S}$ *be the set of subsets of* $X \times X$ *of the form* $\overset{-1}{g}_\iota(V_\iota)$, *where* $\iota \in I$ *and* V_ι *is an entourage of* Y_ι, *and let* $\mathfrak{B}$ *be the set of all finite intersections*

$$\text{(1)} \qquad U(V_{\iota_1}, \ldots, V_{\iota_n}) = \overset{-1}{g}_{\iota_1}(V_{\iota_1}) \cap \cdots \cap \overset{-1}{g}_{\iota_n}(V_{\iota_n})$$

of sets of $\mathfrak{S}$. *Then* $\mathfrak{B}$ *is a fundamental system of entourages of a uniformity* $\mathfrak{U}$ *on* X *which is the initial uniform structure on* X *with respect to the family* (f_ι) (*Set Theory*, Chapter IV, § 2, no. 3), *and in particular* $\mathfrak{U}$ *is the coarsest uniformity on* X *for which all the mappings* f_ι *are uniformly continuous. But otherwise, let* h *be a mapping of a uniform space* Z *into* X; *then* h *is uniformly continuous* (*when* X *is endowed with the uniformity* $\mathfrak{U}$) *if and only if each of the mappings* $f_\iota \circ h$ *is uniformly continuous.*

It is immediately seen that $\mathfrak{B}$ satisfies axioms (B_1) and (U'_I). If $W_\iota = \overset{-1}{g}_\iota(V_\iota)$, then $\overset{-1}{W}_\iota = \overset{-1}{g}_\iota(\overset{-1}{V}_\iota)$ and $\overset{2}{W}_\iota = \overset{-1}{g}_\iota(\overset{2}{V}_\iota)$; hence $\mathfrak{B}$ also satisfies axioms (U'_{II}) and (U'_{III}) and is therefore a fundamental system of entourages of a uniformity $\mathcal{U}$ on X. Furthermore, it follows immediately from the definition of $\mathcal{U}$ and Definition 1 and no. 1 that f_ι is uniformly continuous for each index $\iota \in I$; hence (no. 1, Proposition 2) $f_\iota \circ h$ is uniformly continuous for each $\iota \in I$ if h is. Conversely, suppose that $f_\iota \circ h$ is uniformly continuous for each $\iota \in I$, and consider a set $U(V_{\iota_1}, \ldots, V_{\iota_n})$; by hypothesis, for each k such that $1 \leqslant k \leqslant n$, there is an entourage W_k of Z such that the relation $(z, z') \in W_k$ implies $[f_{\iota_k}(h(z)), f_{\iota_k}(h(z'))] \in V_k$; if

$$W = \bigcap_k W_k,$$

these n relations are simultaneously satisfied whenever z and z' are W-close, so that we have then $(h(z), h(z')) \in U(V_{\iota_1}, \ldots, V_{\iota_n})$, and the proof is complete.

COROLLARY. *The topology on* X *induced by the coarsest uniformity* $\mathcal{U}$ *for which the* f_ι *are uniformly continuous is also the coarsest topology for which the* f_ι *are continuous.*

This is an immediate consequence of the definition of the neighbourhoods of a point in this latter topology (Chapter I, § 2, no. 3, Proposition 4).

The general properties of initial structures (*Set Theory*, Chapter IV, § 2, no. 3, criterion CST 10) imply in particular the following *transitivity* property:

PROPOSITION 5. *Let* X *be a set, let* $(Z_\iota)_{\iota \in I}$ *be a family of uniform spaces, let* $(J_\lambda)_{\lambda \in L}$ *be a partition of* I *and let* $(Y_\lambda)_{\lambda \in L}$ *be a family of sets indexed by* L. *For each* $\lambda \in L$, *let* h_λ *be a mapping of* X *into* Y_λ; *for each* $\lambda \in L$ *and each* $\iota \in J_\lambda$, *let* $g_{\iota\lambda}$ *be a mapping of* Y_λ *into* Z_ι; *and let* $f_\iota = g_{\iota\lambda} \circ h_\lambda$. *Let each* Y_λ *carry the coarsest uniform structure for which the mappings* $g_{\iota\lambda}$ $(\iota \in J_\lambda)$ *are uniformly continuous. Then the coarsest uniform structure on* X *for which the* f_ι *are uniformly continuous is the same as the coarsest uniform structure on* X *for which the* h_λ *are uniformly continuous.*

4. INVERSE IMAGE OF A UNIFORMITY; UNIFORM SUBSPACES

Let X be a set, Y a uniform space, f a mapping of X into Y. The coarsest uniformity $\mathcal{U}$ on X for which f is uniformly continuous is called the *inverse image* under f of the uniform structure of Y. It follows from Proposition 4 of no. 3, and from the formulae which give the inverse image of an intersection, that the inverse images under $g = f \times f$ of the entourages of Y form a fundamental system of entourages for $\mathcal{U}$. The topology induced by $\mathcal{U}$ is the *inverse image* under f of the topology of Y (no. 3, Corollary to Proposition 4).

Remark. If $f: X \to Y$ is *surjective*, then the entourages of Y are the *direct images* under g of the entourages of X.

A mapping f of a uniform space X into a uniform space X′ is uniformly continuous if and only if the inverse image under f of the uniformity of X′ is *coarser* than the uniformity of X.

Let A be a subset of a uniform space X. The *uniformity induced* on A by the uniformity of X is the inverse image of the latter under the canonical injection $A \to X$. By Proposition 4 of no. 3, this is equivalent to the following definition:

DEFINITION 3. *Let* A *be a subset of a uniform space* X. *The uniformity on* A *whose set of entourages is the trace on* $A \times A$ *of the set of entourages of* X *is called the uniformity induced on* A *by the uniformity of* X.

The topology induced by the uniformity induced on A is the same as the topology induced on A by the topology of X; the set A, together with the uniformity and the topology induced by those of X, is called a *uniform subspace of* X.

If A is a subset of a uniform space X and if $f: X \to X'$ is a uniformly continuous mapping, then the restriction $f|A$ is a uniformly continuous mapping of A into X′. If $A' \subset X'$ is such that $f(X) \subset A'$, then the mapping of X into the uniform subspace A′ of X′, having the same graph as f, is again uniformly continuous (no. 3, Proposition 4).

If $B \subset A \subset X$, then the uniform subspace B of X is identical with the uniform subspace B of the uniform subspace A of X (*transitivity* of induced uniform structures; no. 3, Proposition 5).

PROPOSITION 6. *Let* A *be a dense subset of a uniform space* X. *Then the closvres,* in $X \times X$, *of the entourages of the uniform subspace* A *form a fundamental system of entourages of* X.

$A \times A$ is dense in $X \times X$ (Chapter I, § 4, no. 3, Proposition 7). Let V be an open entourage of A; it is the intersection of $A \times A$ with an open entourage U of X. We have $U \subset \overline{V}$ (Chapter I, § 1, no. 6, Proposition 5), and this relation together with $\overline{V} \subset \overline{U}$ establishes the proposition, in view of § 1, no. 2, Corollary 2 of Proposition 2.

5. LEAST UPPER BOUND OF A SET OF UNIFORMITIES

Every family $(\mathfrak{U}_\iota)_{\iota \in I}$ of uniformities on a set X has a *least upper bound* $\mathfrak{U}$ in the ordered set of all uniformities on X; we have only to apply Proposition 4 of no. 3, taking Y_ι to be the set X with the uniformity $\mathfrak{U}_\iota$, and f_ι to be the identity mapping $X \to Y_\iota$. The topology induced by $\mathfrak{U}$ is just the *least upper bound* of the topologies induced by the $\mathfrak{U}_\iota$.

It follows also from Proposition 4 of no. 3 that if X is not empty and if $\mathfrak{U}_\iota$ is the filter of entourages of $\mathcal{U}_\iota$, then the filter of entourages of $\mathcal{U}$ is the *least upper bound* of the filters $\mathfrak{U}_\iota$ (Chapter I, § 6, no. 2).

Example. If ϖ is any finite partition $(A_i)_{1 \leqslant i \leqslant n}$ of a non-empty set X, the set $V_\varpi = \bigcup_i (A_i \times A_i)$ by itself constitutes a fundamental system of entourages of a uniformity $\mathcal{U}_\varpi$ on X (§ 1, no. 1, Example 2); the uniformity of finite partitions on X (no. 2, Remark 1) is then the *least upper bound* of the uniformities $\mathcal{U}_\varpi$.

Remark. A family $(\mathcal{U}_\iota)$ of uniformities on X also has a *greatest lower bound* in the ordered set of all uniformities on X, namely the least upper bound of the set of all uniformities on X which are coarser than each of the $\mathcal{U}_\iota$ (such uniformities exist, since the set of all uniformities on X has a least element). But (supposing X not empty) the filter of entourages of this uniformity is not necessarily the intersection of the filters of entourages of the $\mathcal{U}_\iota$, because this latter filter need not satisfy axiom (U_{III}) (Exercise 4).

6. PRODUCT OF UNIFORM SPACES

DEFINITION 4. *If $(X_\iota)_{\iota \in I}$ is a family of uniform spaces, the product uniform space of this family is the product set*

$$X = \prod_{\iota \in I} X_\iota$$

endowed with the coarsest uniformity for which the projections $\mathrm{pr}_\iota : X \to X_\iota$ are uniformly continuous. This uniformity is called the product of the uniformities of the X_ι, and the uniform spaces X_ι are called the factors of X.

The topology induced by the product uniformity on X is same as the product of the topologies of the X_ι (no. 3, Corollary to Proposition 4).

PROPOSITION 7. *Let $f = (f_\iota)$ be a mapping of a uniform space Y into a product uniform space $X = \prod_{\iota \in I} X_\iota$. Then f is uniformly continuous if and only if each f_ι is uniformly continuous.*

Since $f_\iota = \mathrm{pr}_\iota \circ f$, this is a particular case of Proposition 4 of no. 3.

COROLLARY. *Let $(X_\iota)_{\iota \in I}$, $(Y_\iota)_{\iota \in I}$ be two families of uniform spaces indexed by the same set I. For each $\iota \in I$, let f_ι be a mapping of X_ι into Y_ι. If each of the f_ι is uniformly continuous, then so is the product mapping*

$$f : (x_\iota) \to (f_\iota(x_\iota)).$$

Conversely, if the X_ι are non-empty and f is uniformly continuous, then each f_ι is uniformly continuous.

f can be written as $x \to (f_\iota(\mathrm{pr}_\iota x))$, and the first part of the corollary therefore follows from Proposition 7. The second part is proved by considering a point $a = (a_\iota)$ of $\prod_{\iota \in I} X_\iota$ and repeating the argument of Chapter I, § 4, no. 1, Corollary 1 to Proposition 1, with the phrase "continuous at a (resp. a_κ)" replaced by "uniformly continuous".

The general criterion of transitivity of initial uniformities (no. 3, Proposition 5) shows that, as for the product of topological spaces (Chapter I, § 4, no. 1), the product of uniform spaces is *associative* and that the following is true:

Proposition 8. *Let* X *be a set, let* $(Y_\iota)_{\iota \in I}$ *be a family of uniform spaces, and for each* $\iota \in I$ *let* f_ι *be a mapping of* X *into* Y_ι. *Let* f *be the mapping* $x \to (f_\iota(x))$ *of* X *into* $Y = \prod_{\iota \in I} Y_\iota$, *and let* $\mathcal{U}$ *be the coarsest uniformity on* X *for which the* f_ι *are uniformly continuous. Then* $\mathcal{U}$ *is the inverse image under* f *of the uniformity induced on* $f(X)$ *by the product uniformity on* Y.

Corollary. *For each* $\iota \in I$, *let* A_ι *be a subspace of* Y_ι. *Then the uniformity induced on* $A = \prod_{\iota \in I} A_\iota$ *by the product uniformity on* $\prod_{\iota \in I} Y_\iota$ *is the same as the product of the uniformities of the subspaces* A_ι.

In addition, we see immediately that if X_1, X_2 are two uniform spaces and a_1 is any point of X_1, the mapping $x_2 \to (a_1, x_2)$ is an isomorphism of X_2 onto the subspace $\{a_1\} \times X_2$ of $X_1 \times X_2$; hence:

Proposition 9. *Let* f *be a uniformly continuous mapping of a product uniform space* $X_1 \times X_2$ *into a uniform space* Y; *then every partial mapping*

$$x_2 \to f(x_1, x_2)$$

of X_2 *into* Y *is uniformly continuous.*

In other words, a uniformly continuous function of two arguments is uniformly continuous with respect to each of them separately.

* The example given in Chapter I, § 4, no. 2, Remark 2, shows that the converse of this proposition is false. *

7. INVERSE LIMITS OF UNIFORM SPACES

Let I be a partially ordered set in which the partial ordering is written $\alpha \leqslant \beta$. For each $\alpha \in I$ let X_α be a uniform space, and for each pair of indices α, β such that $\alpha \leqslant \beta$ let $f_{\alpha\beta}$ be a mapping of X_β into X_α.

We shall say that $(X_\alpha, f_{\alpha\beta})$ is an *inverse system of uniform spaces* if (i) $(X_\alpha, f_{\alpha\beta})$ is an *inverse system of sets* (cf. Chapter I, Appendix, no. 1) and (ii) whenever $\alpha \leqslant \beta$, $f_{\alpha\beta}$ is *uniformly continuous*. On $X = \varprojlim X_\alpha$ the *coarsest* uniformity for which the canonical mappings $f_\alpha : X \to X_\alpha$ are uniformly continuous is called the *inverse limit* (with respect to the $f_{\alpha\beta}$) of the uniformities of the X_α, and the set X endowed with this coarsest uniformity is called the *inverse limit of the inverse system of uniform spaces* $(X_\alpha, f_{\alpha\beta})$. All the properties of inverse limits of topological spaces established in Chapter I, § 4, no. 4 (with the exception of Proposition 9) remain valid if we replace "topology" by "uniformity" and "continuous mapping" by "uniformly continuous mapping". In addition:

Proposition 10. *Let* I *be a directed set, let* $(X_\alpha, f_{\alpha\beta})$ *be an inverse system of uniform spaces indexed by* I, *and let* J *be a cofinal subset of* I. *For each* $\alpha \in I$ *let* f_α *be the canonical mapping of* $X = \varprojlim X_\alpha$ *into* X_α, *and let* g_α *denote* $f_\alpha \times f_\alpha$. *Then the family of sets* $\overset{-1}{g}_\alpha(V_\alpha)$, *where* α *runs through* J *and where, for each* $\alpha \in J$, V_α *runs through a fundamental system of entourages of* X_α, *is a fundamental system of entourages of* X.

We leave the proof to the reader; it is a straightforward adaptation of the proof of Proposition 9 of Chapter I, § 4, no. 4.

Finally, the topology on $X = \varprojlim X_\alpha$ induced by the inverse limit of the uniformities of the X_α is the inverse limit of the topologies of the X_α.

3. COMPLETE SPACES

1. CAUCHY FILTERS

Once a set X has been endowed with a uniform structure we can define what is meant by a "small" subset of X (relative to this structure): a "small" subset of X is one in which all the points are "very close" to each other. Precisely:

Definition 1. *If* X *is a uniform space and if* V *is an entourage of* X, *a subset* A *of* X *is said to be* V*-small if every pair of points of* A *are* V*-close* (in other words, if $A \times A \subset V$).

Proposition 1. *In a uniform space* X, *if two sets* A *and* B *are* V*-small and intersect, then their union* $A \cup B$ *is* $\overset{2}{V}$*-small.*

Let x and y be any two points of $A \cup B$, and let $Z \in A \cap B$. Then $(x, z) \in V$ and $(z, y) \in V$, so that $(x, y) \in \overset{2}{V}$.

DEFINITION 2. *A filter* $\mathfrak{F}$ *on a uniform space* X *is a Cauchy filter if for each entourage* V *of* X *there is a subset of* X *which is* V*-small and belongs to* $\mathfrak{F}$.

Here again we may make our language more expressive by the use of the expressions "sufficiently small set" and "a set as small as we please"; thus Definition 2 can be restated by saying that a Cauchy filter is one containing *arbitrarily small sets.*

An infinite sequence (u_n) of points of a uniform space X is said to be a *Cauchy sequence* if the elementary filter associated with the sequence is a Cauchy filter. It comes to the same thing to say that for each entourage V of X there is an integer n_0 such that for all integers $m \geqslant n_0$ and $n \geqslant n_0$ we have $(u_m, u_n) \in V$.

PROPOSITION 2. *On a uniform space* X *every convergent filter is a Cauchy filter.*

If x is any point of X and V is any symmetric entourage of X, then the neighbourhood $V(x)$ of x is $\overset{2}{V}$-small. If $\mathfrak{F}$ is a filter which converges to x, there is a set of $\mathfrak{F}$ contained in $V(x)$, and therefore V-small.

Clearly every filter which is *finer* than a Cauchy filter is a Cauchy filter.

PROPOSITION 3. *Let* $f: X \rightarrow X'$ *be a uniformly continuous mapping. Then the image under* f *of a Cauchy filter base on* X *is a Cauchy filter base on* X'.

Let $g = f \times f$. If V' is an entourage of X', then $\overset{-1}{g}(V')$ is an entourage of X, and the image under f of a $\overset{-1}{g}(V')$-small set is V'-small; hence the result.

It follows in particular that if the uniformity of a uniform space X is replaced by a *coarser* uniformity, then every Cauchy filter with respect to the original uniformity remains a Cauchy filter with respect to the new uniformity.

This fact can be easily remembered in the following form: *the finer the uniformity, the fewer Cauchy filters there are.*

PROPOSITION 4. *Let* X *be a set, let* $(Y_\iota)_{\iota \in I}$ *be a family of uniform spaces, and for each* $\iota \in I$ *let* f_ι *be a mapping of* X *into* Y_ι. *Let* X *carry the coarsest uniformity* $\mathfrak{U}$ *for which the* f_ι *are uniformly continuous. Then in order that a filter base* $\mathfrak{B}$ *on* X *should be a Cauchy filter base it is necessary and sufficient that* $f_\iota(\mathfrak{B})$ *should be a Cauchy filter base on* Y_ι, *for each* $\iota \in I$.

The condition is necessary by Proposition 3. Conversely, suppose that it is satisfied, and let $U(V_{\iota_1}, \ldots, V_{\iota_n})$ be an entourage of the uniformity $\mathfrak{U}$ [§ 2, no. 3, formula (1)]. By hypothesis, for each index k there is a set $M_k \in \mathfrak{B}$ such that $f_{\iota_k}(M_k)$ is V_{ι_k}-small ($1 \leqslant k \leqslant n$). Let M be a set of $\mathfrak{B}$ contained in M_k for $1 \leqslant k \leqslant n$; then for each pair of points x, x' of M we have $[f_{\iota_k}(x), f_{\iota_k}(x')] \in V_{\iota_k}$ for $1 \leqslant k \leqslant n$, so that

$$(x, x') \in U(V_{\iota_1}, \ldots, V_{\iota_n}).$$

This completes the proof.

COROLLARY 1. *If a Cauchy filter on a uniform space X induces a filter on a subset A of X, then this filter is a Cauchy filter on the uniform subspace A.*

COROLLARY 2. *A filter base $\mathfrak{B}$ on a product $\prod_{\iota \in I} X_\iota$ of uniform spaces is a Cauchy filter base if and only if, for each $\iota \in I$, $\mathrm{pr}_\iota(\mathfrak{B})$ is a Cauchy filter base on X_ι.*

2. MINIMAL CAUCHY FILTERS

The minimal elements (with respect to inclusion) of the set of Cauchy filters on a uniform space X are called *minimal Cauchy filters* on X.

PROPOSITION 5. *Let X be a uniform space. For each Cauchy filter $\mathfrak{F}$ on X there is a unique minimal Cauchy filter $\mathfrak{F}_0$ coarser than $\mathfrak{F}$. If $\mathfrak{B}$ is a base of $\mathfrak{F}$ and $\mathfrak{S}$ is a fundamental system of symmetric entourages of X, then the sets $V(M)$ ($M \in \mathfrak{B}$, $V \in \mathfrak{S}$) form a base of $\mathfrak{F}_0$.*

If M, M' are in $\mathfrak{B}$ and V, V' are in $\mathfrak{S}$, then there is a set $M'' \in \mathfrak{B}$ (resp. $V'' \in \mathfrak{S}$) such that $M'' \subset M \cap M'$ (resp. $V'' \subset V \cap V'$); hence $V''(M'') \subset V(M) \cap V'(M')$ and therefore the sets $V(M)$ ($M \in \mathfrak{B}$, $V \in \mathfrak{S}$) indeed form a base of a filter $\mathfrak{F}_0$ on X. Further, if M is V-small, then $V(M)$ is $\overset{3}{V}$-small; hence $\mathfrak{F}_0$ is a Cauchy filter and is clearly coarser than $\mathfrak{F}$. To complete the proof it is enough to show that if $\mathfrak{G}$ is a Cauchy filter coarser than $\mathfrak{F}$, then $\mathfrak{G}$ is finer than $\mathfrak{F}_0$. For each $M \in \mathfrak{B}$ and each $V \in \mathfrak{S}$ there is a set $N \in \mathfrak{G}$ which is V-small; since $N \in \mathfrak{F}$, N meets M; hence $N \subset V(M)$ and so $V(M) \in \mathfrak{G}$.

COROLLARY 1. *For each $x \in X$, the neighbourhood filter $\mathfrak{B}(x)$ of x in X is a minimal Cauchy filter.*

Take $\mathfrak{F}$ in Proposition 5 to be the filter of all subsets of X which contain x, and take $\mathfrak{B}$ to consist of the single element $\{x\}$.

COROLLARY 2. *Every cluster point x of a Cauchy filter $\mathfrak{F}$ is a limit point of $\mathfrak{F}$.*

There is a filter $\mathfrak{G}$ which is finer than both $\mathfrak{F}$ and $\mathfrak{V}(x)$ (Chapter I, § 7, no. 2, Proposition 4); since $\mathfrak{F}$ is a Cauchy filter, so is $\mathfrak{G}$. If $\mathfrak{F}_0$ is the unique minimal Cauchy filter coarser than $\mathfrak{F}$, then both $\mathfrak{F}_0$ and $\mathfrak{V}(x)$ are minimal Cauchy filters coarser than $\mathfrak{G}$. Hence $\mathfrak{F}_0 = \mathfrak{V}(x)$, which shows that $\mathfrak{F}$ converges to x.

COROLLARY 3. *Every Cauchy filter, which is coarser than a filter converging to a point x, also converges to x.*

This is a consequence of Corollary 2.

COROLLARY 4. *If $\mathfrak{F}$ is a minimal Cauchy filter, then every set of $\mathfrak{F}$ has a non-empty interior which also belongs to $\mathfrak{F}$* (in other words, $\mathfrak{F}$ has a base consisting of *open* sets).

Let V be any entourage of X; then there is an open entourage $U \subset V$ (§ 1, no. 2, Corollary 2 to Proposition 2). For each subset M of X, U(M) is open and contained in V(M); hence the result, in view of Proposition 5.

3. COMPLETE SPACES

In a uniform space X, *a Cauchy filter need not have a limit point.*

Examples. 1) Consider the sequence (u_n) on the rational line $\mathbf{Q}$ defined by $u_n = \sum_{p=0}^{n} 2^{-p(p+1)/2}$. If $m > n$ we have

$$|u_m - u_n| \leqslant 2^{-n(n+3)/2} \tag{1}$$

and therefore (u_n) is a *Cauchy sequence.* But this sequence has no limi in $\mathbf{Q}$; for if the rational number a/b were a limit of (u_n), then by (1) we should have for all n

$$|a/b - h_n/2^{n(n+1)/2}| \leqslant 1/2^{n(n+3)/2}$$

where h_n is an integer (depending on n); that is,

$$|a.2^{n(n+1)/2} - bh_n| \leqslant b.2^{-n}$$

for all n. Now the left-hand side of this inequality is an integer for all n, and must therefore be zero whenever n is greater than an integer n_0 such that $b < 2^{n_0}$; we should therefore have $a/b = u_n$ for all $n > n_0$, which is absurd.

2) Let X be an infinite set, and consider the uniformity of finite partitions on X (§ 2, no. 2, Remark 1). Every *ultrafilter* $\mathfrak{F}$ on X is a

Cauchy filter with respect to this uniformity. For if (A_i) is a finite partition of X and

$$V = \bigcup_i (A_i \times A_i)$$

the corresponding entourage, then at least one of the A_i belongs to $\mathfrak{F}$ (Chapter I, § 6, no. 4, Corollary to Proposition 5), and A_i is V-small. On the other hand, X is an infinite discrete space, hence is not compact, and consequently there are ultrafilters on X which do not converge.

DEFINITION 3. *A complete space is a uniform space in which every Cauchy filter converges.*

In a complete space every *Cauchy sequence* (no. 1) is therefore convergent.

Example. On a *discrete* uniform space X a Cauchy filter is a trivial ultrafilter (Chapter I, § 6, no. 4), hence convergent; consequently, X is complete.

From Definitions 2 and 3 of no. 1 and Proposition 2 of no. 1, we deduce the following proposition, known as *Cauchy's criterion*:

PROPOSITION 6. *Let $\mathfrak{F}$ be a filter on a set* X, *and let f be a mapping of* X *into a complete uniform space* X′. *Then f has a limit with respect to $\mathfrak{F}$ if and only if the image of $\mathfrak{F}$ under f is a Cauchy filter base.*

This criterion shows the importance of complete spaces in all questions involving the notion of limit: if a function takes its values in a complete space we can prove the *existence* of a limit *without knowing in advance the value of the limit*; this would be impossible if the definition of limit were the only criterion of convergence at our disposal.

A uniformity which is *finer* than the uniformity of a complete space need not be a uniformity of a complete space (Exercise 2). However, we have the following proposition:

PROPOSITION 7. *Let $\mathfrak{U}_1$, $\mathfrak{U}_2$ be two uniformities on a set* X, *and let $\mathfrak{T}_1$, $\mathfrak{T}_2$ be the topologies induced by these uniformities respectively. Suppose that $\mathfrak{U}_1$ is finer than $\mathfrak{U}_2$, and that there is a fundamental system of entourages for $\mathfrak{U}_1$ which are* closed *in* X × X *in the topology $\mathfrak{T}_2 \times \mathfrak{T}_2$. Then a filter $\mathfrak{F}$ on* X *converges in the topology $\mathfrak{T}_1$ if and only if it is a Cauchy filter in the uniformity $\mathfrak{U}_1$ and converges in the topology $\mathfrak{T}_2$.*

The conditions are clearly necessary, because $\mathfrak{T}_2$ is coarser than $\mathfrak{T}_1$. Conversely, suppose that the conditions are satisfied, and let x be a limit point of $\mathfrak{F}$ with respect to $\mathfrak{T}_2$; we shall show that x is a limit of $\mathfrak{F}$ with respect to $\mathfrak{T}_1$. Let V be a symmetric entourage of $\mathfrak{U}_1$ which is closed in the topology $\mathfrak{T}_2 \times \mathfrak{T}_2$. By hypothesis, $\mathfrak{F}$ contains a set M which is V-small; hence if $x' \in M$ we have $M \subset V(x')$. But $V(x')$ is *closed* in the topology $\mathfrak{T}_2$; hence x, which lies in the closure of M with respect to $\mathfrak{T}_2$, must belong to $V(x')$. It follows that $M \subset \overset{2}{V}(x)$, and the proposition is proved.

COROLLARY. *In the conditions of Proposition 7, if $\mathfrak{U}_2$ is a uniformity of a complete space, then so is $\mathfrak{U}_1$.*

For every Cauchy filter with respect to $\mathfrak{U}_1$ is then a Cauchy filter with respect to $\mathfrak{U}_2$ and therefore converges in the topology $\mathfrak{T}_2$.

Note that the hypotheses of the Corollary to Proposition 7 are satisfied when $\mathfrak{T}_1 = \mathfrak{T}_2$ (§ 1, no. 2, Corollary 2 to Proposition 2).

4. SUBSPACES OF COMPLETE SPACES

PROPOSITION 8. *Every closed subspace of a complete space is complete. Every complete subspace of a* Hausdorff *uniform space (complete or not) is closed.*

Let X be a complete space and let A be a closed subspace of X. If $\mathfrak{F}$ is a Cauchy filter on A, then it is a Cauchy filter base on X (no. 1, Proposition 3) and therefore converges to a point $x \in X$; but since A is closed we have $x \in A$, and therefore $\mathfrak{F}$ converges in the subspace A.

Now let A be a non-closed subset of a *Hausdorff* uniform space X, and let $b \in \overline{A} - A$. The trace $\mathfrak{B}_A$ on A of the neighbourhood filter $\mathfrak{B}$ of b in X is a Cauchy filter on A; but it cannot converge to a point $c \in A$, otherwise c would be a limit point of $\mathfrak{B}$ (no. 2, Proposition 5, Corollary 3) which is absurd since $b \neq c$ and X is Hausdorff.

PROPOSITION 9. *Let* X *be a uniform space and let* A *be a dense subset of* X *such that every Cauchy filter base on* A *converges in* X. *Then* X *is complete.*

It is enough to show that every *minimal* Cauchy filter $\mathfrak{F}$ on X is convergent. Since A is dense and since every set of $\mathfrak{F}$ has a non-empty interior (no. 2, Corollary 4 to Proposition 5), the trace $\mathfrak{F}_A$ of $\mathfrak{F}$ on A is a Cauchy filter on A, hence converges to a point $x_0 \in X$. Since $\mathfrak{F}$ is coarser than the filter on X generated by $\mathfrak{F}_A$, it follows that $\mathfrak{F}$ converges to x_0 (no. 2, Corollary 3 to Proposition 5).

5. PRODUCTS AND INVERSE LIMITS OF COMPLETE SPACES

PROPOSITION 10. *Every product of complete uniform spaces is complete. Conversely, if a product of non-empty uniform spaces is complete, then each of the factors is a complete uniform space.*

The first assertion is a consequence of the characterization of Cauchy filters and convergent filters on a product space (no. 1, Corollary 2 to Proposition 4 and Chapter I, § 7, no. 6, Corollary 1 to Proposition 10). Conversely, suppose

$$X = \prod_{\iota \in I} X_\iota$$

is complete (the X_ι being non-empty) and let $\mathfrak{F}_\varkappa$ be a Cauchy filter on $X_\varkappa$. For each $\iota \neq \varkappa$ let $\mathfrak{F}_\iota$ be a Cauchy filter on X_ι, and consider the product filter (Chapter I, § 6, no. 7)

$$\mathfrak{F} = \prod_{\iota \in I} \mathfrak{F}_\iota \quad \text{on} \quad X;$$

$\mathfrak{F}$ is a Cauchy filter (no. 1, Corollary 2 to Proposition 4), hence is convergent, and therefore so is each $\mathrm{pr}_\varkappa \mathfrak{F} = \mathfrak{F}_\varkappa$ (Chapter I, § 7, no. 6, Corollary 1 to Proposition 10).

Corollary. *Let $(X_\alpha, f_{\alpha\beta})$ be an inverse system of uniform spaces. If the X_α are Hausdorff and complete, then so is $X = \varprojlim X_\alpha$.*

For X is Hausdorff and can be identified with a *closed* subspace of $\prod_{\iota \in I} X_\alpha$ (Chapter I, § 8, no. 2, Corollary 2 to Proposition 7); the corollary therefore follows from Proposition 10 and Proposition 8 (no. 4).

An inverse limit of complete Hausdorff uniform spaces X_α can be *empty*, even if all the X_α are non-empty and all the $f_{\alpha\beta}$ are surjective, as is shown by the case of discrete spaces (*Set Theory*, Chapter III, § 1, Exercise 32). However, we have the following theorem:

Theorem 1 (Mittag-Leffler). *Let $(X_\alpha, f_{\alpha\beta})$ be an inverse system of complete Hausdorff uniform spaces, indexed by a directed set I which has a* countable *cofinal subset; suppose also that, for each $\alpha \in I$, X_α has a* countable *fundamental system of entourages* (*). *Finally, suppose that for each $\alpha \in I$ there is an index $\beta \geqslant \alpha$ satisfying the following condition:*

($ML_{\alpha\beta}$) *For each $\gamma \geqslant \beta$, $f_{\alpha\gamma}(X_\gamma)$ is dense in $f_{\alpha\beta}(X_\beta)$.*

Let $X = \varprojlim X_\alpha$ and let f_α be the canonical mapping $X \to X_\alpha$. Then for each $\alpha \in I$ and each $\beta \geqslant \alpha$ satisfying ($ML_{\alpha\beta}$), $f_\alpha(X)$ is dense in $f_{\alpha\beta}(X_\beta)$ (and consequently X is non-empty if the X_α are all non-empty).

Let (λ_n) be a sequence of indices cofinal in I. Start with an index $\alpha_0 \in I$ and define recursively an increasing sequence (α_n) such that $\alpha_n \geqslant \lambda_n$ and such that ($ML_{\alpha_n, \alpha_{n+1}}$) is true. Clearly the sequence

(*) * This condition signifies that the Hausdorff uniform space X_α is *metrizable*; cf. Chapter IX, § 2, no. 4, Theorem 1. *

(α_n) is cofinal in I. We shall write f_{mn} in place of $f_{\alpha_m \alpha_n}$ for $m \leqslant n$, and we shall put $f_{n,\, n+1}(X_{\alpha_{n+1}}) = Y_n$. Then, if $m \leqslant n$, *$f_{mn}(Y_n)$ is dense in Y_m*; for by definition $f_{m,\, n+1}(X_{\alpha_{n+1}})$ is dense in

$$f_{m,\, m+1}(X_{\alpha_{m+1}}) = Y_m, \quad \text{and} \quad f_{m,\, n+1}(X_{\alpha_{n+1}}) = f_{mn}[f_{n,\, n+1}(X_{\alpha_{n+1}})] = f_{mn}(Y_n).$$

By induction on n and k we can define a fundamental system $(V_{kn})_{k \in \mathbf{N}}$ of closed symmetric entourages of X_{α_n} for each n such that

$$\text{(2)} \qquad \overset{2}{V}_{k+1,\, n} \subset V_{kn},$$

$$\text{(3)} \qquad (f_{n,\, n+1} \times f_{n,\, n+1})(V_{k,\, n+1}) \subset V_{kn}.$$

In effect, let $(U_{kn})_{k \in \mathbf{N}}$ be a fundamental system of entourages of X_{α_n}. If we suppose that the V_{kn} have been defined for a given n and for all $k \in \mathbf{N}$, then since $f_{n,n+1}$ is uniformly continuous we can define the entourage $V_{k,n+1}$ by induction on k so that (3) is satisfied and

$$\overset{2}{V}_{k+1,\, n+1} \subset V_{k,\, n+1} \cap U_{k+1,\, n+1}$$

The assertion follows.

Now let $x_0 \in Y_0$. We shall show that for each integer $k > 0$ there is a point $z \in X$ such that $[x_0, f_{\alpha_0}(z)] \in V_{k-1,0}$; this will prove the theorem. Since $f_{n,n+1}(Y_{n+1})$ is dense in Y_n, we can define by induction a sequence of points $x_n \in Y_n$ such that

$$\text{(4)} \qquad [x_n, f_{n,\, n+1}(x_{n+1})] \in V_{k+n,\, n}.$$

By reason of (3) it follows that if $m \leqslant n$ then

$$\text{(5)} \qquad [f_{mn}(x_n), f_{m,\, n+1}(x_{n+1})] \in V_{k+n,\, m}.$$

From this we conclude that for fixed m the sequence $(f_{mn}(x_n))_{n \geqslant m}$ is a *Cauchy sequence* in X_{α_m} and therefore converges to a point z_m; for by induction it follows from (5) that, for each pair of integers $p \geqslant m$, $q > 0$, we have

$$\text{(6)} \qquad [f_{mp}(x_p), f_{m,\, p+q}(x_{p+q})] \in V_{k+p+q-1,\, m} \circ V_{k+p+q-2,\, m} \circ \cdots \circ V_{k+p,\, m}$$

and by virtue of (2) it is clear that the right-hand side of (6) is contained in $V_{k+p-1,\, m}$. Let q increase indefinitely; we infer in particular that, for $m = p = 0$, we have $(x_0, z_0) \in V_{k-1,0}$, since $V_{k-1,0}$ is *closed*. On the other hand, from the relations $z_m = \lim_{n \to \infty} f_{mn}(x_n)$ and from the continuity of $f_{m,m+1}$, we deduce that $f_{m,m+1}(z_{m+1}) = z_m$ for each $m \geqslant 0$. For each $\gamma \in I$ there is at least one integer n such that $\alpha_n \geqslant \gamma$; putting

$z_\gamma = f_{\gamma, \alpha_n}(z_n)$ we verify immediately that z_γ does not depend on the value of n such that $\alpha_n \geqslant \gamma$, and that the family $(z_\alpha)_{\alpha \in I}$ so defined is a point z of $X = \varprojlim X_\alpha$. Since $f_{\alpha_0}(z) = z_0$, the proof is complete.

COROLLARY 1. *Let* $(X_\alpha, f_{\alpha\beta})$ *be an inverse system of sets indexed by a directed set* I *which has a countable cofinal subset, and suppose that the* $f_{\alpha\beta}$ *are surjective. Then if* $X = \varprojlim X_\alpha$, *the canonical mapping* $f_\alpha : X \to X_\alpha$ *is surjective for each* $\alpha \in I$.

Let each X_α carry the discrete uniformity, and apply Theorem 1.

COROLLARY 2. *Let* I *be a directed set which has a countable cofinal subset. Let* $(X_\alpha, f_{\alpha\beta})$ *and* $(X'_\alpha, f'_{\alpha\beta})$ *be two inverse systems of sets indexed by* I, *and for each* $\alpha \in I$ *let* $u_\alpha : X_\alpha \to X'_\alpha$ *be a mapping, such that the* u_α *form an inverse system of mappings. Let* $u = \varprojlim u_\alpha$. *Let* $x = (x'_\alpha)$ *be an element of*

$$X' = \varprojlim X'_\alpha$$

which satisfies the following condition: for each $\alpha \in I$ *there is an index* $\beta \geqslant \alpha$ *such that for all* $\gamma \geqslant \beta$ *we have* $f_{\alpha\gamma}(\overset{-1}{u}_\gamma(x_\gamma)) = f_{\alpha\beta}(\overset{-1}{u}_\beta(x'_\beta))$. *Then there is an element* $x \in X$ *such that* $u(x) = x'$.

Apply Theorem 1 to the inverse system of sets $\overset{-1}{u}_\alpha(x'_\alpha)$, each carrying the discrete uniformity.

* *Example.* Suppose we are given in **C**: (i) a sequence (a_n) of distinct points such that the sequence $(|a_n|)$ is increasing and tends to $+\infty$; (ii) for each n, a rational function $z \to R_n(z)$ defined in $\mathbf{C} - \{a_n\}$ and having a pole at a_n; (iii) an increasing sequence (B_n) of open discs centred at 0, whose union is **C**, and such that none of the a_k is on the frontier of any of the discs B_n. For each n, let B'_n denote the intersection of $\overline{B}_n$ and the complement in **C** of the set of points a_n; and let X_n denote the set of all mappings

$$z \to S(z) = P(z) + \sum_{a_k \in B_n} R_k(z)$$

of B'_n into **C**, where P is the restriction to B'_n of a function which is continuous in $\overline{B}_n$ and holomorphic in B_n. We define a metric in X_n by putting

$$d_n(S_1, S_2) = \sup_{z \in B'_n} |S_1(z) - S_2(z)|.$$

It is easily verified that X_n is *complete* with respect to this metric. Finally, for $n \leqslant m$, we define a mapping $f_{nm} : X_m \to X_n$ such that if $S \in X_m$ then $f_{nm}(S)$ is the *restriction* of S to B'_n. It is clear that the f_{nm} are *uniformly continuous* and that (X_n, f_{nm}) is an inverse system of uniform spaces. This being so, an element of the *inverse limit* $X = \varprojlim X_n$ can be canonically identified with a *meromorphic function* F in **C**, whose only poles are the points a_n, and which is such that for each n, $F(z) - R_n(z)$ is *holomorphic at* a_n. The classical theorem of Mittag-Leffler asserts that X *is not empty*; by virtue of Theorem 1, we have only to verify the condition (ML_{nn}) for all n. Let

$$S_n = P_n + \sum_{a_k \in B_n} R_k$$

be an element of X_n, where P_n is continuous in $\bar{B}_n$ and holomorphic in B_n; for any $m \geqslant n$, let Q_{mn} be the restriction of

$$\sum_{a_h \in B_m - B_n} R_h \text{ to } B'_n;$$

this latter sum is a holomorphic function in some neighbourhood of $\bar{B}_n$, hence (by Taylor's theorem) for each $\varepsilon > 0$ there is a polynomial P_{mn} such that $|Q_{mn}(Z) - P_{mn}(z)| \leqslant \varepsilon$ in B_n; if S_m is the restriction of $S_n + Q_{mn} - P_{mn}$ to B'_m, we have $S_m \in X_m$ and $|S_m(z) - S_n(z)| \leqslant \varepsilon$ in B'_n. This completes the proof. *

6. EXTENSION OF UNIFORMLY CONTINUOUS FUNCTIONS

The theorem of extension by continuity (Chapter I, § 8, no. 5, Theorem 1) has important additions when the functions in question take their values in a *complete Hausdorff* uniform space.

PROPOSITION 11. *Let* A *be a dense subset of a topological space* X, *and let* f *be a mapping of* A *into a complete Hausdorff uniform space* X′. *Then* f *can be extended by continuity to* X *if and only if, for each* $x \in X$, *the image under* f *of the trace on* A *of the neighbourhood filter of* x *in* X *is a Cauchy filter base in* X′.

This follows from the theorem of extension by continuity (*loc. cit.*) because X′ is *regular* (§ 1, no. 2, Proposition 3) and because on X′ convergent filters are the same as Cauchy filters.

When X is also a *uniform space*, there is the following theorem:

THEOREM 2. *Let* f *be a function defined on a dense subspace* A *of a uniform space* X, *taking its values in a complete Hausdorff uniform space* X′, *and suppose that* f *is uniformly continuous on* A. *Then* f *can be extended to the whole of* X *by continuity, and the extended function* $\bar{f}$ *is uniformly continuous.*

The existence of $\bar{f}$ is an immediate consequence of Propositions 3 and 11 of no. 1. Let us show that $\bar{f}$ is uniformly continuous. Let V′ be a closed symmetric entourage of X′, and let V be an entourage of X such that, when x and y are in A and are V-close, then $f(x)$ and $f(y)$

are V′-close. We may assume that V is the closure in $X \times X$ of an entourage W of A (§ 2, no. 4, Proposition 6). We have $[\bar{f}(x), \bar{f}(y)] \in V'$ when $(x, y) \in W$; since $\bar{f} \times \bar{f}$ is continuous in $X \times X$ (Chapter I, § 4, no. 1, Proposition 1) we have also $[\bar{f}(x), \bar{f}(y)] \in V'$ when $(x, y) \in V = \overline{W}$, since V′ is closed (Chapter I, § 2, no. 1, Theorem 1).

Q.E.D.

Corollary. *Let X_1, X_2 be two complete Hausdorff uniform spaces, and let Y_1, Y_2 be dense subspaces of X_1, X_2 respectively. Then every isomorphism f of Y_1 onto Y_2 extends to an isomorphism of X_1 onto X_2.*

f is uniformly continuous in Y_1, hence (Theorem 2) extends to a uniformly continuous mapping $\bar{f}: X_1 \to X_2$. Likewise the inverse g of f extends to a uniformly continuous mapping $\overset{-1}{f}: X_2 \to X_1$. The function $\bar{g} \circ \bar{f}$ is therefore a continuous mapping of X_1 into itself whose restriction to Y_1 is the identity mapping; by the principle of extension of identities (Chapter I, § 8, no. 1, Corollary 1 to Proposition 2) $\bar{g} \circ \bar{f}$ is therefore the identity mapping of X_1; similarly $\bar{f} \circ \bar{g}$ is the identity map of X_2. Consequently (*Set Theory*, R, § 2, no. 12) $\overset{-1}{f}$ and $\bar{g}$ are bijections and are inverses of each other; they are also uniformly continuous and are therefore isomorphisms (§ 2, no. 1, Proposition 2).

It should be remarked that if f is a *bijective* uniformly continuous mapping of Y_1 onto Y_2, its extension by continuity *need be neither injective nor surjective* (Exercise 3).

7. THE COMPLETION OF A UNIFORM SPACE

Theorem 3. *Let X be a uniform space. Then there exists a complete Hausdorff uniform space $\hat{X}$ and a uniformly continuous mapping $i : X \to \hat{X}$ having the following property:*

(P) *Given any uniformly continuous mapping f of X into a complete Hausdorff uniform space Y, there is a unique uniformly continuous mapping $g : \hat{X} \to Y$ such that $f = g \circ i$.*

If (i_1, X_1) is another pair consisting of a complete Hausdorff uniform space X_1 and a uniformly continuous mapping $i_1 : X \to X_1$ having the property (P), *then there is a unique isomorphism $\varphi : \hat{X} \to X_1$ such that $i_1 = \varphi \circ i$.*

The first statement of the theorem signifies that the pair $(i, \hat{X})$ is the solution of the *universal mapping problem* (*Set Theory*, Chapter IV, § 3, no. 1) in which the Σ-sets are *complete Hausdorff* uniform spaces, the σ-morphisms are uniformly continuous mappings and the α-mappings are uniformly continuous mappings of X into a complete Hausdorff uniform space. The uniqueness of the pair $(i, \hat{X})$ up to a unique isomorphism therefore follows from the general properties of solutions of universal mapping problems (*loc. cit.*). It remains to prove the existence of the pair $(i, \hat{X})$.

1) *Definition of* $\hat{X}$. Let $\hat{X}$ be *the set of minimal Cauchy filters* (no. 2) *on* X. We shall define a uniform structure on $\hat{X}$. For this purpose, if V is any *symmetric* entourage of X, let $\tilde{V}$ denote *the set of all pairs* $(\mathfrak{X}, \mathfrak{Y})$ *of minimal Cauchy filters which have in common a* V-*small set*. We shall show that the sets $\tilde{V}$ form a fundamental system of entourages of a uniform structure on $\hat{X}$:

(i) Since each $\mathfrak{X} \in \hat{X}$ is a Cauchy filter, we have by definition $(\mathfrak{X}, \mathfrak{X}) \in \tilde{V}$ for every symmetric entourage V of X; hence axiom (U'_I) is satisfied.
(ii) If V and V′ are two symmetric entourages of X, then $W = V \cap V'$ is a symmetric entourage, and every set which is W-small is also V-small and V′-small; hence $\tilde{W} \subset \tilde{V} \cap \tilde{V}'$, which prove (B_I).
(iii) The sets $\tilde{V}$ are symmetric by definition, hence (U'_{II}) is satisfied.
(iv) Given a symmetric entourage V of X, let W be a symmetric entourage such that $\overset{2}{V} \subset V$. Consider three minimal Cauchy filters $\mathfrak{X}, \mathfrak{Y}, \mathfrak{Z}$ such that $(\mathfrak{X}, \mathfrak{Y}) \in \tilde{W}$ and $(\mathfrak{Y}, \mathfrak{Z}) \in \tilde{W}$; then there are two W-small sets M, N such that $M \in \mathfrak{X} \cap \mathfrak{Y}$ and $N \in \mathfrak{Y} \cap \mathfrak{Z}$. Since M and N belong to $\mathfrak{Y}$, $M \cap N$ is not empty and therefore (no. 1, Proposition 1) $M \cup N$ is $\overset{2}{W}$-small and hence V-small; since $M \cup N$ belongs to $\mathfrak{X}$ and to $\mathfrak{Z}$ we have $\overset{2}{\tilde{W}} \subset \tilde{V}$; hence (U'_{III}) is satisfied.

We show next that the uniform space $\hat{X}$ is *Hausdorff*. Let $\mathfrak{X}, \mathfrak{Y}$ be two minimal Cauchy filters on X such that $(\mathfrak{X}, \mathfrak{Y}) \in \hat{X}$ for *all* symmetric entourages V of X. It follows immediately that the sets $M \cup N$, where $M \in \mathfrak{X}$ and $N \in \mathfrak{Y}$, form a base of a filter $\mathfrak{Z}$ coarser than $\mathfrak{X}$ and $\mathfrak{Y}$. Now $\mathfrak{Z}$ is a *Cauchy filter*, since for every symmetric entourage V of X there is by hypothesis a V-small set P belonging to both $\mathfrak{X}$ and $\mathfrak{Y}$ and therefore belonging to $\mathfrak{Z}$. By the definition of minimal Cauchy filters, we have $\mathfrak{X} = \mathfrak{Z} = \mathfrak{Y}$, and this shows that $\hat{X}$ is Hausdorff.

2) *Definition of* i; *the uniform structure of* X *is the inverse image under* i *of that of* $\hat{X}$. We know that for each $x \in X$ the neighbourhood filter

$\mathfrak{B}(x)$ of x in X is a minimal Cauchy filter (no. 2, Proposition 5, Corollary 1). So we define $i(x) = \mathfrak{B}(x)$. Let $f = i \times i$; we shall show that for each symmetric entourage V of X we have $\overset{-1}{j}(\tilde{V}) \subset V \cup \overset{-1}{j}[(\overset{3}{V})^{\sim}]$, and this will prove our assertion (§ 2, no. 4). Now, if $[i(x), i(y)] \in \tilde{V}$, there is a V-small set M which is a neighbourhood of each of x and y, hence $(x, y) \in V$. Conversely, if $(x, y) \in V$, it is immediately seen that the set $V(x) \cup V(y)$ is $\overset{3}{V}$-small and is a neighbourhood of each of x and y.

3) $\hat{X}$ *is complete and* $i(X)$ *is dense in* $\hat{X}$. The trace on $i(X)$ of a neighbourhood $\tilde{V}(\mathfrak{X})$ of a point $\mathfrak{X} \in X$ is the set of all $i(x)$ such that

$$(\mathfrak{X}, i(x)) \in \tilde{V}.$$

This relation means that there is a V-small neighbourhood of x in X which belongs to $\mathfrak{X}$, i.e. that x is *an interior point of a* V-*small set of* $\mathfrak{X}$. Let M be the union of the interiors of all V-small sets of $\mathfrak{X}$; then M belongs to $\mathfrak{X}$ (no. 2, Proposition 5, Corollary 4) and from what has been said it follows that $\tilde{V}(\mathfrak{X}) \cap i(X) = i(M)$. We conclude that:

(i) $\tilde{V}(\mathfrak{X}) \cap i(X)$ is not empty, hence $i(X)$ is *dense* in $\hat{X}$.

(ii) The trace of $\tilde{V}(\mathfrak{X})$ on $i(X)$ belongs to the *filter base* $i(\mathfrak{X})$ on X; hence this filter base converges in $\hat{X}$ to the *point* $\mathfrak{X}$.

Now let $\mathfrak{F}$ be a Cauchy filter on $i(X)$; then from 2) above and Proposition 4 of no. 1, $\overset{-1}{i}(\mathfrak{F})$ is a base of a Cauchy filter $\mathfrak{G}$ on X. Let $\mathfrak{X}$ be a minimal Cauchy filter coarser than $\mathfrak{G}$ (no. 2, Proposition 5); then $i(\mathfrak{X})$ is a Cauchy filter base on $i(X)$ (no. 1, Proposition 3), and $\mathfrak{F} = i[\overset{-1}{i}(\mathfrak{F})]$ is finer than the filter whose base is $i(\mathfrak{X})$. Since the latter converges in $\hat{X}$, so does $\mathfrak{F}$, and Proposition 9 of no. 4 therefore shows that $\hat{X}$ is *complete*.

4) *Verification of the property* (P). Let f be a uniformly continuous mapping of X into a *complete Hausdorff* uniform space Y. Let us first show that there is a unique uniformly continuous mapping $g_0 : i(X) \to Y$ such that $f = g_0 \circ i$. Since f is continuous, we have

$$f(x) = \lim f(\mathfrak{B}(x)),$$

hence if we put $g_0(i(x)) = \lim f(\mathfrak{B}(x))$, we have $f = g_0 \circ i$; so it remains to show that g_0 is uniformly continuous in $i(X)$. Let U be an entourage of Y and let V be a symmetric entourage of X such that the relation $(x, x') \in V$ implies $(f(x), f(x')) \in U$; we have seen in 2) that the relation $(i(x), i(x')) \in \tilde{V}$ implies $(x, x') \in V$, hence also is implied

$$(g_0(i(x)), \; g_0(i(x'))) \in U,$$

which proves our assertion.

Let g be the extension of g_0 by continuity to $\hat{X}$ (no. 6, Theorem 2); then $f = g \circ i$, and it is clear that g is the unique continuous mapping of $\hat{X}$ into Y satisfying this relation, since $i(X)$ is dense in $\hat{X}$ (Chapter I, § 8, no. 1, Proposition 2, Corollary 1).

Q.E.D.

Definition 4. *The complete Hausdorff uniform space* $\hat{X}$ *defined in the proof of Theorem 3 is called the Hausdorff completion of* X, *and the mapping* $i : X \to \hat{X}$ *is called the canonical mapping of* X *into its Hausdorff completion.*

We note also the following facts:

Proposition 12. (i) *The subspace* $i(X)$ *is dense in* $\hat{X}$.

(ii) *The graph of the equivalence relation* $i(x) = i(x')$ *is the intersection of the entourages of* X.

(iii) *The uniform structure of* X *is the inverse image under* i *of that of* $\hat{X}$ [*or of that of the subspace* $i(X)$].

(iv) *The entourages of* $i(X)$ *are the images under* $i \times i$ *of the entourages of* X, *and the closures in* $\hat{X} \times \hat{X}$ *of the entourages of* $i(X)$ *form a fundamental system of entourages of* $\hat{X}$.

(i) and (iii) have been proved in the course of the proof of Theorem 3; (iv) is a consequence of (i) and (iii) by virtue of general results proved earlier (§ 2, no. 4, Remark and Proposition 6). The relation

$$i(x) = i(x')$$

means by definition that x and x' have the same neighbourhood filter. But this implies, by definition, that $(x, x') \in V$ for every entourage V of X, and the converse is obvious.

Corollary. *If* X *is a Hausdorff uniform space, then the canonical mapping* $i : X \to \hat{X}$ *is an isomorphism of* X *onto a dense subspace of* $\hat{X}$.

When X is Hausdorff, $\hat{X}$ is said to be the *completion* of X, and we generally *identify* X with a dense subset of $\hat{X}$ by means of i.

Remark. If this identification is made, the minimal Cauchy filters on X are just the traces on X of the neighbourhood filters of points of $\hat{X}$; this follows from the proof of Theorem 3.

The Corollary to Proposition 12 *characterizes* the completion of a Hausdorff uniform space:

PROPOSITION 13. *If* Y *is a complete Hausdorff uniform space and* X *a dense subspace of* Y, *then the canonical injection* $X \to Y$ *extends to an isomorphism of* $\hat{X}$ *onto* Y.

For every uniformly continuous mapping of X into a complete Hausdorff uniform space Z extends uniquely to a uniformly continuous mapping of Y into Z by Theorem 2 of no. 6.

PROPOSITION 14. *Let* X *be a complete Hausdorff uniform space,* $\mathfrak{U}$ *its uniformity, and let* Z *be a dense subspace of* X. *If* $\mathfrak{U}'$ *is a uniformity on* X *which is coarser than* $\mathfrak{U}$ *and which induces the same uniformity as* $\mathfrak{U}$ *on* Z, *then* $\mathfrak{U} = \mathfrak{U}'$.

Let X' denote the set X with the uniformity $\mathfrak{U}'$. The composition of the canonical mapping $X' \to \hat{X}'$ and the identity mapping $X \to X'$ is a uniformly continuous mapping $\varphi : X \to \hat{X}'$. Since Z is Hausdorff for the uniform structure induced by $\mathfrak{U}'$, the restriction of φ to Z is by hypothesis an isomorphism of Z onto the dense subspace $\varphi(Z)$ of $\hat{X}'$; it follows (no. 6, Corollary to Theorem 2) that φ itself is an isomorphism of X onto $\hat{X}'$, hence $X' = \hat{X}'$ and $\mathfrak{U}' = \mathfrak{U}$.

PROPOSITION 15. *Let* X *and* X' *be two uniform spaces. For each uniformly continuous mapping* $f : X \to X'$ *there is a unique uniformly continuous mapping* $\hat{f} : \hat{X} \to \hat{X}'$ *such that the diagram*

$$\begin{array}{ccc} X & \xrightarrow{f} & X' \\ i\downarrow & & \downarrow i' \\ \hat{X} & \xrightarrow[\hat{f}]{} & \hat{X}' \end{array}$$

is commutative (*), *where* $i : X \to \hat{X}$ *and* $i' : X' \to \hat{X}'$ *are the canonical mappings.*

Apply Theorem 3 to the function $i' \circ f : X \to \hat{X}'$.

COROLLARY. *If* $f : X \to X'$ *and* $g : X' \to X''$ *are two uniformly continuous mappings and* $h = g \circ f$, *then* $\hat{h} = \hat{g} \circ \hat{f}$.

This is an immediate consequence of the uniqueness in Proposition 15.

(*) In other words, $i' \circ f = \hat{f} \circ i$.

8. THE HAUSDORFF UNIFORM SPACE ASSOCIATED WITH A UNIFORM SPACE

PROPOSITION 16. *Let* X *be a uniform space and* i *the canonical mapping of* X *into its Hausdorff completion* $\hat{\mathrm{X}}$. *For each uniformly continuous mapping* f *of* X *into a Hausdorff uniform space* Y, *there is a unique uniformly continuous mapping* $h : i(\mathrm{X}) \to \mathrm{Y}$ *such that* $f = h \circ i$.

We may identify Y with a subspace of its completion $\hat{\mathrm{Y}}$ (no. 7, Corollary to Proposition 12), and f can then be considered as a uniformly continuous mapping of X into $\hat{\mathrm{Y}}$. By virtue of Theorem 3, f is then of the form $f = g \circ i$, where g is a uniformly continuous mapping of $\hat{\mathrm{X}}$ into $\hat{\mathrm{Y}}$. If h is the restriction of g to $i(\mathrm{X})$, then clearly $f = h \circ i$, and h maps $i(\mathrm{X})$ into Y. The uniqueness of h is trivial.

The pair $(i, i(\mathrm{X}))$ is therefore the solution of a *universal mapping problem* (*Set Theory*, Chapter IV, § 3, no. 1), where this time we take the Σ-sets to be *Hausdorff* uniform spaces, and the σ-morphisms (resp. α-mappings) to be uniformly continuous mappings (resp. uniformly continuous mappings of X into a Hausdorff uniform space).

DEFINITION 5. *The Hausdorff uniform space* $i(\mathrm{X})$ *defined in the proof of Theorem 3 is called the Hausdorff uniform space associated with* X.

The *Hausdorff completion* of X is thus the completion of the Hausdorff uniform space associated with X.

COROLLARY. *Let* X, Y *be two uniform spaces and* X′, Y′ *the associated Hausdorff spaces. For each uniformly continuous mapping* $f : \mathrm{X} \to \mathrm{Y}$ *there is a unique uniformly continuous mapping* $f' : \mathrm{X}' \to \mathrm{Y}'$ *for which the diagram*

$$\begin{array}{ccc} \mathrm{X} & \xrightarrow{f} & \mathrm{Y} \\ {\scriptstyle i}\downarrow & & \downarrow{\scriptstyle i'} \\ \mathrm{X}' & \xrightarrow[f']{} & \mathrm{Y}' \end{array}$$

is commutative, where i *and* i' *are the canonical mappings.*

Apply Proposition 16 to $i' \circ f : \mathrm{X} \to \mathrm{Y}'$.

The Hausdorff space associated with a uniform space may also be characterized by the following property:

PROPOSITION 17. *Let* X *be a uniform space,* $i(\mathrm{X})$ *its associated Hausdorff space, and let* f *be a mapping of* X *onto a Hausdorff uniform space* X′, *such that the uniformity of* X *is the inverse image under* f *of the uniformity of* X′. *Then the mapping* $g : i(\mathrm{X}) \to \mathrm{X}'$ *such that* $f = g \circ i$ *is an isomorphism.*

By Proposition 16, g is uniformly continuous; also g is obviously surjective, and is also injective because the relation $f(x) = f(y)$ implies by definition that (x, y) belongs to all the entourages of X, and therefore that $i(x) = i(y)$ (no. 7, Proposition 12). Finally, the entourages of X' are the images under $f \times f$ of the entourages of X (§ 2, no. 4, Remark), hence they are also the images under $g \times g$ of the entourages of $i(X)$ (no. 7, Proposition 12); hence the result.

Remark. Let R be the equivalence relation $i(x) = i(x')$ on X. We have seen (no. 7, Proposition 12) that the graph C of R is the intersection of all the entourages of X. It is clear that every open set (and therefore also every closed set) in X is *saturated* with respect to R; taking account of the definition of the inverse image of a topology, we conclude that the canonical bijection of the quotient space X/R onto $i(X)$ induced by i is a *homeomorphism*. The Hausdorff space associated with X can therefore be identified, *qua* topological space, with X/R. The canonical mapping $i : X \to i(X)$ is open and closed, and even proper (Chapter I, § 10, no. 2, Example).

Let X' be another uniform space, C' the intersection of all the entourages of X', and R' the equivalence relation whose graph is C'. Let $f: X \to X'$ be a *continuous* mapping. Since the inverse image under f of any neighbourhood of $f(x)$ is a neighbourhood of x, it follows that the inverse image under f of $C'(f(x))$ contains $C(x)$, and therefore f is *compatible* with R and R', and induces a continuous mapping $X/R \to X'/R'$ (Chapter I, § 3, no. 4, Corollary to Proposition 6). This generalizes the corollary to Proposition 16.

9. COMPLETION OF SUBSPACES AND PRODUCT SPACES

PROPOSITION 18. *Let X be a set, let $(Y_\lambda)_{\lambda \in L}$ be a family of uniform spaces, and for each $\lambda \in L$ let f_λ be a mapping of X into Y_λ. Let X carry the coarsest uniformity $\mathcal{U}$ which makes all the f_λ uniformly continuous. Then the uniformity of the Hausdorff completion $\hat{X}$ of X is the coarsest for which all the mappings $\hat{f}_\lambda : \hat{X} \to \hat{Y}_\lambda$ $(\lambda \in L)$* (no. 7, Proposition 15) *are uniformly continuous. Furthermore, if j_λ is the canonical mapping of Y_λ into $\hat{Y}_\lambda$, and if $g_\lambda = j_\lambda \circ f_\lambda$, then $\hat{X}$ may be identified with the closure in $\prod_{\lambda \in L} \hat{Y}_\lambda$ of the image of X under the mapping $x \to (g_\lambda(x))$.*

Let X' (resp. Y'_λ) be the Hausdorff uniform space associated with X (resp. Y_λ), and let $f'_\lambda : X' \to Y'_\lambda$ be the uniformly continuous mapping which makes the diagram

$$\begin{array}{ccc} X & \xrightarrow{f_\lambda} & Y_\lambda \\ {\scriptstyle i}\downarrow & & \downarrow{\scriptstyle j_\lambda} \\ X' & \xrightarrow[f'_\lambda]{} & Y'_\lambda \end{array}$$

commutative (i being the canonical mapping).

The transitivity of initial uniformities (§ 2, no. 3, Proposition 5) shows on the one hand that $\mathcal{U}$ is the coarsest uniformity for which the mappings $j_\lambda \circ f_\lambda : X \to Y'_\lambda$ are uniformly continuous, and on the other hand that $\mathcal{U}$ is also the inverse image under i of the coarsest uniformity $\mathcal{U}'$ on the set X' for which the f'_λ are uniformly continuous. Now $\mathcal{U}'$ is *Hausdorff*, for if x_1, x_2 are two points of X such that $j_\lambda(f_\lambda(x_1)) = j_\lambda(f_\lambda(x_2))$ for each $\lambda \in L$, then (x_1, x_2) belongs to all the entourages of $\mathcal{U}$ and hence $i(x_1) = i(x_2)$. Proposition 17 of no. 8 therefore shows that $\mathcal{U}'$ is the uniformity of the Hausdorff space X' associated with X.

This being so, the bijection $x' \to (f'_\lambda(x'))$ identifies X with a uniform subspace of the product $\prod_\lambda Y'_\lambda$ (§ 2, no. 6, Proposition 8). Since the Y'_λ are Hausdorff, each Y'_λ can be identified with a dense subspace of its completion $\hat{Y}_\lambda$, and hence $\prod_\lambda Y'_\lambda$ can be identified with a dense subspace of $\prod_\lambda \hat{Y}_\lambda$ (Chapter I, § 4, no. 3, Proposition 7). But $\prod_\lambda \hat{Y}_\lambda$ is Hausdorff and complete (no. 5, Proposition 10); the closure $\overline{X'}$ of X' in $\prod_\lambda \hat{Y}_\lambda$ is therefore a complete Hausdorff subspace (no. 4, Proposition 8) which can be identified with the Hausdorff completion $\hat{X}$ of X; under this identification the mappings $\hat{f}_\lambda$ become the projections onto the factors $\hat{Y}_\lambda$, and the proposition is proved.

Corollary 1. *Let* X *be a uniform space and let* i *be the canonical mapping of* X *into its Hausdorff completion* X ; *let* A *be a subspace of* X *and* $j : A \to X$ *the canonical injection. Then* $\hat{j}: \hat{A} \to \hat{X}$ *is an isomorphism of* $\hat{A}$ *onto the closure of* $i(A)$ *in* $\hat{X}$.

Corollary 2. *Let* $(Y_\lambda)_{\lambda \in L}$ *be a family of uniform spaces. Then the Hausdorff completion of the product space* $\prod_{\lambda \in L} Y_\lambda$ *is canonically isomorphic to the product* $\prod_{\lambda \in L} \hat{Y}_\lambda$.

4. RELATIONS BETWEEN UNIFORM SPACES AND COMPACT SPACES

1. UNIFORMITY OF COMPACT SPACES

Definition 1. *A uniformity on a topological space* X *is said to be compatible with the topology of* X *if the latter coincides with the topology induced by the uniformity.*

A topological space is said to be uniformizable, and its topology is said to be uniformizable, if there exists a uniformity on the space which is compatible with its topology.

There are topological spaces which are not uniformizable, for example any space which does not satisfy axiom (O_{III}) (§ 1, no. 2, Corollary 3 of Proposition 2); hence the question arises of determining under what conditions a topological space is uniformizable.

We shall not give a complete answer to this question until Chapter IX, § 1. In this section we shall examine only one important particular case, that in which X is *compact*. We have then the following theorem:

THEOREM 1. *On a compact space X there is exactly one uniformity compatible with the topology of X; the entourages of this uniformity are all the neighbourhoods of the diagonal Δ in $X \times X$. Furthermore, X endowed with this uniformity is a complete uniform space.*

The last part of the theorem is straightforward; for every Cauchy filter on X has a cluster point [axiom (C)] and is therefore convergent (§ 3, no. 2, Proposition 5, Corollary 2).

Let us show next that if there is a uniformity on X compatible with the topology of X, then the set $\mathfrak{U}$ of entourages of this uniformity is the set of all neighbourhoods of Δ. We know already that every entourage is a neighbourhood of Δ (§ 1, no. 2, Proposition 2), hence we have to show that conversely every neighbourhood of Δ belongs to $\mathfrak{U}$. Suppose that there is a neighbourhood U of Δ which is not in $\mathfrak{U}$; then the sets $V \cap \complement U$, as V runs through $\mathfrak{U}$, form a base of a filter $\mathfrak{G}$ on the compact space $X \times X$; consequently $\mathfrak{G}$ has a cluster point (a, b) not belonging to Δ. Since $\mathfrak{U}$ is a filter coarser than $\mathfrak{G}$, (a, b) is also a cluster point of $\mathfrak{U}$. But the uniformity defined by $\mathfrak{U}$ is Hausdorff by hypothesis; hence the intersection of the closures of the sets of $\mathfrak{U}$ is Δ (§ 1, no. 2, Corollary 2 to Proposition 2 and Proposition 3); thus we arrive at a contradiction.

It remains therefore to show that the set $\mathfrak{V}$ of neighbourhoods of Δ in $X \times X$ is the set of entourages of a uniformity compatible with the topology of X. For this it is enough to show that $\mathfrak{V}$ is the set of entourages of a *Hausdorff* uniformity on X; for then the topology induced by this uniformity will be *coarser* than the topology of X (Chapter I, § 2, no. 2, Proposition 3) and must therefore coincide with the latter (Chapter I, § 9, no. 4, Theorem 2, Corollary 3).

$\mathfrak{V}$ clearly satisfies axioms (F_I) and (F_{II}). Let us show that axioms (U_{II}) and (U_{III}) are also satisfied and that Δ is the intersection of the sets of $\mathfrak{V}$. Take the last point first: every set consisting of a single point $(x, y) \in X \times X$ is closed, since X is Hausdorff; hence if $x \neq y$, the complement of (x, y) in $X \times X$ is a neighbourhood of Δ. Since the symmetry $(x, y) \to (y, x)$ is a homeomorphism of $X \times X$ onto itself, it follows that $V \in \mathfrak{B}$ implies $\overset{-1}{V} \in \mathfrak{B}$, whence (U_{II}). Finally, suppose that (U_{III}) is not satisfied; then there is a set $V \in \mathfrak{B}$ such that for all $W \in \mathfrak{B}$ the set $\overset{2}{W} \cap \complement V$ is not empty; the sets $\overset{2}{W} \cap \complement V$, as W runs through $\mathfrak{B}$, therefore form a filter base on $X \times X$, and this filter base has a cluster point (x, y) not in Δ. Now, since X is *regular* (Chapter I, § 9, no. 2, Corollary to Proposition 1) there are disjoint open neighbourhoods U_1 of x and U_2 of y, and there are *closed* neighbourhoods $V_1 \subset U_1$, $V_2 \subset U_2$ of x and y respectively. Let $U_3 = \complement(V_1 \cup V_2)$, and consider the neighbourhood

$$W = \bigcup_{i=1,2,3} (U_i \times U_i) \text{ of } \Delta \text{ in } X \times X.$$

It follows immediately from these definitions that if $(u, v) \in W$ and $u \in V_1$ (resp. $u \in U_1$), then we must have $v \in U_1$ (resp. $v \in U_1 \cup U_3 = \complement V_2$); hence the neighbourhood $V_1 \times V_2$ of (x, y) in $X \times X$ does not meet $\overset{2}{W}$, and we have a contradiction. This completes the proof.

Remark 1. For every *finite open covering* $\mathfrak{R} = (U_i)_{1 \leqslant i \leqslant n}$ of X, the set

$$V_{\mathfrak{R}} = \bigcup_{i=1}^{n} (U_i \times U_i)$$

is a neighbourhood of Δ in $X \times X$, and these sets $V_{\mathfrak{R}}$ form a *fundamental system of neighbourhoods* of Δ (and therefore a *fundamental system of entourages* of the unique uniformity on X). Let W be any neighbourhood of Δ in $X \times X$; then for each $x \in X$ there is an open neighbourhood U_x of x in X such that $U_x \times U_x \subset W$. Since the U_x $(x \in X)$ form an open covering of X, there exist a finite number of points x_i $(1 \leqslant i \leqslant n)$ such that the U_{x_i} $(1 \leqslant i \leqslant n)$ form a covering $\mathfrak{R}$ of X. We have then $V_{\mathfrak{R}} \subset W$, which proves the assertion.

For this reason the unique uniformity on X is often called the *uniformity of finite open coverings* (cf. Chapter IX, § 4, Exercise 17).

COROLLARY 1. *Every subspace of a compact space is uniformizable.*

COROLLARY 2. *Every locally compact space is uniformizable.*

For by Alexandroff's theorem (Chapter I, § 9, no. 8, Theorem 4) a locally compact space is homeomorphic to a subspace of a compact space.

Remark 2. It can happen that there are several distinct uniformities compatible with the topology of a locally compact space.

For example, we have seen that on an infinite discrete space there is more than one distinct uniformity compatible with the discrete topology (§ 2, no. 2, Remark).

Nevertheless it is *not* the case that the uniqueness of the uniformity compatible with the topology of a uniformizable space is a property which characterizes compact spaces; there exist locally compact spaces which are not compact but whose uniformity is unique (Exercise 4).

THEOREM 2. *Every continuous mapping* f *of a compact space* X *into a uniform space* X′ *is uniformly continuous.*

Let $g = f \times f$, then g is continuous on $X \times X$ (Chapter I, § 4, no. 1, Proposition 1, Corollary 1); hence for each open entourage V′ of X′, $\overset{-1}{g}(V')$ is an open subset of $X \times X$, which evidently contains the diagonal. The theorem thus follows from Theorem 1, since the open entourages of X′ form a fundamental system of entourages (§ 1, no. 2, Proposition 2, Corollary 2).

Under the hypotheses of Theorem 2, the restriction of f to any subspace A of X is uniformly continuous; hence (§ 3, no. 6, Theorem 2):

COROLLARY. *Let* A *be a dense subspace of a compact space* X, *and let* f *be a mapping of* A *into a complete Hausdorff uniform space* X′. *Then* f *can be extended by continuity to the whole of* X *if and only if* f *is uniformly continuous.*

2. COMPACTNESS OF UNIFORM SPACES

DEFINITION 2. *A uniform space* X *is said to be precompact if its Hausdorff completion* $\hat{X}$ *is compact. A subset* A *of a uniform space* X *is said to be a precompact subset if the uniform subspace* A *of* X *is precompact.*

Thus a subset A of a uniform space X is precompact if and only if the closure of $i(A)$ in $\hat{X}$ is compact ($i : X \to \hat{X}$ being the canonical map) (§ 3, no. 9, Proposition 18, Corollary 1).

Example. In any uniform space X, the set of points of a *Cauchy sequence* (x_n) is precompact. Since the images of the x_n in $\hat{X}$ again form a Cauchy sequence, we can assume that X is Hausdorff; the closure in $\hat{X}$ of the set of points x_n then consists of the points x_n and $\lim_{n\to\infty} x_n$ and is therefore compact (Chapter I, § 9, no. 3, Example 2).

THEOREM 3. *A uniform space* X *is precompact if and only if, for each entourage* V *of* X, *there is a* finite *covering of* X *by* V-*small sets.*

We may express this condition more intuitively by saying that X can be covered by a finite number of sets which are *as small as we please.*

Let $i : X \to \hat{X}$ be the canonical mapping, then the entourages of X are the inverse images under $i \times i$ of the entourages of $\hat{X}$ (§ 3, no. 7, Proposition 12). Suppose X is precompact, and let U be any entourage of $\hat{X}$; then there is a symmetric entourage U′ of $\hat{X}$ such that $\overset{2}{U'} \subset U$. Since $\hat{X}$ is compact, there exist a finite number of points $x_j \in \hat{X}$ such that the $U'(x_j)$ (which are U-small) cover $\hat{X}$. If V is the inverse image of U by $i \times i$, the sets $\overset{-1}{i}(U'(x_j))$ are V-small and cover X. Conversely, suppose that for each entourage V of X there is a finite covering of X by V-small sets. We have to show that every ultrafilter $\mathfrak{F}$ on $\hat{X}$ is convergent; since $\hat{X}$ is complete, it is enough to show that $\mathfrak{F}$ is a *Cauchy filter*, i.e. that for each *closed* entourage U of $\hat{X}$ there is a U-small set in $\mathfrak{F}$ (§ 1, no. 2, Proposition 2, Corollary 2). Let V be the inverse image of U under $i \times i$, and let (B_j) be a finite covering of X by V-small sets; then the sets $C_j = i(B_j)$ are U-small and cover $i(X)$, so that

$$\hat{X} = \bigcup_j \overline{C}_j.$$

On the other hand, since $C_j \times C_j \subset U$, and U is closed in $\hat{X} \times \hat{X}$, we have $\overline{C}_j \times \overline{C}_j \subset U$, so that the $\overline{C}_j$ are also U-small. Since $\mathfrak{F}$ is an ultrafilter, one of the $\overline{C}_j$ belongs to $\mathfrak{F}$ (Chapter I, § 6, no. 4, Corollary to Proposition 5).

Q.E.D.

Corollary. *A uniform space X is compact if and only if it is Hausdorff and complete and can be covered by a finite number of* V*-small sets, where* V *is any entourage of* X.

This follows from Theorem 3 and Theorem 1 of no. 1.

Remark 1. A non-Hausdorff *quasi-compact* space is not necessarily uniformizable, since it need not satisfy axiom (O_{III}) (cf. Chapter I, § 9, no. 2); for example most of the non-Hausdorff quasi-compact spaces which appear in algebraic geometry do not satisfy axiom (O_{III}) (cf. Exercise 2).

Proposition 1. *In a uniform space every subset of a precompact set, every finite union of precompact sets and the closure of every precompact set are precompact.*

The first two assertions are immediate consequences of Theorem 3. Let X be a uniform space, A a precompact subset of X, and let $i : X \to \hat{X}$ be the canonical mapping. $i(\overline{A})$ is contained in the closure of $i(A)$ in $\hat{X}$ (Chapter I, § 2, no. 1, Theorem 1), hence the closure of $i(\overline{A})$ in $\hat{X}$ is contained in a compact set and is therefore compact.

Remark 2. In a uniform space X a *relatively compact* set A is *precompact*, since A is contained in a compact set. On the other hand, even if X is Hausdorff, a precompact set *need not be relatively compact* in X, as is shown by the case where X itself is precompact but not compact.

PROPOSITION 2. *Let* $f: X \to Y$ *be a uniformly continuous mapping. If* A *is any precompact subset of* X, *then* $f(A)$ *is a precompact subset of* Y.

For if $i: X \to \hat{X}$ and $j: Y \to \hat{Y}$ are the canonical mappings, we have $j(f(A)) = \hat{f}(i(A))$ (§ 3, no. 7, Proposition 15) and hence $j(f(A))$ is relatively compact in $\hat{Y}$ (Chapter I, § 9, no. 4, Theorem 2, Corollary 1).

PROPOSITION 3. *Let* X *be a set, let* $(Y_\lambda)_{\lambda \in L}$ *be a family of uniform spaces, and for each* $\lambda \in L$ *let* f_λ *be a mapping of* X *into* Y_λ. *Let* X *carry the coarsest uniformity for which the* f_λ *are uniformly continuous. Then a subset* A *of* X *is precompact if and only if* $f_\lambda(A)$ *is a precompact subset of* Y_λ *for each* $\lambda \in L$.

The condition is necessary by virtue of Proposition 2. Sufficiency follows from the characterization of the Hausdorff completion of X given in § 3, no. 9, Proposition 18, and Tychonoff's theorem (Chapter I, § 9, Theorem 3, Corollary).

3. COMPACT SETS IN A UNIFORM SPACE

The following proposition for an arbitrary uniform space is a sharper form of Proposition 2 of Chapter I, § 9, no. 2 for compact spaces.

PROPOSITION 4. *In a uniform space* X, *let* A *be a compact set and* B *a closed set such that* $A \cap B = \varnothing$. *Then there is an entourage* V *of* X *such that* $V(A)$ *and* $V(B)$ *do not intersect.*

If the proposition were false, then none of the sets $A \cap \overset{2}{V}(B)$, where V runs through the set of symmetric entourages of X, would be empty; hence these sets would form a filter base on A, which would have a cluster point $x_0 \in A$. Hence for each symmetric entourage V of X, $\overset{3}{V}(x_0)$ would meet B and therefore, as B is closed, we should have $x_0 \in B$, contrary to hypothesis.

COROLLARY. *Let* A *be a compact set in a uniform space* X; *then as* V *runs through the set of entourages of* X, *the sets* $V(A)$ *form a fundamental system of neighbourhoods of* A.

Let U be any open neighbourhood of A, then $B = \complement U$ is closed and does not meet A; hence, by Proposition 4, there is an entourage V such that $V(A) \cap V(B) = \varnothing$, and therefore $V(A) \subset U$.

4. CONNECTED SETS IN A COMPACT SPACE

DEFINITION 3. *Let* V *be a symmetric entourage of a uniform space* X. *A finite sequence* $(x_i)_{0 \leqslant i \leqslant n}$ *of points of* X *is said to be a* V-*chain if* x_i *and* x_{i+1} *are* V-*close for* $0 \leqslant i < n$. *The points* x_0 *and* x_n *are called the ends of the* V-*chain, and they are said to be joined by the* V-*chain.*

Given a symmetric entourage V, the relation "there is a V-chain joining x and y" is an equivalence relation between x and y in X. Let $A_{x,V}$ be the equivalence class of x for this relation, i.e. the set of all $y \in X$ which can be joined to x by a V-chain. Clearly if $y \in A_{x,V}$ then $V(y) \subset A_{x,V}$; hence $A_{x,V}$ is open in X; and the complement of $A_{x,V}$, being a union of equivalence classes, is also open. Hence:

PROPOSITION 5. *In a uniform space* X, *the set* $A_{x,V}$ *of points which can be joined by a* V-*chain to a given point* x *is both open and closed in* X.

For each $x \in X$, let A_x denote the intersection of the sets $A_{x,V}$ as V runs through the set of symmetric entourages of X; A_x is the equivalence class of x for the equivalence relation "for all symmetric entourages V, there is a V-chain joining x and y".

PROPOSITION 6. *In a compact space* X, *the component of* x, *the set* A_x, *and the intersection of the neighbourhoods of* x *which are both open and closed, are all three identical.*

It is enough to show that A_x is *connected:* for in any uniform space X the component of x is contained in the intersection of the neighbourhoods of x which are both open and closed, and this intersection is contained in A_x by Proposition 5.

Suppose A_x is not connected. Since A_x is *closed*, there are two non-empty disjoint closed sets B and C such that $B \cup C = A_x$. By Proposition 4 of § 3 there is an entourage U of X such that $U(B) \cap U(C)$ is empty.

Let W be an *open* entourage such that $\overset{2}{W} \subset U$, and let H be the *closed* set which is the complement of $W(B) \cup W(C)$ in X. Suppose for example that $x \in B$, and let y be a point of C. Then for each symmetric entourage $V \subset W$ one sees immediately, by induction on i, that every V-chain $(x_i)_{0 \leqslant i \leqslant n}$ joining x and y in X must have a point in H, by the choice of W. Since by hypothesis x and y can be joined by a V-chain for each symmetric entourage V, we see that $H \cap A_{x,V}$ is not empty if $V \subset W$. On the other hand, if $V' \subset V$ then clearly $A_{x,V'} \subset A_{x,V}$; thus it follows that, as V runs through the set of symmetric entourages of X, the sets $H \cap A_{x,V}$ form a filter base of *closed* sets in the compact space H. Hence all these sets have a common point and therefore H meets A_x; but this contradicts the definition of H.

COROLLARY. *Let* X *be a locally compact space and let* K *be a* compact *component of* X. *Then the neighbourhoods of* K *which are both open and closed form a fundamental system of neighbourhoods of* K.

Let V be a relatively compact open neighbourhood of K in X (Chapter I, § 9, no. 7, Proposition 10) and let F be its frontier. Let $U \subset \overline{V}$ be a set which is both open and closed *with respect to* $\overline{V}$. Then U is closed in X, and if in addition U does not meet F, then U is open in X (for then $U \subset V$ and U is open in V). Hence it is enough to show that there is a subset of $\overline{V}$ which contains K, does not meet F and is both open and closed with respect to $\overline{V}$.

Suppose this is not the case: then the intersections of F with the subsets of $\overline{V}$ which contain K and are open and closed in $\overline{V}$ form a filter base of closed sets in F; F is compact, so these sets have a common point $y \in F$. But this is absurd; since $\overline{V}$ is compact, K is a component of $\overline{V}$, and by virtue of Proposition 6, the intersection of the sets which contain K and are both open and closed in $\overline{V}$ is just K. The Corollary is thus proved.

PROPOSITION 7. *Let* X *be a compact space and let* R *be the equivalence relation on* X *whose classes are the components of* X. *Then the quotient space* X/R *is compact and totally disconnected.*

We know from Chapter I, § II, no. 5, Proposition 9 that X/R is totally disconnected; hence we have to show that X/R is *Hausdorff* (Chapter I, § 10, no. 4, Proposition 8.) Let A and B be two distinct components of X. By Proposition 6 there is a symmetric entourage U of X such that no point of A can be joined to any point of B by a U-chain. The set V (resp. W) of points of X which can be joined to a point $x \in A$ (resp. $y \in B$) by a U-chain is both open and closed in X (Proposition 5) and contains A (resp. B); these sets are therefore open neighbourhoods of A and B respectively, are saturated with respect to R and do not intersect. This completes the proof.

EXERCISES

§ 1

1) Let X be an infinite set and let $\mathfrak{F}$ be an ultrafilter on X such that the sets of $\mathfrak{F}$ have empty intersection. For each $A \in \mathfrak{F}$, let V_A denote the subset $\Delta \cup (A \times A)$ of $X \times X$. Show that, as A runs through $\mathfrak{F}$, the sets V_A form a fundamental system of entourages for a uniformity $\mathfrak{U}(\mathfrak{F})$ on X, and that the topology induced by this uniformity is discrete.

*2) On the real line **R**, carrying the additive uniformity, show that as V runs through the filter of entourages of **R** the sets $V(\mathbf{Z})$ do not form a fundamental system of neighbourhoods of the set **Z** of rational integers.* (Cf. § 4, no. 3, Corollary to Proposition 4).

*3) Let V be that entourage of the additive uniformity on **R** which consists of all pairs (x, y) such that either $|x - y| \leqslant 1$ or $xy \geqslant 1$. Show that V is closed in $\mathbf{R} \times \mathbf{R}$, but that if A denotes the (closed) subset of **R** consisting of all integers $n \geqslant 2$, then $V(A)$ is not closed in **R**.*

4) Show that if a uniform space is a Kolmogoroff space (Chapter I, § 1, Exercise 2) then it is Hausdorff.

¶ 5) *a*) Let X be a uniform space, $\mathfrak{U}$ its uniformity; for each entourage V of X, let $\tilde{V}$ be the subset of $\mathfrak{P}(X) \times \mathfrak{P}(X)$ consisting of all pairs (M, N) of subsets of X such that *both* $M \subset V(N)$ and $N \subset V(M)$. Show that the sets $\tilde{V}$ form a fundamental system of entourages of a uniformity $\tilde{\mathfrak{U}}$ on $\mathfrak{P}(X)$.

b) On the set $\mathfrak{P}_0(X)$ of non-empty subsets of X, show that the topology induced by the topology $\mathfrak{T}(\tilde{\mathfrak{U}})$ induced by $\tilde{\mathfrak{U}}$ is finer than the topology $\mathfrak{T}_\Phi$ (Chapter I, § 2, Exercise 7).

c) If X has at least two points and if $\mathfrak{U}$ is Hausdorff, show that the topology induced by $\mathfrak{T}(\tilde{\mathfrak{U}})$ on $\mathfrak{P}_0(X)$ is never coarser than $\mathfrak{T}_\Omega$. The topology induced by $\mathfrak{T}(\tilde{\mathfrak{U}})$ on the set $\mathfrak{F}(X)$ of non-empty closed subsets of X is finer than the topology induced by $\mathfrak{T}_\Omega$ on this set if and only if, for each closed subset A of X, the sets V(A) form a fundamental system of neighbourhoods of A as V runs through the set of entourages of X.

**d*) Show that on the quotient set X/R defined in Chapter I, § 11, Exercise 18, the quotient topology, and the topologies defined by $\mathfrak{T}_\Omega$, $\mathfrak{T}_\Phi$ and $\mathfrak{T}(\tilde{\mathfrak{U}})$ (where $\mathfrak{U}$ is the usual uniformity on $\mathbf{R}^2$) are all four distinct.$_*$

§ 2

*1) Show that on the real line **R** (with the additive uniformity) the function $|x|$ is uniformly continuous; the function $1/x$ is uniformly continuous in every interval $[a, +\infty[$ where $a > 0$; and that this function is continuous, but not uniformly continuous, in the interval $]0, +\infty[$.$_*$

2) Show that the topologies induced on **Z** by the p-adic uniformities are not comparable for different primes p.

3) Let $\mathfrak{F}_1$, $\mathfrak{F}_2$ be two distinct ultrafilters on an infinite set X. Show that the uniformities $\mathfrak{U}(\mathfrak{F}_1)$ and $\mathfrak{U}(\mathfrak{F}_2)$ (§ 1, Exercise 1) are not comparable, and that their least upper bound is the discrete uniformity. What is their greatest lower bound? Deduce that the set of Hausdorff uniformities on X is equipotent to $\mathfrak{P}(\mathfrak{P}(X))$ [cf. Chapter I, § 8, Exercise 6 *a*)].

4) *a*) Let $\mathfrak{U}$ and $\mathfrak{U}'$ be the filters of entourages of two uniformities on the same non-empty set X. Show that the intersection of the filters $\mathfrak{U}$ and $\mathfrak{U}'$ is the filter of entourages of a uniformity on X if and only if, for each $V \in \mathfrak{U}$ and $V' \in \mathfrak{U}'$, there exist $W \in \mathfrak{U}$ and $W' \in \mathfrak{U}'$ such that $WW' \subset V \cup V'$.

b) Give an example of two filters of entourages $\mathfrak{U}_{\varpi}$, $\mathfrak{U}_{\varpi'}$, each defined by a finite partition of X (no. 5, Example), which do not satisfy the condition of *a*).

5) Let $(X_\iota)_{\iota \in I}$ be a family of uniform spaces and let $\mathfrak{c}$ be an infinite cardinal. For each family $(V_\iota)_{\iota \in H}$, where $\operatorname{card}(H) < \mathfrak{c}$ and V_ι is an entourage of X_ι for each $\iota \in H$, let $U((V_\iota))$ be the set of pairs (x, y) of points of $X = \prod_{\iota \in I} X_\iota$ such that $(\mathrm{pr}_\iota x, \mathrm{pr}_\iota y) \in V_\iota$ for each $\iota \in H$. Show that the sets $U((V_\iota))$ form a fundamental system of entourages

of a uniformity on X, and that the topology induced by this uniformity is that defined in Exercise 9 of Chapter I, § 4.

6) Let X be a uniform space, $\mathfrak{U}$ its uniformity, $\tilde{\mathfrak{U}}$ the corresponding uniformity on $\mathfrak{P}(X)$ (§ 1, Exercise 5).

a) Show that the mapping $x \to \{x\}$ is an isomorphism of the uniform space X onto a subspace of the uniform space $\mathfrak{P}(X)$.

b) Show that the uniformity induced by $\tilde{\mathfrak{U}}$ on the set $\mathfrak{F}(X)$ of non-empty closed subsets of X is Hausdorff, and that $\tilde{\mathfrak{U}}$ is the inverse image under the mapping $M \to \overline{M}$ of the uniformity induced by $\tilde{\mathfrak{U}}$ on $\mathfrak{F}(X)$.

c) Show that the mapping $(M, N) \to M \cup N$ of $\mathfrak{P}(X) \times \mathfrak{P}(X)$ into $\mathfrak{P}(X)$ is uniformly continuous.

d) Let Y be another uniform space. If $f: X \to Y$ is a uniformly continuous mapping, show that the mapping $M \to f(M)$ of $\mathfrak{P}(X)$ into $\mathfrak{P}(Y)$ is uniformly continuous.

e) Let $\mathfrak{B}$ be a compact subset of the set $\mathfrak{F}(X)$ of non-empty closed subsets of X, $\mathfrak{F}(X)$ carrying the topology induced by the topology $\mathfrak{T}(\tilde{\mathfrak{U}})$ induced by $\tilde{\mathfrak{U}}$. Show that the set $\bigcup_{M \in \mathfrak{B}} M$ is closed in X.

§ 3

*1) Let $\mathfrak{U}$ denote the additive uniformity on the real line **R**, and let $\mathfrak{U}'$ denote the inverse image of $\mathfrak{U}$ under the mapping $x \to x^3$ of **R** onto itself. Show that $\mathfrak{U}'$ is strictly finer than $\mathfrak{U}$, but that the Cauchy filters are the same for both uniformities.$_*$

2) *a*) Let X be an infinite set and let $\mathfrak{F}$ be a non-trivial ultrafilter on X. Show that if X carries the uniformity $\mathfrak{U}(\mathfrak{F})$ (§ 1, Exercise 1) then the complement of X in its completion $\hat{X}$ consists of a single point, and that the topological space $\hat{X}$ can be identified with the space associated with the ultrafilter $\mathfrak{F}$ (Chapter I, § 6, no. 5, Example).

**b*) Deduce from *a*) an example of a uniformity on **R** which is finer than the additive uniformity and with respect to which **R** is not complete.$_*$

*3) *a*) Let $\mathfrak{U}$ denote the additive uniformity on the real line **R**, let $\mathfrak{U}_1$ denote the uniformity induced on **R** by that of the extended line $\overline{\mathbf{R}}$, and let $\mathfrak{U}_2$ denote the uniformity induced on **R** by that of the one-point compactification $\tilde{\mathbf{R}} = \mathbf{P}_1(\mathbf{R})$ of **R**. Show that $\mathfrak{U}$ is strictly finer than $\mathfrak{U}_1$, that $\mathfrak{U}_1$ is strictly finer than $\mathfrak{U}_2$, and that the topologies induced by these three uniformities are the same; also that the completions of **R** with respect to these three uniformities are respectively **R**, $\overline{\mathbf{R}}$

and $\tilde{\mathbf{R}}$. The identity mapping of $\mathbf{R}$ extends by continuity to an injective but not surjective mapping $\mathbf{R} \to \overline{\mathbf{R}}$ and to a surjective but not injective mapping $\overline{\mathbf{R}} \to \tilde{\mathbf{R}}$.

b) Deduce from *a*) an example of two Hausdorff uniform spaces X, Y and a uniformly continuous bijection $u : X \to Y$ whose extension by continuity $\bar{u} : \hat{X} \to \hat{Y}$ is neither injective nor surjective.$_*$

¶ 4) With the notation of Exercise 5 of § 2, suppose that all the uniform spaces X_ι are complete. Show that the space X is complete with respect to the uniformity there defined (argue by *reductio ad absurdum*, using § 2, Proposition 5, Corollary 2).

5) Let X be a Hausdorff uniform space such that X is the intersection of a countable family of open sets of $\hat{X}$. Show that if Y is any Hausdorff uniform space which contains X as a uniform subspace, then X is the intersection of a closed subset of Y and a countable family of open sets of Y.

6) Let X be a uniform space and let $i : X \to \hat{X}$ be the canonical mapping of X into its Hausdorff completion. Let X_0 be the sum of X and $\hat{X} - i(X)$, and let $j : X_0 \to \hat{X}$ be the mapping which agrees with i on X and with the identity mapping on $\hat{X} - i(X)$. Show that X_0 is complete with respect to the uniformity which is the inverse image under j of the uniformity of $\hat{X}$, also that if Y is any complete uniform space such that X is a uniform subspace of Y, then the identity mapping $X \to X$ can be extended uniquely to a continuous mapping $X_0 \to Y$.

¶ 7) Let X be a Hausdorff uniform space, $\mathfrak{U}$ its uniformity, $\mathfrak{U}_0$ the uniformity induced on $\mathfrak{F}(X)$ by the uniformity $\tilde{\mathfrak{U}}$ (§ 2, Exercise 6 *b*), $\mathfrak{U}_{00}$ the uniformity induced on $\mathfrak{F}(\mathfrak{F}(X))$ by the uniformity $\tilde{\mathfrak{U}}_0$. Show that the canonical mapping $x \to \{\{x\}\}$ of X into $\mathfrak{F}(\mathfrak{F}(X))$ extends to an isomorphism of $\hat{X}$ onto a closed uniform subspace of $\mathfrak{F}(\mathfrak{F}(X))$ (carrying the uniformity $\mathfrak{U}_{00}$) (use Proposition 9 of no. 4).

§ 4

¶ 1) Let $\mathfrak{R}$ be an open covering of a compact space X. Show that there is an entourage V of the uniformity of X such that, for each $x \in X$, $V(x)$ is contained in a set belonging to $\mathfrak{R}$ [remark that for each $x \in X$ there is an entourage W_x such that $\overset{2}{W}_x(x)$ is contained in a set of $\mathfrak{R}$, and cover X by a finite number of sets $W_x(x)$].

2) Prove that a quasi-compact space X is uniformizable if and only if its topology is the inverse image of the topology of a compact space Y under a surjective mapping $f : X \to Y$.

3) Let f be a continuous mapping of a compact space X into a compact space Y, and let V be an open entourage of X such that the relation $f(x) = f(y)$ implies $(x, y) \in V$. Show that there is an entourage W of Y such that the relation $(f(x), f(y)) \in W$ implies $(x, y) \in V$.

4) Let X be the locally compact space defined in Chapter I, § 9, Exercise 11. Show that every neighbourhood of the diagonal Δ in $X \times X$ contains a set of the form $[x, b[\times [x, b[$. Deduce that there is only one uniformity on X compatible with the topology of X, and that the completion of X for this uniformity can be identified with the interval $X' = [a, b]$ carrying the topology $\mathfrak{T}v(X)$. (Observe that every ultrafilter on X must be a Cauchy filter with respect to any uniformity compatible with the topology of X).

5) Show that a uniform space X is precompact if and only if every ultrafilter on X is a Cauchy filter. (Remark that if X is not precompact then there is an entourage V of X such that the sets $\complement V(x)$, as x runs through X, generate a filter on X).

6) Let X be a uniform space in which every sequence of points has at least one cluster point. Show that X is precompact (use *reductio ad absurdum*).

7) A subset A of a uniform space X is said to be *bounded* if for each entourage V of X there is a finite set F and an integer $n > 0$ such that $A \subset \overset{n}{V}(F)$.

a) The union of two bounded subsets is bounded. The closure of a bounded subset is bounded. Every precompact subset is bounded.

b) If $f : X \to Y$ is uniformly continuous, then the image under f of every bounded subset of X is a bounded subset of Y.

c) In a product of non-empty uniform spaces, a set is bounded if and only if each of its projections is bounded.

d) For each symmetric entourage V of X, let R_V denote the union of the sets $\overset{n}{V}$ as n runs through the integers > 0. Show that R_V is both open and closed in $X \times X$ and that, in the notation of no. 4,

$$R_V(x) = A_{x, V}$$

for each $x \in X$. Suppose that $R_V = X \times X$ for each symmetric entourage V of X (this will be the case in particular if X is connected); show that a set $A \subset X$ is bounded if and only if, for each symmetric entourage V of X and each $x_0 \in X$, there is an integer $n > 0$ such that $A \subset \overset{n}{V}(x_0)$.

8) Let X be a uniform space, V a closed entourage of X, A a compact subset of X. Show that $V(A)$ is closed in X (cf. Chapter I, § 10, no. 2, Theorem 1, Corollary 5).

¶ 9) Let X be a locally compact space which carries a uniformity compatible with its topology and is such that there is an entourage V with the property that $V(x)$ is relatively compact in X for all $x \in X$.

a) Show that X is complete with respect to this uniformity.

b) Let U be a symmetric entourage of X such that $\overset{2}{U} \subset V$. Show that if A is relatively compact in X, then so is $U(A)$.

c) Show that X is paracompact [use *b*), Exercise 7 and Chapter I, § 9, no. 10, Theorem 5].

d) Let W be a closed symmetric entourage of X such that $\overset{2}{W} \subset V$. Show that if A is closed in X then so is $W(A)$ (use Exercise 8).

10) Let X be a Hausdorff uniform space which has an entourage V such that $V(x)$ is precompact for each $x \in X$. Show that the completion $\hat{X}$ is locally compact and satisfies the conditions of Exercise 9. [If W is a symmetric open entourage of X such that $\overset{3}{W} \subset V$, show that the closure $\overline{W}$ of W in $\hat{X} \times \hat{X}$ is such that $\overline{W}(x)$ is compact for each $x \in \hat{X}$].

¶ 11) Let X be a Hausdorff uniform space, $\mathfrak{U}$ its uniformity, $\mathfrak{F}(X)$ the set of all non-empty closed subsets of X. Let $\mathfrak{F}(X)$ carry the uniformity induced by the uniformity $\tilde{\mathfrak{U}}$ defined in § 1, Exercise 5 *a*).

a) Show that if X' is precompact then so is $\mathfrak{F}(X)$. [If (A_i) is a finite covering of X by V-small sets (V symmetric), show that the sets $\tilde{V}(B_k)$ cover $\mathfrak{F}(X)$, where the B_k are all the unions of sets A_i].

b) A point $A \in \mathfrak{F}(X)$ is such that every neighbourhood of A in the topology induced by $\mathfrak{T}(\tilde{\mathfrak{U}})$ on $\mathfrak{F}(X)$ contains a neighbourhood of A in the topology induced by $\mathfrak{T}_\Theta$ (Chapter I, § 8, Exercise 12) on $\mathfrak{F}(X)$, if and only if A is precompact.

c) Show that the topologies $\mathfrak{T}(\tilde{\mathfrak{U}})$ and $\mathfrak{T}_\Theta$ induce the same topology on the set $\mathfrak{K}(X)$ of compact subsets of X [use *b*) and Exercise 5 of § 1].

¶ 12) *a*) Let R be a Boolean algebra, and identify R with a Boolean algebra of subsets of the set Ω of maximal prefilters on R (Chapter I, § 6, Exercise 20). For each finite partition $\varpi = (A_i)$ of Ω, formed of sets belonging to R, let C_{ϖ} denote the subset $\bigcup_i (A_i \times A_i)$ of $\Omega \times \Omega$. Show that the sets C_{ϖ} form a fundamental system of entourages of a Hausdorff uniformity on Ω. The completion, $\hat{\Omega}$, of Ω with respect to this uniformity is a totally disconnected compact space. A subset of $\hat{\Omega}$ is both open and closed in $\hat{\Omega}$ if and only if it is of the form $\overline{A}$, where $A \in R$. Show that $A \to \overline{A}$ is a bijective mapping of R onto the set of subsets of $\hat{\Omega}$ which are both open and closed, and that $\overline{\complement A} = \complement\overline{A}$, $\overline{A \cup B} = \overline{A} \cup \overline{B}$ and $\overline{A \cap B} = \overline{A} \cap \overline{B}$. Deduce that the canonical mapping $\Omega \to \hat{\Omega}$ is bijective.

b) Let X be a Hausdorff topological space in which the sets that are both open and closed form a base for the topology. Take R to be the Boolean algebra formed by subsets of X which are both open and closed; then the topology induced on X by the topology of the completion $\hat{X}$ of X with respect to the uniformity $\mathfrak{U}$ defined in *a*) is the given topology on X. Further, the maximal elements of the set of filters on X having a base formed of open sets, and the maximal elements of the set of filters on X having a base formed of closed sets are Cauchy filters with respect to the uniformity $\mathfrak{U}$. Deduce that there are continuous surjections $\varphi : \tilde{X} \to \hat{X}$, $\psi : X' \to \hat{X}$, where $\tilde{X}$, X' are the spaces defined in Chapter I, § 9, Exercises 25 and 24. * Show that if X is the rational line **Q**, then ψ is not bijective.* If X is extremally disconnected (Chapter I, § 11, Exercise 21), then ψ is bijective, $\hat{X}$ is extremally disconnected and can be identified with the semi-regular space associated with X' (Chapter I, § 8, Exercise 20); thus extremally disconnected spaces can be characterized as dense subspaces of extremally disconnected compact spaces. Hence give an example of an extremally disconnected compact space which has no isolated points [(cf. Chapter I, § 11, Exercise 21 *f*)].

In particular, if we take X to be a discrete space, then $\hat{X}$ can be identified with the ultrafilter space of X (Chapter I, § 9, Exercise 26).

13) Let X be a locally compact space and U an open subset of X. Let B be the union of U and the relatively compact components of $\complement U$. Show that B is open and is the union of U and those subsets of $\complement U$ which are both open and compact in $\complement U$. (Embed X in its one-point compactification $X' = X \cup \{\omega\}$, observe that in $X' - U$ the component of ω is $\{\omega\} \cup \complement B$, and use Proposition 6 of no. 4).

14) Let X be a compact space, and let $\mathfrak{B}$ be a filter base on X consisting of connected closed sets. Show that the (non-empty) intersection of the sets of $\mathfrak{B}$ is closed and connected (argue by *reductio ad absurdum*, using Proposition 4 of no. 3 and Chapter I, § 9, no. 1, Theorem 1).

15) Let X be a compact space. Then the space $\mathfrak{F}(X)$ of non-empty closed subsets of X, with the topology induced by $\mathfrak{T}(\tilde{\mathfrak{U}})$ (Exercise 11), is compact (Chapter I, § 9, Exercise 13). Consider an ultrafilter Ψ on $\mathfrak{F}(X)$, and suppose that for each entourage V of the uniformity of X,

and each $\mathfrak{X} \in \Psi'$, there exists $M \in \mathfrak{X}$ such that any two points of M can be joined by a V-chain contained in M. Show that the limit point A of Ψ' in $\mathfrak{F}(X)$ is a connected subset of X (use Proposition 6 of no. 4).

¶ 16) Let X be a connected compact space and let A, B be two disjoint non-empty closed subsets of X. Show that there is a component of $\complement(A \cup B)$ whose closure meets both A and B. (Let U be an entourage of the uniformity of X such that $\overset{2}{U}(A) \cap \overset{2}{U}(B) = \emptyset$. Show first that the set $\mathfrak{M}_U$ of components of $\complement(\overline{U(A)} \cup \overline{U(B)})$ which meet both $\overline{U(A)}$ and $\overline{U(B)}$ is not empty, by proving that for each symmetric entourage $W \subset U$ there exist $x \in U(A)$ and $y \in U(B)$ and a W-chain which joins x to y and all of whose points other than x and y belong to $\complement(U(A) \cup U(B))$, and then applying Exercise 15. For each entourage $V \subset U$, show that every $K \in \mathfrak{M}_V$ contains a set $H \in \mathfrak{M}_U$, and that the set $\mathfrak{N}_{U,V}$ of sets $H \in \mathfrak{M}_U$ which are contained in a set $K \in \mathfrak{M}_V$ is closed in $\mathfrak{F}(X)$; and conclude that as V runs through the set of entourages contained in U, the intersection of the sets $\mathfrak{N}_{U,V}$ is not empty).

17) Let X be a connected locally compact space. Let K be any non-empty compact subset of X distinct from X. Show that every component of K meets the frontier of K in X (use the Corollary to Proposition 6 of no. 4). Deduce that if A is any non-empty relatively compact open subset of X distinct from X, then the closure of every component of A meets $\complement A$.

¶ 18) *a*) Show that a compact connected space X cannot be the union of a countable infinity of mutually disjoint non-empty closed sets [by *reductio ad absurdum:* if (F_n) is a countably infinite partition of X into closed sets, show with the help of Exercise 17 that there is a compact connected set K which does not meet F_1 but meets an infinite number of sets F_n; in order to do this consider the component of a point of F_2 with respect to a compact neighbourhood of F_2 not meeting F_1. Then use induction].

b) Extend the result of *a*) to a locally compact connected space X under one or the other of the following two supplementary conditions: one of the F_n is compact and connected, or X is locally connected. [In the first case, reduce to *a*) by means of Exercise 17].

**c*) Let X be the subspace of $\mathbf{R}^3$ which is the union of the following subspaces: A_n is the half-line $x > 0$, $y = 1/n$, $z = 0$ $(n \geqslant 1)$; B_n is the interval $2n < x < 2n + 2$, $y = 0$, $z = 0$ $(n \geqslant 0)$; C_n is the set defined by the relations

$$x = 2n + 1, \qquad 0 \leqslant y \leqslant \frac{1}{n+1}, \qquad z = y\left(y - \frac{1}{n+1}\right),$$

where $n \geqslant 0$.

Show that X is connected and locally compact and is the union of a countable infinite family of mutually disjoint connected non-empty closed sets. *

¶ 19) A compact connected space X is said to be *irreducible* between two of its points x, y if there is no compact connected subset of X other than X itself which contains both x and y.

a) If x and y are two distinct points of a compact connected space X, show that X has a compact connected subspace K which is irreducible between x and y (use Exercise 14 and Zorn's Lemma).

b) Show that a compact connected space having at least two points cannot be irreducible between every pair of its points (use Exercise 17).

c) Let X be a compact connected space, and suppose that there is a point $a \in X$ which has a connected closed neighbourhood $V \neq X$. If x and y are any two points of $\complement V$, show that there is a compact connected subset of X which contains x and one of the points a, y but not the other (use Exercise 17). Under the same hypotheses, show that if x, y, z are any three distinct points of X, then X cannot be simultaneously irreducible between x and y, between y and z, and between x and z.

¶ 20) Let X be a compact connected space and let L be the set of all points of X which have a fundamental system of connected neighbourhoods. The points of $S = \complement L$ are called the *singular* points of X. The points of the interior $\mathring{L}$ of L and the components of $\overline{S}$ are called the *prime constituents* of X.

a) Let $A \neq X$ be a non-empty open subset of X, let K be a component of $\overline{A}$ and let F be the closure of $\overline{A} \cap \complement K$. Show that if $A \cap K \cap F$ is not empty then all its points are singular. If $x \in A \cap F \cap K$, the component Q of x in F meets $\complement A$ [consider a point of $\complement K$ contained in $V(x)$ and its component in $\overline{A}$, and show by using Exercise 17 that the set of points of F which can be joined to x by a V-chain in F meets $\complement A$]. Deduce that the prime constituent P which contains x meets $\complement A$ (remark that P contains the component of x in $Q \cap A$, and use Exercise 17 applied to the compact connected space Q and the open subset $Q \cap A$ of Q).

b) Deduce from *a*) that if P is a prime constituent of X and U is a neighbourhood of P, then P has a connected neighbourhood contained in U (by *reductio ad absurdum*).

c) Let V be a symmetric entourage of X, and let V' denote the set of pairs (x, y) of points of X which can be joined by a V-chain, all of whose points are singular, except perhaps for x and y. As V runs through the symmetric entourages of X, the corresponding sets V′ form a fundamental system of entourages of a uniformity (in general not Hausdorff) on X. Let X′ be the associated Hausdorff space, and show that the inverse images of the points of X′ under the canonical mapping $X \to X'$ are the prime constituents of X. The (compact connected) space X′ is called the *space of prime constituents* of X. Show that X′ is *locally connected* [use *b*) and Chapter I, § 10, no. 4, Proposition 8].

**d*) Let (r_n) be the sequence of all the rational numbers contained in $]0,1[$, arranged in some order. For each irrational $x \in [0,1]$, let

$$f(x) = \sum_n 2^{-n} \sin 1/(x - r_n).$$

Let X be the closure in $\mathbf{R}^2$ of the graph of *f*. Show that X is connected and irreducible between its points with abscissas 0 and 1, but that X has only one prime constituent.*

*21) In the locally compact space X defined in Exercise 20 *b*) of Chapter I, § 10, let S be the equivalence relation whose classes are the components of X. Show that the quotient space X/S is not Hausdorff.*

22) Show that every totally disconnected compact space X is homeomorphic to the inverse limit of an inverse system of finite discrete spaces. (Consider the finite partitions of X into open sets and use the Corollary to Proposition 6 of no. 4).

¶ 23) Let $f: X \to Y$ be a continuous open mapping, X locally compact and Y Hausdorff.

a) If K is any compact connected subset of Y and C is any compact component of $\overset{-1}{f}(K)$, show that $f(C) = K$ (reduce to the case $K = Y$ and use the Corollary to Proposition 6 of no. 4). *Give an example of a non-compact component C′ of $\overset{-1}{f}(K)$ such that $f(C') \neq K$.*

b) Suppose in addition that Y is locally compact and locally connected. Show that if U is any connected open subset of Y and R is any relatively compact component of $\overset{-1}{f}(U)$ in X, then $f(R) = U$. [Observe that in a locally compact locally connected space, a connected open set U is the union of the compact connected sets contained in U and containing a given point $y_0 \in U$; then apply *a*).]

c) Suppose in addition that X is locally connected. Show that if K is any connected subset of Y which is either open or compact with a non-empty interior, and if H is any compact subset of X, then there is only a finite number of components of $\overset{-1}{f}(K)$ contained in H. [Use *a*) and *b*)].

HISTORICAL NOTE

(Numbers in brackets refer to the bibliography at the end of this note.)

The principal concepts and results on uniform spaces have emerged gradually from the theory of real variables, and it is only in recent years that they have been systematically studied. Cauchy, in his attempt to put the theory of series on a rigorous basis (cf. the Historical Notes to Chapters I and IV) took as starting point a principle which he seems to have thought self-evident, namely that a necessary and sufficient condition for the convergence of a sequence (a_n) is that $|a_{n+p} - a_n|$ is as small as we please whenever n is sufficiently large (see for example [2]). Together with Bolzano [1] he was undoubtedly one of the first to state this principle explicitly and to recognize its importance; hence the name of "Cauchy sequences" given to sequences of real numbers which satisfy the condition in question, and by extension to sequences (x_n) of points in a metric space (Chapter IX) such that the distance from x_{n+p} to x_n is as small as we please whenever n is large enough; and hence finally the name of "Cauchy filter" given to the generalization of Cauchy sequences which has been studied in this chapter.

When later the intuitive notion of real number ceased to be regarded as adequate, and attempts were made to define real numbers in terms of rational numbers in order to provide a solid basis for Analysis, it was Cauchy's principle which provided the most fertile of the definitions put forward in the second half of the nineteenth century. This definition, due to Cantor [3] (and developed along the lines of Cantor's ideas by Heine [5] and independently by Méray, amongst others) consists in making a real number correspond to every Cauchy sequence of rational numbers ("fundamental sequence" in Cantor's terminology); the same real number corresponds to two Cauchy sequences (a_n) and (b_n) of rational numbers if and only if $|a_n - b_n|$ tends to zero. The essential idea here is that, from a certain point of view (in fact, the point of view of the "uniform structure "defined in this chapter, § 1, no. 1, Example 1), the set **Q** of rational numbers is "incomplete", and that the set of real numbers is the "complete" set which one gets from **Q** by "completion".

On the other hand, Heine was the first to define uniform continuity for real-valued functions of one or more real variables, in work largely inspired by the ideas of Weierstrass and Cantor [4], and he proved that every real-valued function which is continuous on a bounded closed interval of **R** is uniformly continuous on this interval: this is "Heine's theorem". By Theorem 2 of § 4 this result is related to the compactness of a bounded closed interval of **R** ("Borel-Lebesgue theorem", Chapter IV, § 2, Theorem 2; cf. the Historical Notes to Chapters I and IV), and Heine's proof of his theorem can also serve, with some modifications, to prove the Borel-Lebesgue theorem (and this has appeared as sufficient reason to some authors for calling the latter theorem the "Heine-Borel theorem").

These ideas became extended to more general spaces when metric spaces came to be studied, first in particular cases and then in general (cf. Chapter IX); in a metric space a distance is given (i.e. a real-valued function of pairs of points, satisfying certain axioms) which defines both a topology and a uniformity. Fréchet, who was the first to give a general definition of these spaces, recognized the importance of Cauchy's principle [6] and also proved for metric spaces a theorem equivalent to Theorem 3 of § 4 ([6] and [7]). Hausdorff developed considerably the theory of metric spaces in his "Mengenlehre" ([8]; cf. also [8 *a*]) and in particular realized that one can apply Cantor's construction, described earlier, to these spaces and hence obtain a "complete" metric space from a "non-complete" metric space (i.e. one in which Cauchy's principle is not valid).

Metric spaces are a particular type of "uniform spaces"; the latter were defined in full generality by A. Weil at a fairly recent date [9]. Previous to this, the notions and results on "uniform structure" could be used only in connection with metric spaces, and this fact explains the important part played by metric spaces and metrizable spaces (and in particular by compact metrizable spaces) in many modern works on topology, in questions where distance has no real usefulness. Once one has the definition of a uniform space, there is no difficulty (especially if one also has the notion of a filter at one's disposal) in extending to these spaces almost the whole of the theory of metric spaces as given e.g. by Hausdorff (and similarly in extending, for example, to arbitrary compact spaces the results on compact metric spaces given in Alexandroff-Hopf's *Topologie* [10]). This is what we have done in this chapter; in particular, the theorem on completion of uniform spaces (§ 3, Theorem 3) is no more than a transposition, without any essential modifications, of Cantor's construction of the real numbers.

BIBLIOGRAPHY

[1] B. Bolzano, *Rein analytischer Beweis des Lehrsatzes, dass zwischen je zwei Werthen, die ein entgegengesetztes Resultat gewähren, wenigstens eine reelle Wurzel liegt*, Ostwald's Klassiker, no. 153, Leipzig, 1905.

[2] A.-L. Cauchy, Sur la convergence des séries [*Exercices d'Analyse*, 2e Année, Paris, 1827, p. 221 = *Oeuvres* (II), vol. 7, Paris (Gauthier-Villars) 1889, p. 267].

[3] G. Cantor, *Gesammelte Abhandlungen*, Berlin (Springer), 1932.

[4] E. Heine, Über trigonometrische Reihen, *Crelle's Journal*, **71** (1870), pp. 353-365.

[5] E. Heine, Die Elemente der Functionenlehre, *Crelle's Journal*, **74** (1872), pp. 172-188.

[6] M. Fréchet, Sur quelques points du calcul fonctionnel, *Rend. Palermo*, **22** (1906), pp. 1-74.

[7] M. Fréchet, Les ensembles abstraits et le calcul fonctionnel, *Rend. Palermo*, **30** (1910), pp. 1-26.

[8] F. Hausdorff, *Grundzüge der Mengenlehre*, Leipzig (Veit), 1914.

[8 *a*] F. Hausdorff, *Mengenlehre*, Berlin (de Gruyter), 1927.

[9] A. Weil, Sur les espaces à structure uniforme et sur la topologie générale, *Act. Scient. et Ind.*, no. 551, Paris (Hermann), 1937.

[10] P. Alexandroff and H. Hopf, *Topologie I*, Berlin (Springer), 1935.

CHAPTER III

Topological Groups

(Elementary Theory)

1. TOPOLOGIES ON GROUPS

1. TOPOLOGICAL GROUPS

In the first four sections of this chapter the law of composition of a group will generally be written *multiplicatively*, and e shall denote the identity element; translation of results into additive notation (which, we recall, is reserved exclusively to *commutative* groups) is usually left to the reader.

DEFINITION 1. *A topological group is a set* G *which carries a group structure and a topology and satisfies the following two axioms :*

(GT_I). *The mapping* $(x, y) \to xy$ *of* $G \times G$ *into* G *is continuous.*

(GT_{II}). *The mapping* $x \to x^{-1}$ *of* G *into* G (the *symmetry* of the group G) *is continuous.*

A group structure and a topology on a set G are said to be *compatible* if they satisfy (GT_I) and (GT_{II}).

Examples. 1) The discrete topology on a group G is compatible with the group structure. A topological group whose topology is discrete is called a *discrete* group.

Again, the *coarsest* topology (Chapter I, § 2, no. 2) on G is compatible with the group structure of G.

* 2) In Chapter IV we shall see that the topology of the rational line **Q** (resp. the real line **R**) is compatible with the additive group structure of **Q** (resp. **R**). *

3) If G is a topological group, its topology is compatible with the structure of the group G^0 which is the opposite of G; G^0, with this topology, is said to be the topological group *opposite* to G.

Axioms (GT_I) and (GT_{II}) are equivalent to the following:

(GT'). *The mapping* $(x, y) \to xy^{-1}$ *of* $G \times G$ *into* G *is continuous.*

Clearly (GT_I) and (GT_{II}) together imply (GT'). Conversely, (GT') implies (GT_{II}), for $x \to ex^{-1} = x^{-1}$ is then continuous; and (GT') and (GT_{II}) together imply (GT_I), for $(x, y) \to x(y^{-1})^{-1} = xy$ is then continuous.

If a is any element of G, the *left translation* $x \to ax$ (resp. the *right translation* $x \to xa$) is continuous, by (GT_I), and is therefore a *homeomorphism* of G onto G. The mappings $x \to axb$, as a and b run through G, thus form a *group of homeomorphisms* of G; the mappings $x \to axa^{-1}$ (resp. $x \to ax$, $x \to xa$), where a runs through G, form a subgroup of this group of homeomorphisms. Again, since the symmetry $x \to x^{-1}$ is an involutory permutation of G, axiom (GT_{II}) shows that this mapping is a *homeomorphism* of G onto G.

If A is an open (resp. closed) subset of G, and if x is any point of G, then the sets $x.A$, $A.x$ and A^{-1} (*) are open (resp. closed), for they are the transforms of A under one of the preceding homeomorphisms. If A is *open* and B is any subset of G, then AB and BA are *open*, because they are unions of open sets [axiom (O_I)]. If V is a neighbourhood of e in G, then VA and AV are *neighbourhoods* of A; for if W is an open neighbourhood of e contained in V, then WA and AW are open and contain A.

On the other hand, AB need not be closed when A is closed, even if B is closed too (cf. § 4, no. 1, Corollary 1 to Proposition 1).

* For example, in the additive group of the real line **R**, the subgroup **Z** of the rational integers is closed, and so is the subgroup $\theta\mathbf{Z}$ consisting of all integer multiples $n\theta$ of an *irrational* number θ; but the subgroup $\mathbf{Z} + \theta\mathbf{Z}$ of **R**, which is the set of all real numbers $m + n\theta$ (where m and n take all integer values) is not closed in **R**, as we shall see in Chapter V, § 1.

Again, let A be the subset of the additive group of $\mathbf{R} \times \mathbf{R}$ which consists of all (x, y) pairs such that $x \geqslant 0$ and $0 \leqslant y \leqslant 1 - \frac{1}{x+1}$; and let B be the set of all pairs $(x, 0)$, as x runs through **R**. A and B are closed, but $A + B$ is the set of all pairs (x, y) such that $0 \leqslant y < 1$, and is not closed in $\mathbf{R} \times \mathbf{R}$. *

Let X be a topological space and let f and g be two mappings of X into a topological group G. If f and g are continuous at a point x_0

(*) We recall that if A and B are two subsets of a group G, then A.B or AB denotes the set of all products xy where $x \in A$ and $y \in B$; A^{-1} denotes the set of all elements x^{-1} where $x \in A$. If B consists of a single element x, we write $x.A$ or xA (resp. $A.x$ or Ax) in place of $\{x\}.A$ (resp. $A.\{x\}$).

of X, then so are $\overset{-1}{f}$ and fg (*), by Theorem 2 of Chapter I, § 2, no. 1. In particular, the continuous mappings of X into G form a *subgroup* of the group G^X of all mappings of X into G.

Again, let f and g be two mappings of a set X, *filtered* by a filter $\mathfrak{F}$, into a *Hausdorff* topological group G. If $\lim_{\mathfrak{F}} f$ and $\lim_{\mathfrak{F}} g$ exist, then so do $\lim_{\mathfrak{F}} \overset{-1}{f}$ and $\lim_{\mathfrak{F}} fg$, and we have (Chapter I, § 7, no. 4, Proposition 9, Corollary 1)

$$\lim_{\mathfrak{F}} \overset{-1}{f} = (\lim_{\mathfrak{F}} f)^{-1}, \tag{1}$$

$$\lim_{\mathfrak{F}} fg = (\lim_{\mathfrak{F}} f)\,(\lim_{\mathfrak{F}} g). \tag{2}$$

When G is a *commutative* group, written *additively*, the axiom (GT′) indicates that $(x, y) \to x - y$ is a continuous mapping. If f and g are mappings of a topological space X into G, continuous at a point x_0, then $f - g$ is continuous at this point. The formulas (1) and (2) can be transcribed similarly.

2. NEIGHBOURHOODS OF A POINT IN A TOPOLOGICAL GROUP

Let $\mathfrak{B}$ be the neighbourhood filter of the identity element e in a topological group G, and let a be any point of G. Since $x \to ax$ and $x \to xa$ are homeomorphisms, it follows that the neighbourhood filter of a is the family $a.\mathfrak{B}$ of sets $a.V$, where V runs through $\mathfrak{B}$, and is also the family $\mathfrak{B}.a$ of sets $V.a$. Thus we know the neighbourhood filter of *any* point of a topological group as soon as we know the neighbourhood filter of the *identity element* e of the group.

If we say that xy and x^{-1} are continuous at $x = y = e$, we obtain (Chapter I, § 2, no. 1):

(GV_I). *Given any* $U \in \mathfrak{B}$, *there exists* $V \in \mathfrak{B}$ *such that* $V.V \subset U$.

(GV_{II}). *Given any* $U \in \mathfrak{B}$, *we have* $\overset{-1}{U} \in \mathfrak{B}$.

Every filter $\mathfrak{B}$ on G which satisfies (GV_I) and (GV_{II}) also satisfies (GV_a). *Given any* $U \in \mathfrak{B}$, *there exists* $V \in \mathfrak{B}$ *such that* $V.V^{-1} \subset U$. For by (GV_I), there exists $W \in \mathfrak{B}$ such that $W.W \subset U$, and by (GV_{II}) there exists $V \in \mathfrak{B}$ such that $V \subset W \cap W^{-1}$; hence $V^{-1} \subset W$ and therefore $V.V^{-1} \subset W.W \subset U$.

Conversely, if a filter $\mathfrak{B}$ on G satisfies (GV_a), it follows first of all that e *belongs to every set* $U \in \mathfrak{B}$; for if $V \in \mathfrak{B}$ is such that $V.V^{-1} \subset U$,

(*) We recall that $\overset{-1}{f}$ is the mapping $x \to (f(x))^{-1}$ and fg the mapping $x \to f(x)\,g(x)$; they should not be confused with $\overset{-1}{f}$ and $f \circ g$ (when these are defined) (*Set Theory*, R, § 2, nos. 6 and 11).

then, since V is not empty, we have $x.x^{-1} = e \in \mathrm{U}$ for every $x \in \mathrm{V}$. The condition (GV_a) therefore implies that $\mathrm{V}^{-1} \subset \mathrm{V}.\mathrm{V}^{-1} \subset \mathrm{U}$, so that $\mathrm{U}^{-1} \in \mathfrak{B}$ whenever $\mathrm{U} \in \mathfrak{B}$. Finally, if $\mathrm{V} \in \mathfrak{B}$ is such that $\mathrm{V}.\mathrm{V}^{-1} \subset \mathrm{U}$, and $\mathrm{W} \in \mathfrak{B}$ is such that $\mathrm{W} \subset \mathrm{V} \cap \mathrm{V}^{-1}$, we have $\mathrm{W}.\mathrm{W} \subset \mathrm{U}$. We see thus that (GV_a) is *equivalent* to the conjunction of $(\mathrm{GV}_{\mathrm{I}})$ and $(\mathrm{GV}_{\mathrm{II}})$.

Finally, since $x \to axa^{-1}$ is a homeomorphism which leaves e fixed, $\mathfrak{B}$ has the following property:

$(\mathrm{GV}_{\mathrm{III}})$. *For all* $a \in \mathrm{G}$ *and all* $\mathrm{V} \in \mathfrak{B}$, *we have* $a.\mathrm{V}a^{-1} \in \mathfrak{B}$.

These three properties of the filter $\mathfrak{B}$ are *characteristic*:

PROPOSITION 1. *Let* G *be a group and let* $\mathfrak{B}$ *be a filter on* G *satisfying the axioms* $(\mathrm{GV}_{\mathrm{I}})$, $(\mathrm{GV}_{\mathrm{II}})$ *and* $(\mathrm{GV}_{\mathrm{III}})$. *Then there is a unique topology on* G, *compatible with the group structure of* G, *for which* $\mathfrak{B}$ *is the neighbourhood filter of the identity element* e. *For this topology the neighbourhood filter of any point* $a \in \mathrm{G}$ *is the same as each of the two filters* $a.\mathfrak{B}$ *and* $\mathfrak{B}.a$.

If there is a topology with the required properties, then by what has been said above the neighbourhood filter of a coincides with each of the filters $a.\mathfrak{B}$ and $\mathfrak{B}.a$; hence the topology is unique, if it exists. Its existence will be established if we show 1) that the filters $a.\mathfrak{B}$ are the neighbourhood filters of a topology on G, and 2) that this topology is compatible with the group structure of G.

1) The filter $a.\mathfrak{B}$ satisfies axiom $(\mathrm{V}_{\mathrm{III}})$ (see Chapter I § 1, no. 2) by reason of $(\mathrm{GV}_{\mathrm{I}})$ and $(\mathrm{GV}_{\mathrm{II}})$, as we have already seen; hence to show that $a.\mathfrak{B}$ is the neighbourhood filter of a in a topology on G, we have to verify axiom $(\mathrm{V}_{\mathrm{IV}})$. Let then V be any set of $\mathfrak{B}$, and W a set of $\mathfrak{B}$ such that $\mathrm{W}.\mathrm{W} \subset \mathrm{V}$; then for any $x \in a.\mathrm{W}$ we have $x.\mathrm{W} \subset a.\mathrm{W}.\mathrm{W} \subset a.\mathrm{V}$, so that $a.\mathrm{V}$ belongs to the filter $x.\mathfrak{B}$; hence $(\mathrm{V}_{\mathrm{IV}})$ is satisfied.

2) Let us now show that the topology defined by the neighbourhood filters $a.\mathfrak{B}$ satisfies (GT'). Let a, b be any two points of G; if we put $x = au$ and $y = bv$, then we have to show that xy^{-1} is as near as we please to ab^{-1} whenever u and v are close enough to e. Now $(ab^{-1})^{-1}(xy^{-1}) = buv^{-1}b^{-1}$; let U be any neighbourhood of e, then we shall have $buv^{-1}b^{-1} \in \mathrm{U}$ if $uv^{-1} \in b^{-1}\mathrm{U}b = \mathrm{V}$, and $\mathrm{V} \in \mathfrak{B}$ by reason of $(\mathrm{GV}_{\mathrm{III}})$. But by $(\mathrm{GV}_{\mathrm{I}})$ and $(\mathrm{GV}_{\mathrm{II}})$ there exists $\mathrm{W} \in \mathfrak{B}$ such that $\mathrm{W}.\mathrm{W}^{-1} \subset \mathrm{V}$; hence it is enough to take $u \in \mathrm{W}$ and $v \in \mathrm{W}$ in order to have $xy^{-1} \in (ab^{-1})\,\mathrm{U}$. This completes the proof.

A common method of defining a topology compatible with a group structure on G consists in giving a filter satisfying the axioms $(\mathrm{GV}_{\mathrm{I}})$, $(\mathrm{GV}_{\mathrm{II}})$ and $(\mathrm{GV}_{\mathrm{III}})$. The corresponding conditions for a *filter base* $\mathfrak{B}$ are as follows:

$(\mathrm{GV}'_{\mathrm{I}})$. *Given any* $\mathrm{U} \in \mathfrak{B}$, *there exists* $\mathrm{V} \in \mathfrak{B}$ *such that* $\mathrm{V}.\mathrm{V} \subset \mathrm{U}$.

(GV'_{II}). *Given any* $U \in \mathfrak{B}$, *there exists* $V \in \mathfrak{B}$ *such that* $V^{-1} \subset U$.

(GV'_{III}). *Given any* $a \in G$ *and any* $U \in \mathfrak{B}$, *there exists* $V \in \mathfrak{B}$ *such that* $V \subset a.U.a^{-1}$.

A neighbourhood of e which coincides with its image under the symmetry $x \to x^{-1}$ is said to be *symmetric*. If V is any neighbourhood of e, then $V \cup V^{-1}$, $V \cap V^{-1}$ and $V.V^{-1}$ are symmetric neighbourhoods. By (GV_{II}), the symmetric neighbourhoods form a *fundamental system of neighbourhoods* of e. Also it follows from (GV_I) that when V runs through a fundamental system of neighbourhoods of e, the sets V^n (where n is a fixed integer $\neq 0$) form a fundamental system of neighbourhoods of e.

Remark. If G is *commutative*, we have $x.A.x^{-1} = A$ for every subset A of G and every $x \in G$, and therefore (GV_{III}) [resp. (GV'_{III})] is automatically satisfied for every filter (resp. filter base) on G. On the other hand, if G is not abelian, then (GV_{III}) is not a consequence of (GV_I) and (GV_{II}) [see Exercise 5].

If G is a commutative group, written *additively*, the axioms which characterize the filter $\mathfrak{B}$ of neighbourhoods of the origin for a topology compatible with the group structure of G are therefore the following:

(GA_I). *Given any* $U \in \mathfrak{B}$, *there exists* $V \in \mathfrak{B}$ *such that* $V + V \subset U$.

(GA_{II}). *Given any* $U \in \mathfrak{B}$, *we have* $-U \in \mathfrak{B}$.

PROPOSITION 2. *A topological group* G *is Hausdorff if and only if the set* $\{e\}$ *is closed.*

Clearly, if G is Hausdorff, then $\{e\}$ is closed. Conversely, if $\{e\}$ is closed, then the diagonal Δ of $G \times G$ is closed, because it is the inverse image of $\{e\}$ under the continuous mapping $(x, y) \to xy^{-1}$; hence (Chapter I, § 8, no. 1, Proposition 1) G is Hausdorff.

COROLLARY. *A topological group* G *is Hausdorff if and only if the intersection of the neighbourhoods of* e *consists only of the point* e.

The condition is clearly necessary. Conversely, if the intersection of all the neighbourhoods of e is just $\{e\}$, then given any $x \neq e$ there is a neighbourhood V of e such that $x^{-1} \notin V$ and therefore $e \notin xV$. This shows that x is not in the closure of $\{e\}$, and thus $\{e\}$ is closed, so that G is Hausdorff.

Example. Definition of a topology on a group by means of a set of subgroups.
If $\mathfrak{B}$ is a *filter base* on a group G, formed of *subgroups* of G, then it is immediately seen that $\mathfrak{B}$ satisfies axioms (GV'_I) and (GV'_{II}), since $H.H^{-1} = H$ for any subgroup H of G. Hence the set $\mathfrak{B}$ will be a *fundamental system of neighbourhoods* of e in a topology compatible with the group structure of G, provided that $\mathfrak{B}$ satisfies (GV_{III}); this will in

particular be the case if all the subgroups in $\mathfrak{B}$ are normal, hence always if G is *commutative*. The topology thus defined is *Hausdorff*, by Proposition 2, if and only if *the intersection of all the subgroups in* $\mathfrak{B}$ *consists only of* e. The most interesting cases are those in which the subgroup $\{e\}$ *is not in* $\mathfrak{B}$ (otherwise the topology defined by $\mathfrak{B}$ is the *discrete* topology): if $\{e\} \notin \mathfrak{B}$, the topology defined by $\mathfrak{B}$ is Hausdorff only if $\mathfrak{B}$ is an *infinite* set.

Since the intersection of two subgroups is a subgroup, we can define a topology on G, compatible with its group structure, starting from *any* set $\mathfrak{F}$ of subgroups of G: let $\mathfrak{G}$ be the set of all subgroups $a.\mathrm{H}.a^{-1}$, where $\mathrm{H} \in \mathfrak{F}$ and $a \in \mathrm{G}$, and let $\mathfrak{B}$ be the set of all *finite* intersections of subgroups belonging to $\mathfrak{G}$. Then $\mathfrak{B}$ is a filter base and satisfies $(\mathrm{GV}'_{\mathrm{III}})$.

Consider in particular the additive group of a *ring* A. Every set $\mathfrak{F}$ of *ideals* of A defines a topology compatible with this additive group structure. This topology is Hausdorff if the intersection of all the ideals of $\mathfrak{F}$ is the zero ideal, and it is not discrete if no finite intersection of the ideals of $\mathfrak{F}$ is the zero ideal. Topologies defined in this way play a large part in the theory of numbers (see the Exercises of §§ 6 and 7 of this chapter).

3. ISOMORPHISMS AND LOCAL ISOMORPHISMS

In accordance with the general definitions (*Set Theory*, Chapter IV, § 1, no. 5) an *isomorphism* f of a topological group G *onto* a topological group G′ is a bijective mapping of G onto G′ which is simultaneously an *isomorphism of the group structure of* G onto that of G′, and a *homeomorphism of* G onto G′. In other words, f is an isomorphism of G onto G′ if and only if: 1) f is bijective; 2) $f(xy) = f(x)f(y)$ for all $x, y \in \mathrm{G}$; and 3) f is bicontinuous.

For example, if a is any point of G, the mapping $x \to axa^{-1}$ is an isomorphism of G onto G, that is (*loc. cit.*), an *automorphism* of the topological group G. It is called an *inner automorphism*.

If a topology $\mathcal{T}$ is compatible with the group structure of a group G, then $\mathcal{T}$ is also compatible with the group structure *opposite* to that of G. If G^0 denotes the topological group obtained by giving the group opposite to G the topology $\mathcal{T}$, then the symmetry $x \to x^{-1}$ is an *isomorphism* of the topological group G onto the topological group G^0.

Definition 2. *If* G *and* G′ *are two topological groups, a local isomorphism of* G *with* G′ *is a homeomorphism* f *of a neighbourhood* V *of the identity element of* G *onto a neighbourhood* V′ *of the identity element of* G′ *which satisfies the following conditions*:

1) *For each pair* x, y *of points of* V *such that* $xy \in \mathrm{V}$,

$$f(xy) = f(x)f(y).$$

2) *If* g *is the mapping inverse to* f, *then for each pair of points* x', y' *of* V' *such that* $x'y' \in V'$, *we have* $g(x'y') = g(x')\,g(y')$.
The mapping g *is then a local isomorphism of* G' *with* G.
Two topological groups G, G' *are said to be locally isomorphic if there exists a local isomorphism of* G *with* G'.

Isomorphic topological groups are evidently locally isomorphic. The converse is false.

* For example, we shall see in Chapter V, § 1, that the topological groups **R** and **T** are locally isomorphic but not isomorphic. *

If f is a local isomorphism of G with G', then every *restriction* of f to a neighbourhood of the identity element of G is again a local isomorphism of G with G'.

A local isomorphism of G with G is called a *local automorphism* of G.

In general, if f is a homeomorphism of a neighbourhood V of the identity element of G onto a neighbourhood V' of the identity element of G' which satisfies condition 1) of Definition 2, f does not necessarily satisfy condition 2) (see Exercise 7). Nevertheless, G and G' are in fact *locally isomorphic*.

PROPOSITION 3. *Let* G *and* G' *be two topological groups, and let* f *be a homeomorphism of a neighbourhood* V *of the identity element of* G *onto a neighbourhood* V' *of the identity element of* G', *which satisfies condition* 1) *of Definition* 2. *Then* f *is an extension of a local isomorphism of* G *with* G'.

For it is not hard to see that if W is a neighbourhood of the identity element of G such that $W.W \subset V$, then the restriction of f to W is a local isomorphism of G with G'.

2. SUBGROUPS, QUOTIENT GROUPS, HOMOMORPHISMS, HOMOGENEOUS SPACES, PRODUCT GROUPS

1. SUBGROUPS OF A TOPOLOGICAL GROUP

Let G be a topological group and let H be a subgroup of G. By (GT'), the topology induced on H by the topology of G is compatible with the group structure of H. The structure of a *topological group* thus defined on H is said to be *induced* by that of G. Whenever we consider a subgroup H of G as a topological group, it is always this induced structure which is under consideration, unless the contrary is expressly stated.

PROPOSITION 1. *The closure* $\overline{H}$ *of a subgroup* H *of a topological group* G *is a subgroup of* G. *If* H *is a normal subgroup of* G, *then so is* $\overline{H}$.

If $a, b \in \overline{H}$ then $ab^{-1} \in \overline{H}$, because the mapping $(x, y) \to xy^{-1}$ is continuous on $G \times G$ and transforms $H \times H$ into H (Chapter I, § 2, no. 1, Theorem 1). In the same way, the continuity of the mapping $x \to axa^{-1}$ shows that if H is normal then $\overline{H}$ is normal.

In particular, the closure N of the set $\{e\}$ consisting only of the identity element of G, is a *normal subgroup* of G; and $N = \{e\}$ if and only if G is *Hausdorff* (§ 1, no. 2, Proposition 2).

PROPOSITION 2. *If* G *is a* Hausdorff *topological group, then the closure of a commutative subgroup of* G *is a commutative subgroup of* G.

By reason of Proposition 1, we may limit ourselves to the case where H is dense in G. The continuous functions xy and yx are equal on $H \times H$, and are therefore equal on $G \times G$, by virtue of the principle of extension of identities (Chapter I, § 8, no. 1, Proposition 2, Corollary 1).

PROPOSITION 3. *Let* G *be a Hausdorff topological group and let* M *be any subset of* G. *Then the set* M′ *of elements of* G *which commute with each element of* M *is a closed subgroup of* G. *In particular, the centre of* G *is closed in* G.

For M′ is the intersection of the sets F_m $(m \in M)$, where F_m is the set of all $x \in G$ such that $xm = mx$; and F_m is closed (Chapter I, § 8, no. 1, Proposition 2).

PROPOSITION 4. *Let* G *be a topological group and let* H *be a subgroup of* G *which is locally closed at one point of* H (Chapter I, § 3, no. 3, Definition 2). *Then* H *is closed in* G.

By translation, H is locally closed at each of its points, i.e. H is locally closed in G. Let V be a symmetric open neighbourhood of e in G such that $V \cap H$ is closed in V. If $x \in \overline{H}$, then xV meets H; if $y \in xV \cap H$, we have $x \in yV$, and $y(V \cap H) = (yV) \cap H$ is closed in yV. But x lies in the closure of $(yV) \cap H$, hence $x \in H$.

COROLLARY. *A subgroup of a topological group is open if and only if it has an interior point. Every open subgroup is closed.*

If a subgroup H has an interior point, then by translation all the points of H are interior and so H is open. The second assertion of the corollary is a particular case of Proposition 4.

PROPOSITION 5. *A subgroup* H *of a topological group* G *is discrete if and only if* H *has an isolated point. Every discrete subgroup of a* Hausdorff *group is closed.*

If H is discrete, every point of H is isolated. Conversely if H has an isolated point, then by translation every point of H is isolated and therefore H is discrete. If H is discrete and G is Hausdorff, then there is a neighbourhood V of e such that $V \cap H = \{e\}$; $\{e\}$ is closed in G and therefore *a fortiori* in V, so that H is locally closed at e. Hence H is closed in G by Proposition 4.

> *Remark.* Let H be any subgroup of a topological group G. For each $x \in \overline{H}$ we have $\overline{xH} = x.\overline{H} = \overline{H}$, since translation by x is a homeomorphism of G onto G. In other words, for each $x \in \overline{H}$, xH is *dense* in $\overline{H}$. It follows that if H is not closed, then $\overline{H} \cap \complement H$ is *dense* in $\overline{H}$.

2. COMPONENTS OF A TOPOLOGICAL GROUP

Let V be a *symmetric* neighbourhood of e in G. The subgroup generated by V, which is denoted by V^∞ consists of all products $\prod_{i=1}^{n} x_i$ of finite sequences of elements of V. V^∞ is *open*, since it has e as an interior point, and is therefore *closed* by Proposition 4 of no. 1. It follows that:

PROPOSITION 6. *A connected topological group is generated by every neighbourhood of the identity element.*

> The converse of this proposition is in general false, as we shall see in Chapter IV (§ 2, no. 5). If a topological group G is generated by each neighbourhood of the identity element, the most we can say is that G contains *no open subgroup* other than G.
>
> * As an example of a non-connected group G which has an open subgroup distinct from G, we may cite the *multiplicative group* $\mathbf{R}^*$ of non-zero real numbers, in which the subgroup $\mathbf{R}^*_+$ of real numbers > 0 is both open and closed (see Chapter IV, § 3, no. 2). *

PROPOSITION 7. *In a topological group* G, *the component* K *of the identity element* e *is a closed normal subgroup. The component of any point* $x \in G$ *is the coset* $x.K = K.x$.

If $a \in K$, then $a^{-1}K$ is connected and contains e; hence $K^{-1}K \subset K$, which shows that K is a subgroup of G. This subgroup is invariant under all automorphisms of G, and in particular under all inner automorphisms, therefore K is *normal* in G; also K is closed (Chapter I, § 11, no. 5, Proposition 9). Finally, the left translation $y \to xy$ is a homeomorphism of G which sends e to x, and hence the component of x is $x.K$.

The component of the identity element e of G is called the *identity component* of G.

3. DENSE SUBGROUPS

The following Proposition generalizes Proposition 1 of no. 1:

PROPOSITION 8. *Let* H *be a dense subgroup of a topological group* G, *and let* K *be a normal subgroup of* H. *Then the closure* $\overline{K}$ *of* K *in* G *is a normal subgroup of* G.

For the mapping $(z, x) \to zxz^{-1}$ is continuous on $G \times G$, and maps $H \times K$ into K; hence (Chapter I, § 2, no. 1, Theorem 1) it maps $G \times \overline{K} = \overline{H} \times \overline{K}$ into $\overline{K}$.

PROPOSITION 9. *Let* H *be a dense subgroup of a topological group* G. *If* H *is generated by every neighbourhood of the identity element in* H, *then* G *is generated by every neighbourhood of the identity element in* G.

Let V be any symmetric neighbourhood of e in G. Then $V \cap H$ is a neighbourhood of e in H, and hence generates H. It follows that V generates a subgroup H′ which contains H; but H′ is open and closed (no. 1, Corollary to Proposition 4), and therefore contains $\overline{H} = G$.

4. SPACES WITH OPERATORS

Let X be a topological space and let G be a topological group. G is said to *operate continuously* on X if the following conditions are satisfied:

1) X has G as a group of operators; in other words X is endowed with an external law of composition $(s, x) \to s.x$ for which G is the set of operators, and which is such that $s.(t.x) = (st).x$ and $e.x = x$ for all $s, t \in G$ and all $x \in X$.

2) The mapping $(s, x) \to s.x$ of $G \times X$ into X is *continuous*.

LEMMA 1. *If a topological group* G *operates continuously on a topological space* X, *then for each* $s \in G$ *the mapping* $x \to s.x$ *is a homeomorphism of* X *onto* X.

For this mapping is a continuous bijection whose inverse $x \to s^{-1}.x$ is also continuous.

We recall that for each $x \in X$ the set $G.x$ of transforms $s.x$ of x by the elements s of G is called the *orbit* of x (with respect to the group of operators G), and that the set of all $s \in G$ such that $s.x = x$ is a subgroup of G called the *stabilizer* of x. The relation $R(x, y)$: "y belongs to the orbit of x" is an equivalence relation on X, called the equivalence relation *defined by* G; the equivalence classes with respect to this relation are the orbits of the points of X. The topological space X/R is called the *orbit space* of X (with respect to G), or the *quotient space of* X *by the group* G, and is denoted by X/G; and the topology of X/G is said to be the *quotient of the topology of* X *by* G.

LEMMA 2. *If a topological group* G *operates continuously on a topological space* X, *then the equivalence relation* R *defined by* G *is open.*

For the saturation with respect to R of an open subset U of X is the set $\bigcup_{s \in G} s.U$, and each $s.U$ is open by Lemma 1.

Examples. 1) Let H be a subgroup of a topological group G. H operates continuously on G by the external law $(s, x) \to sx$. H also operates continuously on G by the external law $(x, s) \to sxs^{-1}$.

2) *If K is a topological division ring (§ 6, no. 7), the multiplicative group K* operates continuously on K by the external law $(s, x) \to sx$. *

3) Let G be a topological group, X a topological space. Then the mapping $(s, x) \to x$ of $G \times X$ into X is an external law of composition on X, and G operates continuously on X with respect to this law; G is then said to operate *trivially* on X.

Remark. Instead of saying that a topological group G operates continuously on a topological space X, it is often said that G operates continuously *on the left* on X. When the topological group G^0 opposite to G operates continuously on X, we say that G operates continuously *on the right* on X. It comes to the same thing to say that X has a continuous external law of composition $(s, x) \to s.x$ with G as set of operators, such that $s.(t.x) = (ts).x$ and $e.x = x$. Such a law is often written *on the right* : $(s, x) \to x.s$ (whence the terminology), and we have then $(x.t).s = x.(ts)$. If G operates continuously on the right on X by the law $(s, x) \to x.s$, then also G operates continuously on the left on X according to the external law $(s, x) \to x.s^{-1}$, by virtue of axiom (GT_{II}).

Let X (resp. X′) be a set with a group of operators G (resp. G′) and let f: G $\to$ G′ be a homomorphism and g : X $\to$ X′ a mapping. f and g are said to be *compatible* if $g(s.x) = f(s).g(x)$ for all $s \in$ G and all $x \in$ X. If X″ is a third set with a group of operators G″, if f' : G′ $\to$ G″ is a homomorphism, g' : X′ $\to$ X″ a mapping, and if f' and g' are compatible, then $f' \circ f$ and $g' \circ g$ are compatible. When X, X′ are topological spaces and G, G′ are topological groups operating continuously on X, X′ respectively, (f, g) is said to be a *morphism* of the space with operators X into the space with operators X′, provided that f and g are *continuous* and *compatible*. Passing to the quotients, g then induces a continuous mapping X/G $\to$ X′/G′ (Chapter I, § 3, no. 4, Corollary to Proposition 6).

Let G be a topological group operating continuously on a topological space X, and let φ be the canonical mapping of X onto the orbit space X/G. Let A be any subset of X, and let A′ be the subspace of X which is the saturation of A with respect to the equivalence relation R defined by G (thus A′ is the union of the orbits of points of A, and is said to be the *saturation of* A *with respect to* G). G operates continuously on A′ by the restriction of $(s, x) \to s.x$ to G $\times$ A′. Moreover, since R is open (Lemma 2) and A′ is saturated, it follows from Proposition 4 of Chapter I, § 5, no. 2, and from the relation $\varphi(A) = \varphi(A')$, that:

PROPOSITION 10. *The canonical bijection of the subspace* $\varphi(A)$ *of* X/G *onto the orbit space* A′/G *is a homeomorphism.*

Now let S be an equivalence relation on X such that, for each $s \in$ G, the mapping $x \to s.x$ is *compatible* with S [in other words, such that the relation $x \equiv y \pmod{S}$ implies $s.x \equiv s.y \pmod{S}$]; for the sake of brevity we shall say that the relation S is *compatible with the group* G. If ψ is the canonical mapping of X onto X/S, and if $s.\psi(x)$ denotes the class mod S of $s.x$, then X/S has G as groups of operators with respect to the external law $(s, \psi(x)) \to s.\psi(x) = \psi(s.x)$. Moreover:

PROPOSITION 11. *If the equivalence relation* S *on* X *is open and compatible with* G, *then* G *operates continuously on* X/S.

Since the relation of equality on G and the relation S on X are open, it is enough to show that the mapping $(s, x) \to s.\psi(x) = \psi(s.x)$ of G $\times$ X into X/S is continuous (Chapter I, § 5, no. 3, Corollary to Proposition 8); but this follows from the continuity of ψ and of the mapping

$$(s, x) \to s.x.$$

Remark. Let G′ be another topological group operating continuously on X, and suppose that $s.(s'x) = s'(s.x)$ for all $s \in$ G, $s' \in$ G′ and

$x \in X$. Then the equivalence relation S defined by G′ is compatible with G, and since S is open (Lemma 2) it follows that G operates continuously on X/G′. Similarly, G′ operates continuously on X/G. In these circumstances the two operations of the two groups G and G′ on X are said to commute.

5. HOMOGENEOUS SPACES

Let G be a topological group and H a subgroup of G. H operates continuously *on the right* on G according to the external law $(t, x) \to xt$, and the orbit of a point $x \in G$ is the left coset xH. The set of orbits is therefore what we have called in algebra the *homogeneous space* G/H. Whenever we speak of G/H as a topological space, we shall always mean the orbit space of G (with respect to H) unless the contrary is expressly stated; i.e. the quotient space of G by the equivalence relation $x^{-1}y \in H$. In conformity with the general definitions, we say that the topology of this space is the *quotient by* H *of the topology of* G.

Proposition 12. *The group* G *operates continuously on every homogeneous space* G/H.

Since the equivalence relation $x^{-1}y \in H$ is open (no. 4, Lemma 2) this is a particular case of Proposition 11 of no. 4.

Proposition 13. *Let* G *be a topological group and let* H *be a subgroup of* G. *Then the homogeneous space* G/H *is Hausdorff if and only if* H *is closed in* G.

H is an equivalence class for the relation $x^{-1}y \in H$ and therefore, if G/H is Hausdorff, H is closed in G. Conversely, if H is closed, then the graph of this relation is closed in $G \times G$, since it is the inverse image of H under the continuous mapping $(x, y) \to x^{-1}y$. Since the relation $x^{-1}y \in H$ is open, it follows from Chapter I, § 8, no. 3, Proposition 8 that G/H is Hausdorff.

Proposition 14. *Let* G *be a topological group and let* H *be a subgroup of* G. *Then the homogeneous space* G/H *is discrete if and only if* H *is open in* G.

For the inverse images in G of the points of G/H under the canonical mapping are the cosets xH $(x \in G)$; and these sets are open in G if and only if H is open in G.

Let X be a topological space on which a topological group G operates continuously and *transitively;* X is then (in the algebraic sense) a *homo-*

geneous space of G. Let x be a point of X, H_x its stabilizer. The continuous surjection $s \to s.x$ of G onto X factorizes canonically as

$$G \xrightarrow{f_x} G/H_x \xrightarrow{g_x} X$$

where f_x is the canonical mapping of G onto the homogeneous space G/H_x, and g_x is the bijection $s.H_x \to s.x$ of G/H_x onto X; moreover (Chapter I, § 3, no. 4, Proposition 6) g_x is a *continuous* mapping. But g_x is not necessarily a homeomorphism of G/H_x onto X (Exercise 29). If g_x is a homeomorphism for each $x \in X$, then we say that X is a *topological homogeneous space* (of the topological group G); for this to be so it is necessary and sufficient that, for each $x \in E$, the mapping $s \to s.x$ of G into X should be *open*.

Proposition 15. *Let* X *be a topological space on which a topological group* G *operates continuously and transitively. For* X *to be a topological homogeneous space (relative to* G) *it is sufficient that for some point* $x_0 \in X$ *the mapping* $s \to s.x_0$ *transforms each neighbourhood of* e *in* G *into a neighbourhood of* x_0 *in* X.

Every $x \in X$ can be written as $x = t.x_0$ for some $t \in G$. If V is a neighbourhood of e, then $V.x = (Vt).x_0$ is a neighbourhood of x, for we can write $(Vt).x_0 = t((t^{-1}Vt).x_0)$, and the assertion follows from the facts that $t^{-1}Vt$ is a neighbourhood of e in G and that $y \to t.y$ is a homeomorphism of X onto itself (no. 4, Lemma 1). It follows that if U is any open subset of G and if x is any point of X, then $U.x$ is open in X; for if $t \in U$, then $t^{-1}U$ is a neighbourhood of e, hence $(t^{-1}U).x$ is a neighbourhood of x, and $t.((t^{-1}U).x) = U.x$ is a neighbourhood of $t.x$. Hence $U.x$ is open in X, so that the mapping $s \to s.x$ of G into X is open. This completes the proof.

6. QUOTIENT GROUPS

Proposition 16. *Let* G *be a topological group and let* H *be a normal subgroup of* G. *Then the quotient by* H *of the topology of* G *is compatible with the group structure of* G/H.

If $x \to \dot{x}$ is the canonical mapping of G onto G/H, then we have to show that $(\dot{x}, \dot{y}) \to \dot{x}\dot{y}^{-1}$ is a continuous mapping of $(G/H) \times (G/H)$ into G/H. Since the equivalence relation $x^{-1}y \in H$ is open (no. 4, Lemma 2), it is enough to show that $(x, y) \to \dot{x}\dot{y}^{-1}$ is a continuous mapping of $G \times G$ into G/H (Chapter I, § 5, no. 3, Corollary to Proposition 8, and § 3, no. 4, Proposition 6). But $(x, y) \to \dot{x}\dot{y}^{-1}$ is the composition of the continuous mappings $x \to \dot{x}$ and $(x, y) \to xy^{-1}$, hence is continuous.

Whenever in the sequel we consider a quotient group G/H of a topological group G as a topological group, it is always to be understood that the topology of G/H is the quotient by H of the topology of G, unless the contrary is expressly stated.

PROPOSITION 17. *Let* φ *be the canonical mapping of a topological group* G *onto a quotient group* G/H. *If* $\mathfrak{B}$ *is a fundamental system of neighbourhoods of* e *in* G, *then* $\varphi(\mathfrak{B})$ *is a fundamental system of neighbourhoods of the identity element* $\varphi(e)$ *of* G/H.

This is a particular case of Proposition 5 of Chapter I, § 5, no. 3.

Propositions 13 and 14 give in particular, for quotient groups:

PROPOSITION 18. *Let* G *be a topological group and let* H *be a normal subgroup of* G.

a) *The quotient group* G/H *is Hausdorff if and only if* H *is closed in* G.

b) *The quotient group* G/H *is discrete if and only if* H *is open in* G.

If G is a topological group and N is the closure of $\{e\}$ in G, then N is a closed normal subgroup of G (no. 1, Proposition 1), hence G/N is Hausdorff; G/N is called the *Hausdorff group associated with* G.

PROPOSITION 19. *If* H *is a discrete normal subgroup of a topological group* G, *then* G/H *is locally isomorphic to* G.

Let V be a neighbourhood of e in G which does not contain any point of H other than e, and let W be a symmetric open neighbourhood of e in G such that $W^2 \subset V$. Then the restriction to W of the canonical mapping φ of G onto G/H is *injective*; for if $x, y \in W$ and $\varphi(x) = \varphi(y)$, then $x^{-1}y \in W^2 \subset V$ and $x^{-1}y \in H$, so that $x = y$. By Proposition 17 it follows that the restriction of φ to W is a homeomorphism of W onto $\varphi(W)$; since moreover $\varphi(xy) = \varphi(x)\varphi(y)$ for all $x, y \in W$, we conclude that G and G/H are locally isomorphic (§ 1, no. 3, Proposition 3).

7. SUBGROUPS AND QUOTIENT GROUPS OF A QUOTIENT GROUP

Let G be a topological group, let H be a normal subgroup of G, and let $\varphi : G \to G/H$ be the canonical mapping. We know that if A′ is a subgroup of G/H, then $\overset{-1}{\varphi}(A')$ is a subgroup of G which contains H. Conversely, if A is a subgroup of G, then $\varphi(A)$ is a subgroup of G/H; moreover, there is a canonical bijection of the quotient group $A/(A \cap H)$ onto the subgroup $\varphi(A)$ of G/H, and a canonical bijection of $\varphi(A)$ onto the quotient group AH/H, and both these bijections are isomorphisms *for the group structures*.

PROPOSITION 20. *Let* A *be a subgroup of a topological group* G, *let* H *be a normal subgroup of* G, *and let* φ *denote the canonical mapping of* G *onto* G/H. *Then the canonical bijection of* $\varphi(A)$ *onto* AH/H *is an isomorphism of topological groups.*

This follows from the preceding remarks, and from Proposition 10 of no. 4.

The canonical bijection of $A/(A \cap H)$ onto $\varphi(A)$ is a continuous homomorphism, since it arises from the restriction of φ to A by passing to the quotients; but in general the topological groups $A/(A \cap H)$ and AH/H *are not isomorphic* (cf. § 4, no. 1, Proposition 1, Corollary 3).

* For example, take G to be the additive group **R** of real numbers, take H to be the group **Z** of integers, and take A to be the group $\theta\mathbf{Z}$ of integer multiples of an irrational number θ. Then $A \cap H = \{0\}$, so that $A/(A \cap H)$ is a discrete group, isomorphic to **Z**; on the other hand, $A + H$ is dense in **R** (as we shall see in Chapter V, § 1, no. 1, Proposition 1), hence $(A + H)/H$, which is locally isomorphic to $A + H$ (no. 6, Proposition 19), is not a discrete group and consequently is not isomorphic to $A/(A \cap H)$. *

Nevertheless, we have the following proposition:

PROPOSITION 21. *Let* G *be a topological group,* G_0 *a dense subgroup of* G, H_0 *a normal subgroup of* G_0, H *the closure of* H_0 *in* G *and* φ *the canonical mapping* $G \to G/H$. *Then the canonical bijection* $G_0/H_0 \to \varphi(G_0)$ *is an isomorphism of the topological group* G_0/H_0 *onto a dense subgroup of* G/H.

Since $H_0 = H \cap G_0$, it is enough to show that if U_0 is any open subset of G_0 which is saturated with respect to the relation $x^{-1}y \in H_0$, then U_0 is the intersection of G_0 and an open subset of G which is saturated with respect to the relation $x^{-1}y \in H$ (Chapter I, § 3, no. 6, Proposition 10). Let U be an open subset of G such that $U_0 = U \cap G_0$. Since $U_0 = U_0H_0$, it is easily seen that $U_0 = UH_0 \cap G_0$; but UH_0 is open in G, so we may suppose that $U = UH_0$. The set UH is open in G and is saturated with respect to the relation $x^{-1}y \in H$; hence the proposition will be proved if we show that $UH \cap G_0 = U_0$. Now, if $u \in U$ and $h \in H$ are such that $uh \in G_0$, then there is a symmetric neighbourhood V of e in G such that $uV \subset U$; since Vh is a neighbourhood of h in G, there exists $z \in V$ such that $zh \in H_0$. But then we have $uz^{-1} \in U$, and $uh = (uz^{-1})(zh)$; therefore $UH \cap G_0 \subset UH_0$. But $UH_0 = U$, hence $UH \cap G_0 \subset U \cap G_0 = U_0$. This completes the proof.

Let G be a topological group operating continuously on a topological space X, and let K be a normal subgroup of G which is contained in the stabilizer of *each* point of X. For each $x \in X$, the relation $s \equiv t \pmod{K}$ therefore implies $s.x = t.x$, and passing to the quotient we define a mapping $\dot{s} \to \dot{s}.x$ of G/K into X. It is immediately verified that with respect to the external law $(\dot{s}, x) \to \dot{s}.x$, X has G/K as a group of operators. Moreover, G/K operates *continuously* on X with respect to this law; for the relation of equality on X and the relation $s \equiv t \pmod{K}$ on G are both open equivalence relations, and therefore the result follows from the continuity of the mapping

$$(s, x) \to \dot{s}.x = s.x$$

of $G \times X$ into X (Chapter I, § 5, no. 3, Corollary to Proposition 8, and § 3, no. 4, Proposition 6).

Now let G be a topological group operating continuously on a topological space X, and let H be any *normal* subgroup of G; then H operates continuously on X. Let S be the equivalence relation defined by H on X; then S is open (no. 4, Lemma 2). The relation S is *compatible with the group* G (no. 4); for if $y \equiv x \pmod{S}$ then there exists $t \in H$ such that $y = t.x$; hence for all $s \in G$ we have $s.y = (sts^{-1}).s.x$, and $sts^{-1} \in H$ since H is normal in G; thus $s.y \equiv s.x \pmod{S}$. If ψ is the canonical mapping of X onto X/S, the group G therefore operates continuously on X/S with respect to the external law

$$(s, \psi(x)) \to \psi(s.x)$$

(no. 4, Proposition 11). Moreover, the group H is contained in the stabilizer of each of the points of X/S; as we have seen above, G/H *operates continuously on* X/S = X/H with respect to the external law $(\dot{s}, \psi(x)) \to \psi(s.x)$. If R denotes the equivalence relation on X defined by G, then the relation S implies R, and the equivalence relation R/S on X/S is that defined by the group G/H. Hence (Chapter I, § 3, no. 4, Proposition 7):

Proposition 22. *Let* G *be a topological group operating continuously on a topological space* X, *and let* H *be a normal subgroup of* G. *Then the canonical bijection of* X/G *onto* (X/H)/(G/H) *is a homeomorphism.*

Corollary. *Let* G *be a topological group,* H *a normal subgroup of* G, *and* K *a normal subgroup of* G *which contains* H. *Then the canonical bijection of* G/K *onto* (G/H)/(K/H) *is an isomorphism of topological groups.*

We know already that this bijection is an isomorphism of groups, and Proposition 22 (applied to the group K operating on the right on G) shows that it is a homeomorphism.

8. CONTINUOUS HOMOMORPHISMS AND STRICT MORPHISMS

Proposition 23. *A homomorphism* f *of a topological group* G *into a topological group* G′ *is continuous on* G *if and only if it is continuous at one point of* G.

Suppose f is continuous at a point $a \in G$; then if V' is any neighbourhood of $f(a)$, $V = \overset{-1}{f}(V')$ is a neighbourhood of a. Hence if x is any point of G, we have

$$f(xa^{-1}V) = f(x)[f(a)]^{-1}f(V) \subset f(x)[f(a)]^{-1}V',$$

and therefore f is continuous at x.

A continuous homomorphism of a topological group G into a topological group G′ is also called a *morphism* of G into G′ for the topological group structures.

Let f be a continuous homomorphism of a topological group G into a topological group G′; the inverse image $H = \overset{-1}{f}(e')$ of the identity element e' of G′ is a *normal subgroup* of G, and $f(G)$ is a *subgroup* of G′. Consider the canonical factorization $f = \psi \circ \dot{f} \circ \varphi$, where φ is the canonical mapping $G \to G/H$, ψ is the canonical injection $f(G) \to G'$ and lastly $\dot{f}$ is a *continuous bijective homomorphism* of the quotient group G/H onto the subgroup $f(G)$ (Chapter I, § 3, no. 5); $\dot{f}$ is said to be the bijective homomorphism *associated* with f. In general, $\dot{f}$ is not an isomorphism of topological groups.

For example, let G′ be a non-discrete topological group, and G the topological group obtained by giving G′ the discrete topology; then the identity mapping of G into G′ is a continuous bijective homomorphism, but is not bicontinuous.

Definition 1. *A continuous homomorphism of a topological group* G *into a topological group* G′ *is said to be a strict morphism of* G *into* G′ *if the bijective homomorphism* $\dot{f}$ *of* $G/\overset{-1}{f}(e')$ *onto* $f(G)$, *associated with* f, *is an isomorphism of topological groups (in other words, if* $\dot{f}$ *is bicontinuous).*

An isomorphism of a topological group G onto a topological group G′ is therefore a bijective strict morphism of G onto G′.

Proposition 24. *Let* f *be a continuous homomorphism of a topological group* G *into a topological group* G′. *Then the following three statements are equivalent:*

a) f *is a strict morphism.*

b) *The image under* f *of every open set in* G *is an open set in* $f(G)$.

c) *The image under* f *of every neighbourhood of the identity element in* G *is a neighbourhood of the identity element in* G′.

In view of Lemma 2 of no. 4, the equivalence of a) and b) follows immediately from the definitions (Chapter I, § 5, no. 3, Proposition 5). The equivalence of b) and c) is a particular case of Proposition 15 of no. 5, if we observe that G operates continuously on $f(G)$ by the external law $(s, f(t)) \to f(st)$.

Remarks. 1) From condition b) of Proposition 24 it follows that every continuous homomorphism of a topological group into a *discrete* group is a *strict morphism.*

If G is *compact* and $f(G)$ is *Hausdorff,* then the bijective homomorphism $\dot{f}$ associated with f is bicontinuous (Chapter I, § 10, no. 2, Theorem 1, Corollary 2, and no. 1, Proposition 5, Corollary 4). Hence *every continuous homomorphism of a compact group into a Hausdorff group is a strict morphism.*

2) Let f be a strict morphism of G into G′ and let g be a strict morphism of G′ into G″. If f is *surjective* or if g is *injective,* it follows immediately from Proposition 24 that $g \circ f$ is a strict morphism of G into G″. But this conclusion is no longer necessarily valid if neither of the two preceding conditions is satisfied, even if f is injective and g surjective (Exercise 19).

3) Let f be a continuous homomorphism of a topological group G into a topological group G′, and let H be a normal subgroup of G. f induces a homomorphism g of the group G/H onto the quotient group $f(G)/f(H)$. This homomorphism g is *continuous.* Moreover, if f is a *strict morphism* of G into G′, then g is a *strict morphism* of G/H onto $f(G)/f(H)$; for if U is open in G/H, and if φ (resp. φ') denotes the canonical mapping of G onto G/H [resp. of $f(G)$ onto $f(G)/f(H)$], then we have $g(u) = \varphi'(f(\overset{-1}{\varphi}(u)))$, and since $\overset{-1}{\varphi}(u)$ is open in G, it follows that $g(u)$ is open in $f(G)/f(H)$, which proves our assertion.

9. PRODUCTS OF TOPOLOGICAL GROUPS

Let $(G_\iota)_{\iota \in I}$ be a family of topological groups. Then we can define a group structure on the product set

$$G = \prod_{\iota \in I} G_\iota$$

(the *product* of the group structures of the G_ι) by defining $(x_\iota) . (y_\iota) = (x_\iota y_\iota)$. If e_ι is the identity element of G_ι, then $e = (e_\iota)$ is the identity element of G, and we have $(x_\iota)^{-1} = (x_\iota^{-1})$. The *product topology* (Chapter I, § 2, no. 3) of the topologies of the G_ι is *compatible* with this group structure. For the mapping $((x_\iota), (y_\iota)) \to (x_\iota y_\iota^{-1})$ of G × G into G is the composition of the mapping $((x_\iota, y_\iota)) \to (x_\iota y_\iota^{-1})$ of

$$\prod_{\iota \in I} G_\iota \times G_\iota \quad \text{into} \quad G,$$

and the canonical mapping $((x_\iota), \quad (y_\iota)) \to ((x_\iota, \quad y_\iota))$ of

$$G \times G \quad \text{onto} \quad \prod_{\iota \in I} (G_\iota \times G_\iota);$$

and both these mappings are continuous (Chapter I, § 4, no. 1, Corollary 1 of Proposition 1, and Proposition 2).

DEFINITION 2. *The topological group obtained by giving the product set*

$$G = \prod_{\iota \in I} G_\iota$$

the group structure which is the product of the group structures of the G_ι *and the topology which is the product of the topologies of the* G_ι *is called the product of the topological groups* G_ι.

If $(J_\varkappa)_{\varkappa \in K}$ is a *partition* of I, then G is isomorphic to the product of the topological groups $\prod_{\iota \in J_\varkappa} G_\iota$ (associativity of the product).

If H_ι is a subgroup of G_ι, then the product of the topological groups H_ι is isomorphic to the subgroup $\prod_\iota H_\iota$ of $\prod_\iota G_\iota$. In particular, if J is any subset of I, and $J' = \complement J$, then the topological group $\prod_{\iota \in J} G_\iota$ is isomorphic to the normal subgroup $G'_J = (\prod_{\iota \in J} G_\iota) \times (\prod_{\iota \in J'} \{e_\iota\})$ of G. Since the projection of every open set is an open set, the projection pr_J of G onto $\prod_{\iota \in J} G_\iota$ is a *strict morphism*, and consequently the quotient group G/G' is isomorphic to $G'_{J'}$; G is isomorphic to the product $G'_J \times (G/G'_J)$.

PROPOSITION 25. *Let* $(G_\iota)_{\iota \in I}$ *be a family of topological groups, and let* H *be the normal subgroup of* $G = \prod_{\iota \in I} G_\iota$ *consisting of all* $x = (x_\iota)$ *such that the* x_ι *are equal to the identity element* e_ι *of* G_ι *except for a finite number of indices. Then the subgroup* H *is dense in* G.

This is a particular case of Chapter I, § 4, no. 3, Proposition 8.

Let $(X_\iota)_{\iota \in I}$ be a family of topological spaces, and for each $\iota \in I$ let G_ι be a topological group operating continuously on X_ι. It is clear that the product group $G = \prod_{\iota \in I} G_\iota$ then operates continuously on the product space $X = \prod_{\iota \in I} X_\iota$ according to the law

$$((s_\iota), (x_\iota)) \to (s_\iota . x_\iota)$$

(Chapter I, § 4, no. 1, Proposition 1, Corollary 1, and Proposition 2). Moreover the orbit under G of a point $x = (x_\iota)$ of X is the product of the orbits of the x_ι (with respect to the groups G_ι). Let φ_ι be the canonical mapping of X_ι onto X_ι/G_ι, and let $\varphi = (\varphi_\iota)$ be the product mapping of X onto $\prod_{\iota \in I} (X_\iota/G_\iota)$; then the preceding remark shows that the bijection canonically associated with φ maps the orbit space X/G onto $\prod_{\iota \in I} (X_\iota/G_\iota)$. Moreover:

PROPOSITION 26. *The bijection of* X/G *onto* $\prod_{\iota \in I} (X_\iota/G_\iota)$ *canonically associated with* (φ_ι) *is a homeomorphism.*

For since the φ_ι are surjective and open, it follows that $\varphi = (\varphi_\iota)$ is an open mapping (Chapter I, § 5, no. 3, Corollary to Proposition 8).

COROLLARY. *Let* $(G_\iota)_{\iota \in I}$ *be a family of topological groups, and for each* $\iota \in I$ *let* H_ι *be a normal subgroup of* G_ι; *let* φ_ι *denote the canonical mapping of* G_ι *onto* G_ι/H_ι. *Let* $G = \prod_{\iota \in I} G_\iota$, $H = \prod_{\iota \in I} H_\iota$. *Then the bijective homomorphism of* G/H *onto* $\prod_{\iota \in I} (G_\iota/H_\iota)$ *associated with the continuous homomorphism* $(x_\iota) \to (\varphi_\iota(x_\iota))$ *is an isomorphism of topological groups.*

For this homomorphism is an isomorphism of group structures.

Remark. If G is a *commutative* topological group, written additively, then the mapping $(x, y) \to x + y$ of $G \times G$ onto G is a *strict morphism.* For since $(x + x') + (y + y') = (x + y) + (x' + y')$ it is a *homomorphism* of $G \times G$ onto G; also it is continuous, and the image of a neighbourhood $V \times V$ of the origin in $G \times G$ under this mapping is the neighbourhood $V + V$ of the origin in G.

10. SEMI-DIRECT PRODUCTS

Let L, N be two subgroups of a group G, such that LN = NL; then LN is a *subgroup* of G, since

$$(LN)(LN)^{-1} = LNN^{-1}L^{-1} = LNL = LLN = LN$$

Moreover, the mapping $\varphi : (x, y) \to xy$ of $N \times L$ into G is *injective* if and only if $N \cap L = \{e\}$. For it is clear that if φ is injective, then $N \cap L = \{e\}$; and conversely, the relation $x'y' = xy$, where $x, x' \in N$ and $y, y' \in L$, implies that $x^{-1}x' = yy'^{-1} \in N \cap L$; hence if $N \cap L = \{e\}$, then φ is injective. Thus φ is *bijective* if and only if LN = G and $N \cap L = \{e\}$.

If N is a *normal* subgroup of G (or more generally is normal in some subgroup of G containing $N \cup L$), the condition LN = NL is automatically satisfied. Moreover, for each $y \in L$ the mapping $\sigma_y : x \to yxy^{-1}$ is an *automorphism* of the group N, and for any two elements u, v of L we have

(1) $$\sigma_{uv} = \sigma_u \circ \sigma_v;$$

and for any x, x' in N and y, y' in L we have also

(2) $$(xy)(x'y') = (x\sigma_y(x'))(yy').$$

Conversely:

PROPOSITION 27. *Let* N *and* L *be two groups,* e' *and* e'' *their respective identity elements. Suppose that we are given a homomorphism* $y \to \sigma_y$ *of* L *into the group* Γ *of automorphisms of* N. *Then:*

1) *On the product set* $S = N \times L$, *the internal law of composition*

(3) $$(x, y)(x', y') = (x\sigma_y(x'), yy')$$

defines a group structure, for which $j_1 : x \to (x, e'')$ *is an isomorphism of* N *onto a normal subgroup of* S, $j_2 : y \to (e', y)$ *is an isomorphism of* L *onto a subgroup of* S, *and* $\mathrm{pr}_2 : S \to L$ *is a surjective homomorphism whose kernel is* $j_1(N)$ *and which is such that* $\mathrm{pr}_2 \circ j_2$ *is the identity automorphism of* L.

2) *Let* $f : N \to G$, $g : L \to G$ *be two homomorphisms into a group* G, *such that*

(4) $$f(\sigma_y(x)) = g(y)f(x)g(y^{-1})$$

for all $x \in N$ *and all* $y \in L$. *Then there is a unique homomorphism* $h : S \to G$ *such that* $f = h \circ j_1$ *and* $g = h \circ j_2$.

If x, x', x'' are elements of N and y, y', y'' are elements of L, then

$$((x, y)(x', y'))(x'', y'') = (x\sigma_y(x'), yy')(x'', y'') = (x\sigma_y(x')\sigma_{yy'}(x''), yy'y'')$$

and

$$(x, y)((x', y')(x'', y'')) = (x, y)(x'\sigma_{y'}(x''), y'y'') = (x\sigma_y(x'\sigma_{y'}(x'')), yy', y'')$$

and therefore the associativity of the law (3) follows from the facts that $y \to \sigma_y$ is a homomorphism of L into Γ and that σ_y is an automorphism of N. Clearly (e', e'') is the identity element of (3), and finally

$$(x, y)(\sigma_{y^{-1}}(x^{-1}), y^{-1}) = (\sigma_{y^{-1}}(x^{-1}), y^{-1})(x, y) = (e', e'')$$

so that (x, y) has an inverse in S. The other assertions of 1) are clear. On the other hand, since $(x, y) = (x, e'')(e', y)$, a homomorphism h which satisfies the conditions of 2) must necessarily satisfy

$$h(x, y) = f(x)g(y),$$

hence is unique if it exists; moreover it is immediate from (4) that

$$f(x\sigma_y(x'))g(yy') = f(x)g(y)f(x')g(y^{-1})g(y)g(y') = f(x)g(y)f(x')g(y')$$

which shows that $(x, y) \to f(x)g(y)$ is indeed a homomorphism of S into G satisfying the conditions of 2).

Corollary. *The homomorphism* h *defined in* 2) *of Proposition* 27 *is injective if and only if* f *and* g *are injective and* $f(\mathrm{N}) \cap g(\mathrm{L}) = \{e\}$; *and* h *is surjective if and only if* $f(\mathrm{N})g(\mathrm{L}) = \mathrm{G}$.

Since $h(x, y) = f(x)g(y)$, the second assertion is obvious; moreover it follows from (4) that $f(\mathrm{N})g(\mathrm{L}) = g(\mathrm{L})f(\mathrm{N})$, and the first assertion follows from the remarks at the beginning of this sub-section.

The group S defined in Proposition 27 is said to be the *external semi-direct product* of N and L (relative to σ); we shall generally identify N (resp. L) with the normal subgroup $j_1(\mathrm{N})$ [resp. the subgroup $j_2(\mathrm{L})$] of S. If σ_y is the identity element of Γ for all $y \in \mathrm{L}$, we retrieve the usual notion of the *product* of two groups.

Now let G be a group, and let L and N be two subgroups of G such that $\mathrm{LN} = \mathrm{NL}$ and such that N is *normal* in NL, so that for each $y \in \mathrm{L}$, $\sigma_y : x \to yxy^{-1}$ is an automorphism of N, and $y \to \sigma_y$ is a homomorphism of L into the automorphism group Γ of N. It follows then from Proposition 27 that if S is the external semi-direct product of N and L (relative to σ), then $h : (x, y) \to xy$ is a *homomorphism* of S into G. h is bijective if and only if we have $\mathrm{N} \cap \mathrm{L} = \{e\}$ and $\mathrm{NL} = \mathrm{G}$ (Corollary to Proposition 27); G is then said to be the *semi-direct product* of its normal subgroup N and its subgroup L, and we often identify G with S by means of h.

Proposition 28. *Let* L, N *be two topological groups, let* $y \to \sigma_y$ *be a homomorphism of* L *into the group of automorphisms* Γ *of the (non-topological) group structure of* N; *and suppose that the mapping* $(x, y) \to \sigma_y(x)$ *of* $\mathrm{N} \times \mathrm{L}$ *into* N *is continuous. Then :*

1) *On the external semi-direct product* S *of* N *and* L, *relative to* σ, *the product of the topologies of* N *and* L *is compatible with the group structure; the canonical injections* $j_1 : \mathrm{N} \to \mathrm{S}$ *and* $j_2 : \mathrm{L} \to \mathrm{S}$ *are isomorphisms of the topological groups* N *and* L *respectively onto the subgroups* $j_1(\mathrm{N})$ *and* $j_2(\mathrm{L})$ *of* S, *and* pr_2 *is a strict morphism of* S *onto* L.

2) *Let* $f : \mathrm{N} \to \mathrm{G}$ *and* $g : \mathrm{L} \to \mathrm{G}$ *be two continuous homomorphisms into a topological group* G, *satisfying* (4); *then the homomorphism*

$$(x, y) \to f(x)g(y)$$

of S *into* G *is continuous.*

This is an immediate consequence of the definitions and of the properties of the product topology.

The topological group S thus defined is said to be the *external topological semi-direct product* of N and L (relative to σ); notice that the condition imposed on σ implies that L *operates continuously on the left* on N according to the external law $(x, y) \to \sigma_y(x)$ [no. 4].

Now let G be a topological group and let N and L be two subgroups of G such that G is the *semi-direct product* of N and L, *qua* non-topological group; it is then clear that the mapping $(x, y) \to \sigma_y(x)$ is *continuous* on $N \times L$, and that the canonical bijective homomorphism

$$h : (x, y) \to xy$$

of S onto G is *continuous*. But this homomorphism is not necessarily *bicontinuous*; when it is bicontinuous, G is said to be the *topological semi-direct product* of N and L. For this to be so it is necessary and sufficient that, if $p : G \to N$ and $q : G \to L$ are the mappings which make correspond to $z \in G$ the unique elements $p(z) \in N$ and $q(z) \in L$ such that $z = p(z)q(z)$, then one of the mappings p, q is *continuous* (in which case both are continuous). It comes to the same thing to say that the restriction to L of the canonical mapping $G \to G/N$ is an isomorphism of the topological group L onto the topological group G/N.

3. UNIFORM STRUCTURES ON GROUPS

1. THE RIGHT AND LEFT UNIFORMITIES ON A TOPOLOGICAL GROUP

In a topological group G we can perceive the possibility of defining a notion of " sufficiently near points " and hence a uniform structure, by operating as follows: if x and y are any two points of G, we apply to *both* points the translation which sends one of them, say x, to the identity element e; the " proximity " of x and y is then evaluated in some sense by the neighbourhood V of e into which y is translated. This translation, which consists in multiplying both x and y by x^{-1}, can be carried out on the *right* or on the *left*, and we shall see that in either case we obtain effectively a uniformity on G which is *compatible* with the topology of G. Let us take the case in which the translations are performed on the *right*; then to each neighbourhood V of e there corresponds the set V_d of pairs $(x, y) \in G \times G$ such that $yx^{-1} \in V$. Let $\mathfrak{G}_d$ be the family of sets V_d, as V runs through the neighbourhood filter $\mathfrak{B}$ of e. Then $\mathfrak{G}_d$ is a *fundamental system of entourages* (Chapter II, § 1, no. 1). For since $e \in V$, the diagonal Δ of $G \times G$ is contained in V_d for each $V \in \mathfrak{B}$, hence $\mathfrak{G}_d$ is a filter base and satisfies axiom (U'_I); since the relations $yx^{-1} \in V$ and $xy^{-1} \in V^{-1}$ are equivalent, we have $\overset{-1}{V}_d = (V^{-1})_d$, hence $\overset{-1}{V}_d \in \mathfrak{G}_d$ by (GV_{II}), so that (U'_{II}) is satisfied; and finally, the relations $zx^{-1} \in V$ and $yz^{-1} \in V$ imply $yx^{-1} \in V.V$; hence $V_d \circ V_d$ is contained in $(V.V)_d$, and (GV_I) shows that $\mathfrak{G}_d$ satisfies (U'_{III}).

The uniformity defined by $\mathfrak{G}_d$ is compatible with the topology of G, for the relations $y \in V_d(x)$ and $y \in V.x$ are equivalent by definition; in other words $V_d(x) = V.x$.

The argument is analogous when the translations are on the *left*, and we may therefore make the following definition:

Definition 1. *The right* (resp. *left*) *uniformity on a topological group* G *is the uniformity for which a fundamental system of entourages is obtained by making correspond to each neighbourhood* V *of the identity element* e, *the set* V_d (resp. V_s) *of pairs* (x, y) *such that* $yx^{-1} \in V$ (resp. $x^{-1}y \in V$).

If V runs through a fundamental system of neighbourhoods of e, the sets V_d (resp. V_s) form a fundamental system of entourages of the right (resp. left) uniformity.

To each proposition on the topology of a uniform space there corresponds a proposition on the topology of a group; the translation is made according to Definition 1 and the formulae $V_d(x) = V.x$, $V_d(A) = V.A$, $V_s(x) = x.V$, $V_s(A) = A.V$, which are immediate consequences of the definition. For example, if A is any non-empty subset of G then we have (Chapter II, § 1, no. 2, Corollary 1 to Proposition 2)

$$\bar{A} = \bigcap_{V \in \mathfrak{B}} V.A = \bigcap_{V \in \mathfrak{B}} A.V. \tag{1}$$

Again (Chapter II, § 1, no. 2, Corollary 3 to Proposition 2), *every Hausdorff group is regular.*

The right and left uniformities on a topological group are in general *distinct* (see Exercise 4). Obviously they coincide if the group is *commutative*, for then $V_d = V_s$; also they coincide if the group is *compact* (Chapter II, § 4, no. 1, Theorem 1).

In general, we shall denote by G_s (resp. G_d) the *uniform space* obtained by giving the set G its left (resp. right) uniformity.

Proposition 1. *The left and right translations are isomorphisms of the right uniformity onto itself.*

As to the right translations, the result is clear, since the relation $yx^{-1} \in V$ is equivalent to $(ya)(xa)^{-1} \in V$ [in other words, the mapping $(x, y) \to (xa, ya)$ leaves V_d fixed]. For the left translations, the result follows from (GV_{III}); for $yx^{-1} \in V$ if and only if $(ay)(ax)^{-1} \in aVa^{-1}$; hence $x \rightarrow ax$ is uniformly continuous on G_d.

Similarly, the right and left translations are isomorphisms of the left uniformity onto itself.

Every *inner automorphism* $x \rightarrow axa^{-1}$ of G is therefore an automorphism for the group structure of G, for the topology of G, and for both the uniformities of G.

PROPOSITION 2. *The symmetry* $x \rightarrow x^{-1}$ *is an isomorphism of the right uniformity onto the left uniformity.*

This is an immediate consequence of Definition 1.

The reader should beware of supposing that the mapping $(x, y) \rightarrow xy$ of the uniform space $G_d \times G_d$ into the uniform space G_d is in general uniformly continuous. Similarly, the symmetry $x \rightarrow x^{-1}$, considered as a mapping of G_d onto G_d, is not in general uniformly continuous (see Exercises 3 and 4).

PROPOSITION 3. *Every continuous homomorphism f of a topological group G into a topological group G′ is uniformly continuous when considered as a mapping of* G_d *into* G'_d (or of G_s into G'_s).

For if V' is a neighbourhood of the identity element in G', and $V = \overset{-1}{f}(V')$, then the relation $yx^{-1} \in V$ implies

$$f(y)(f(x))^{-1} = f(yx^{-1}) \in V'.$$

2. UNIFORMITIES ON SUBGROUPS, QUOTIENT GROUPS AND PRODUCT GROUPS

If H is a subgroup of a topological group G, then the uniformity induced on H by the right uniformity of G is none other than the right uniformity of the topological group H.

If H is a normal subgroup of G, and if φ is the canonical mapping of G onto G/H, we obtain a fundamental system of entourages of the right uniformity of the quotient group G/H by associating with each neighbourhood V of the identity element in G, the set of all pairs $(\dot{x}, \dot{y})$ of G/H such that $\dot{x}\dot{y}^{-1} \in \varphi(V)$ (§ 2, no. 6, Proposition 17). This condition means that there is at least one point $x \in \dot{x}$ and at least one point $y \in \dot{y}$ such that $yx^{-1} \in V$ [i.e. such that $(x, y) \in V_d$]. In particular, if N is the closure of the subset $\{e\}$ of G, then the right uniformity on G/N is isomorphic to the Hausdorff uniformity *associated* with the right uniformity on G (cf. Chapter II, § 3, no. 8).

Finally, on a product of a family (G_ι) of topological groups, the right uniformity is the *product* of the right uniformities of the G_ι (cf. Chapter II, § 2, no. 6).

There are analogous results for the left uniformity.

The left and right uniformities on the product group $\prod_{\iota \in I} G_\iota$ are identical if and only if the left and right uniformities on each factor G_ι coincide. This will always be the case if some of the G_ι are commutative and the others are compact.

3. COMPLETE GROUPS

Definition 2. *A topological group is said to be complete if its left and right uniformities are structures of complete spaces.*

From Proposition 2 of no. 1, for a group to be complete it is sufficient that *one* of its uniformities is a structure of a complete space. G is complete if and only if its associated Hausdorff group (§ 2, no. 6) is complete.

Every *closed* subgroup of a complete group is complete (Chapter II, § 3, no. 4, Proposition 8). Every product of complete groups is complete (Chapter II, § 3, no. 5, Proposition 10).

On the other hand, if G is a complete group and H is a closed normal subgroup of G, then the quotient group G/H is not necessarily complete (however, see Chapter IX, § 3, no. 1, Proposition 4).

Proposition 4. *If, in a topological group* G, *there is a neighbourhood* V *of* e *which is complete with respect to either the right or the left uniformity, then* G *is complete.*

Suppose for example that V is complete with respect to the right uniformity, and let $\mathfrak{F}$ be a Cauchy filter on G_d; then $\mathfrak{F}$ contains a V_d-small set M, and if $x_1 \in M$ we therefore have $M \subset Vx_1$. Hence the trace of $\mathfrak{F}$ on the complete subspace Vx_1 of G_d is a Cauchy filter, which converges to a point x_0; since x_0 is a cluster point of $\mathfrak{F}$, it is a limit of $\mathfrak{F}$ (Chapter II, § 3, no. 2, Corollary 2 to Proposition 5).

Corollary 1. *A locally compact group is complete.*

For every compact space is complete with respect to its unique uniformity (Chapter II, § 4, no. 1, Theorem 1).

Corollary 2. *Every locally compact subgroup of a Hausdorff topological group* G *is closed in* G.

For every complete subspace of a Hausdorff uniform space is closed (Chapter II, § 3, no. 4, Proposition 8).

Proposition 5. *Let* G_1 *be a topological group, let* G_2 *be a complete Hausdorff topological group, and let* H_1 (resp. H_2) *be a dense subgroup of* G_1 (resp. G_2). *Then every continuous homomorphism* u *of* H_1 *into* H_2 *can be uniquely extended to a continuous homomorphism* $\bar{u}$ *of* G_1 *into* G_2. *Furthermore, if* G_1 *is Hausdorff and complete, and if* u *is an isomorphism of* H_1 *onto* H_2, *then* $\bar{u}$ *is an isomorphism of* G_1 *onto* G_2.

$\bar{u}$ is uniformly continuous with respect to the right uniformities of H_1 and H_2 (no. 1, Proposition 3), hence admits a unique extension to a mapping $\bar{u}$ of G_1 into G_2 which is uniformly continuous with respect to the right uniformities of these groups (Chapter II, § 3, no. 6, Theorem 2). Moreover, by virtue of the principle of extension of identities (Chapter I, § 8, no. 1, Corollary 1 to Proposition 2), $\bar{u}$ is a homomorphism of G_1 into G_2, whence the first assertion of the Proposition. To prove the second assertion, it is enough to consider the isomorphism v of H_2 onto H_1, which is the inverse of u, and its extension $\bar{v}$ to a continuous homomorphism of G_2 into G_1; by reason of the uniqueness of the extension, $\bar{v} \circ \bar{u}$ and $\bar{u} \circ \bar{v}$ are the identity mappings of G_1 and G_2 respectively, and therefore (*Set Theory*, R, § 2, no. 12) $\bar{u}$ is bijective.

Remark. If the continuous homomorphism u is bijective, it does not follow in general that $\bar{u}$ is either injective or surjective (cf. Exercise 12); but see no. 5, Proposition 9.

4. COMPLETION OF A TOPOLOGICAL GROUP

Let G be a Hausdorff topological group. The *uniform space* G_d may be considered as a dense *subspace* of its *completion* $\hat{G}_d$. We shall investigate whether G can be considered as a dense *subgroup* of a *complete Hausdorff group* G′. If so, then the *uniform space* G'_d must be isomorphic to $\hat{G}_d$ (Chapter II, § 3, no. 6, Corollary to Theorem 2), and we must therefore be able to define on $\hat{G}_d$ a *topological group structure* which *induces* the given topological group structure on G. Consequently we have to examine: 1) Whether we can *extend by continuity* the functions xy and x^{-1} to $\hat{G}_d \times \hat{G}_d$ and $\hat{G}_d$ respectively; 2) whether the functions thus extended do indeed define a *group* structure on $\hat{G}_d$ (they will then necessarily define a topological group structure on $\hat{G}_d$ inducing the given structure on G). We have next to establish that 3) when the preceding operations are possible, the topological group which they define is *complete*. Finally, we shall see that 4) if there is a complete group satisfying the given conditions, then it is *unique* up to isomorphism.

1) *Extension of xy and x^{-1} by continuity.* Since the functions xy and x^{-1} are not in general uniformly continuous, we cannot apply the theorem of extension of uniformly continuous functions (Chapter II, § 3, no. 6, Theorem 2). Nevertheless, we can extend xy, by virtue of Proposition 11 of Chapter II, § 3, no. 6 and the following Proposition:

PROPOSITION 6. *Let $\mathfrak{F}$ and $\mathfrak{G}$ be two Cauchy filters on G_d. Then the image of the filter $\mathfrak{F} \times \mathfrak{G}$ under the mapping $(x, y) \to xy$ is a Cauchy filter base on G_d.*

Let us evaluate the "proximity" of xy and $x'y'$ in G_d; in other words, let us form the product $(x'y')(xy)^{-1} = x'y'y^{-1}x^{-1}$. For each $a \in G$, we can also write $(x'y')(xy)^{-1} = (x'a^{-1})(ay'y^{-1}a^{-1})(ax^{-1})$. We shall see that by suitable choice of a, each of the three factors of this product is very small whenever the pairs (x, y) and (x', y') belong to a sufficiently small set of $\mathfrak{F} \times \mathfrak{G}$. Let V be any neighbourhood of e in G; then there is a V_d-small set $A \in \mathfrak{F}$. Choose a in A, then if x and x' are any two points of A, we have $x'a^{-1} \in V$ and $ax^{-1} \in V$. On the other hand, the relation $ay'y^{-1}a^{-1} \in V$ is equivalent to

$$y'y^{-1} \in a^{-1}Va = W,$$

and since W is a neighbourhood of e, there is a W_d-small set $B \in \mathfrak{G}$. Hence for all (x, y) and (x', y') in $A \times B$, we have $(x'y')(xy)^{-1} \in V^3$, and this completes the proof.

In order that x^{-1} can be extended by continuity to $\hat{G}_d$ it is necessary and sufficient that *the image, under the symmetry $x \to x^{-1}$, of a Cauchy filter on G_d is a Cauchy filter on G_d* (Chapter II, § 3, no. 6, Proposition 11). There are examples of topological groups in which this condition is not satisfied (cf. Chapter X, § 3, Exercise 16); we shall suppose that it is satisfied for the remainder of this proof.

2) *The extended functions xy and x^{-1} define a group structure on $\hat{G}_d$.* For if we apply the principle of extension of identities (Chapter I, § 8, no. 1, Proposition 2, Corollary 1) to the functions $x(yz)$ and $(xy)z$, defined on $\hat{G}_d \times \hat{G}_d \times \hat{G}_d$ and equal on the dense subspace $G_d \times G_d \times G_d$, we see that the law of composition $(x, y) \to xy$ is *associative* on $\hat{G}_d$. For the same reason, the functions x, ex, xe are identical on $\hat{G}_d$, and the functions e, xx^{-1}, $x^{-1}x$ are identical on $\hat{G}_d$.

3) *The topological group $\hat{G}_d$ is complete.* Let $\mathfrak{U}_d$ be its *right uniformity*, and let $\mathfrak{U}$ be the uniformity on $\hat{G}_d$ obtained by *completing* the right uniformity of G. Then $\mathfrak{U}$ and $\mathfrak{U}_d$ induce the *same* uniformity on G, and therefore every Cauchy filter base $\mathfrak{B}$ *on* G with respect to $\mathfrak{U}_d$ is also a Cauchy filter base with respect to $\mathfrak{U}$. Now $\mathfrak{B}$ converges in $\hat{G}_d$,

since $\mathcal{U}$ is a complete uniformity; and since $\mathcal{U}$ and $\mathcal{U}_d$ induce the same topology on $\hat{G}_d$, it follows (Chapter II, § 3, no. 4, Proposition 9) that $\mathcal{U}_d$ is a complete uniformity. This conclusion also shows that $\mathcal{U}$ and $\mathcal{U}_d$ *coincide* (Chapter II, § 3, no. 6, Corollary to Theorem 2).

4) *Uniqueness.* This follows from Proposition 5 of no. 3.

To sum up, we have proved the following theorem:

THEOREM 1. *A Hausdorff topological group* G *is isomorphic to a dense subgroup of a complete group* $\hat{G}$ *if and only if the image, under the symmetry* $x \to x^{-1}$, *of a Cauchy filter with respect to the right uniformity of* G *is a Cauchy filter with respect to this uniformity. The complete group* $\hat{G}$ (which is called the *completion* of G) *is then unique* (up to isomorphism).

PROPOSITION 7. *Let* G *be a Hausdorff topological group which has a completion* $\hat{G}$. *Then the closures in* $\hat{G}$ *of the neighbourhoods of the identity element in* G *form a fundamental system of neighbourhoods of the identity element in* $\hat{G}$.

Since $\hat{G}$ is regular, every neighbourhood of the identity element in $\hat{G}$ contains the closure V of an open neighbourhood U of e in $\hat{G}$, and V is also the closure of the trace of U on G.

Let G be a group which is not necessarily Hausdorff; let $N = \overline{\{e\}}$, and let $G' = G/N$ be the Hausdorff group associated with G (§ 2, no. 6). If G' has a completion $\hat{G}'$, this completion is called the *Hausdorff completion* of G and is denoted by $\hat{G}$; $\hat{G}'_d$ (resp. $\hat{G}'_s$) is then the *Hausdorff completion* (Chapter II, § 3, no. 7) of the uniform space G_d (resp. G_s).

PROPOSITION 8. *Let* G *be a topological group which has a Hausdorff completion* $\hat{G}'$. *Then every continuous homomorphism* u *of* G *into a complete Hausdorff group* H *can be uniquely factorized into* $u = v \circ \varphi$, *where* v *is a continuous homomorphism of* $\hat{G}'$ *into* H *and* φ *is the canonical mapping of* G *into* $\hat{G}'$ (*the composition of the canonical injection of* G' *into* $\hat{G}'$ *and the canonical homomorphism* ψ *of* G *onto* $G/N = G'$).

Since the kernel of u is closed and contains e, it contains N, and hence u can be written as $u = w \circ \psi$, where w is a continuous homomorphism of G' into H; now apply Proposition 5 of no. 3 to w.

5. UNIFORMITY AND COMPLETION OF A COMMUTATIVE TOPOLOGICAL GROUP

We have already remarked that the left and right uniformities coincide on a commutative topological group G; whenever we speak of *the uniformity* of G, it is this unique uniformity to which we refer.

THEOREM 2. *Let* G *be a commutative topological group. Then the functions* x^{-1} *and* xy *are uniformly continuous on* G *and* G × G *respectively. Moreover* G *admits a Hausdorff completion* $\hat{G}$, *and* $\hat{G}$ *is commutative.*

The uniform continuity of x^{-1} follows from Proposition 2 of no. 1, and that of xy from Proposition 3 of no. 1, since $(x, y) \to xy$ is a continuous homomorphism of G × G into G. If G is Hausdorff, it satisfies the condition of Theorem 1 of no. 4 (as does every Hausdorff group whose left and right uniformities coincide); moreover the functions xy and yx are equal on $\hat{G} \times \hat{G}$ by the principle of extension of identities; hence the second part of the theorem, by considering in the general case the Hausdorff group associated with G.

It follows in particular from this theorem that if f and g are two uniformly continuous mappings of a uniform space X into a commutative group G, written additively, then the functions $-f$ and $f + g$ are uniformly continuous.

PROPOSITION 9. *Let* G *be a commutative group, and let* $\mathfrak{T}_1$, $\mathfrak{T}_2$ *be two Hausdorff topologies compatible with the group structure of* G. *Suppose that* $\mathfrak{T}_1$ *is finer than* $\mathfrak{T}_2$ *and that there is a fundamental system of neighbourhoods of* 0 *for* $\mathfrak{T}_1$ *which are closed for* $\mathfrak{T}_2$. *Let* G_1, G_2 *be the completions of* G *with respect to the topologies* $\mathfrak{T}_1$, $\mathfrak{T}_2$ *respectively, and let* $f: G_1 \to G_2$ *be the continuous homomorphism which extends the identity mapping of* G (no. 3, Proposition 5). *Then* f *is injective.*

Suppose that G is written additively. Let $\mathfrak{U}_1$ be the uniformity on G corresponding (no. 1) to the topology $\mathfrak{T}_1$: it will suffice to show that if $\mathfrak{F}$ and $\mathfrak{F}'$ are two minimal Cauchy filters (Chapter II, § 3, no. 2) with respect to $\mathfrak{U}_1$, which converge in G_2 to the same point a, then $\mathfrak{F} = \mathfrak{F}'$ (Chapter II, § 3, no. 7). For this, it is enough to show that $\mathfrak{F} \cap \mathfrak{F}'$ is a Cauchy filter with respect to $\mathfrak{U}_1$. Let V be a neighbourhood of 0 in G with respect to $\mathfrak{T}_1$, such that V is closed in $\mathfrak{T}_2$, and let W be a symmetric neighbourhood of 0 in G with respect to $\mathfrak{T}_1$, such that $W + W \subset V$. By hypothesis, there is a W_a-small set M (resp. M') in $\mathfrak{F}$ (resp. $\mathfrak{F}'$); if $x \in M$ and $y \in M$ we have $y - x \in W$, i.e. $y \in x + W$. If $\overline{W}$ and $\overline{V}$ are the closures of W and V in G_2, it follows that $y \in x + \overline{W}$, and therefore, since a is in the closure of M, that $a \in x + \overline{W}$ for each $x \in M$. Similarly, $a \in x' + \overline{W}$ for each $x' \in M'$, and hence $x - x' \in \overline{W} + \overline{W}$; but since $(x, y) \to x + y$ is a continuous mapping of $G_2 \times G_2$ into G_2, we have $\overline{W} + \overline{W} \subset \overline{W + W} \subset \overline{V}$. It follows that if $x \in M$ and $x' \in M'$, then $x - x' \in \overline{V} \cap G = V$, because V is *closed* in $\mathfrak{T}_2$; and this completes the proof.

COROLLARY 1. *Under the hypotheses of Proposition* 9, *if* A *is a subset of* G *which is a complete subspace with respect to the uniformity* $\mathfrak{U}_2$ *corresponding to* $\mathcal{T}_2$, *then* A *is also a complete subspace with respect to the uniformity* $\mathfrak{U}_1$ *corresponding to* $\mathcal{T}_1$.

If A_1 is the closure of A in G_1, then $f(A_1)$ is contained in the closure of A in G_2, which by hypothesis is equal to A. Since $f(A) = A$ by definition and f is injective, we have $A_1 = A$.

COROLLARY 2. *Let* G *be a commutative group and let* $\mathcal{T}_1, \mathcal{T}_2$ *be two topologies compatible with the group structure of* G. *Suppose that* $\mathcal{T}_1$ *is finer than* $\mathcal{T}_2$ *and that there is a fundamental system* $\mathfrak{B}$ *of neighbourhoods of* 0 *with respect to* $\mathcal{T}_1$ *which are complete with respect to the uniformity* $\mathfrak{U}_2$ *corresponding to* $\mathcal{T}_2$. *Then* G *is complete with respect to the uniformity* $\mathfrak{U}_1$ *corresponding to* $\mathcal{T}_1$.

The sets of $\mathfrak{B}$ are closed in the topology $\mathcal{T}_2$, hence complete for the uniformity $\mathfrak{U}_1$ by Corollary 1; the result therefore follows from Proposition 4 of no. 3.

4. GROUPS OPERATING PROPERLY ON A TOPOLOGICAL SPACE; COMPACTNESS IN TOPOLOGICAL GROUPS AND SPACES WITH OPERATORS

1. GROUPS OPERATING PROPERLY ON A TOPOLOGICAL SPACE

DEFINITION 1. *Let* G *be a topological group operating continuously on a topological space* X. G *is said to operate properly on* X *if the mapping*

$$\theta : (s, x) \to (x, s.x) \text{ of } G \times X \text{ into } X \times X$$

is proper (Chapter I, § 10, no. 1, Definition 1).

Let $\Gamma \subset G \times X \times X$ be the graph of the mapping $\rho : (s, x) \to s.x$. Since ρ is continuous, the mapping $\sigma : (s, x) \to (s, x, s.x)$ is a homeomorphism of $G \times X$ onto Γ, and the composite mapping

$$G \times X \xrightarrow{\sigma} \Gamma \xrightarrow{\mathrm{pr}_{23}} X \times X$$

is just θ. Definition 1 is therefore equivalent to saying that the restriction of pr_{23} to Γ is a *proper* mapping of Γ into $X \times X$.

Theorem 1 of Chapter I, § 10, no. 2 shows that G operates properly on X if and only if the following condition is satisfied:

For each set A filtered by an ultrafilter $\mathfrak{F}$, and each mapping

$$\alpha \to (s_\alpha, x_\alpha)$$

of A into $G \times X$, if the mapping $\alpha \to (s_\alpha . x_\alpha, x_\alpha)$ has a limit (b, a) with respect to $\mathfrak{F}$, then $\alpha \to s_\alpha$ has a limit $t \in G$ with respect to $\mathfrak{F}$, such that $t.a = b$.

Examples. 1) Let H be a *closed* subgroup of a topological group G. If G operates properly on X, then so does H, since $H \times X$ is closed in $G \times X$ (Chapter I, § 10, no. 1, Proposition 5, Corollary 1). If for example we take $X = G$, with G operating on itself by left translations, then the mapping $G \times X \to X \times X$ is a homeomorphism, hence proper; thus H *operates properly on* G *by left translations.*

2) If G operates properly on X, then it operates properly on every subspace X′ of X which is a *union of orbits of points of* X (in other words, X′ is saturated with respect to the equivalence relation defined by G). For the inverse image of $X' \times X'$ in $G \times X$ is $G \times X'$, and we can apply Proposition 3 of Chapter I, § 10, no. 1.

PROPOSITION 1. *Let* G *be a topological group operating continuously on a topological space* X, *and let* K *be a quasi-compact subset of* G. *Then the mapping* $\rho : (s, x) \to s.x$ *of* $K \times X$ *into* X *is proper.*

ρ factorizes into $K \times X \to K \times X \xrightarrow{pr_2} X$, where $\alpha(s, x) = (s, s.x)$. α is a homeomorphism, for $\alpha^{-1} : (s, y) \to (s, s^{-1}.y)$ is continuous. Since K is quasi-compact, pr_2 is proper (Chapter I, § 10, no. 2, Theorem 1, Corollary 5); hence ρ is proper (Chapter I, § 10, no. 1, Proposition 5).

COROLLARY 1. *If* A *is a closed* (resp. *compact*) *subset of* X, *then* K.A *is closed in* X (resp. *compact if* X *is Hausdorff*).

The assertion concerning closed sets follows from Proposition 1 and the fact that a proper mapping is closed (Chapter I, § 10, no. 1, Proposition 1). The assertion concerning compact sets is trivial.

It should be noted that if L is a compact subset of X, and F a closed subset of G, then F. L is not necessarily closed in X (§ 2, Exercise 29; cf. no. 5, Corollary to Proposition 12).

COROLLARY 2. *If* K *is a quasi-compact subgroup of a topological group* G, *then the equivalence relation* $x^{-1}y \in K$ *is closed, and the canonical mapping* $\varphi : G \to G/K$ *is proper.*

The first assertion follows from Corollary 1 applied to G acting by right translations on itself, and the second assertion then follows from Chapter I, § 10, no. 2, Theorem 1.

COROLLARY 3. *Let* K *be a quasi-compact normal subgroup of a topological group* G, *and let* φ *be the canonical mapping* $G \to G/K$. *Then for each closed subgroup* A *of* G *the canonical bijection of* $A/A \cap K$ *onto* $\varphi(A)$ *is an isomorphism of topological groups.*

Since $x^{-1}y \in K$ is a closed equivalence relation (Corollary 2), the Corollary follows from Chapter I, § 5, no. 2, Proposition 4.

PROPOSITION 2. *Let* K *be a compact group operating continuously on a Hausdorff space* X. *Then:*

a) K *operates properly on* X.

b) *The mapping* $(s, x) \to s.x$ *of* $K \times X$ *into* X *is proper.*

c) *The canonical mapping of* X *onto* X/K *is proper.*

b) is a consequence of Proposition 1. As to a), since K is compact, $pr_2 : (s, x) \to x$ is proper (Chapter I, § 10, no. 2, Theorem 1, Corollary 5); hence, X being Hausdorff, $(s, x) \to (x, s.x)$ is proper (Chapter I, § 10, no. 1, Proposition 5, Corollary 3). It remains to prove c). By Corollary 1 to Proposition 1, the canonical mapping $\varphi : X \to X/K$ is closed. If Z is any topological space and if we make K operate trivially on Z, then K operates continuously on $X \times Z$ and hence the canonical mapping $X \times Z \to (X \times Z)/K$ is closed. But $(X \times Z)/K$ can be canonically identified with $(X/K) \times Z$ (§ 2, no. 4, Lemma 2 and Chapter I, § 5, no. 3, Corollary to Proposition 8); hence the canonical mapping $X \times Z \to (X \times Z)/K$ can be identified with $\varphi \times 1$, and since it is closed for all Z it follows that φ is proper.

COROLLARY 1. *Under the hypotheses of Proposition* 2, X *is compact* (resp. *locally compact*) *if and only if* X/K *is compact* (resp. *locally compact*).

This follows from the fact that the canonical mapping $X \to X/K$ is proper, in view of Proposition 9 of Chapter I, § 10, no. 4.

COROLLARY 2. *Let* G *be a Hausdorff topological group and let* K *be a compact subgroup of* G. *Then* G *is compact* (resp. *locally compact*) *if and only if* G/K *is compact* (resp. *locally compact*).

Apply Corollary 1 to K operating on G by right translations.

2. PROPERTIES OF GROUPS ACTING PROPERLY

PROPOSITION 3. *If a topological group* G *operates properly on a topological space* X, *then the orbit space* X/G *is Hausdorff. If also* G *is Hausdorff, then* X *is Hausdorff.*

Let $C \subset X \times X$ be the graph of the equivalence relation R defined by G on X; then C is the image of $G \times X$ under the mapping $\theta : (s, x) \to (x, s.x)$. Since θ is proper, C is closed in $X \times X$ (Chapter I, § 10, no. 1, Proposition 1). Since the relation R is open (§ 2, no. 4, Lemma 2), it follows that X/G is Hausdorff (Chapter I, § 8, no. 3, Proposition 8).

Now suppose that G is Hausdorff. Then the mapping $x \to (e, x)$ of X into $G \times X$ is a homeomorphism of X onto a closed subspace of $G \times X$, and is therefore proper (Chapter I, § 10, no. 1, Proposition 2). If we compose this mapping with the mapping $(s, x) \to (x, s.x)$ of $G \times X$ into $X \times X$, which by hypothesis is proper, we obtain a proper mapping of X into $X \times X$, namely the diagonal mapping $x \to (x, x)$. Hence the diagonal Δ of $X \times X$ is closed in X, and therefore X is Hausdorff.

PROPOSITION 4. *Let* G *be a topological group operating properly on a topological space* X, *and let* x *be a point of* X. *Let* $G.x$ *denote the orbit of* x, *and let* K_x *denote the stabilizer of* x. *Then*:

a) *The mapping* $s \to s.x$ *of* G *into* X *is proper.*

b) K_x *is quasi-compact.*

c) *The canonical mapping of* G/K_x *onto* $G.x$ *is a homeomorphism.*

d) *The orbit* $G.x$ *is closed in* X.

The inverse image of $\{x\} \times X$ under $\theta : (s, x) \to (x, s.x)$ is $G \times \{x\}$; hence, by Proposition 3 of Chapter I, § 10, no. 1, the restriction of θ to $G \times \{x\}$ is a proper mapping of $G \times \{x\}$ into $\{x\} \times X$, whence a) follows. Since K_x is the inverse image of x under $s \to s.x$, b) follows from Chapter I, § 10, no. 2, Theorem 1. c) and d) are consequences of a), by virtue of Chapter I, § 10, no. 1, Propositions 2 and 5 b).

Remark. Proposition 4 shows that if a topological group G operates properly on a homogeneous space G/H, then the subgroup H is quasi-compact (and therefore compact if G is Hausdorff). It can be shown that this is also a sufficient condition for G to operate properly on G/H (Exercise 3).

PROPOSITION 5. *Let* G (resp. G′) *be a topological group operating continuously on a topological space* X (resp. X′). *Let* φ *be a continuous homomorphism of* G *into* G′ *and let* ψ *be a continuous mapping of* X *into* X′ *compatible with* φ (§ 2, no. 4). *Then:*

(i) *If* φ *is surjective and* ψ *is surjective and proper, and if* G *operates properly on* X, *then* G′ *operates properly on* X′.

(ii) *If* φ *is proper, if* G′ *operates properly on* X′ *and if* X *is Hausdorff, then* G *operates properly on* X.

To prove (i), consider the commutative diagram

$$\begin{array}{ccc} G \times X & \xrightarrow{\theta} & X \times X \\ \alpha \downarrow & & \downarrow \beta \\ G' \times X' & \xrightarrow{\theta'} & X' \times X' \end{array}$$

where $\alpha = \varphi \times \psi$ and $\beta = \psi \times \psi$. By hypothesis, θ is proper; so is β [Chapter I, § 10, no. 1, Proposition 4 a)]; hence $\beta \circ \theta = \theta' \circ \alpha$ is proper [Chapter I, § 10, no. 1, Proposition 5 a)]. Since α is surjective it follows that θ' is proper [Chapter I, § 10, no. 1, Proposition 5 *b*)].

To prove (ii) consider an ultrafilter $\mathfrak{U}$ on G × X, such that the mappings

$$(s, x) \to s.x \qquad \text{and} \qquad (s, x) \to x$$

converge with respect to $\mathfrak{U}$ to y_0 and x_0 respectively. It follows that $(s, x) \to \varphi(s).\psi(x)$ and $(s, x) \to \psi(x)$ converge with respect to $\mathfrak{U}$. Since G′ operates properly on X′, this implies (no. 1) that $(s, x) \to \varphi(s)$ converges with respect to $\mathfrak{U}$ to a point $s'_0 \in G'$. Since φ is proper, we deduce (Chapter I, § 10, no. 2, Theorem 1) that $(s, x) \to s$ converges with respect to $\mathfrak{U}$ to a point $s_0 \in G$. The uniqueness of the limit in X then shows that $y_0 = s_0 x_0$, and hence that G operates properly on X (no. 1).

3. GROUPS OPERATING FREELY ON A TOPOLOGICAL SPACE

Definition 2. *Let* G *be a group operating on a set* X. G *is said to operate freely on* X *if the stabilizer of every element of* X *is* $\{e\}$, *in other words if the relations* $s.x = x$, $s \in G$, $x \in X$ *imply* $s = e$.

Example. Let G be a group and let H be a subgroup of G. Then H operates freely by (left or right) translations on G.

Let G be a group operating freely on a set X, let R be the equivalence relation defined by G on X, and let $C \subset X \times X$ be the graph of R. If $(x, y) \in C$, then there exists $s \in G$ such that $s.x = y$; and s is *unique*, since $s.x = s'.x$ implies $s'^{-1}s.x = x$, and therefore $s'^{-1}s = e$ (as G operates freely). If we make correspond to $(x, y) \in C$ the unique $s \in G$ such that $s.x = y$, we define a mapping φ: C → G, which we shall call the *canonical mapping* of C into G. With this notation:

PROPOSITION 6. *Let* G *be a topological group operating continuously on a topological space* X, *and suppose that* G *operates freely on* X. *Then* G *operates properly on* X *if and only if the following condition is satisfed:*

(FP) *The graph* C *of the equivalence relation defined by* G *is closed in* $X \times X$, *and the canonical mapping* $\varphi : C \to G$ *is continuous.*

The set C is the image of the mapping $\theta : (s, x) \to (x, s.x)$ of $G \times X$ into $X \times X$. We know (Chapter I, § 10, no. 1, Proposition 2) that θ is proper if and only if C is closed in $X \times X$ and (if θ' denotes the mapping θ considered as a mapping of $G \times X$ into C) θ' is a homeomorphism. Now the hypothesis implies that θ' is bijective and that its inverse is the mapping $(x, y) \to (\varphi(x, y), x)$. Hence θ' is a homeomorphism if and only if φ is continuous.

4. LOCALLY COMPACT GROUPS OPERATING PROPERLY

PROPOSITION 7. *Let* G *be a locally compact group operating continuously on a Hausdorff space* X. *Then* G *operates properly on* X *if and only if, for each pair of points* x, y *of* X, *there is a neighbourhood* V_x *of* x *and a neighbourhood* V_y *of* y *such that the set* K *of all* $s \in G$ *for which* $s.V_x \cap V_y \neq \emptyset$ *is relatively compact in* G.

Let F be the compact space obtained by adjoining a point at infinity ω to G, and let Γ be the graph of $\rho : (s, x) \to s.x$ considered as a subset of $F \times X \times X$. Let us show that if the restriction of pr_{23} to Γ is proper, then Γ is *closed in* $F \times X \times X$. Indeed, this hypothesis implies that the map $u: (t, s, x, y) \to (t, x, y)$ of $F \times \Gamma$ into $F \times X \times X$ is closed. If Γ' is the set of points (s, s) in $F \times G$, where $s \in G$, then Γ' is closed in $F \times G$, since it is the graph of the canonical injection $G \to F$ (Chapter I, § 8, no. 1, Corollary 2 to Proposition 2); hence the intersection $(\Gamma' \times X \times X) \cap (F \times \Gamma)$ is closed in $F \times \Gamma$, and it is immediately seen that its image under u is the set Γ considered as a subset of $F \times X \times X$; hence Γ is closed in $F \times X \times X$. Now, we have $(\{\omega\} \times X \times X) \cap \Gamma = \emptyset$. From the definition of F, therefore, for every point $(x, y) \in X \times X$ there is a neighbourhood W of (x, y) in $X \times X$ and a compact subset K of G such that

$$((G - K) \times W) \cap \Gamma;$$

is empty; and since we may take W to be a neighbourhood $V_x \times V_y$, where V_x and V_y are neighbourhoods of x and y respectively in X, the statement "$((G - K) \times W) \cap \Gamma = \emptyset$" becomes "if $s \notin K$, then $s.V_x \cap V_y = \emptyset$". We have thus proved the necessity of the condition stated in the proposition. Conversely, suppose this condition is satisfied; let A be a set filtered by an ultrafilter $\mathfrak{F}$, and let $\alpha \to (s_\alpha, x_\alpha)$ be a mapping of A into $G \times X$ such that $\lim_{\mathfrak{F}} x_\alpha = x$ and $\lim_{\mathfrak{F}} s_\alpha . x_\alpha = y$. Suppose that K, V_x and V_y satisfy the condition of the proposition. By hypothesis there is a set $M \in \mathfrak{F}$ such that if $\alpha \in M$ then $x_\alpha \in V_x$ and $s_\alpha . x_\alpha \in V_y$, hence $s_\alpha \in K$. This shows that $\alpha \to s_\alpha$ converges with respect to $\mathfrak{F}$, and the proof is complete.

If G is compact, the condition of Proposition 7 is trivially satisfied; we thus retrieve Proposition 2 a).

Proposition 7 shows in particular that a *discrete* group G, operating continuously on a Hausdorff space X, operates properly on X if and only if, for each pair (x, y) of points of X, there is a neighbourhood V_x of x and a neighbourhood V_y of y such that the set of points $s \in G$ for which $s.V_x \cap V_y \neq \emptyset$ is *finite*.

PROPOSITION 8. *Let* G *be a discrete group operating properly on a Hausdorff space* X. *Let* x *be a point of* X *and let* K_x *be the stabilizer of* x. *Then:*

a) *The subgroup* K_x *is finite and there is an open set* $U \subset X$, *containing* x, *which is stable under* K_x, *and on which the equivalence relation induced by the relation defined by* G *is the equivalence relation defined by* K_x.

b) *The canonical mapping* $U/K_x \to X/G$ *is a homeomorphism of* U/K_x *onto an open neighbourhood of the class of* x *in* X/G.

By Proposition 7, K_x is finite. To construct an open set U which satisfies the required conditions, notice first that by Proposition 7 there is an open set U_0 containing x and such that the set K of $s \in G$ for which $s.U_0 \cap U_0 \neq \emptyset$ is finite. Clearly $K_x \subset K$; let $s_1, \ldots, s_n$ be the elements of $K - K_x$. If we put $x_i = s_i . x$ $(1 \leqslant i \leqslant n)$, then $x_i \neq x$ for each i; since X is Hausdorff, there is for each index i an open neighbourhood V_i of x and an open neighbourhood V'_i of $s_i . x$ such that $V_i \cap V'_i = \emptyset$. Let $U_i = V_i \cap s_i^{-1}.V'_i$; then U_i is clearly open and contains x, and we have $U_i \cap s_i . U_i \subset V_i \cap V'_i = \emptyset$. Let $U' = U_0 \cap U_1 \cap \ldots \cap U_n$; U' is open, contains x and is such that $U' \cap s.U' = \emptyset$ for $s \notin K_x$. Putting $U = \bigcap_{t \in K_x} t.U'$ we finally obtain an open set, stable under K_x, containing x and such that $U \cap s.U = \emptyset$ for $s \notin K_x$: U is the open set required.

The fact that the canonical mapping $U/K_x \to X/G$ is a homeomorphism of U/K_x onto an open set in X/G follows from Chapter I, § 5, no. 2, Proposition 4, since U is open and the equivalence relation defined by G is open (§ 2, no. 4, Lemma 2).

COROLLARY. *If we suppose in addition that* $K_x = \{e\}$, *then the point* x *has an open neighbourhood* U *such that the restriction to* U *of the canonical mapping* $X \to X/G$ *is a homeomorphism of* U *onto an open subset of* X/G.

5. GROUPS OPERATING CONTINUOUSLY ON A LOCALLY COMPACT SPACE

PROPOSITION 9. *Let* G *be a topological group operating continuously on a locally compact space* X. *Then if* X/G *is Hausdorff it is locally compact.*

Since the equivalence relation on X defined by G is open (§ 2, no. 4, Lemma 2), the proposition results from Chapter I, § 10, no. 4, Proposition 10.

PROPOSITION 10. *Let* G *be a topological group operating continuously on a locally compact space* X, *and suppose that* X/G *is Hausdorff. Let* φ *be the canonical mapping of* X *onto* X/G. *Then if* K' *is any compact subset of* X/G, *there is a compact subset* K *of* X *such that* $\varphi(K) = K'$.

Since the equivalence relation defined by G is open (§ 2, no. 4, Lemma 2), the proposition is a particular case of Proposition 10 of Chapter I, § 10, no. 4.

PROPOSITION 11. *Let* G *be a Hausdorff topological group operating properly on a non-empty space* X. *If* X *is compact* (resp. *locally compact*) *then so are* G *and* X/G.

By hypothesis, the mapping $\theta : (s, x) \to (x, s.x)$ of $G \times X$ into $X \times X$ is proper; if $X \times X$ is compact (resp. locally compact) then the Corollary to Proposition 9 of Chapter I, § 10, no. 4 shows that $G \times X$ is also compact (resp. locally compact), and therefore so is G since $X \neq \emptyset$. Since X/G is Hausdorff (no. 2, Proposition 3), the compactness (resp. local compactness) of X implies the compactness (resp. local compactness) of X/G [Chapter I, § 10, no. 4, Proposition 8 (resp. Proposition 9)] (see § 2, Exercise 29).

We shall now give criteria which allow us to assert that a Hausdorff topological group G operates properly on a locally compact space X. For each pair of subsets K, L of X we denote by P(K, L) the set of all $s \in G$ such that $s.K \cap L \neq \emptyset$.

THEOREM 1. *Let* G *be a Hausdorff topological group which operates continuously on a topological space* X. *Let* K *be a compact subset of* X *and let* L *be a closed subset of* X. *Then*:

a) *The set* P(K, L) *is closed in* G.

b) *If* G *operates properly on* X *and if* L *is compact, then* P(K, L) *is compact.*

c) *Conversely, if* X *is locally compact and if, for each pair* K, L *of compact subsets of* X, P(K, L) *is relatively compact in* G [*and therefore compact by* a)]; *then* G *operates properly on* X (*and if* X *is not empty,* G *is locally compact by Proposition* 11).

The mapping $(s, x) \to s.x$ of $G \times K$ into X is continuous, and the inverse image L′ of L under this mapping is therefore closed. Since K is compact, the projection $pr_1 : G \times K \to G$ is proper (Chapter I, § 10, no. 2, Corollary 5 of Theorem 1) and the image of L′ under pr_1 is therefore closed. This image is P(K, L), hence a) is proved.

To prove b) : X is Hausdorff (no. 2, Proposition 3). By hypothesis the mapping $\theta : (s, x) \to (x, s.x)$ of $G \times X$ into $X \times X$ is proper; since $K \times L$ is compact, so is $\overset{-1}{\theta}(K \times L)$ since X is Hausdorff (Chapter I, § 10, no. 2, Proposition 6). Hence the projection P(K, L) of $\overset{-1}{\theta}(K \times L)$ into G is a compact set.

To prove c) : since $K \times L$ is closed in $X \times X$, it follows that $\overset{-1}{\theta}(K \times L)$ is closed in $P(K, L) \times K$ and is therefore compact under the hypotheses of c). Since every compact subset of $X \times X$ is contained in a compact subset of the form $K \times L$, it follows that the inverse image of any compact subset of $X \times X$ under θ is compact, and since $X \times X$ is locally compact, this shows that θ is proper (Chapter I, § 10, no. 3, Proposition 7) [see Chapter IV, § 1, Exercise 4c)].

Remark. Clearly, we have $P(K, L) \subset P(K \cup L, K \cup L)$; therefore for G to operate properly on a locally compact space X, it is sufficient that, for each compact subset K of X, the set P(K, K) is relatively compact in G. In particular, a *discrete* group G operates properly on a locally compact space X if and only if, for each compact subset K of X, the set of $s \in G$ such that $s.K \cap K \neq \emptyset$ is *finite.*

Example. * Let X be a complex analytic manifold, analytically isomorphic to a bounded open subset of $\mathbf{C}^n$, and let G be the group of analytic automorphisms of X. The topology of compact convergence is compatible with the group structure of G, and it can be shown that G operates properly on X. In particular, every discrete subgroup of G operates properly on X.

Take for example X to be the upper half-plane $\mathfrak{J}(z) > 0$, which is analytically isomorphic to an open disc in **C**. Then G is the group of all transformations $z \to (az + b)/(cz + d)$, where a, b, c, d are real and $ad - bc \neq 0$. The subgroup H of G which consists of all such transformations for which a, b, c, d are integers and $ad - bc = 1$ is a discrete subgroup of G, called the *modular group.* By what has been said above, it operates properly on the upper half-plane $\mathfrak{J}(z) > 0$. *

PROPOSITION 12. *Let* G *be a Hausdorff topological group which operates continuously on a topological space* X. *Let* K *be a compact subset of* X, *and let* ρ_K *be the mapping* $(s, x) \to s.x$ *of* $G \times K$ *into* X. *Then*:

a) *If* G *operates properly on* X, ρ_K *is proper.*

b) *If* X *is locally compact and* ρ_K *is proper for each compact subset* K *of* X, G *operates properly on* X.

The mapping ρ_K factorizes as $G \times K \xrightarrow{\theta_K} K \times X \xrightarrow{pr_2} X$, where θ_K is the restriction to $G \times K$ of the mapping $\theta : (s, x) \to (x, s.x)$ of $G \times X$ into $X \times X$. Since $\overset{-1}{\theta}(K \times X) = G \times K$, θ_K is proper if θ is proper (Chapter I, § 10, no. 1, Proposition 3). On the other hand, since K is compact, the projection $pr_2 : K \times X \to X$ is proper (Chapter I, § 10, no. 2, Theorem 1, Corollary 5), hence ρ_K is proper (*ibid.*, no. 1, Proposition 5).

Suppose conversely that ρ_K is proper for each compact $K \subset X$. If L is a compact subset of X, then $\overset{-1}{\rho_K}(L)$ is a compact subset of $G \times K$, whose projection into G is P(K, L); therefore P(K, L) is compact. Hence if X is locally compact it follows from Theorem 1 that G operates properly on X.

COROLLARY. *Let* G *be a Hausdorff topological group operating properly on a topological space* X. *If* K *is any compact subset of* X *and if* F *is any closed subset of* G, *then* F.K *is closed in* X.

This follows from Proposition 12 and from Chapter I, § 10, no. 1, Proposition 1.

6. LOCALLY COMPACT HOMOGENEOUS SPACES

PROPOSITION 13. *Let* G *be a locally compact group and let* H *be a closed subgroup of* G. *Then the homogeneous space* G/H *is locally compact and paracompact.*

Since G/H is Hausdorff (§ 2, no. 5, Proposition 13) it is locally compact, by Proposition 9 of no. 5 applied to H operating on G on the right. Thus it remains to show that G/H is paracompact. Let V be a symmetric compact neighbourhood of e in G, and let $G_0 = V^\infty$ be the subgroup of G generated by V. G_0 is open (§ 2, no. 1, Corollary to Proposition 4) and operates continuously on G/H (§ 2, no. 5, Proposition 12). If we can show that each of the orbits $G_0.z$ ($z \in G/H$) is an *open* subset of G/H and a countable union of compact sets, then it will follow that G/H is the *topological sum* of the distinct orbits $G_0.z$ and is therefore paracompact (Chapter I, § 9, no. 10, Theorem 5). That $G_0.z$ is open in G/H follows from the facts that G_0 is open in G and the equivalence relation $x^{-1}y \in H$ is open (§ 2, no. 4, Lemma 2). On the other hand, $G_0 z$ is the union of the sets $V^n.z$ ($n \geqslant 1$), and since V^n is compact in G and G/H is Hausdorff, $V^n.z$ is compact. This completes the proof.

PROPOSITION 14. *In a locally compact group* G, *the identity component* C *is the intersection of the open subgroups of* G.

C is a closed normal subgroup of G (§ 2, no. 2, Proposition 7), and hence G/C is locally compact (Proposition 13) and totally disconnected (Chapter I, § 11, no. 5, Proposition 9). Since the inverse image of an open subgroup of G/C, under the canonical mapping of G onto G/C, is an open subgroup of G containing C, we see that we can restrict ourselves to proving the proposition for the group G/C. In other words, we may assume that G is totally disconnected. We know then (Chapter II, § 4, no. 4, Corollary to Proposition 6) that every compact neighbourhood V of e contains a neighbourhood U of e which is both open and closed. Since U is compact and $B = \complement U$ is closed, there is a symmetric open neighbourhood W of e such that $W \subset U$ and $UW \cap BW = \emptyset$ (§ 3, no. 1 and Chapter II, § 4, no. 3, Proposition 4), and *a fortiori* $UW \subset U$. By induction on n it follows that $W^n \subset U$ for every integer $n > 0$. Hence the subgroup $W^\infty = \bigcup_{n>0} W^n$, generated by W, is contained in U; but W^∞ is open in G (§ 2, no. 1, Corollary to Proposition 4). This completes the proof.

We have also proved:

COROLLARY 1. *If* G *is a totally disconnected locally compact group, then every neighbourhood of* e *in* G *contains an open subgroup of* G.

COROLLARY 2. *A locally compact group is connected if it is generated by each neighbourhood of the identity element.*

COROLLARY 3. *Let* G *be a locally compact group, let* H *be a closed subgroup of* G, *and let* φ *be the canonical mapping of* G *onto* G/H. *Then the components of* G/H *are the closures of the images under* φ *of the components of* G.

Let C be the identity component of G. The components of G are the sets sC, where $s \in G$ (§ 2, no. 2, Proposition 7); $\varphi(sC)$ is clearly connected, hence so is $\overline{\varphi(sC)}$ (Chapter I, § 11, no. 1, Proposition 1). But $\varphi(sC) = \varphi(sCH)$, and since sCH is saturated with respect to the equivalence relation defined by H, and since this equivalence relation is open

(§ 2, no. 4, Lemma 2), we have $\overline{\varphi(s\mathrm{CH})} = \varphi(\overline{s\mathrm{CH}}) = \varphi(s.\overline{\mathrm{CH}})$ (Chapter I, § 5, no. 3, Proposition 7).

Put $\mathrm{L} = \overline{\mathrm{CH}}$. L is a closed subgroup of G which contains C and H; hence to prove that the sets $\varphi(s.\mathrm{L}) = s.\varphi(\mathrm{L})$ are the components of G/H it is enough to show that the quotient space of G/H by the equivalence relation whose classes are the sets $s.\varphi(\mathrm{L})$ is totally disconnected. Now this quotient space is homeomorphic to the homogeneous space G/L (Chapter I, § 3, no. 4, Proposition 7); we are thus reduced to proving that when $\mathrm{C} \subset \mathrm{H}$, G/H is *totally disconnected*. Since G/H may be identified with (G/C)/(H/C) (§ 2, no. 7, Proposition 22), we may even assume that G itself is totally disconnected. Every neighbourhood of $\varphi(e)$ in G/H contains a neighbourhood of the form $\varphi(\mathrm{V})$, where V is a neighbourhood of e in G, and therefore (Corollary 1) contains a neighbourhood of the form $\varphi(\mathrm{K})$, where K is a *compact open* subgroup of G. $\varphi(\mathrm{K})$ is therefore both open and closed in G/H, and this shows that the component of $\varphi(e)$ in G/H consists of $\varphi(e)$ alone. By translation, the same is true for the component of every point of G/H, and the corollary is proved.

5. INFINITE SUMS IN COMMUTATIVE GROUPS

1. SUMMABLE FAMILIES IN A COMMUTATIVE GROUP

In this section we shall be concerned only with *Hausdorff commutative topological groups*, whose law of composition is written *additively*. Only the most important results will be translated into multiplicative notation.

Let G be a Hausdorff commutative group, let I be any index set and let $(x_\iota)_{\iota \in \mathrm{I}}$ be a family of points of G, indexed by I. With each *finite* subset J of I we associate the element $s_\mathrm{J} = \sum_{\iota \in \mathrm{J}} x_\iota$ of G, which we call the *finite partial sum of the family* $(x_\iota)_{\iota \in \mathrm{I}}$, corresponding to the set J. If $\mathfrak{F}(\mathrm{I})$ denotes the *set of finite subsets* of I, we have thus a mapping $\mathrm{J} \to s_\mathrm{J}$ of $\mathfrak{F}(\mathrm{I})$ into G. Now $\mathfrak{F}(\mathrm{I})$ is a *directed* set with respect to the relation $\subset$; for if J and J′ are two elements of $\mathfrak{F}(\mathrm{I})$, then $\mathrm{J} \subset \mathrm{J} \cup \mathrm{J}'$ and $\mathrm{J}' \subset \mathrm{J} \cup \mathrm{J}'$, and $\mathrm{J} \cup \mathrm{J}'$ is a finite subset of I. Let Φ be the *section filter* of the directed set $\mathfrak{F}(\mathrm{I})$ (Chapter I, § 6, no. 3).

DEFINITION 1. *Let* $(x_\iota)_{\iota\in I}$ *be a family of points of a Hausdorff commutative group* G; *let* $\mathfrak{F}(I)$ *be the set of finite subsets of the index set* I, *and for each finite subset* J *of* I, *let* s_J *be the sum of those* x_ι *such that* $\iota \in J$. *The family* $(x_\iota)_{\iota\in I}$ *is said to be summable if the mapping* $J \to s_J$ *has a limit with respect to the section filter* Φ *of the set* $\mathfrak{F}(I)$ *directed by the relation* $\subset$; *this limit is then said to be the sum of the family* $(x_\iota)_{\iota\in I}$ *and is denoted by* $\sum_{\iota\in I} x_\iota$ (or simply $\sum_i x_\iota$, or even $\sum x_\iota$, when there is no risk of ambiguity).

Definition 1 is equivalent to the following: *the family* (x_ι) *is summable and its sum is* s *if, for each neighbourhood* V *of the origin in* G, *there is a finite subset* J_0 *of* I *such that for each finite subset* $J \supset J_0$ *of* I *we have* $s_J \in s + V$.

If G is written *multiplicatively*, and if we put $p_J = \prod_{\iota\in J} x_\iota$ for each finite subset J of I, the family (x_ι) will be said to be *multipliable* if the mapping $J \to p_J$ has a limit with respect to the filter Φ; this limit is called the *product* of the family (x_ι), and is denoted by $\prod_{\iota\in I} x_\iota$.

Remarks. 1) When I is *finite*, Definition 1 reduces to the ordinary definition of the sum of a finite family. More generally, if I is arbitrary and $x_\iota = 0$ except when the index ι belongs to a *finite* subset J of I, then the sum $\sum_{\iota\in I} x_\iota$ is equal to $\sum_{\iota\in J} x_\iota$.

2) The definition of a summable family does not involve any *ordering* of the index set I, and we may therefore say that the notion of a sum thus defined is *commutative*. More precisely, we have the following property: let $(x_\iota)_{\iota\in I}$ be a summable family and let φ be a bijective mapping of an index set K onto the set I; then if we put $y_\varkappa = x_{\varphi(\varkappa)}$, the family $(y_\varkappa)_{\varkappa\in K}$ is summable and has the same sum as (x_ι). For if $s = \sum_{\iota\in I} x_\iota$, and if $\sum_{\iota\in J} x_\iota \in s + V$ for every finite subset J containing the finite subset J_0, then we shall have $\sum_{\varkappa\in L} y_\varkappa \in s + V$ for every finite subset L of K containing $\overset{-1}{\varphi}(J_0)$.

3) Definition 1 applies, more generally, to every family of points in a *Hausdorff topological space* X on which has been defined an *associative and commutative law of composition*, written additively; for it does not make use of the axioms of topological groups.

2. CAUCHY'S CRITERION

Let $(x_\iota)_{\iota\in I}$ be a summable family in G. Then for each neighbourhood V of the origin in G there is a finite subset J_0 of I such that, for each finite subset K of I *which does not meet* J_0, we have $s_K \in V$. For $J = J_0 \cup K$ is an arbitrary finite subset containing J_0; let $s = \sum_{\iota \in J} x_\iota$ and let W be a symmetric neighbourhood of 0 such that $W + W \subset V$; then, by Definition 1, we may choose J_0 such that $s_J \in s + W$ and $s_{J_0} \in S + W$, which means that $s_K = s_J - s_{J_0} \in W + W \subset V$.

Conversely, suppose that the family (x_ι) has this property. Then the image of the filter Φ, under the mapping $J \to s_J$, is a *Cauchy filter base* in G. For if J is a finite subset containing J_0, and if K denotes $J \cap \complement J_0$, then $K \cap J_0 = \varnothing$ and $s_K = s_J - s_{J_0}$, so that $s_J \in s_{J_0} + V$. If J' is another finite subset containing J_0, then $s_J - s_{J'} \in V + V$, and the result follows. Consequently:

THEOREM 1 (Cauchy's criterion). *In a Hausdorff commutative group* G, *in order that a family* $(x_\iota)_{\iota \in I}$ *should be summable it is necessary that, for each neighbourhood* V *of the origin in* G, *there is a finite subset* J_0 *of* I *such that* $\sum_{\iota \in K} x_\iota \in V$ *for all finite subsets* K *of* I *which do not meet* J_0. *This necessary condition is also sufficient if* G *is complete.*

Thus by taking away a (sufficiently large) finite number of terms from the family (x_ι), every *finite partial sum* of the surviving subfamily has to be as *close to* 0 *as we please.*

An immediate consequence of the first part of Theorem 1 is the following proposition:

PROPOSITION 1. *If the family* (x_ι) *is summable, then every neighbourhood of* 0 *contains all the* x_ι *except for a finite subfamily* (in other words, if I is infinite, we have $\lim x_\iota = 0$ with respect to the *filter of complements of finite subsets of* I).

This *necessary* condition for a family (x_ι) to be summable is *by no means sufficient* in general, even if G is complete; we shall see many examples later (see Chapter IV, § 7).

COROLLARY. *Let* $(x_\iota)_{\iota \in I}$ *be a summable family in a commutative group whose identity element has a countable fundamental system of neighbourhoods. Then the set of indices* ι *such that* $x_\iota \neq 0$ *is countable.*

Let (V_n) be a countable fundamental system of neighbourhoods of 0. If H_n is the set of all indices ι such that $x_\iota \notin V_n$, then the set H of indices ι such that $x_\iota \neq 0$ is the union of the sets H_n, and each of the H_n is *finite* by Proposition 1.

This corollary is no longer necessarily valid if we do not assume that the origin has a countable fundamental system of neighbourhoods. * Consider, for example, the product group $\mathbf{R}^{\mathbf{R}}$ [the additive group of all finite real-valued functions of a real variable, with the topology of pointwise convergence (cf. Chapter X, § 1, no. 3)], and let f_a be the element of $\mathbf{R}$ such that $f_a(a) = 1$ and $f_a(x) = 0$ if $x \neq a$; then the family $(f_a)_{a\in\mathbf{R}}$ is summable and its sum is the function whose value at each point of $\mathbf{R}$ is 1. *

Remark. When G is written multiplicatively, Cauchy's criterion takes the following form : for the family $(x_\iota)_{\iota\in I}$ to be multipliable it is necessary that, for each neighbourhood V of the identity element, there should exist a finite subset J_0 of I such that, for each finite subset K of I which does not meet J_0, we have $\prod_{\iota\in K} x_\iota \in V$; and this condition is sufficient provided that G is complete. We deduce that if I is infinite and (x_ι) is multipliable, then $\lim x_\iota = 1$ with respect to the filter of complements of finite subsets of I; if in addition the identity element has a countable fundamental system of neighbourhoods, then the set of indices ι such that $x_\iota \neq 1$ is countable.

3. PARTIAL SUMS; ASSOCIATIVITY

PROPOSITION 2. *In a complete group* G, *every subfamily of a summable family is summable.*

For if Cauchy's criterion is satisfied by a family $(x_\iota)_{\iota\in I}$, then it is trivially satisfied by every subfamily.

If $(x_\iota)_{\iota\in I}$ is summable, it follows therefore that the sum $\sum_{\iota\in J} x_\iota$ is defined for every subset J of I, finite or not : it is again called the *partial sum* of the family (x_ι), corresponding to the subset J of the index set. The set of partial sums of a summable family is evidently contained in the *closure* of the set of *finite* partial sums.

THEOREM 2 (*Associativity* of the sum). *Let* $(x_\iota)_{\iota\in I}$ *be a summable family in a complete group* G, *and let* $(I_\lambda)_{\lambda\in L}$ *be any partition of* I. *If* s_λ denotes $\sum_{\iota\in I_\lambda} x_\iota$, *then the family* $(s_\lambda)_{\lambda\in L}$ *is summable and has the same sum as the family* $(x_\iota)_{\iota\in I}$.

Thus if we have a summable family in a complete group, we can *associate* its terms arbitrarily into subfamilies and form the sum of each subfamily thus obtained; the family of these partial sums is again summable and its sum is equal to that of the given family.

Let $s = \sum_{\iota \in I} x_\iota$, and let V be any *closed* neighbourhood of 0 in G. Then there is a finite subset J_0 of I such that, for each finite subset J of I which contains J_0, we have $\sum_{\iota \in J} x_\iota \in s + V$. Let K_0 be the subset of L consisting of indices λ such that $J_\lambda = I_\lambda \cap J_0$ is *not empty*: K_0 is clearly finite. Let K be any finite subset of L which contains K_0; we shall show that $\sum_{\lambda \in K} s_\lambda \in s + V$, which will establish the theorem. Now s_λ is very close to a finite partial sum of (x_ι), whose indices all belong to I_λ; more precisely, given any symmetric neighbourhood W of 0, there exists for each $\lambda \in K$ a finite subset H_λ of I_λ, *containing* J_λ and such that $s_\lambda - \sum_{\iota \in H_\lambda} x_\iota \in W$. Let $J = \sum_{\lambda \in K} H_\lambda$; then J is a finite subset of I containing J_0, and we have

$$\sum_{\iota \in J} x_\iota = \sum_{\iota \in \bigcup_{\lambda \in K} H_\lambda} x_\iota = \sum_{\lambda \in K} \Big(\sum_{\iota \in H_\lambda} x_\iota \Big)$$

by the associativity of finite sums. Hence by reason of the choice of J_0 and the H_λ, we have

$$\sum_{\lambda \in K} s_\lambda \in s + V + nW$$

where n is the number of elements in K; this relation holds for each W, hence also $\sum_{\lambda \in K} s_\lambda \in s + V$, since V (being closed) is the intersection of the neighbourhoods $V + nW$ [§ 3, no. 1, formula (1)].

Q.E.D.

We can thus write the *associativity formula* for sums:

$$\sum_{\lambda \in L} \Big(\sum_{\iota \in I_\lambda} x_\iota \Big) = \sum_{\iota \in \bigcup_{\lambda \in L} I_\lambda} x_\iota, \tag{1}$$

which is valid whenever the family (I_λ) is a *partition* of its union and the right-hand side is defined. In particular, if the index set is a *product* $I = L \times M$, and if the "double" family $(x_{\lambda\mu})_{(\lambda, \mu) \in L \times M}$ is *summable*, we have the *formula of change of order of summation*

$$\sum_{(\lambda, \mu) \in L \times M} x_{\lambda\mu} = \sum_{\lambda \in L} \Big(\sum_{\mu \in M} x_{\lambda\mu} \Big) = \sum_{\mu \in M} \Big(\sum_{\lambda \in L} x_{\lambda\mu} \Big). \tag{2}$$

It should be remarked that the *left-hand side* of (1) can have a meaning, without the right-hand side being defined. Consider for example the case where $I = L \times \{1, 2\}$ and L is infinite, and I_λ consists of the two elements $(\lambda, 1)$ and $(\lambda, 2)$; if we take $x_{\lambda,1} = a$, $x_{\lambda,2} = -a$, where a is any non-zero element of G, then all the partial sums corresponding to the I_λ are zero, and therefore the left-hand side of (1) is defined and equal to 0, whereas the right-hand side of (1) has no meaning, as Proposition 1 shows.

Likewise, it can happen that the left-hand side of (2) is not defined but that each of the two "double sums" in (2) has a meaning; and the elements of G which they represent need not be equal (see Chapter IV, § 7, Exercise 17).

Thus although it is always possible to "associate" the terms of a sum, it is not possible, on the other hand, to "dissociate" into their elements those terms of a sum which appear themselves as sums. Nevertheless, this operation is legitimate whenever the number of these "dissociable" terms is *finite*.

PROPOSITION 3. *Let* $(x_\iota)_{\iota \in I}$ *be a family of points of a group* G, *and let* $(I_\lambda)_{\lambda \in L}$ *be a* finite *partition of* I. *If each of the subfamilies* $(x_\iota)_{\iota \in I_\lambda}$ *is summable, then the family* $(x_\iota)_{\iota \in I}$ *is summable and the formula* (1) *is valid.*

It is enough to prove the proposition when $L = (1,2)$; having done so, we then proceed by induction on the number of elements of L. Let $s_1 = \sum_{\iota \in I_1} x_\iota$ and $s_2 = \sum_{\iota \in I_2} x_\iota$. For each neighbourhood V of the origin, there is a finite subset J_1 (resp. J_2) of I_1 (resp. I_2) such that, for each finite subset H_1 (resp. H_2) of I_1 (resp. I_2) containing J_1 (resp. J_2), we have $\sum_{\iota \in H_1} x_\iota \in S_1 + V$ (resp. $\sum_{\iota \in H_2} x_\iota \in s_2 + V$). If we put $J_0 = J_1 \cup J_2$, it follows that, for each finite subset H of I which contains J_0, we have $\sum_{\iota \in H} x_\iota \in S_1 + s_2 + V + V$, and the result follows.

4. SUMMABLE FAMILIES IN A PRODUCT OF GROUPS

PROPOSITION 4. *Let* $G = \prod_{\lambda \in L} G_\lambda$ *be the product of a family of Hausdorff commutative groups. Then a family* $(x_\iota)_{\iota \in I}$ *of points of* G *is summable if and only if, for each* $\lambda \in L$, *the family* $(\mathrm{pr}_\lambda\, x_\iota)_{\iota \in I}$ *is summable; and if* s_λ *is the sum of this family, then* $s = (s_\lambda)$ *is the sum of the family* (x_ι).

This follows immediately from the condition for convergence, with respect to a filter, of a function which takes its values in a product space (Chapter I, § 1, no. 6, Corollary 1 to Proposition 10); in effect, for each finite subset J of I, we have

$$\mathrm{pr}_\lambda\Big(\sum_{\iota \in J} x_\iota\Big) = \sum_{\iota \in J} \mathrm{pr}_\lambda x_\iota.$$

5. IMAGE OF A SUMMABLE FAMILY UNDER A CONTINUOUS HOMOMORPHISM

PROPOSITION 5. *Let f be a continuous homomorphism of a commutative group* G *into a commutative group* G′. *If* (x_ι) *is a summable family in* G, *then* $(f(x_\iota))$ *is a summable family in* G′, *and we have*

$$\Sigma f(x_\iota) = f(\Sigma x_\iota). \tag{3}$$

If J is any finite subset of the index set, then $f\Big(\sum_{\iota \in J} x_\iota\Big) = \sum_{\iota \in J} f(x_\iota)$, and the image under f of a convergent filter base is a convergent filter base (Chapter I, § 7, no. 4, Corollary 1 to Proposition 9).

PROPOSITION 6. *Let* (x_ι), (y_ι) *be two summable families, with the same index set, in a group* G. *Then the families* $(-x_\iota)$, (nx_ι) $(n \in \mathbf{Z})$, $(x_\iota + y_\iota)$ *are summable, and we have*

$$\Sigma(-x_\iota) = -\Sigma x_\iota, \tag{4}$$

$$\Sigma(nx_\iota) = n\Sigma x_\iota, \tag{5}$$

$$\Sigma(x_\iota + y_\iota) = \Sigma x_\iota + \Sigma y_\iota. \tag{6}$$

For $x \to -x$ and $x \to nx$ are continuous homomorphisms of G into G; on the other hand, if (x_ι) and (y_ι) are summable, then the family (x_ι, y_ι) is summable in G × G, and since $(x, y) \to x + y$ is a continuous homomorphism of G × G into G, we deduce (6).

Remark. Propositions 4 and 5 again apply to the case, mentioned earlier, of summable families in a topological space X which has an associative and commutative law of composition; the same holds for Proposition 3 and formula (6) if we suppose in addition that $x + y$ is continuous on X × X.

6. SERIES

Consider a sequence of points $(x_n)_{n \in \mathbf{N}}$ in a Hausdorff commutative group, and form the sequence of *partial sums* $s_n = \sum_{p=0}^{n} x_p$ $(n \in \mathbf{N})$. The mapping $(x_n) \to (s_n)$ is a *bijection* of the set $G^{\mathbf{N}}$ of sequences (x_n) of points of G, onto itself; for if the sequence (s_n) is given, the sequence (x_n) is determined by the relations $x_0 = s_0$, $x_n = s_n - s_{n-1}$ $(n \geqslant 1)$.

The *series defined by the sequence* (x_n), or the *series whose general term* is x_n [or simply the *series* (x_n), by abuse of language, if there is no risk of confusion] is defined to be the *pair* of sequences (x_n) and (s_n) thus associated. The series defined by the sequence (x_n) is said to be *convergent* if the sequence (s_n) is convergent; the limit of this sequence is called *the sum of the series* and is written $\sum_{n=0}^{\infty} x_n$ (or $\sum_{n=0}^{\infty} x_n$, by abuse of notation).

If the series whose general term is x_n is *convergent*, we shall sometimes allow ourselves, by abuse of language, to refer to it as "the series $\sum_{n=0}^{\infty} x_n$" or "the series $x_0 + x_1 + \cdots + x_n + \cdots$".

A *necessary* condition for the convergence of the series whose general term is x_n is that the sequence (s_n) should be a *Cauchy sequence*, that is to say, that for each neighbourhood V of the origin in G there is an integer n_0 such that for each pair of integers $n \geqslant n_0$, $p > 0$, we have

$$s_{n+p} - s_n = \sum_{i=n+1}^{n+p} x_i \in \mathrm{V}.$$

If G is *complete*, this condition is also *sufficient* (*Cauchy's criterion for series*).

If the series whose general term is x_n is convergent, we have in particular $\lim_{n \to \infty} (s_n - s_{n-1}) = \lim_{n \to \infty} x_n = 0$; but this *necessary* condition for convergence is by no means sufficient in general, even if G is complete (see Chapter IV, § 7).

Proposition 7. *If the series defined by the sequences (x_n) and (y_n) are convergent, then so are the series defined by the sequences $(-x_n)$ and $(x_n + y_n)$, and we have*

$$\sum_{n=0}^{\infty} (-x_n) = -\sum_{n=0}^{\infty} x_n, \tag{7}$$

$$\sum_{n=0}^{\infty} (x_n + y_n) = \sum_{n=0}^{\infty} x_n + \sum_{n=0}^{\infty} y_n. \tag{8}$$

This is an obvious consequence of the continuity of $-x$ on G, and of $x + y$ on $\mathrm{G} \times \mathrm{G}$.

Corollary. *If (x_n), (y_n) are two sequences of points of G such that $x_n = y_n$ except for a finite number of indices, and if the series whose general term is x_n converges, then so does the series whose general term is y_n.*

For the series whose general term is $x_n - y_n$ has all its terms zero from a certain index onwards.

This corollary may be put in the form that *we may alter arbitrarily a finite number of terms of a convergent series without it ceasing to be convergent.*

In particular, if $y_n = 0$ for $n < m$, and $y_n = x_n$ for $n \geqslant m$, we see that the series whose general term is y_n converges if and only if the series whose general term is x_n converges; its sum is denoted by $\mathop{\mathsf{S}}\limits_{n=m}^{\infty} x_n$ and is called the *residue* of index m of the series (x_n). Since

$$\mathop{\mathsf{S}}\limits_{n=m}^{\infty} x_n = \mathop{\mathsf{S}}\limits_{n=0}^{\infty} x_n - s_{m-1},$$

the residue of index m of a convergent series *tends to* 0 as m tends to $+\infty$.

If a sequence $(x_n)_{n\in \mathrm{I}}$ has as index set an infinite subset I of **N**, and if φ denotes the *strictly order-preserving* bijection of **N** onto I, then the series defined by the sequence $(x_{\varphi(n)})_{n\in\mathbf{N}}$ is called, by abuse of language, the *series* defined by the sequence $(x_n)_{n\in\mathrm{I}}$; if this series is convergent, its sum is denoted by $\mathop{\mathsf{S}}\limits_{n\in\mathrm{I}} x_n$. It is immediately verified that this series converges if and only if the series whose general term is (z_n) converges, where $z_n = x_n$ if $n \in \mathrm{I}$ and $z_n = 0$ if $n \in \complement \mathrm{I}$.

It is important to notice that if the series defined by a sequence $(x_n)_{n\in\mathbf{N}}$ converges, there can exist infinite subsets I of **N** such that the series defined by the subsequence $(x_n)_{n\in\mathrm{I}}$ *does not converge* (see Exercise 5 and Chapter IV, § 7).

Propositions 4 and 5 extend immediately to series, and we leave their formulation in this case to the reader.

Proposition 8 (Restricted associativity of series). *Let (k_n) be a strictly increasing sequence of integers $\geqslant 0$. If the series whose general term is x_n converges, and if we put $u_n = \sum\limits_{p=k_{n-1}}^{k_n-1} x_p$, then the series whose general term is u_n is convergent, and we have $\mathop{\mathsf{S}}\limits_{n=0}^{\infty} u_n = \mathop{\mathsf{S}}\limits_{n=0}^{\infty} x_n$.*

For the sequence of partial sums of the series (u_n) is a *subsequence* (s_{k_n-1}) of the sequence (s_n) of partial sums of the series (x_n).

7. COMMUTATIVELY CONVERGENT SERIES

Let (x_n) be a *summable* sequence in G, and let $s = \sum\limits_{n\in\mathbf{N}} x_n$ be its sum. Then for each neighbourhood V of 0, there exists $\mathrm{J}_0 \in \mathfrak{F}(\mathbf{N})$ such that $s_\mathrm{J} \in s + \mathrm{V}$ whenever $\mathrm{J} \in \mathfrak{F}(\mathbf{N})$ and $\mathrm{J}_0 \subset \mathrm{J}$. Let m be the largest integer in J_0; then if $n \geqslant m$ we have $s_n \in s + \mathrm{V}$, and therefore the *series* (x_n) is convergent and its sum is s. But the converse is *false*:

the sequence of terms of a convergent series can very well fail to be summable (see Chapter IV, § 7).

Moreover, the definition of a convergent series essentially involves the order structure of **N**. If the series (x_n) is convergent, and if σ is a *permutation* of **N**, then the series $(x_{\sigma(n)})$ is not necessarily convergent (cf. Chapter IV, § 7, Exercise 15).

DEFINITION 2. *A series defined by a sequence* (x_n) *is said to be commutatively convergent if, for each permutation* σ *of* **N**, *the series defined by the sequence* $(x_{\varphi(n)})$ *is convergent.*

PROPOSITION 9. *The series defined by the sequence* (x_n) *is commutatively convergent if and only if the sequence* (x_n) *is summable; and then, for each permutation* σ *of* **N**, *we have*

$$\mathop{\mathsf{S}}_{n=0}^{\infty} x_\sigma = \sum_{n\in\mathbf{N}} x_n.$$

If the sequence is summable, then clearly the series is commutatively convergent. To prove the reverse implication we shall argue by *reductio ad absurdum* and suppose that the series (x_n) is commutatively convergent but that the sequence (x_n) is not summable. The image of the filter Φ under the mapping $H \to s_H$ therefore cannot be a Cauchy filter base in G, otherwise this filter base would converge, since by hypothesis it has a cluster point (Chapter II, § 3, no. 2, Proposition 5, Corollary 2). Hence there is a neighbourhood V of 0 in G such that, for each finite subset J of **N**, there is a finite subset H of **N** which does not meet J and is such that $\sum_{n\in H} x_n \notin V$. Hence, by induction, we can define a partition of **N** into *finite* subsets H_k $(k \in \mathbf{N})$ such that $\sum_{n\in H_k} x_n \notin V$ for an infinite number of indices k. Clearly there is a permutation σ of **N** such that for each k, the values of n for which $\sigma(n) \in H_k$ are consecutive. If σ is such a permutation, then the series whose general term is $x_{\sigma(n)}$ cannot be convergent, and we have a contradiction.

If the group G is written *multiplicatively*, the *infinite product defined by a sequence* (x_n) of points of G (or *infinite product whose general term is* x_n, or even *product* x_n if there is no possibility of confusion) is defined to be the pair formed by the sequence (x_n) and the sequence of partial products $p_n = \prod_{k=0}^{n} x_k$. The infinite product is said to *converge* if the sequence (p_n) converges, and the limit of this sequence is denoted by $\mathop{\mathsf{P}}_{n=0}^{\infty} x_n$ (or $\prod_{n=0}^{\infty} x_n$ by abuse of notation). We leave to the reader the task of transcribing into multiplicative notation the properties of series which we have established.

6. TOPOLOGICAL GROUPS WITH OPERATORS; TOPOLOGICAL RINGS, DIVISION RINGS AND FIELDS

1. TOPOLOGICAL GROUPS WITH OPERATORS

On a set G, a structure of a *group with operators* and a topology are said to be *compatible* if the topology and the group structure of G are compatible (§ 1, no. 1) and if in addition the endomorphisms of G produced by the operators are *continuous*. The set G, together with the given topology and structure of group with operators, is then said to be a *topological group with operators*.

If H is a stable subgroup of a topological group G with operators, then the topology induced on H by the topology of G is compatible with the structure of group with operators on H. Furthermore:

PROPOSITION 1. *If* H *is a stable subgroup of a topological group* G *with operators, then the closure* $\overline{H}$ *of* H *in* G *is a stable subgroup of* G.

We know already (§ 2, no. 1, Proposition 1) that $\overline{H}$ is a subgroup of G. Also if α is any operator on G, the image of H under the continuous mapping $x \to x^{\alpha}$ is contained in H, and therefore the image of $\overline{H}$ is contained in $\overline{H}$ (Chapter I, § 2, no. 1, Theorem 1).

Let H be a stable normal subgroup of a topological group G with operators. Then for each operator α on G, the mapping of G/H into itself induced by $x \to x^{\alpha}$ is continuous (§ 2, no. 8, Remark 3), and the structure of group with operators on G/H is therefore compatible with the quotient topology on G/H.

Let $(G_{\iota})_{\iota \in I}$ be a family of topological groups with operators, where each G_{ι} is assumed to have the same operator set Ω. For each $\alpha \in \Omega$, the mapping $x \to ((\mathrm{pr}_{\iota} x)^{\alpha})$ of $G = \prod_{\iota \in I} G_{\iota}$ into itself is continuous (Chapter I, § 4, no. 1, Proposition 1), and the structure of groups with operators on G is therefore compatible with the product topology on G.

If G is a Hausdorff topological group with operators and if G has a completion $\hat{G}$ (§ 3, no. 4), then every endomorphism $x \to x^{\alpha}$ of G defined by an operator on G can be extended by continuity to an endomorphism of $\hat{G}$ (§ 3, no. 3, Proposition 5). Hence $\hat{G}$ has the structure of a topological group with operators, and the operator set is the same for $\hat{G}$ as for G.

2. TOPOLOGICAL DIRECT SUM OF STABLE SUBGROUPS

Since the study of commutative groups with operators is equivalent to the study of modules we shall sometimes allow ourselves to use the terminology proper to modules for arbitrary commutative groups with operators; thus we may speak of *linear mappings* instead of homomorphisms of commutative groups with operators, and we may use the word *projector* to denote an idempotent endomorphism of a commutative group with operators.

If a commutative topological group E with operators (written additively) is the direct sum of a *finite* family $(M_i)_{1 \leqslant i \leqslant n}$ of stable subgroups, then the canonical bijection $(x_i) \to \sum_{i=1}^{n} x_i$ of the product group $\prod_{i=1}^{n} M_i$ onto E is *continuous*, but is *not necessarily a homeomorphism.*

Definition 1. *Let* E *be a commutative topological group with operators, and let* $(M_i)_{1 \leqslant i \leqslant n}$ *be a finite family of stable subgroups of* E, *such that* E *is the direct sum of the* M_i. *Then* E *is said to be the topological direct sum of the* M_i *if the canonical mapping* $(x_i) \to \sum_{i=1}^{n} x_i$ *of the product group* $\prod_{i=1}^{n} M_i$ *onto* E *is a homeomorphism* (and therefore an isomorphism of topological groups with operators).

Proposition 2. *Let* E *be a commutative topological group with operators which is the direct sum of stable subgroups* M_i $(1 \leqslant i \leqslant n)$. *Let* $(p_i)_{1 \leqslant i \leqslant n}$ *be the family of projectors associated with the decomposition* $E = \sum_{i=1}^{n} M_i$. *Then* E *is the topological direct sum of the* M_i *if and only if the* p_i *are continuous.*

For the mapping $x \to (p_i(x))$ is the inverse of the mapping $(x_i) \to \sum_{i=1}^{n} x_i$.

Since $1 = \sum_{i=1}^{n} P_i$ (where 1 denotes the identity mapping of E) it is sufficient for $n - 1$ of the projectors p_i to be continuous in order that the n th should also be continuous.

If E is the topological direct sum of two stable subgroups M, N, then N is said to be a *topological complement* of M in E; this is the case if and only if the canonical mapping of E/M onto is an *isomorphism* of topological groups with operators.

COROLLARY. *Let* E *be a commutative topological group with operators, and let* M *be a stable subgroup of* E. *Then the following conditions are equivalent:*

a) M *has a topological complement in* E.

b) *There is a continuous projector* p *of* E *into* E *such that* $p(E) = M$.

c) *The identity mapping of* M *can be extended to a continuous linear mapping of* E *onto* M.

It follows from Proposition 2 that a) implies b), and it is clear that b) implies c). Finally, if p is a continuous linear mapping of E onto M which extends the identity mapping of M, then p is a continuous projector, and the projectors p and $1 - p$ are associated with the direct sum decomposition $E = M + N$, where $N = \overset{-1}{p}(0)$.

Remarks. 1) To avoid confusion, we shall sometimes say that a stable subgroup of E which is a complement of M (in the sense of the structure of group with operators, without a topology) is an *algebraic complement* of M.

2) If a *Hausdorff* commutative topological group with operators is the topological direct sum of a family $(M_i)_{1 \leqslant i \leqslant n}$ of stable subgroups, then each of the subgroups M_i is *closed* in E, for M_i is the set of all $x \in E$ such that $p_i(x) = x$ (Chapter I, § 8, no. 1, Proposition 2).

PROPOSITION 3. *Let* E, F *be two commutative topological groups with operators, and let* u *be a continuous linear mapping of* E *into* F. *In order that there should exist a continuous linear mapping* v *of* F *into* E *such that* $u \circ v$ *is the identity mapping of* F (in which case u is said to be *right invertible* and v is said to be a *right inverse* of u) *it is necessary and sufficient that* u *is a strict morphism* (§ 2, no. 8) *of* E *onto* F *and that* $\overset{-1}{u}(0)$ *has a topological complement in* E.

The conditions are *necessary*. For we have then $u(v(F)) = F$ and *a fortiori* $u(E) = F$; furthermore, if $p = v \circ u$, then p is a continuous linear mapping of E into itself such that $p^2 = p$; therefore (Corollary to Proposition 2) $p(E) = v(u(E)) = v(F)$ has a topological complement $\overset{-1}{p}(0)$ in E; but since $u(p(x))) = u(x)$ by hypothesis, we have $\overset{-1}{u}(0) = \overset{-1}{p}(0)$. Finally, the bijective mapping of $E/\overset{-1}{p}(0)$ onto F, associated with u, is the composition of the bijective mapping of $E/\overset{-1}{u}(0)$ onto $v(F)$, associated with p, and the restriction of u to $v(F)$; since v is continuous, both these mappings are isomorphisms, and therefore u is a strict morphism of E onto F.

The conditions are *sufficient*. For if φ is the canonical homomorphism of E onto $E/\overset{-1}{u}(0)$, to say that $\overset{-1}{u}(0)$ has a topological complement M in E is to say that the restriction of φ to M is an isomorphism of M onto $E/\overset{-1}{u}(0)$. Since on the other hand $u = w \circ \varphi$, where w is an isomorphism of $E/\overset{-1}{u}(0)$ onto F, we see that the restriction of u to M is an isomorphism of M onto F, and the inverse isomorphism v is therefore such that $u \circ v$ is the identity mapping of F onto itself.

PROPOSITION 4. *Let* E, F *be two commutative topological groups with operators, and let* u *be a continuous linear mapping of* E *into* F. *In order that there should exist a continuous linear mapping* v *of* F *into* E *such that* $v \circ u$ *is the identity mapping of* E *onto itself* (in which case u is said to be *left invertible* and v is said to be a *left inverse* of u) *it is necessary and sufficient that* u *is a (topological) isomorphism of* E *onto* $u(E)$, *and that* $u(E)$ *has a topological complement in* F.

The conditions are *sufficient*, for if they are satisfied we obtain a left inverse v of u by taking the composition of the isomorphism of $u(E)$ onto E which is the inverse of u, with a continuous projector of F onto $u(E)$.

The conditions are *necessary*. For the relation $v(u(x)) = x$ shows that $\overset{-1}{u}(0) = \{0\}$; u is therefore a bijection of E onto $u(E)$, and since the restriction of v to $u(E)$ is continuous, it follows that u is an isomorphism of E onto $u(E)$. On the other hand, if we put $q = u \circ v$, then q is a continuous linear mapping of F onto $u(E)$ such that $q^2 = q$, which proves (Corollary to Proposition 2) that $u(E)$ has a topological complement in F.

3. TOPOLOGICAL RINGS

DEFINITION 2. *A topological ring is a set* A *which carries a ring structure and a topology satisfying the following axioms:*

(AT_I). *The mapping* $(x, y) \to x + y$ *of* A × A *into* A *is continuous.*

(AT_{II}). *The mapping* $x \to -x$ *of* A *into* A *is continuous.*

(AT_{III}). *The mapping* $(x, y) \to xy$ *of* A × A *into* A *is continuous.*

The first two axioms express that the topology of A is compatible with its *additive group* structure (§ 1, no. 1).

If a ring structure and a topology are given on a set A, they are said to be *compatible* if they satisfy axioms (AT_I), (AT_{II}) and (AT_{III}).

Examples. 1) On any ring A, the *discrete* topology is compatible with the ring structure. A topological ring whose topology is discrete is called a *discrete* ring.

* 2) We shall see in Chapter IV that the topology of the rational line **Q** (resp. the real line **R**) is compatible with the ring structure of **Q** (resp. **R**). *

In a topological ring every left *homothety* $x \to ax$ (resp. every right homothety $x \to xa$) is continuous (and is a homeomorphism if a is a unit of A).

Since we can write identically

$$xy - x_0 y_0 = (x - x_0)(y - y_0) + (x - x_0)y_0 + x_0(y - y_0)$$

the axiom (AT_{III}) [in view of (AT_I) and (AT_{II})] is equivalent to the conjunction of the following two axioms :

$(AT_{III\,a})$. *Given any* $x_0 \in A$, *the mappings* $x \to x_0 x$ *and* $x \to x x_0$ *are continuous at the point* $x = 0$.

$(AT_{III\,b})$. *The mapping* $(x, y) \to xy$ of $A \times A$ *into* A *is continuous at the point* $(0, 0)$.

From this we can deduce a necessary and sufficient set of conditions which the *filter* $\mathfrak{V}$ *of neighbourhoods of* 0 in a ring A must satisfy in order to define a topology on A compatible with its ring structure : $\mathfrak{V}$ must satisfy axioms (GA_I) and (GA_{II}) of § 1, and also the following two axioms :

(AV_I). *For all* $x_0 \in A$ *and all* $V \in \mathfrak{V}$, *there exists* $W \in \mathfrak{V}$ such *that* $x_0 W \subset V$ *and* $W x_0 \subset V$.

(AV_{II}). *For all* $V \in \mathfrak{V}$, *there exists* $W \in \mathfrak{V}$ *such that* $WW \subset V$.

Remark. In analysis one fairly often meets rings which satisfy axioms (AT_I), (AT_{II}) and $(AT_{III\,a})$, but not $(AT_{III\,b})$. * An example is the ring of measures on a compact group, where multiplication is convolution and the topology is the vague topology. *

Example 3). Let $\mathfrak{V}$ be a *filter base* on a ring A, consisting of *two-sided ideals.* $\mathfrak{V}$ is a fundamental system of neigbourhoods of 0 for a topology compatible with the additive group structure of A, and it follows immediately from (AV_I) and (AV_{II}) that this topology is compatible with the *ring* structure of A.

Let X be a topological space, and let f and g be two mappings of X into a topological ring A. If f and g are continuous at a point $x_0 \in X$, then $f + g$, $-f$ and fg are continuous at this point. It follows that the continuous mappings of X into A form a *subring* of the ring A^X of all mappings of X into A. We see also that, if A is *commutative*, then every *polynomial in* n *variables*, with coefficients in A and defined on A^n, is *continuous* on A^n. Again, let f and g be two mappings of a set X, *filtered* by a filter $\mathfrak{F}$, into a *Hausdorff* topological ring A; if $\lim_{\mathfrak{F}} f$ and $\lim_{\mathfrak{F}} g$ exist, then so do $\lim_{\mathfrak{F}} (f + g)$, $\lim_{\mathfrak{F}}(-f)$ and $\lim_{\mathfrak{F}} (fg)$, and we have (Chapter I, § 7, no. 4, Proposition 9, Corollary 1, and § 8, no. 1, Proposition 1)

(1) $$\lim_{\mathfrak{F}}(f + g) = \lim_{\mathfrak{F}} f + \lim_{\mathfrak{F}} g,$$

(2) $$\lim_{\mathfrak{F}} (-f) = -\lim_{\mathfrak{F}} f,$$

(3) $$\lim_{\mathfrak{F}} (fg) = (\lim_{\mathfrak{F}} f)(\lim_{\mathfrak{F}} g).$$

4. SUBRINGS; IDEALS; QUOTIENT RINGS; PRODUCTS OF RINGS

If H is a subring of a topological ring A, then the topology induced on H by that of A is compatible with the ring structure of H. The topological ring structure thus defined on H is said to be *induced* by that of A.

PROPOSITION 5. *Let* H *be a dense subring of a topological ring* A, *and let* K *be a subring* (resp. *left ideal, right ideal, two-sided ideal*) *of* H. *Then the closure* $\overline{K}$ *of* K *in* A *is a subring* (resp. *left ideal, right ideal, two-sided ideal*) *of* A.

The proof is the same as for Proposition 8 of § 2, n. 3 : if for example K is a left ideal in H, then the mapping $(z, x) \to zx$ is continuous on $A \times A$ and maps $H \times K$ into K; hence it maps $A \times \overline{K} = \overline{H} \times \overline{K}$ into $\overline{H}$.

Let H be a *two-sided* ideal in a topological ring A. By the same argument as for quotient groups, we see that the *quotient* of the topology of A by the relation $x - y \in H$ is compatible with the ring structure of A/H. In particular, if A is not Hausdorff, then the closure N of $\{0\}$ in A is a *closed two-sided ideal*, by Proposition 5; the quotient ring, which is Hausdorff (§ 2, no. 5, Proposition 13) is called the *Hausdorff ring associated with* A.

Let $(A_\iota)_{\iota \in I}$ be a family of topological rings. On the set $A = \prod_{\iota \in I} A_\iota$, the *product* of the topologies of the A_ι is compatible with the product of the ring structures of the A_ι (same proof as for product groups); the topological ring A thus defined is called the *product* of the topological rings A_ι.

5. COMPLETION OF A TOPOLOGICAL RING

When we speak of *the* uniformity of a topological ring, we always mean the uniformity of its *additive group*, unless the contrary is expressly stated; in particular, A is said to be a *complete* ring if the additive group of A is complete.

Let A be a *Hausdorff* topological ring; as an additive group A can be considered as a dense subgroup of a *complete Hausdorff commutative* group $\hat{A}$, which is determined up to isomorphism (§ 3, no. 5, Theorem 2). In order that we should be able to consider A as a *subring* of a *complete ring*, it is necessary to be able to extend the function xy by continuity to the space $\hat{A} \times \hat{A}$. The possibility of this extension will follow from the following more general theorem :

THEOREM 1. *Let* E, F, G *be three complete Hausdorff commutative groups; let* A *be a dense subgroup of* E *and let* B *be a dense subgroup of* F. *If* f *is a* **Z** *continuous* **Z**-*bilinear* (*) *mapping of* A × B *into* G, *then* f *can be extended by continuity to a continuous* **Z**-*bilinear mapping of* E × F *into* G.

Let (x_0, y_0) be an arbitrary point of E × F, and let $\mathfrak{U}$, $\mathfrak{V}$ be the traces on A, B respectively of the neighbourhood filters of x_0, y_0 ($\mathfrak{U}$, $\mathfrak{V}$ are filters, by hypothesis). To show that f can be extended by continuity, it is enough to show that $f(\mathfrak{U} \times \mathfrak{V})$ is a *Cauchy filter-base* on G (Chapter II, § 3, no. 6, Proposition 11). Consider the identity

$$f(x', y') - f(x, y) = f(x' - x, y_1) + f(x_1, y' - y) + f(x' - x, y' - y_1).$$

We shall show that by taking (x, y) and (x', y') in a sufficiently small set of $\mathfrak{U} \times \mathfrak{V}$, and by choosing x_1 and y_1 suitably, we can make each of the terms on the right-hand side very small. Let W be any neighbourhood of 0 in G; since f is continuous at $(0, 0) \in A \times B$, there is a set $U \in \mathfrak{V}$ and a set $V \in \mathfrak{V}$ such that $f(x' - x, y' - y) \in W$ whenever $x \in U$, $x' \in U$, $y \in V$, $y' \in V$. Take a point $x_1 \in U$ and a point $y_1 \in V$; then for all x, x' in U and all y, y' in V we shall have

$$f(x' - x, y' - y_1) + f(x - x_1, y' - y) \in W + W.$$

On the other hand, the partial mapping $x \to f(x, y_1)$ is continuous on A; hence there is a set $U' \subset U$, belonging to $\mathfrak{U}$, and such that, whenever x and x' belong to U', we have $f(x' - x, y_1) \in W$. Likewise there exists $V' \subset V$ belonging to $\mathfrak{V}$ such that, whenever y and y' belong to V', we have $f(x_1, y' - y) \in W$. Consequently, if (x, y) and (x', y') are any two points of $U' \times V'$, then

$$f(x', y') - f(x, y) \in W + W + W + W;$$

this proves the existence of the extension $\bar{f}$ of f. The fact that $\bar{f}$ is **Z**-*bilinear* is an immediate consequence of the principle of extension of identities (Chapter I, § 8, no. 1, Corollary 1 to Proposition 2).

Q.E.D.

In the application of this theorem to a Hausdorff topological ring A, we take E, F and G to be $\hat{A}$, A and B to be the ring A, and f to be

---

(*) f is said to be **Z**-bilinear if, for all elements x, x' of A and all elements $y, y' \in B$ we have

$$f(x + x', y) = f(x, y) + f(x', y)$$

and

$$f(x, y + y') = f(x, y) + f(x, y').$$

the **Z**-bilinear mapping $(x, y) \to xy$, which by hypothesis is continuous. We denote again by xy the value of the extended function on $\hat{A} \times \hat{A}$; this function is a law of composition on $\hat{A}$, and to say that it is **Z**-bilinear means that it is distributive on both sides with respect to addition; and it is also *associative*, by the principle of extension of identities. Consequently:

PROPOSITION 6. *A Hausdorff topological ring* A *is isomorphic to a dense subring of a complete Hausdorff ring* $\hat{A}$, *which is determined up to isomorphism* (and is called the *completion* of A).

If A is *commutative* (resp. *has an identity element*) then the same is true of $\hat{A}$ (principle of extension of identities).

Let A be a topological ring, not necessarily Hausdorff; let N be the closure of $\{0\}$ in A, and let $A' = A/N$ be the Hausdorff ring associated with A. Then $\hat{A}'$, the completion of A′, is called the *Hausdorff completion* of A and is also denoted by $\hat{A}$. One shows as in § 3, no. 4, Proposition 8 that every continuous ring homomorphism u of A with a *complete Hausdorff* topological ring C can be uniquely factorized as $u = v \circ \varphi$, where v is a continuous homomorphism of $\hat{A}$ into C and φ is the canonical mapping of A into $\hat{A}$. If A, B are two topological rings, and $u : A \to B$ is a continuous homomorphism, there is therefore a unique continuous homomorphism $\hat{u} : \hat{A} \to \hat{B}$ such that the diagram

$$\begin{array}{ccc} A & \xrightarrow{u} & B \\ {\scriptstyle\varphi}\downarrow & & \downarrow{\scriptstyle\psi} \\ \hat{A} & \xrightarrow{\hat{u}} & \hat{B} \end{array}$$

is commutative (φ and ψ being the canonical mappings); we have only to apply the preceding result to $\psi \circ u$.

6. TOPOLOGICAL MODULES

DEFINITION 3. *Given a topological ring* A *with an identity element, a topological left* A*-module is a set* E, *together with:*

1) *a left* A*-module structure;*

2) *a topology compatible with the additive group structure of* E, *and satisfying the following axiom:*

(MT) *The mapping* $(\lambda, x) \to \lambda x$ *of* $A \times E$ *into* E *is continuous.*

We define similarly the notion of a topological *right* A-module; since every right A-module can be considered as a left A°-module, where A° is the opposite ring of A, and since the topology of A is compatible with the ring structure of A°, there is no need to distinguish between topological right A-modules and topological left A°-modules.

Examples. * 1) A topological vector space over **R** (resp. **C**) is a topological **R** (resp. **C**) module.

2) Let A be a ring and let $\mathfrak{B}$ be a filter base on A consisting of two-sided ideals of A; let E be a left A-module. If we give A the topology (compatible with its ring structure) for which $\mathfrak{B}$ is a fundamental system of neighbourhoods of o (no. 3, Example 3), and E the topology (compatible with its additive group structure) in which the sets $\mathfrak{a}E$, as $\mathfrak{a}$ runs through $\mathfrak{B}$, form a fundamental system of neighbourhoods of o (§ 1, no. 2, Example), it is immediately verified that E is a topological A-module.

Remark. Given a topological ring A, consider, on a left A-module E, a topology compatible with the additive group structure of E. By virtue of the identity

$$\lambda x - \lambda_0 x_0 = (\lambda - \lambda_0)x_0 + \lambda_0(x - x_0) + (\lambda - \lambda_0)(x - x_0)$$

the axiom (MT) is equivalent to the conjunction of the following three axioms :

(MT$_I'$). *For each* $x_0 \in$ E, *the mapping* $\lambda \to \lambda x_0$ *is continuous at the point* $\lambda = $ o.

(MT$_{II}'$). *For each* $\lambda_0 \in$ A, *the mapping* $x \to \lambda_0 x$ *is continuous at the point* $x = $ o.

(MT$_{III}'$). *The mapping* $(\lambda, x) \to \lambda x$ *is continuous at the point* (o, o).

We deduce from this a necessary and sufficient set of conditions that the *filter* $\mathfrak{V}$ *of neighbourhoods of* o in an A-module E must satisfy in order to define a topology on E compatible with its module structure; $\mathfrak{V}$ must satisfy the axioms (GA$_I$) and (GA$_{II}$) of § 1, no. 2, and in addition must satisfy the following three axioms :

(MV$_I$). *For each* $x_0 \in$ E *and* $V \in \mathfrak{V}$, *there is a neigbourhood* S *of* o *in* A *such that* $S.x_0 \subset V$.

(MV$_{II}$). *For each* $\lambda_0 \in$ A *and* $V \in \mathfrak{V}$, *there exists* $W \in \mathfrak{V}$ *such that* $\lambda_0 W \subset V$.

(MV$_{III}$). *For each* $V \in \mathfrak{V}$ *there exists* $U \in \mathfrak{V}$ *and a neighbourhood* T *of* o *in* A *such that* $T.U \subset V$.

Every commutative topological group is a topological **Z**-module when the ring **Z** is given the discrete topology.

If M is a submodule of a topological A-module E, it is clear that the topology induced on M by the topology of E is compatible with the module structure of M. Moreover, on the quotient A-module E/M, the topology which is the quotient by M of the topology of E is compatible with the A-module structure. To see this it is enough to show that the mapping $(\lambda, \dot{x}) \to \lambda\dot{x}$ of $A \times (E/M)$ onto E/M is continuous (where $x \to \dot{x}$ denotes the canonical mapping of E onto E/M). Now, since we may identify the additive topological groups $A \times (E/M)$ and $(A \times E)/(\{0\} \times M)$ (§ 2, no. 9, Corollary to Proposition 26), it is enough to show that the mapping $(\lambda, x) \to \lambda\dot{x}$ of $A \times E$ into E/M is continuous; and this is immediate, since the mapping in question is the composition of $x \to \dot{x}$ and $(\lambda, x) \to \lambda x$.

Let $(E_\iota)_{\iota \in I}$ be an arbitrary family of topological A-modules, and let $E = \prod_{\iota \in I} E_\iota$ be the A-module which is the product of the E_ι. Then the product topology on E is compatible with the A-module structure of E. To prove this it is enough to show that the mapping $(\lambda, x) \to (\lambda . \mathrm{pr}_\iota x)_{\iota \in I}$ of $A \times E$ into E is continuous, or (by Proposition 1 of Chapter I, § 4, no. 1) that for each index $\varkappa \in I$ the mapping $(\lambda, x) \to \lambda . \mathrm{pr}_\varkappa x$ is a continuous mapping of $A \times E$ into $E_\varkappa$; but this mapping is the composition of $(\lambda, x_\varkappa) \to \lambda x_\varkappa$ and $(\lambda, x) \to (\lambda, \mathrm{pr}_\varkappa x)$, both of which are continuous.

Let A be a Hausdorff topological ring and E a Hausdorff topological A-module. Let $\hat{E}$ be the additive group which is the completion of the commutative topological group E (§ 3, no. 5, Theorem 2). The **Z**-bilinear mapping $(\lambda, x) \to \lambda x$ of the product $A \times E$ of the additive groups A, E into the additive group E can be extended by continuity to a **Z**-bilinear mapping of $\hat{A} \times \hat{E}$ into $\hat{E}$ (no. 5, Theorem 1), and this mapping we continue to denote by $(\lambda, x) \to \lambda x$. By virtue of the principle of extension of identities, we have $\lambda(\mu x) = (\lambda\mu)x$ for $\lambda \in \hat{A}$, $\mu \in \hat{A}$ and $x \in \hat{E}$, and $1 . x = x$ for all $x \in \hat{E}$; the external law $(\lambda, x) \to \lambda x$ therefore defines an $\hat{A}$-module structure on $\hat{E}$ compatible with its topology. The topological $\hat{A}$-module $\hat{E}$ thus defined is called the *completion* of the topological A-module E.

Let E be a topological module over a topological ring A, where neither A nor E is necessarily Hausdorff. Let N (resp. F) be the closure of $\{0\}$ in A (resp. E). N is a two-sided ideal of A (no. 4, Proposition 5) and F is a sub-A-module of E (no. 1, Proposition 1); furthermore, by continuity we have $\lambda x \in F$ whenever $\lambda \in N$ or $x \in F$. We can therefore define, by passing to the quotients, a mapping $(\dot{\lambda}, \dot{x}) \to \dot{\lambda}\dot{x}$ of $(A/N) \times (E/F)$ into E/F; it is easily verified (by use of the Corollary to Proposition 26 of § 2, no. 9) that this mapping is continuous, and therefore defines a structure of a topological (A/N)-module on E/F. If we put $B = A/N$ and $L = E/F$, then the B-module L is called the Hausdorff module *associated* with E; its completion $\hat{L}$ is a topological module over the Hausdorff completion $\hat{A}$ (equal by definition to $\hat{B}$) of A (no. 5), and this module $\hat{L}$ is called the *Hausdorff completion* of E and is denoted by $\hat{E}$. We see as in § 3, no. 4, Proposition 8 that every continuous homomorphism $u : E \to G$ of E into a complete Hausdorff $\hat{A}$-module G factorizes uniquely into $u = v \circ \varphi$, where v is a continuous homomorphism of $\hat{E}$ into G and φ is the canonical mapping of E into $\hat{E}$. We conclude that if E, E′ are two topological A-modules and $u : E \to E'$ is a continuous homomorphism, then there is a unique continuous homomorphism $\hat{u} : \hat{E} \to \hat{E}'$ such that the diagram

$$\begin{array}{ccc} E & \xrightarrow{u} & E' \\ {\scriptstyle\varphi}\downarrow & & \downarrow{\scriptstyle\varphi'} \\ \hat{E} & \xrightarrow{\hat{u}} & \hat{E}' \end{array}$$

is commutative, φ and φ' being the canonical mappings.

The definitions and results of this section apply equally well to *pseudo-modules* over an arbitrary topological ring, by deleting all mention of the identity element of the ring.

7. TOPOLOGICAL DIVISION RINGS AND FIELDS

In what follows, and in Chapters IV and V, if K is a *division ring* we shall denote by K^* the *multiplicative group* of non-zero elements of K.

DEFINITION 4. *A topological division ring is a set* K *carrying a division ring structure and a topology compatible with the ring structure of* K, *and satisfying in addition the following axiom:*

(KT) *The mapping* $x \to x^{-1}$ *of* K^* *into* K^* *is continuous.*

A commutative topological division ring is called a topological field.

A division ring structure and a topology on a set K are said to be *compatible* if the corresponding ring structure and the topology are compatible and if in addition axiom (KT) is satisfied.

Examples. 1) On any division ring K, the *discrete* topology is compatible with the division ring structure. A topological division ring whose topology is discrete is called a *discrete* division ring.

* 2) The topology of the rational line **Q** (resp. the real line **R**) is compatible with the field structure of **R** (resp. **R**) (see Chapter IV, § 3). *

Definition 4 shows that, if K is a topological division ring, then the topology *induced* by that of K on the multiplicative group K^* is compatible with the group structure of K^*.

If $a \neq 0$, the homotheties $x \to ax$ and $x \to xa$ are homeomorphisms of K onto itself; and so is the mapping $x \to ax + b$ for all $b \in K$. Note that the homotheties $x \to ax$ and $x \to xa$ are *automorphisms* of the (topological) *additive group* of K if $a \neq 0$. If V is any neighbourhood of 0 in K, it follows therefore that aV and Va are neighbourhoods of 0 for all $a \neq 0$.

Let X be a topological space, and let f be a mapping of X into a topological division ring K. If f is continuous at a point $x_0 \in X$ and if $f(x_0) \neq 0$, then f^{-1} is continuous at x_0. In particular, if K is a topological *field*, then every *rational function* in n variables with coefficients in K is continuous at every point of K^n where its denominator does not vanish.

Likewise, if f is a mapping of a set X, filtered by a filter $\mathfrak{F}$, into a Hausdorff topological division ring K, and if $\lim_{\mathfrak{F}} f$ exists and is not 0, then $\lim_{\mathfrak{F}} f^{-1}$ exists and we have

$$(4) \qquad \lim_{\mathfrak{F}} f^{-1} = (\lim_{\mathfrak{F}} f)^{-1}.$$

If H is a *division subring* of a topological division ring K, then the topology induced on H by the topology of K is compatible with the division ring structure of H. The structure of a topological division ring thus defined on H is said to be *induced* by that of K. Furthermore, $\overline{H}$ is also a *division subring* of K (the proof is analogous to that of Proposition 5).

In a topological division ring K, the closure of the set $\{0\}$ is a two-sided ideal, by Proposition 5, and hence must be either $\{0\}$ or K. In other words, if the topology of K is not the coarsest topology (Chapter I, § 2, no. 2) then it is Hausdorff (§ 1, no. 2, Proposition 2).

8. UNIFORMITIES ON A TOPOLOGICAL DIVISION RING

If K is a topological division ring it is necessary to distinguish between: 1) the uniformity of the *additive group* of K, which is defined on K and is called the *additive uniformity* of K; and 2) the left and right uniformities of the *multiplicative group* K^*, which are defined on K^* and are called (by abuse of language) the *multiplicative* uniformities of K.

The uniformity *induced* on K^* by the additive uniformity of K is in general *distinct* from the multiplicative uniformities of K (see Exercise 17).

By Proposition 6, a Hausdorff topological division ring K can be considered as a *dense subring* of a *complete Hausdorff ring* $\hat{K}$. In order that $\hat{K}$ should be a *topological division ring* it is necessary that the mapping

$x \to x^{-1}$ be *extendable by continuity* to $(\hat{K})^*$; and this necessary condition is also sufficient, for the functions xx^{-1}, $x^{-1}x$ and 1 are then equal on $(\hat{K})^*$ by reason of the principle of extension of identities, and therefore the value of the extended function is the *inverse* of x for each $x \neq 0$ in $\hat{K}$. In other words (cf. Chapter II, § 3, no. 6, Proposition 11):

PROPOSITION 7. *The completion $\hat{K}$ of a Hausdorff topological division ring K is a topological division ring if and only if the image under the mapping $x \to x^{-1}$ of every Cauchy filter (with respect to the additive structure) which does not have a cluster point at 0, is a Cauchy filter (with respect to the additive structure).*

There are topological division rings in which this condition is not satisfied and in which the ring $\hat{K}$ has zero divisors (see Exercise 26). Moreover, even if the completion $\hat{K}$ is a topological division ring, there is no *a priori* reason to assume that the *multiplicative* structures of $\hat{K}$ are structures of a *complete* space. However, this will be the case for division rings K such that $\hat{K}$ is *locally compact* (see Chapter I, § 9, no. 7, Proposition 13, and Chapter III, § 3, no. 3, Proposition 4) and for topological *fields;* for the latter, we have the following proposition:

PROPOSITION 8. *If the additive uniform structure of a topological field K is a structure of a complete Hausdorff space, then the multiplicative structure on K^* is a structure of a complete space.*

We shall show that if $\mathfrak{F}$ is a Cauchy filter with respect to the *multiplicative* structure on K^*, then $\mathfrak{F}$ is a Cauchy filter with respect to the *additive* structure on K, and does not converge to 0; this will establish the result. Let U be any neighbourhood of 0 in K, let V be a closed neighbourhood of 0 such that $V \subset U$, $VV \subset U$ [axiom (AV_{II})] and $-1 \notin V$; then by hypothesis there is a set $A \in \mathfrak{F}$ such that, for all $x \in A$ and $y \in A$, we have $x^{-1}y \in 1 + V$. Let $a \in A$, then $A \subset a + aV$, and $a + aV$ is a closed set which does not contain 0; hence 0 is not in the closure of A and is therefore not a cluster point of $\mathfrak{F}$. Let W be a neighbourhood of 0 such that $aW \subset V$ [axiom (AV_I)]; then by hypothesis there is a set $B \in \mathfrak{F}$ such that $B \subset A$ and, for all $x \in B$ and $y \in B$, we have $x^{-1}y \in 1 + W$; hence $y - x \in xW \subset AW \subset aW + aVW$; but K is commutative, and therefore $aVW = aWV \subset VV \subset U$; consequently $y - x \in U + U$ and the Proposition is proved.

The same proof shows that Proposition 8 can be extended to the case in which every Cauchy filter with respect to *one* of the multiplicative structures of K is also a Cauchy filter with respect to the other multiplicative structure.

7. INVERSE LIMITS OF TOPOLOGICAL GROUPS AND RINGS

Throughout this section, I *denotes a non-empty directed set* (*) *and* $\alpha \leqslant \beta$ *denotes the partial order relation in* I. *Unless the contrary is expressly stated, all the inverse systems considered are indexed by* I.

1. INVERSE LIMITS OF ALGEBRAIC STRUCTURES

Let $(X_\alpha, f_{\alpha\beta})$ be an inverse system of sets, and suppose that each X_α is endowed with an internal law of composition, everywhere defined, and written multiplicatively. Suppose also that the $f_{\alpha\beta}$ are homomorphisms with respect to these internal laws. Since

$$f_{\alpha\beta}(x_\beta \cdot y_\beta) = f_{\alpha\beta}(x_\beta) \cdot f_{\alpha\beta}(y_\beta)$$

whenever $\alpha \leqslant \beta$ and x_β and y_β belong to X_β, it is clear that $X = \varprojlim X_\alpha$ is a *stable subset* of the product $\prod_\alpha X_\alpha$, with respect to the internal law $(x_\alpha) \cdot (y_\alpha) = (x_\alpha \cdot y_\alpha)$. Let $(\Lambda_\alpha, \varphi_{\alpha\beta})$ be another inverse system of sets, indexed by I, and suppose that each X_α carries an external law of composition, everywhere defined, written multiplicatively and having Λ_α as a set of operators, such that whenever $\alpha \leqslant \beta$ we have

$$f_{\alpha\beta}(\lambda_\beta \cdot x_\beta) = \varphi_{\alpha\beta}(\lambda_\beta) \cdot f_{\alpha\beta}(x_\beta)$$

for $\lambda_\beta \in \Lambda_\beta$ and $x_\beta \in X_\beta$. Then we can define an external law on $\prod_\alpha X_\alpha$, having $\prod_\alpha \Lambda_\alpha$ as a set of operators, by defining $(\lambda_\alpha) \cdot (x_\alpha) = (\lambda_\alpha \cdot x_\alpha)$; restricting the set of operators to $\Lambda = \varprojlim \Lambda_\alpha$, we have an external law on $\prod_\alpha X_\alpha$ with respect to which X is again a *stable subset.* The internal (resp. external) law thus defined on X is said to be the *inverse limit* of the internal (resp. external) laws of the X_α. In the case of external laws, it may happen that all the Λ_α are identical with the same set Λ_0 and

(*) The reader may easily verify that most of the definitions and results which precede Proposition 1 in this section remain valid if I is merely *partially ordered.*

that all the $\varphi_{\alpha\beta}$ are identity mappings; then, since I is a directed set, Λ can be identified with Λ_0.

It is immediately verified that the usual properties of associativity, commutativity, existence of an identity element for an internal law (provided that the $f_{\alpha\beta}$ map identity element to identity element), distributivity of an external law with respect to an internal law, etc., are preserved under passage to the inverse limit.

Let Σ be a species of algebraic structure and let Σ_0 be the *impoverished* structure corresponding to Σ. Whenever we speak of an inverse system of sets $(X_\alpha, f_{\alpha\beta})$ endowed with structures of species Σ, we shall always suppose that the $f_{\alpha\beta}$ are *homomorphisms* for these structures. If we endow $X = \varprojlim X_\alpha$ with the internal and external laws which are the respective inverse limits of the internal and external laws of the X_α, then X carries an algebraic structure of *species* Σ_0. Naturally it remains to be seen in each particular case whether or not this structure is of species Σ.

For example, if $(G_\alpha, f_{\alpha\beta})$ is an inverse system of groups (resp. rings), then $\varprojlim G_\alpha$ is a subgroup (resp. subring) of $\prod_\alpha G_\alpha$, and is called the *inverse limit* of the system $(G_\alpha, f_{\alpha\beta})$ of groups (resp. rings).

Let $(X_\alpha, g_{\alpha\beta})$ be an inverse system of sets and let $(G_\alpha, f_{\alpha\beta})$ be an inverse system of groups; suppose that each X_α has G_α as a group of operators and that whenever $\alpha \leqslant \beta$ we have

$$(1) \qquad g_{\alpha\beta}(s_\beta . x_\beta) = f_{\alpha\beta}(s_\beta) . g_{\alpha\beta}(x_\beta)$$

for $x_\beta \in X_\beta$ and $s_\beta \in G_\beta$. Then $X = \varprojlim X_\alpha$ has $G = \varprojlim G_\alpha$ as group of operators. It follows from (1) that, if $\alpha \leqslant \beta$, the mappings $f_{\alpha\beta}$ and $g_{\alpha\beta}$ are compatible (§ 2, no. 4) and hence define a mapping $\varphi_{\alpha\beta} : X_\beta/G_\beta \to X_\alpha/G_\alpha$ of the quotient sets, such that $(X_\alpha/G_\alpha, \varphi_{\alpha\beta})$ is an inverse system. Moreover, if $f_\alpha : G \to G_\alpha$ and $g_\alpha : X \to X_\alpha$ are the canonical mappings, then f_α and g_α are compatible and consequently define a mapping $h_\alpha : X/G \to X_\alpha/G_\alpha$ of the quotient sets; it is clear that the h_α form an inverse system of mappings, whose inverse limit is therefore a mapping $h : X/G \to \varprojlim X_\alpha/G_\alpha$, *which is not necessarily either injective or surjective* (Exercise 1).

Again, let $(A_\alpha, \varphi_{\alpha\beta})$ be an inverse system of rings and let $(M_\alpha, f_{\alpha\beta})$ be an inverse system of commutative groups; and suppose that each M_α carries a left A_α-module structure in such a way that whenever $\alpha \leqslant \beta$ we have

$$(2) \qquad f_{\alpha\beta}(\lambda_\beta . x_\beta) = \varphi_{\alpha\beta}(\lambda_\beta) . f_{\alpha\beta}(x_\beta)$$

for $x_\beta \in M_\beta$ and $\lambda_\beta \in A_\beta$; then $\varprojlim M_\alpha$ has a structure of a left module over $\varprojlim A_\alpha$. If we suppose in addition that, for each α, A_α is commutative and that each M_α has an A_α-algebra structure, and finally that $(M_\alpha, f_{\alpha\beta})$ is an inverse system of *rings*, then $\varprojlim M_\alpha$ has a structure of an *algebra* over $\varprojlim A_\alpha$.

Let $(G_\alpha, f_{\alpha\beta})$ be an inverse system of groups, and for each α let H_α be a subgroup of G_α. If $f_{\alpha\beta}(H_\beta) \subset H_\alpha$ whenever $\alpha \leqslant \beta$, then the inverse system of subsets H_α of G_α is an inverse system of groups with respect to the restrictions of the $f_{\alpha\beta}$, and $H = \varprojlim H_\alpha$ is a *subgroup* of $G = \varprojlim G_\alpha$. If each H_α is a normal subgroup of G_α, then H is a normal subgroup of G. If $(G'_\alpha, f'_{\alpha\beta})$ is another inverse system of groups and, for each α, $u_\alpha : G_\alpha \to G'_\alpha$ is a homomorphism such that the u_α form an inverse system of mappings, then, if H_α is the kernel of u_α, we have $f_{\alpha\beta}(H_\beta) \subset H_\alpha$ whenever $\alpha \leqslant \beta$; $u = \varprojlim u_\alpha$ is a homomorphism of G into $G' = \varprojlim G'_\alpha$, and $H = \varprojlim H_\alpha$ is the kernel of u. If we put $K_\alpha = u_\alpha(G_\alpha)$, then we have $f'_\alpha(K_\alpha) \subset K_\alpha$ whenever $\alpha \leqslant \beta$, so that the K_α form an inverse system of subgroups of the G'_α; but $K = \varprojlim K_\alpha$ *is not necessarily the image of* G *under* u [Exercise 1 *c*)].

We obtain analogous results by replacing "group" by "ring", "subgroup" by "ideal" (left or right); we leave to the reader the task of stating the analogous results for modules and algebras.

2. INVERSE LIMITS OF TOPOLOGICAL GROUPS AND SPACES WITH OPERATORS

We shall say that an inverse system $(G_\alpha, f_{\alpha\beta})$ is an *inverse system of topological groups* if the G_α are topological groups and the $f_{\alpha\beta}$ are *continuous* homomorphisms. Then $G = \varprojlim G_\alpha$ is a subgroup of the product group $\prod_\alpha G_\alpha$; if we endow G with the topological group structure induced by that of $\prod_\alpha G_\alpha$, then the topological group so obtained is called the *inverse limit* of the inverse system of topological groups $(G_\alpha, f_{\alpha\beta})$. If the G_α are Hausdorff (resp. Hausdorff and complete) then G is Hausdorff and closed in $\prod_\alpha G_\alpha$ (resp. Hausdorff and complete) (Chapter I, § 8, no. 2, Proposition 7, Corollary 2 and Chapter II, § 3, no. 5, Corollary to Proposition 10).

If $(G'_\alpha, f'_{\alpha\beta})$ is another inverse system of topological groups and if, for each α, $u_\alpha : G_\alpha \to G'_\alpha$ is a continuous homomorphism such that the u_α form an inverse system of mappings, then $u = \varprojlim u_\alpha$ is a continuous homomorphism of G into $G' = \varprojlim G'_\alpha$ (Chapter I, § 4, no. 4). The same results are valid when "topological group" is replaced by "topological ring"; we leave to the reader the task of stating the analogous results for topological modules (§ 6, no. 6).

Let $(X_\alpha, g_{\alpha\beta})$ be an inverse system of topological spaces, and let $(G_\alpha, f_{\alpha\beta})$ be an inverse system of topological groups. Suppose that each G_α *operates continuously* on X_α (§ 2, no. 4) and that the relations (1) hold for all $x_\beta \in X_\beta$, $s_\beta \in G_\beta$ and $\alpha \leqslant \beta$. We have seen (no. 1) that $X = \varprojlim X_\alpha$ has $G = \varprojlim G_\alpha$ as group of operators; furthermore, G operates *continuously* on X. For if g_α (resp. f_α) is the canonical mapping $X \to X_\alpha$ (resp. $G \to G_\alpha$), then by definition

$$g_\alpha(s.x) = f_\alpha(s).g_\alpha(x)$$

and therefore the mappings $(s.x) \to g_\alpha(s.x)$ are continuous on $X \times G$, which shows the continuity of $(s, x) \to (s.x)$ (Chapter I, § 4, no. 4).

The mapping $h_\alpha : X/G \to X_\alpha/G_\alpha$ induced by f_α and g_α is therefore continuous (§ 2, no. 4), and so is the mapping $h: X/G \to \varprojlim X_\alpha/G_\alpha$ defined by the h_α (Chapter I, § 2, no. 3, Proposition 4).

PROPOSITION 1. *Suppose that the* X_α *and the* G_α *satisfy the above hypotheses.*
a) *If the stabilizer of each point of* X_α *is a compact subgroup of* G_α *for each* $\alpha \in I$, *then the stabilizer of each point* $x = (x_\alpha)$ *of* X *is a compact subgroup of* G; *the orbit of* x (*with respect to* G) *is canonically homeomorphic to the inverse limit of the orbits of the* x_α (*with respect to* G_α); *and the canonical mapping* $h : X/G \to \varprojlim X_\alpha/G_\alpha$ *is injective.*
b) *If, for each* $\alpha \in I$, *every orbit of a point of* X_α (*with respect to* G_α) *is compact, then every orbit of a point of* X (*with respect to* G) *is relatively compact, and* h *is surjective. If also* h *is bijective, then every orbit of a point of* X *is compact.*

Let $x = (x_\alpha) \in X$, and for each $\alpha \in I$ let $X'_\alpha = G_\alpha.x_\alpha$ be the orbit of x_α. If $\alpha \leqslant \beta$, it follows from (1) and the relation $g_{\alpha\beta}(x_\beta) = x_\alpha$ that $g_{\alpha\beta}(X'_\beta) \subset X'_\alpha$, and hence that (X'_α) is an inverse system of subsets of the X_α. For each $\alpha \in I$ let $u_\alpha : G_\alpha \to X'_\alpha$ be the continuous mapping $s_\alpha \to s_\alpha.x_\alpha$; the u_α form an inverse system of mappings, and $u = \varprojlim u_\alpha$ is the continuous mapping $s \to s.x$ of G into the subspace $X' = \varprojlim X'_\alpha$ of X. The hypothesis of a) implies that $\overset{-1}{u}_\alpha(y_\alpha)$ is compact for each $y_\alpha \in X'_\alpha$. Since also u_α is surjective, the conditions of Chapter I, § 9, no. 6, Proposition 8, Corollary 2 are satisfied, and the first two assertions of a) are therefore established. Hence, if $x = (x_\alpha)$ and $y = (y_\alpha)$ are such that x_α and y_α belong to the same orbit with respect to G_α for each $\alpha \in I$, then x and y belong to the same orbit with respect to G, and therefore h is injective.

Likewise, the hypothesis of b) implies that the inverse system of canonical mappings $v_\alpha : X_\alpha \to X_\alpha/G_\alpha$ satisfies the conditions of Chapter I, § 9, no. 6, Proposition 8, Corollary 2, and therefore its inverse limit $v = \varprojlim v_\alpha : X \to \varprojlim X_\alpha/G_\alpha$ is surjective, and the inverse image under

v of every point of $\varprojlim X_\alpha/G_\alpha$ is compact. Since v factorizes into

$$X \xrightarrow{\psi} X/G \xrightarrow{h} \varprojlim X_\alpha/G_\alpha,$$

where ψ is the canonical mapping, the assertions of b) follow.

Corollary 1. *If the* G_α *are compact and the* X_α *Hausdorff, then the conclusions of* a) *and* b) *are valid.*

For the hypotheses of a) and b) are satisfied, since every closed subgroup of G_α is compact and $u_\alpha : s_\alpha \to s_\alpha . x_\alpha$ is a continuous mapping of a compact space into a Hausdorff space.

Corollary 2. *If, for each* $\alpha \in I$, *the group* G_α *operates transitively on the space* X_α, *and if the stabilizer of each point of* X_α *is a compact subgroup of* G_α, *then* G *operates transitively on* X *and the stabilizer of each point of* X *is a compact subgroup of* G.

For hypothesis *a*) is satisfied, and $X'_\alpha = X_\alpha$ for each α.

Corollary 3. *Suppose that the* G_α *are Hausdorff. For each* $\alpha \in I$, *let* K_α *be a compact subgroup of* G_α, *such that* $f_{\alpha\beta}(K_\beta) \subset K_\alpha$ *whenever* $\alpha \leqslant \beta$. *Then, if* $K = \varprojlim K_\alpha$, *the canonical mapping* h *of the homogeneous space* G/K *into* $\varprojlim G_\alpha/K_\alpha$ *is a homeomorphism.*

The fact that h is bijective follows from Corollary 1, applied by replacing X_α by G_α, and G_α by K_α operating by *right* translations (§ 2, no. 5). Let φ be the canonical mapping $G \to G/K$, and for each α let f_α be the canonical mapping $G \to G_\alpha$. If, for each α, V_α runs through a fundamental system of open neighbourhoods of the identity element e_α of G_α, then the sets $V = \overset{-1}{f}_\alpha(V_\alpha)$ (α and V_α variable) form a fundamental system of neighbourhoods of the identity element e in G (Chapter I, § 4, no. 4, Proposition 9), and the sets $\varphi(V.K)$ form a fundamental system of neighbourhoods of $\varphi(e)$ in G/K. We have to show that the image of $\varphi(V.K)$ under h contains a neighbourhood of $h(\varphi(e))$, that is that there exists $\beta \geqslant \alpha$ and a neighbourhood W_β of e_β in G_β such that $\overset{-1}{f}_\beta(W_\beta . K_\beta) \subset V.K$. Now, the relation $x \in V.K$ is equivalent to the existence of y in K such that $f_\alpha(xy^{-1}) \in V_\alpha$, i.e. $f_\alpha(x) \in V_\alpha . f_\alpha(K)$; so that $V.K = \overset{-1}{f}_\alpha(V_\alpha . f_\alpha(K))$. Let $U_\alpha = V_\alpha . f_\alpha(K)$; we shall see that there exists $\beta \geqslant \alpha$ such that if we put $U_\beta = \overset{-1}{f}_{\alpha\beta}(U_\alpha)$, we have $K_\beta \subset U_\beta$; it will then follow that there is a neighbourhood W_β of e_β in G_β such that $W_\beta . K_\beta \subset U_\beta$ (Chapter II, § 4, no. 3, Corollary to Proposition 4), and this will establish the desired relation

$$\overset{-1}{f}_\beta(W_\beta . K_\beta) \subset \overset{-1}{f}_\beta(U_\beta) = V.K.$$

We proceed by *reductio ad absurdum:* for each $\beta \geqslant \alpha$, let M_β denote $K_\beta \cap \complement U_\beta$; since $\overset{-1}{f}_{\beta\gamma}(U_\beta) = U_\gamma$ if $\alpha \leqslant \beta \leqslant \gamma$, the M_β form an inverse system of compact subsets of the G_β (for $\beta \geqslant \alpha$). If they were all non-empty, then their inverse limit M would also be non-empty (Chapter I, § 9, no. 6, Proposition 8). It is clear that $M \subset K$ and $f_\alpha(M) \subset M_\alpha$; but this is absurd since $f_\alpha(K) \subset U_\alpha$, and the proof is therefore complete.

3. APPROXIMATION OF TOPOLOGICAL GROUPS

Let G be a group and let $(H_\alpha)_{\alpha \in I}$ be a family of normal subgroups of G such that $H_\alpha \supset H_\beta$ whenever $\alpha \leqslant \beta$. For each $\alpha \in I$ let $G_\alpha = G/H_\alpha$, and for $\alpha \leqslant \beta$ let $f_{\alpha\beta}$ be the canonical homomorphism $G/H_\beta \to G/H_\alpha$, which therefore maps a coset T of H_β in G to the coset TH_α of H_α in G. Clearly $(G_\alpha, f_{\alpha\beta})$ is an inverse system of groups; the elements of $\tilde{G} = \varprojlim G_\alpha$ are families $(T_\alpha)_{\alpha \in I}$, where T_α is a coset of H_α in G for each α, and $T_\alpha \supset T_\beta$ whenever $\alpha \leqslant \beta$. The mapping $i : s \to (sH_\alpha)$ of G into $\tilde{G}$ is the inverse limit of the canonical homomorphisms $G \to G/H_\alpha$ and is therefore a homomorphism of G into $\tilde{G}$, and the inverse image under i of an element $(T_\alpha) \in \tilde{G}$ is equal to $\bigcap_{\alpha \in I} T_\alpha$. The kernel of i is therefore $\bigcap_{\alpha \in I} H_\alpha$, and the image of i consists of all families $(T_\alpha) \in \tilde{G}$ whose intersection is *non-empty*.

Now suppose that G is a *topological group;* if we give each $G_\alpha = G/H_\alpha$ the quotient topology, it is clear that $(G_\alpha, f_{\alpha\beta})$ is an inverse system of topological groups, and that $i : G \to \tilde{G}$ is a continuous homomorphism.

PROPOSITION 2. *Let* G *be a topological group and let* $(H_\alpha)_{\alpha \in I}$ *be a family of normal subgroups of* G *such that* $H_\alpha \supset H_\beta$ *whenever* $\alpha \leqslant \beta$, *and which satisfy the following condition:*

(AP) *For each* $\alpha \in I$, H_α *is closed in* G *and every neighbourhood of the identity element* e *in* G *contains one of the* H_α (in other words, the filter base formed by the H_α converges to e).

Then the mapping $i : G \to \tilde{G} = \varprojlim G/H_\alpha$ *is a strict morphism of* G *onto* $i(G)$; $\tilde{G}$ *is Hausdorff and* $i(G)$ *is dense in* $\tilde{G}$; *and finally the kernel of* i *is the closure of* $\{e\}$ *in* G. *If in addition one of the* H_α *is complete, then* i *is surjective.*

Clearly the $G_\alpha = G/H_\alpha$ are Hausdorff (§ 2, no. 5, Proposition 13), hence so is $\tilde{G}$ (since it is a subspace of $\prod_{\alpha \in I} G_\alpha$). The kernel H of i is the intersection of the H_α and is therefore a closed subgroup of G. Since each neighbourhood of e contains some H_α, it contains H and therefore [§ 3, no. 1, formula (1)] H is the closure of $\{e\}$. Let us next show that $i(G)$ is dense in $\tilde{G}$. Let f_α be the canonical mapping $\tilde{G} \to G_\alpha$, which is the restriction to $\tilde{G}$ of the projection pr_α; $\varphi_\alpha = f_\alpha \circ i$ is the canonical mapping $G \to G/H_\alpha$. If U is any non-empty open set of $\tilde{G}$, then there is an index $\alpha \in I$ and a non-empty open set U_α in G_α such that $\overset{-1}{f}_\alpha(U_\alpha) \subset U$ (Chapter I, § 4, no. 4, Proposition 9); therefore

$$\overset{-1}{i}(U) \supset \overset{-1}{\varphi}_\alpha(U_\alpha);$$

but since φ_α is surjective, $\overset{-1}{i}(U)$ is not empty, and therefore $i(G) \cap U \neq \emptyset$.

To see that i is a strict morphism of G onto $i(G)$, consider a neighbourhood V of e in G. There is a neighbourhood W of e in G such that $W^2 \subset V$, and an index $\alpha \in I$ such that $H_\alpha \subset W$; it follows that V contains $WH_\alpha = \overset{-1}{\varphi}_\alpha(\varphi_\alpha(W)) = \overset{-1}{i}(\overset{-1}{f}_\alpha(\varphi_\alpha(W)))$. Since $\overset{-1}{f}_\alpha(\varphi_\alpha(W))$ is a neighbourhood of the identity element in $\tilde{G}$, the result follows.

Finally, suppose that there is an index $\gamma \in I$ such that H_γ is complete. To show that i is surjective it is enough to prove that every family $(T_\alpha)_{\alpha \in I} \in \tilde{G}$ has a non-empty intersection. Since T_γ is obtained by translation from H_γ, it is a complete subspace of G (with respect to both the right and the left uniformities). Moreover, since every neighbourhood U of e in G contains one of the H_α, the corresponding set T_α is U_d- (or U_s-) small, and hence the set of T_α contained in T_γ is a *Cauchy* filter base; therefore it converges in T_γ, and since the T_α are closed in G (since they are translations of the H_α), their intersection is not empty.

Q.E.D.

COROLLARY 1. *If the condition* (AP) *is satisfied and if in addition the (Hausdorff) groups* G/H_α *are complete, then the group* G *has a Hausdorff completion which can be identified with* $\tilde{G}$; *and the mapping* $i : G \to \tilde{G}$ *is then identified with the canonical mapping* (§ 3, no. 3, Proposition 5).

For $\tilde{G}$ is then complete (no. 2), and Proposition 2 shows that $i(G)$ is isomorphic with the Hausdorff group associated with G; since it is dense in $\tilde{G}$, the corollary follows (§ 3, no. 3, Proposition 5). In particular:

COROLLARY 2. *Let* G *be a group and let* (H_α) *be a family of normal subgroups of* G, *directed with respect to the relation* $H_\alpha \supset H_\beta$. *If we endow* G *with the group topology for which the* H_α *form a fundamental system of neighbourhoods of the identity element* e, *then the Hausdorff group associated with* G *is isomorphic to* $G_1 = G/\left(\bigcap_\alpha H_\alpha\right)$; G_1 *has a completion* $\hat{G}_1 = \tilde{G}$; *and the canonical mapping* $G_1 \to \hat{G} = \varprojlim G/H_\alpha$ *extends to an isomorphism of* $\hat{G} = \hat{G}_1$ *onto* $\tilde{G}$.

The subgroup H_α of G is open and therefore also closed (§ 2, no. 1, Corollary to Proposition 4), and G/H_α is discrete (§ 2, no. 6, Proposition 18); hence the conditions of Corollary 1 are satisfied.

For the remainder of this sub-section we shall assume that G is *Hausdorff* and that (H_α) is a family of *compact* normal subgroups of G, which is directed with respect to the relation $H_\alpha \supset H_\beta$ and which satisfies the condition (AP); by virtue of Proposition 2, the mapping

$$i : G \rightarrow \tilde{G} = \varprojlim G/H_\alpha$$

is then an *isomorphism of topological groups* which permits us to identify G and $\tilde{G}$. We denote by f_α the canonical mapping $G \rightarrow G/H_\alpha$.

Lemma 1. *Under the hypotheses of Proposition 2, if* E *is any closed subset of* G *we have* $E = \bigcap_\alpha EH_\alpha$.

For E is the intersection of the sets EV where V runs through the neighbourhood filter of e [§ 3, no. 1, formula (1)], and every neighbourhood of e contains an H_α; whence the result, since $E \subset EH_\alpha$.

Proposition 3. *Suppose that* G *is Hausdorff and that the* H_α *are compact and satisfy* (AP).

a) *Let* L *be a closed subgroup of* G; *then, for each* $\alpha \in I$, *the subgroup* $L_\alpha = f_\alpha(L)$ *of* $G_\alpha = G/H_\alpha$ *is closed, and the isomorphism* i *of* G *onto* $\varprojlim G_\alpha$ *gives by restriction an isomorphism of* L *onto* $\varprojlim L_\alpha$. *If also* L *is normal in* G, *then* L_α *is normal in* G_α *for each* $\alpha \in I$, *and by passing to the quotients,* i *induces an isomorphism of* G/L *onto* $\varprojlim G_\alpha/L_\alpha$.

b) *Conversely, for each* $\alpha \in I$ *let* L_α *be a closed subgroup of* G_α, *such that* $L_\alpha = f_{\alpha\beta}(L_\alpha)$ *whenever* $\alpha \leqslant \beta$. *Then there is a unique closed subgroup* L *of* G *such that* $L_\alpha = f_\alpha(L)$ *for each* $\alpha \in I$; *and if in addition* L_α *is normal in* G_α *for each* $\alpha \in I$, *then* L *is normal in* G.

a) Since H_α is compact, LH_α is closed in G (§ 4, no. 1, Proposition 1, Corollary 1) and therefore L_α is closed in G_α. Since i identifies the topological groups G and $\varprojlim G_\alpha$, and since $\varprojlim L_\alpha$ may be identified with a (topological) subgroup of $\varprojlim G_\alpha$, i identifies the subgroup $\bigcap_\alpha LH_\alpha$ of G with $\varprojlim L_\alpha$, and to prove the first assertion it is enough to remark that $L = \bigcap_\alpha LH_\alpha$ by Lemma 1. On the other hand, if L is normal, then for each $\alpha \in I$ the mapping $f'_\alpha : G/L \rightarrow G_\alpha/L_\alpha$ induced by f_α is a surjective strict morphism (§ 2, no. 8, Remark 3), whose kernel is the *compact* normal subgroup $H_\alpha L/L$ of G/L, the canonical image of the compact subgroup H_α of G. Since the subgroups $H_\alpha L/L$ of G/L satisfy condition (AP) (§ 2, no. 8, Proposition 24), and since G/L is Hausdorff, the last assertion of a) is a consequence of Proposition 2.

b) Let $f'_{\alpha\beta}$ be the restriction of $f_{\alpha\beta}$ to L_β ($\alpha \leqslant \beta$). Then $(L_\alpha, f'_{\alpha\beta})$ is an inverse system of topological groups, whose inverse limit L can be identified with the subgroup $G \cap \prod_\alpha L_\alpha$ of G. By hypothesis, $f'_{\alpha\beta}$ is surjective and its kernel is the compact subgroup $f_\beta(H_\alpha) \cap L_\beta$ of L_β; consequently (Chapter I, § 9, no. 6, Corollary 1 to Proposition 8) we have $L_\alpha = f_\alpha(L)$ for all $\alpha \in I$. If L' is another closed subgroup of G such that $f_\alpha(L') = L_\alpha$ for all $\alpha \in I$, then $L'H_\alpha = \overset{-1}{f}_\alpha(L_\alpha)$, whence (Lemma 1) $L' = \bigcap_\alpha L'H_\alpha = \bigcap_\alpha \overset{-1}{f}_\alpha(L_\alpha) = L$. Finally, the last assertion of b) follows from the formula $L = \bigcap_\alpha \overset{-1}{f}_\alpha(L_\alpha)$, since the $\overset{-1}{f}_\alpha(L_\alpha)$ are now normal subgroups of G.

PROPOSITION 4. *Suppose that* G *is Hausdorff, the* H_α *are compact and that* (AP) *is satisfied. If* C_α *is the identity component of* $G_\alpha = G/H_\alpha$, *then the identity component* C *of* G *can be identified with* $\varprojlim C_\alpha$, *and we have* $f_\alpha(C) = C_\alpha$.

The proposition is a consequence of the following lemma :

LEMMA 2. *Let* G *be a Hausdorff topological group and let* H *be a compact normal subgroup of* G, *and* φ *the canonical mapping* $G \to G/H$. *If* C *is the identity component of* G, *then* $\varphi(C)$ *is the identity component of* $G' = G/H$.

Once this lemma is established, we shall have $f_\alpha(C) = C_\alpha$ for each $\alpha \in I$, and since C is a closed subgroup of G (§ 2, no. 2, Proposition 7) it suffices to apply Proposition 3 a).

To prove the lemma, notice first that if C' is the component of the identity element e' in G', then $\varphi(C) \subset C'$ since $\varphi(C)$ is connected. Suppose that $\varphi(C) \neq C'$. Since C is a closed normal subgroup of G (§2, no. 2, Proposition 7), $\varphi(C)$ is a normal subgroup of G'; if ψ is the canonical mapping $G' \to G'/\varphi(C)$, then $\psi(C')$ is connected and does not consist of the identity element alone; hence the identity component of $G'/\varphi(C)$ does not consist of the identity element alone. But $G'/\varphi(C)$ is isomorphic to $(G/H)/(HC/H)$, hence to G/HC, and consequently also to $(G/C)/(HC/C)$ (§ 2, no. 7, Corollary to Proposition 22, and Proposition 20). Now, G/C is Hausdorff and totally disconnected (Chapter I, § 11, no. 5, Proposition 9), and HC/C, being the canonical image of the compact normal subgroup H of G, is a compact subgroup of G/C. It is therefore sufficient to prove Lemma 2 under the additional hypothesis that G is *totally disconnected*, i.e. $C = \{e\}$.

Suppose then that $C' \neq \{e'\}$; replacing G by its subgroup $\overset{-1}{f}(C')$, which is totally disconnected and contains H, we may suppose that G' is *connected* and does not consist of a single point.

Let $\mathfrak{M}$ be the set of closed subgroups L of G such that $LH = G$. We shall show that the set $\mathfrak{M}$, ordered by the relation $\supset$, is *inductive*. Indeed, if $\mathfrak{T}$ is a linearly ordered subset of $\mathfrak{M}$, then for each $x \in G$ the set of sets $xH \cap L$ with $L \in \mathfrak{T}$ is a filter base formed of closed sets in the compact space xH; hence the intersection of these sets is not empty, which shows that the intersection of the subgroups $L \in \mathfrak{T}$ belongs to $\mathfrak{M}$. Applying Zorn's lemma, we see therefore that $\mathfrak{M}$ has a *minimal* element L_0. Since H is compact, $G/H = L_0H/H$ is isomorphic to $L_0/(L_0 \cap H)$ (§ 4, no. 1, Proposition 1, Corollary 3); since L_0 is totally disconnected and $L_0 \cap H$ is compact, we see that we may replace G by L_0; in other words, we may suppose in addition that there is *no* closed subgroup $L \neq G$ such that $LH = G$.

Now let F be the intersection of the neighbourhoods of the identity element which are both open and closed in G, and let us show that F is a closed subgroup of G. Clearly F is closed in G; hence it is enough to show that $F^{-1}.F \subset F$. But if $x \in F$ and if V is an open and closed neighbourhood of e in G, then so is xV; for otherwise e would belong to the complement W of xV in G, which is again open and closed, and we should have $x \notin W$; hence $x \notin F$, contrary to hypothesis. It follows that xF, the intersection of the sets xV as V runs through the open and closed neighbourhoods of e in G, contains F; in other words, $x^{-1}F \subset F$, which proves our assertion. Since G is totally disconnected and $G \neq \{e\}$, we have $F \neq G$. But if V is an open and closed neighbourhood of e in G, then VH is also open and closed in G (§ 4, no. 1, Corollary 1 to Proposition 1), hence $\varphi(V)$ is both open and closed in G/H, and this implies that $\varphi(V) = G/H$ by virtue of the hypothesis. We shall conclude from this that $FH = G$, which will give us the desired contradiction and complete the proof of the lemma. Indeed, for each $x \in G$, xH meets every neighbourhood V of e which is both open and closed, hence also meets the intersection F of these neighbourhoods, since the sets $Vx \cap H$ form a filter base consisting of closed sets in the compact space xH.

Q.E.D.

Remark. If the subgroup H_α is compact for *some* $\alpha \in I$, then H_β is compact for all $\beta \geqslant \alpha$, because it is a closed subgroup of H_α. Since the set of all $\beta \in I$ such that $\beta \geqslant \alpha$ is cofinal in I, it makes essentially no difference, in the study of the group G, whether we suppose that one of the H_α is compact or that *all* the H_α are compact.

4. APPLICATION TO INVERSE LIMITS

PROPOSITION 5. *Let* $(G_\alpha, f_{\alpha\beta})$ *be an inverse system of Hausdorff topological groups such that the* $f_{\alpha\beta}$ *are surjective strict morphisms with compact kernels. Then for each* $\alpha \in \mathrm{I}$ *the canonical mapping* f_α *of* $\mathrm{G} = \varprojlim \mathrm{G}_\alpha$ *into* G_α *is a surjective strict morphism whose kernel is compact.*

The facts that f_α is surjective and that its kernel is compact are consequences of Chapter I, § 9, no. 6, Corollary 1 to Proposition 8. It remains to show that f_α is a strict morphism. Let e (resp. e_α) denote the identity element of G (resp. G_α). Each neighbourhood V of e in G contains a set of the form $\overset{-1}{f}_\beta(\mathrm{V}_\beta)$, where V_β is a neighbourhood of e_β in G_β, and we may suppose that $\beta \geqslant \alpha$; since $f_{\alpha\beta}$ is a surjective strict morphism, $f_{\alpha\beta}(\mathrm{V}_\beta)$ is a neighbourhood of e_α in G_α, and since f_β is surjective, we have $\mathrm{V}_\beta \subset f_\beta(\mathrm{V})$, whence

$$f_\alpha(\mathrm{V}) = f_{\alpha\beta}(f_\beta(\mathrm{V})) \supset f_{\alpha\beta}(\mathrm{V}_\beta);$$

this shows that $f_\alpha(\mathrm{V})$ is a neighbourhood of e_α in G_α.

If $\mathrm{H}_\alpha = \overset{-1}{f}_\alpha(e_\alpha)$, then each H_α is a compact normal subgroup of G; the H_α clearly satisfy the condition (AP) of no. 3; and G_α can be identified with $\mathrm{G}/\mathrm{H}_\alpha$. In particular Propositions 3 and 4 of no. 3 apply to G and the H_α.

Corollary 1. *Let* $(\mathrm{G}_\alpha, f_{\alpha\beta})$ *be an inverse system of topological groups which satisfies the hypotheses of Proposition* 5; *let* $(\mathrm{G}'_\alpha, f'_{\alpha\beta})$ *be an inverse system of topological groups, and for each* α *let* $u_\alpha : \mathrm{G}_\alpha \to \mathrm{G}'_\alpha$ *be a surjective strict morphism with compact kernel, such that the* u_α *from an inverse system of mappings. Then* $u = \varprojlim u_\alpha$ *is a strict morphism of* $\mathrm{G} = \varprojlim \mathrm{G}_\alpha$ *onto* $\mathrm{G}' = \varprojlim \mathrm{G}'_\alpha$, *and its kernel is compact.*

Let N_α be the kernel of u_α. $\mathrm{L}_\alpha = \overset{-1}{f}_\alpha(\mathrm{N}_\alpha)$ is then the kernel of the surjective strict morphism $v_\alpha = u_\alpha \circ f_\alpha : \mathrm{G} \to \mathrm{G}'_\alpha$; since $\mathrm{L}_\alpha/\mathrm{H}_\alpha$ is isomorphic to N_α (§ 2, no. 7, Proposition 20), L_α is a *compact* normal subgroup of G (§ 4, no. 1, Corollary 2 to Proposition 2). The kernel L of u is the intersection of the L_α. Let φ denote the canonical mapping $\mathrm{G} \to \mathrm{G}/\mathrm{L}$; then we can write $v_\alpha = w_\alpha \circ \varphi$, where w_α is a strict morphism of G/L onto G'_α, whose kernel is $\mathrm{L}_\alpha/\mathrm{L}$. Since the intersection of the $\mathrm{L}_\alpha/\mathrm{L}$ is the identity element of G/L, and since the $\mathrm{L}_\alpha/\mathrm{L}$ form a filter base and are compact, this filter base converges to the identity element of G/L (Chapter I, § 9, no. 1, Corollary to Theorem 1). Proposition 2 of no. 3 then shows that $w = \varprojlim w_\alpha$ is an *isomorphism* of G/L onto G′; it follows that $w \circ \varphi$ is a strict morphism of G onto G′, with kernel L. But it is clear that $u = w \circ \varphi$, and thus the corollary is proved.

Corollary 2. *Let* $(\mathrm{G}_\alpha, f_{\alpha\beta})$ *be an inverse system of topological groups which satisfies the conditions of Proposition* 5, *and let* G′ *be a topological group in which there is a neighbourhood* V′ *of the identity element* e' *which contains no subgroup of* G' *other than* $\{e'\}$. *Then if* $v : G \to G'$ *is any continuous homomorphism, there is an index* $\alpha \in I$ *and a continuous homomorphism* $v_\alpha : G_\alpha \to G'$ *such that* $v = v_\alpha \circ f_\alpha$.

For since $\overset{-1}{v}(V')$ is a neighbourhood of e in G, there is an index α and a neighbourhood V_α of e_α in G_α such that $\overset{-1}{f}_\alpha(V_\alpha) \subset \overset{-1}{v}(V')$. Hence $v(H_\alpha) \subset V'$, and since $v(H_\alpha)$ is a subgroup of G', it follows that $v(H_\alpha) = \{e'\}$. Since f_α may be identified with the canonical mapping $G \to G/H_\alpha$, the corollary is a consequence of the canonical factorization of a continuous homomorphism (§ 2, no. 8).

EXERCISES

§ 1

1) Every topology compatible with the group structure of a *finite* group G may be obtained by taking as neighbourhoods of the identity element the sets containing a normal subgroup H of G.

2) A *semi-topological* group is a group G endowed with a topology such that, for each $a \in G$, the translations $x \to ax$ and $x \to xa$ are continuous on G, and such that the symmetry $x \to x^{-1}$ is continuous on G (*).

a) A filter $\mathfrak{B}$ on a group G is the neighbourhood filter of e in a topology which makes G a semi-topological group if and only if $\mathfrak{B}$ satisfies axioms (GV_{II}) and (GV_{III}) and the family of filters

$$\mathfrak{B}(x) = x . \mathfrak{B} = \mathfrak{B} . x \qquad (x \in G)$$

satisfies axiom (V_{IV}) of Chapter I, § 1, no. 2.

b) If G is an infinite group, show that the topology in which the open sets are $\varnothing$ and the complements of finite subsets of G makes G a non-Hausdorff semi-topological group in which $\{e\}$ is closed, but is not compatible with the group structure of G.

c*) Define a topology on **R, finer than the usual topology, for which **R** is a semi-topological but not topological group [consider a decreasing sequence (r_n) of numbers > 0, tending to 0, and take the intersections of symmetric intervals $]-a, a[$ and of the complement of the set of points $\pm r_n$). *

(*) If G is *locally compact* these conditions imply that the topology of G is compatible with its group structure (Chapter X, § 3, Exercise 25).

3) Let A be a subset of a semi-topological (resp. topological) group G. If V runs through the neighbourhood filter of e in G, then the intersection of the sets AV, or of the sets VA (or of the sets VAV), is the closure $\overline{A}$ of A.

4) A *paratopological* group is a group G endowed with a topology satisfying axiom (GT_I) (*).

a) A filter $\mathfrak{B}$ on a group G is the neighbourhood filter of e for a topology which makes G a paratopological group if and only if $\mathfrak{B}$ satisfies axioms (GV_I) and (GV_{III}) and $e \in V$ for all $V \in \mathfrak{B}$. The corresponding topology is then unique; it is Hausdorff if and only if the intersection of the sets $V.V^{-1}$, where V runs through $\mathfrak{B}$, consists of the point e alone.

b) On the group **Z** of integers, let V_n be the set consisting of 0 and all integers $m \geqslant n$, for each $n > 0$. Show that the V_n form a fundamental system of neighbourhoods of 0 for a non-Hausdorff topology on **Z**, and that with this topology **Z** is a paratopological group in which $\{0\}$ is closed.

c) A paratopological group G is a topological group if and only if, for each subset A of G, the intersection of the sets AV (or of the sets VA) is the closure $\overline{A}$ of A (as V runs through the neighbourhood filter of e) (to see that the condition is sufficient, take A to be the complement of an open neighbourhood of e).

5) Let $\mathfrak{B}$ be a filter on a group G, satisfying axioms (GV_I) and (GV_{II}).

a) There is a unique topology $\mathcal{T}_s$ (resp. $\mathcal{T}_d$) on G such that, for each $a \in G$, the neighbourhood filter of a with respect to $\mathcal{T}_s$ (resp. $\mathcal{T}_d$) is $a.\mathfrak{B}$ (resp. $\mathfrak{B}.a$).

Let G_s (resp. G_d) be the topological space obtained by endowing G with the topology $\mathcal{T}_s$ (resp. $\mathcal{T}_d$). For each $a \in G$, the left translation $x \to ax$ (resp. the right translation $x \to xa$) is a continuous mapping of G_s into G_s (resp. of G_d into G_d). The symmetry $x \to x^{-1}$ is a continuous mapping of G_s into G_d and of G_d into G_s.

b) The following conditions are equivalent: 1) V satisfies (GV_{III}); 2) for each $a \in G$, $x \to xa$ is a continuous mapping of G_s into G_s; 3) $x \to x^{-1}$ is a continuous mapping of G_s into G_s.

**c*) Let G be the group $\mathbf{GL}(2, \mathbf{Q}_p)$ of 2×2 square matrices over the p-adic field $\mathbf{Q}_p$. For each integer $n > 0$, let H_n be the set of

(*) If G is *locally compact*, this condition implies that the topology of G is compatible with its group structure (Chapter X, § 3, Exercise 25).

matrices of the form $I + p^n U$, where U is a 2×2 matrix whose elements are p-adic integers. Show that the H_n are subgroups of G whose intersection consists only of the identity element. Deduce that if $\mathfrak{B}$ is the filter which has the H_n as a base, then the corresponding topologies T_s and T_d on G are Hausdorff. Show that these topologies are distinct.$_*$

6) Let G be a topological group and let V be an open neighbourhood of e in G. Show that the set of $x \in G$ such that $x^2 \in V$ is the union of the neighbourhood W of e such that $W^2 \subset V$.

*7) Let φ be the canonical mapping of **R** onto $\mathbf{T} = \mathbf{R}/\mathbf{Z}$, and let f be the mapping $x \to \varphi(3x)$, restricted to the neighbourhood $V = [-1/8, +1/8]$ of 0 in **R**. Show that f is a homeomorphism of V onto $f(V)$ and satisfies condition 1) of Definition 2 of no. 3, but is not a local isomorphism from **R** to **T**. $_*$

¶ 8) Let G be a Hausdorff topological group, and let $s \neq e$ be a point of G. Show that there is a symmetric neighbourhood V of e such that $s \notin V^2$, and that G can be covered by at most 17 sets of the form $a_k V b_k$ ($1 \leqslant k \leqslant 17$). (Consider a maximal element in the set of symmetric neighbourhoods V of e such that $s \notin V^2$. We are thus led to consider the set S of $x \in G$ such that $x^2 = s$; observe that if x, y are two elements of S such that $xy^{-1} \in S$, then we must have $x^2 = e$, and if z is a third element of S such that $xz^{-1} \in S$ and $yz^{-1} \in S$, then z must commute with xy^{-1}, and $xy^{-1}z^{-1} \notin S$.) If G is commutative, the number 17 may be replaced by 5.

9) Show that on a group G, the least upper bound of a family of topologies compatible with the group structure of G is compatible with the group structure of G.

§ 2

1) Extend Propositions 3, 6, 7 and 9, and the corollary to Proposition 4, to semi-topological groups (§ 1, Exercise 2) and paratopological groups (§ 1, Exercise 4).

2) *a*) Extend Propositions 1, 2 and 4 to semi-topological groups (§ 1, Exercise 2). (To prove that the closure $\overline{H}$ of a subgroup H is a subgroup start by remarking that $S\overline{H} \subset \overline{H}$ for all $s \in H$).

b) Let G be the group $\mathbf{Z}^{(\mathbf{N})}$, and let e_i ($i \in \mathbf{N}$) be the element of G all of whose coordinates are zero except the ith, and whose ith coordinate is 1. For each $n \in \mathbf{N}$, let V_n be the set of all $z = (z_k) \in G$ such that $z = 0$ for $k \leqslant n$ and $z_k \geqslant 0$ for $k > n$. The V_n form a fundamental system of neighbourhoods of 0 for a topology which makes G into a Hausdorff paratopological group. Let H be the subgroup of G generated by the elements $e_0 + e_n$ $(n \geqslant 1)$ Show that H is discrete but not closed in G, and that $\overline{H}$ is not a subgroup of G.

3) Give an example of a non-Hausdorff topological group in which the centre is not closed and the closure of the centre is a non-commutative subgroup (cf. § 1, Exercise 1). Give an example of a semi-topological group having the same properties and in which every set consisting of a single point is closed [cf. § 1, Exercise 2 *b*)].

¶ 4) *a*) In a semi-topological group G, the intersection of the neighbourhoods of e which are both open and closed is a closed subgroup H which is invariant under all automorphisms of G (remark that there can be no partition of G into two open and closed subsets, each of which meets H).

b) On the set $\mathfrak{G}$ of all subgroups of G, define a mapping $\varphi : \mathfrak{G} \to \mathfrak{G}$ as follows: if H is any subgroup of G, then $\varphi(H)$ is the subgroup of H which is the intersection of all the simultaneously open and closed neighbourhoods of e in H. In the set $\mathfrak{G}$, ordered by the relation $\supset$, consider the *chain* Γ of G with respect to the mapping φ (*Set Theory*, Chapter III, § 2, Exercise 6); show that the smallest subgroup of Γ is the connected component of e in G (remark that this smallest subgroup must be connected).

5) A semi-topological group G is *quasi-topological* if, for each $a \in G$, the mapping $x \to xax^{-1}$ of G into G is continuous. Every commutative semi-topological group is quasi-topological.

a) Show that the normalizer of a closed subgroup of a quasi-topological group is closed.

b) Let G be a quasi-topological group which is generated by each neighbourhood of e (resp. which is connected). Show that every discrete (resp. totally disconnected) normal subgroup D of G is contained in the centre of G (if $a \in D$, show that there is a neighbourhood V of e such that $xax^{-1} = a$ for each $x \in V$).

**c*) Let G be the group of matrices $\begin{pmatrix} 1 & 0 \\ y & x \end{pmatrix}$, where x and y are real numbers and $x > 0$. If we identify G with a subset of $\mathbf{R}^2$ by means of the mapping $(x, y) \to \begin{pmatrix} 1 & 0 \\ y & x \end{pmatrix}$, then the topology $\mathcal{T}_0$ induced on G by the topology of $\mathbf{R}^2$ is compatible with the group structure of G. Let H be the normal subgroup of G consisting of all matrices $\begin{pmatrix} 1 & 0 \\ y & 1 \end{pmatrix}$, and let H* be the complement of the identity element in H. Define a topology $\mathcal{T}$, finer than $\mathcal{T}_0$, by taking as a fundamental system of neighbourhoods of e in G the intersections of neighbourhoods of e in the topology $\mathfrak{T}_0$ with the complement of H^* in G. With respect to this topology $\mathfrak{T}$, show that G is a connected semi-topological group which is not quasi-topological and which has H as a discrete subgroup not contained in the centre. $_*$

* 6) Let G be the orthogonal group $\mathbf{O}_2(\mathbf{Q})$, identified with a subspace of the space $\mathbf{Q}^4$ of all 2×2 matrices over $\mathbf{Q}$. Show that the topology induced on G by the topology of $\mathbf{Q}^4$ is compatible with the group structure of G. Show that the commutator subgroup of G is not closed and that its closure in G is the group $\mathbf{SO}_2(\mathbf{Q})$. $_*$

7) *a*) Show that if G is a connected quasi-topological group (Exercise 5) then the commutator subgroup of G is connected. (Let P_k be the set of all products of k commutators $x^{-1}y^{-1}xy$; show that P_k is a union of connected sets each of which meets P_{k-1}).

b) Let G be a non-commutative infinite group whose commutator subgroup is finite (for example, the product of a non-commutative finite group and an infinite commutative group). If G carries the topology defined in Exercise 2 *b*) of § 1, show that G is a connected semi-topological group whose commutator subgroup is not connected.

¶ 8) Let G be a quasi-topological group (Exercise 5).

a) Extend Proposition 8 to G (show first that $x\overline{K}x^{-1} \subset \overline{K}$ if $x \in H$). Give an example of semi-topological group which is not quasi-topological and for which Proposition 8 is not valid [cf. § 1, Exercice 2 *b*)].

b) Let H, K be two subgroups of G such that $H \supset K$ and K contains the commutator subgroup of H. Show that $\overline{K}$ contains the commutator subgroup of $\overline{H}$ (analogous method).

c) Deduce from *b*) that if every subset of G consisting of a single point is closed, then the closure of every solvable (resp. commutative) subgroup in G is solvable (resp. commutative) (argue on the length of the derived series of the subgroup under consideration).

9) Let G be a quasi-topological group, and let H be a closed normal subgroup of G which contains the commutator subgroup of G. Show that if the component K of e in H is solvable, then so is the component L of e in G [show by use of Exercise 7 *a*) that K contains the commutator subgroup of L].

10) Let G be a quasi-topological group in which every subset consisting of a single point is closed and whose identity component C is such that G/C is finite. Show that if G/C has k elements, then the set of conjugates xax^{-1} of an element $a \in G$ is either infinite or else has at most k

elements (consider the mapping $x \to xax^{-1}$ of G onto this set). In particular, if G is connected, the set of conjugates of every element not in the centre of G is infinite.

11) Let G be a group, topologized or not, which operates on a topological space X in such a way that each of the mappings $x \to s.x$ $(s \in G)$ of X into X is continuous. Extend the results of no. 4 to this situation.

12) Let G be a semi-topological (resp. paratopological, quasi-topological) group and let H be a normal subgroup of G.

a) Show that if G/H is endowed with the quotient topology of G by H, then G/H is a semi-topological (resp. paratopological, quasi-topological) group.

b) Extend the results of no. 7 to semi-topological (resp. paratopological, quasi-topological) groups.

13) Let G be a semi-topological (resp. paratopological) group and let H be a dense subgroup of G. What is the topology of the homogeneous space G/H?

14) Let G be a semi-topological group and let H be a normal subgroup of G. Show that the semi-topological group $G/\overline{H}$ is isomorphic to the Hausdorff group associated with G/H.

¶ 15) Let G be a quasi-topological group in which every set consisting of a single point is closed. Show that if G is solvable then there is a composition series of G consisting of *closed* subgroups such that the successive quotient groups of the series are commutative [argue by induction on the length of the derived series of G, using Exercise 8 *c*)].

¶ 16) Let H be a subgroup of a semi-topological group G, contained in the component K of e. Show that the components of the space G/H are the images of the components of G under the canonical mapping f of G onto G/H [if L is a component of G/H, show that $\overset{-1}{f}(L)$ is connected, by *reductio ad absurdum*]. Show that K is the smallest of the closed subgroups H of G such that G/H is totally disconnected.

* 17) Let G be the additive group of mappings $n \to f(n)$ of $\mathbf{N}$ into $\mathbf{Q}$ such that $\lim_{n\to\infty} f(n)$ exists in $\mathbf{R}$. For each $\alpha > 0$, let V_α be the set of $f \in G$ such that $|f(n)| < \alpha$ for all $n \in \mathbf{N}$. Show that the V_α satisfy axioms (GV_I) and (GV_{II}) and that with the topology defined by the V_α, the group G is Hausdorff and totally disconnected. Let H be the subgroup of all $f \in G$ such that $\lim_{n\to\infty} f(n) = 0$. Show that H is closed and that G/H is isomorphic to $\mathbf{R}$ and therefore connected. *

18) A continuous homomorphism f of a topological group G into a topological group G′ is a strict morphism if and only if the image under f of every open set in G has at least one interior point with respect to $f(G)$.

19) Let G, G′, G″ be three topological groups; let f be a strict morphism of G into G′, and let g be a strict morphism of F′ into G″. Show that if $f(G)$ contains the kernel $\overset{-1}{g}(e'')$ of g (e'' being the identity element of G″), then $g \circ f$ is a strict morphism of G into G″ (cf. no. 7, Corollary to Proposition 21). Give an example in which f is injective and the preceding condition is not satisfied, and $g \circ f$ is not a strict morphism of G into G″ (cf. no. 7, Remark following Proposition 20).

20) Let H, H′ be two subgroups of a semi-topological group G, such that $H' \subset H$. Show that if f, f' are the canonical mappings of G onto G/H and G/H′ respectively, then there is a unique mapping φ of G/H′ onto G/H such that $f = \varphi \circ f'$; φ is continuous and open. If moreover H′ is open in H, then every point x of G/H′ has a neighbourhood V such that the restriction of φ to V is a homeomorphism of V onto a neighbourhood of $\varphi(x)$ in G/H.

21) Let G be a topological group and let H, K be two subgroups of G such that $G = HK$ and $H \cap K = \{e\}$. Each $x \in G$ can be written uniquely in the form $x = f(x)g(x)$, where $f(x) \in H$ and $g(x) \in K$. The mapping $(y, z) \to yz$ of $H \times K$ onto G is a homeomorphism if and only if one of the mappings f, g is continuous. Show that this condition is always satisfied if one of the subgroups H, K is compact and closed, and the other is closed (remark that if H and K are closed, f and g can have no cluster point other than e when x tends to e in G).

22) Let G be the product of a family $(G_\iota)_{\iota \in I}$ of topological groups, and let G′ be a topological group such that there is a neighbourhood V′ of the identity element e' in G′ which contains no normal subgroup of G′ other than $\{e'\}$. If f is a continuous homomorphism of G into G′, show that there is a subset J of I whose complement J′ is finite, such that f is equal to e' on the subgroup $\prod_{\iota \in I} G_\iota \times \prod_{\iota \in J'} \{e_\iota\}$ of G.

23) Consider, on the product set of a family $(G_\iota)_{\iota \in I}$ of topological groups, the group structure which is the product of the group structures of the G_ι and the topology generated by the set of products $\prod_{\iota \in I} A_\iota$, where A_ι is open in G_ι for each $\iota \in I$ and where the set of $\iota \in I$ such that $A_\iota \neq G_\iota$ has cardinal $< \mathfrak{c}$, where $\mathfrak{c}$ is a given infinite cardinal (Chapter I, § 4, Exercise 9). Show that this topology is compatible with

the group structure of G. Hence construct an example of a non-discrete Hausdorff topological group in which every compact subset is finite (cf. Chapter I, § 9, Exercise 4).

¶ 24) Let G and G′ be two locally isomorphic connected groups.

a) Let f be a local isomorphism of G with G′, defined on a neighbourhood V of the identity element of G. Let H be the subgroup of the product $G \times G'$ generated by the set of points $(x, f(x))$ where $x \in V$. Consider the filter base on H consisting of the images of neighbourhoods of e contained in V under the mapping $x \to (x, f(x))$. Show that this filter base is a fundamental system of neighbourhoods of the identity element of H for a topology $\mathcal{T}$ compatible with the group structure of H.

b) Show that there exist two discrete normal subgroups K, K′ contained in the centre of H, such that G is isomorphic to H/K and G′ is isomorphic to H/K′ [use Exercise 5 *b*)].

* *c*) Take G and G′ each to be the group **T**; let φ be the canonical homomorphism of **R** onto $\mathbf{T} = \mathbf{R}/\mathbf{Z}$, and let f be the local isomorphism which maps each point $\varphi(x)$, where $-1/4 \leqslant x \leqslant 1/4$, to $\varphi(\theta x)$, θ being an irrational number such that $0 < \theta < 1$. Show that for this local isomorphism, the topology $\mathcal{T}$ defined on H is distinct from the topology induced on H by the product topology on $G \times G'$ *.

¶ 25) Let G be a topological group, K a normal subgroup of G, φ the canonical mapping $G \to G/K$. Let f be a continuous homomorphism of a topological group H into G, such that every element of $f(H)$ commutes with every element of K. Then $(y, z) \to f(y)z$ is a continuous homomorphism g of $H \times K$ into G. g is a strict morphism of $H \times K$ into G if and only if $\varphi \circ f$ is a strict morphism of H into G/K.

26) Let $(G_\iota)_{\iota \in I}$ be a family of topological groups, and for each $\iota \in I$ let K_ι be an *open* normal subgroup of G_ι. Let $\mathfrak{V}$ denote the neighbourhood filter of the identity element in the product topological group $K = \prod_{\iota \in I} K_\iota$. Show that in the product group $G = \prod_{\iota \in I} G_\iota$, $\mathfrak{V}$ is a base of the neighbourhood filter of e for a topology compatible with the group structure of G. The group G, endowed with this topology, is called the *local product* of the G_ι (relative to the K_ι); K is then an open normal subgroup of G, and G/K is isomorphic to the (discrete) product of the groups G_ι/K_ι. If, for each $\iota \in I$, K'_ι is another open normal subgroup of G_ι, then the local product topologies on G, defined by the families (K_ι) and (K'_ι), coincide if and only if $K'_\iota = K_\iota$ for all but a finite number of indices.

For each $\varkappa \in I$, let $G'_\varkappa$ be the normal subgroup of G consisting of all $x = (x_\iota)$ such that $x_\iota = e_\iota$ whenever $\iota \neq \varkappa$. Then $G'_\varkappa$, with the topology induced by a local product topology on G, is canonically isomorphic to $G_\varkappa$. The closure in G of the subgroup generated by the G'_ι is the normal subgroup G_0 of G which consists of all $x = (x_\iota)$ such that $x_\iota \in K_\iota$ for all but a finite number of indices. G_0 is said to be the *local direct product* of the G_ι (relative to the K'); G_0/K is then a discrete group isomorphic to the direct product of the G_ι/K_ι.

¶ 27) A group G is said to have the property $P(n)$ (n an integer > 0) if, for each system $(s_1, \ldots, s_n)$ of n elements of G, there is a permutation $\sigma \in \mathfrak{S}_n$, other than the identity permutation (but depending on $s_1, \ldots, s_n$) such that $s_{\sigma(1)} \ldots s_{\sigma(n)} = s_1 \ldots s_n$.

a) Show that the non-commutative group $\mathfrak{S}_3$ satisfies $P(4)$.

b) Let G be a connected Hausdorff topological group. Show that if G satisfies $P(n)$ for some integer $n > 1$, then G is commutative. [Show that G satisfies $P(n-1)$: given $n - 1$ elements $s_1, \ldots, s_{n-1}$ of G, reduce to the case where there is a neighbourhood U of e such that, for each $x \in U$, we have $s_1 \ldots s_{n-1}x = s_1 \ldots s_i x s_{i+1} \ldots s_{n-1}$ for some index $i = i(x)$; deduce that the centralizer in G of one of the elements $s_{j+1} \ldots s_{n-1}$ $(0 \leqslant j \leqslant n-2)$ is both open and closed and therefore equal to G. Hence the result if $j \geqslant 1$; if $j = 0$, show that

$$s_1 s_2 \ldots s_{n-1} = s_2 \ldots s_{n-1} s_1].$$

¶ 28) *a*) Let S be a stable subset of a topological group G. Show that $\mathring{S}$ and $\overline{S}$ are stable and that $S\mathring{S} \subset \mathring{S}$ and $\mathring{S}S \subset \mathring{S}$. If also S is open and $e \in \overline{S}$, then $S = \mathring{\overline{S}}$ (if $x \in \overline{S} - S$, show that x is a frontier point of S).

b) From now on in this exercise, G is a commutative topological group, written additively. For every non-empty subset A of G, let $s(A)$ denote the set of all $x \in G$ such that $x + A \in A$, i.e. the intersection of the sets $A - y$ as y runs through A; let $b(A) = s(A) \cap s(-A)$. The set $s(A)$ is a stable subset of G, and $b(A)$ is a subgroup of G consisting of all $x \in G$ such that $x + A = A$. Show that if F is any non-empty closed subset of G, then $s(F)$ is closed. For every non-empty subset A of G, we have $s(A) \subset s(\overline{A})$ and $s(A) \subset s(\mathring{A})$; and if $A = \mathring{\overline{A}}$, then $s(\mathring{\overline{A}}) = s(\overline{A})$.

c) Let $\mathfrak{S}$ be the set of all stable open subsets S of G such that $0 \notin S$, and suppose that $\mathfrak{S} \neq \varnothing$. Then the set $\mathfrak{S}$, ordered by inclusion, is inductive; let $\mathfrak{M} \neq \varnothing$ be the set of its maximal elements. Show that if $M \in \mathfrak{M}$, then $M = \mathring{M}$ [use *a*)] and that, for each integer $n > 0$, M is equal to the set of all $x \in G$ such that $nx \in M$. In addition, the subgroup $b(M)$ is equal to the complement of $M \cup (-M)$ in G (if $x \notin M$ and $-x \notin M$, consider the union of the sets $kx + M$ for integers $k \geqslant 0$); deduce that $G = M - M$.

A closed subgroup B of G is said to be *residual* if there exists $M \in \mathfrak{M}$ such that $B = b(M)$.

d) The intersection of all the residual subgroups of G is called the *radical* of G and is denoted by T_G. If $T_G = G$ (resp. $T_G = \{0\}$) then G is said to be a *radical* group (resp. a group *without radical*). For x to belong to T_G it is necessary and sufficient that every stable open subset of G which contains x should also contain 0. Deduce that if G is discrete, then T_G is the torsion subgroup of G. If f is any continuous homomorphism of G into a commutative topological group G', then $f(T_G) \subset T_{G'}$. If H is a subgroup of G contained in T_G, then $T_{G/H} = T_G/H$. The subgroup T_G is the smallest of the subgroups H of G such that G/H is without radical.

e) Show that, among the subgroups H of G such that $T_H = H$, there is a largest, $H_0 \subset T_G$, which is closed.

f) If $T_G \neq G$, then T_G is open if and only if there is a residual open subgroup in G [show that if $b(M)$ is open for some $M \in \mathfrak{M}$, then $T_G \supset b(M)$, by using *e*)].

* 29) Let G be the topological group $\mathbf{R}$, let φ be the canonical mapping of $\mathbf{R}$ onto its quotient group $\mathbf{T} = \mathbf{R}/\mathbf{Z}$, and let θ be an irrational number. The group G operates continuously on $E = \mathbf{T}^2$ by the rule $(s, (x, y)) \to (x + \varphi(s), y + \varphi(\theta s))$. Show that the stabilizer of each $z \in E$ consists of 0 alone, that the orbit of z is dense in E and that it is not a topological homogeneous space relative to G. *

30) A topological group G is said to have *no arbitrarily small subgroups* if there is a neighbourhood V of e in G such that $\{e\}$ is the only subgroup of G contained in V.

a) Let G be a topological group, H a normal subgroup of G. Show that if H and G/H have no arbitrarily small subgroups, then neither does G.

b) Deduce from *a*) that if H_1, H_1 are two normal subgroups of a topological group G such that G/H_1 and G/H_2 have no arbitrarily small subgroups, then neither does $G/(H_1 \cap H_2)$.

¶ 31) Let G be a connected topological group, H a subgroup of G, U an open subset of G such that $H = UH$. Let $V = H \cap (U^{-1}U)$; show that $H = V^{\infty}$ [show that if $A = H \cap \complement(V^{\infty})$, then the sets $U.V^{\infty}$ and $U.A$ do not intersect].

§ 3

1) On a topological group G, the right uniformity is the only uniformity compatible with the topology of G which has a fundamental system of entourages each of which is invariant under all translations

$$(x,y) \to (xa, ya).$$

2) The definition of the left and right uniformities on a topological group G uses only the properties (GV_I) and (GV_{II}) of the neighbourhood filter $\mathfrak{B}$ of e in G. If we suppose that $\mathfrak{B}$ is a filter satisfying these two axioms, but not necessarily satisfying axiom (GV_{III}), show that the left and right uniformities are respectively compatible with the topologies $\mathcal{T}_s$ and $\mathcal{T}_d$ defined in Exercise 5 *a*) of § 1.

3) Show that the following conditions on a topological group G are equivalent:

α) The left and right uniformities on G are the same.

β) If V is any neighbourhood of e in G, there is a neighbourhood W of e such that $xWx^{-1} \subset V$ for all $x \in G$.

γ) The symmetry $x \to x^{-1}$ is a uniformly continuous mapping of G_d onto G_d (or of G_s onto G_s).

δ) The mapping $(x,y) \to xy$ is a uniformly continuous mapping of one of the four spaces $G_d \times G_d$, $G_s \times G_s$, $G_d \times G_s$, $G_s \times G_d$ into one of the two spaces G_d, G_s.

* 4) Let $G = \mathbf{GL}(2, \mathbf{R})$ be the multiplicative group of invertible square 2×2 matrices with real elements. For each integer $n > 0$, let V_n be the set of matrices $X = \begin{pmatrix} x & y \\ z & t \end{pmatrix} \in G$ such that $|x - 1| \leqslant 1/n$, $|y| \leqslant 1/n$, $|z| \leqslant 1/n$, $|t - 1| \leqslant 1/n$. Show that the family of the V_n is a fundamental system of neighbourhoods of the identity element for a topology compatible with the group structure of G. With respect to this topology, G is locally compact and the left and right uniformities on G are distinct. *

5) Let G be a topological group and let K be a normal subgroup of G. Considering G/K as a subset of $\mathfrak{P}(G)$, show that the right uniformity on the topological group G/K is induced by the uniformity on $\mathfrak{P}(G)$ constructed from the right uniformity on G by the procedure of Exercise 5 of Chapter II, § 1.

6) Show that the least upper bound of the right and left uniformities on a topological group G (Chapter II, § 2, no. 5) is a uniformity compatible with the topology of G: it is called the *two-sided uniformity* on G. Show that every Hausdorff topological group is isomorphic to a dense

subgroup of a topological group which is complete with respect to its two-sided uniformity.

¶ 7) *a*) Let G be a Hausdorff topological group. If there is a neighbourhood V_0 of e in G such that the symmetry $x \to x^{-1}$, considered as a mapping of G_d into G_d, is uniformly continuous on V_0, show that G has a completion (show that the condition of Theorem 1 of § 4 is satisfied).

* *b*) Show that if G is an infinite product of groups each identical with the topological group defined in Exercise 4, then G is complete but there is no neighbourhood of e on which $x \to x^{-1}$, considered as a mapping of G_d into G_d, is uniformly continuous $_*$.

8) A Hausdorff topological group G is said to be *locally precompact* if there is a neighbourhood V_0 of e which is precompact with respect to the right (or left) uniformity on G. Show that every locally precompact group has a locally compact completion (use Exercise 7).

9) Let G be a Hausdorff topological group and let K be a closed normal subgroup of G. Show that if the topological groups K and G/K are complete, then so is G [consider a minimal Cauchy filter on G with respect to the left (or right) uniformity on G, and its image in G/K].

10) Let $(G_\iota)_{\iota \in I}$ be a family of topological groups and let $G = \prod_{\iota \in I} G_\iota$ be the product group endowed with the topology defined in Exercise 23 of § 2. Show that if the G_ι are complete, then so is G.

11) With the hypotheses and notation of Exercise 26 of § 2, suppose that each of the groups G_ι is Hausdorff and has a completion. Show that the local product G of the G_ι, relative to a family (K_ι) of open normal subgroups, has a completion $\hat{G}$ which is isomorphic to the local product of the $\hat{G}_\iota$, relative to the closures $\overline{K}_\iota$ of the K_ι.

12) *a*) Let G′ be a complete Hausdorff group; let G_0 be a dense subgroup of G′, distinct from G′; and let G be the topological group obtained by giving G_0 the discrete topology. The identity mapping $G \to G_0$ is then a bijective continuous homomorphism, but its continuous extension $\hat{G} \to \hat{G}_0$ is not surjective.

* *b*) Let G be the group $\mathbf{Q}^2$, θ an irrational number, u the continuous homomorphism $(x, y) \to x + \theta y$ of G into $\mathbf{R}$, and let $G' = u(G)$. Then $u : G \to G'$ is bijective, but its continuous extension $\hat{G} \to \hat{G}'$ is not injective.

c) From *a*) and *b*) construct an example of a bijective continuous homomorphism $G \to G'$ of Hausdorff commutative groups whose continuous extension $\hat{G} \to \hat{G}'$ is neither injective nor surjective $_*$.

§ 4

1) *a*) A subgroup H of a topological group G operates properly on G according to the external law $(s, x) \rightarrow sxs^{-1}$ if and only if G is Hausdorff and H is compact (use Propositions 2 and 4).

b) Let G be a topological group and let H be a subgroup of G. Then the mapping $(x, y) \rightarrow xy$ of $H \times G$ into G is proper if and only if H is quasi-compact (consider the inverse image of e under this mapping).

2) Give an example of a compact group operating continuously (and hence not properly, cf. Proposition 3) on a non-Hausdorff space (cf. § 2, Exercise 13).

¶ 3) A topological group G operates properly on a homogeneous space G/K if and only if K is quasi-compact. (To show that the condition is sufficient, show that the canonical mapping of G onto G/K is proper and use Chapter I, § 10, no. 1, Proposition 5.)

4) Give an example of a Hausdorff topological group G operating properly on a Hausdorff topological space X, such that the mapping $(s, x) \rightarrow s.x$ is not closed and the canonical mapping $X \rightarrow X/G$ is not closed [cf. Exercise 1 *b*)].

5) A subgroup H of a topological group G operates properly on G by left translations if and only if H is closed in G.

* 6) For each real number $a \geqslant 1$, put

$$f_a(t) = \left(t + \frac{a(a+2)}{a+1}, -\frac{a}{a+1}\right) \quad \text{if} \quad t < -\frac{a}{a+1},$$

$$f_a(t) = (a, t) \quad \text{if} \quad -\frac{a}{a+1} \leqslant t \leqslant \frac{a}{a+1},$$

$$f_a(t) = \left(-t + \frac{a(a+2)}{a+1}, \frac{a}{a+1}\right) \quad \text{if} \quad t > \frac{a}{a+1}.$$

Let C_a denote the set of all points $f_a(t)$ as t runs through **R**, and let $X \subset \mathbf{R}^2$ be the union of the sets C_a for $a \geqslant 1$ and the lines D' and D'', where D' (resp. D'') is the set of points $(t, -1)$ (resp. $(t, 1)$) for $t \in \mathbf{R}$.

a) The group **R** operates on X according to the law $(s, z) \rightarrow s.z$ such that:

(i) $s.(t, -1) = (s + t, -1)$;

(ii) $s.(t, 1) = (t - s, 1)$;

(iii) $s.f_a(t) = f_a(s + t)$ for $a \geqslant 1$.

Show that $\mathbf{R}$ operates continuously on X and that the four conditions of Proposition 4 are satisfied, but that $X/\mathbf{R}$ is not Hausdorff.

b) Let S be the equivalence relation on X whose classes consist of a single point z if $z \notin D' \cup D''$, and are of the form $(z, -z)$ if $z \in D'$ or $z \in D''$. Let X′ be the quotient space X/S. The relation S is compatible with the group $\mathbf{R}$ operating on X (§ 2, no. 4); show that $\mathbf{R}$ operates continuously on X′, that the four conditions of Proposition 4 are satisfied and that $X'/\mathbf{R}$ is Hausdorff, but that $\mathbf{R}$ does not operate properly on X′. $_*$

7) Let G be a Hausdorff group operating continuously on a locally compact space X, and suppose that (i) for every compact subset K of X, the restriction to $G \times K$ of the mapping $(s, x) \to s.x$ is a closed mapping of $G \times K$ into X; (ii) for each $x \in X$, $s \to s.x$ is a proper mapping of G into X. Show that G operates properly on X (use Proposition 12).

8) Let G_0 be a non-discrete Hausdorff topological group in which every compact subset is finite (§ 2, Exercise 23). Let G be the group G_0 with the discrete topology. Show that G operates continuously but not properly on G_0 by left translation, and is such that P(K, L) is compact for all compact subsets K, L of G_0 (Theorem 1).

¶ 9) Let G be a topological group operating continuously on two spaces X and X′, and let $f : X \to X'$ be a continuous mapping compatible with the identity mapping of G (§ 2, no. 4). Let $\bar{f} : X/G \to X'/G$ be the continuous mapping induced by f on the quotient spaces.

a) Show that if f is open (resp. closed, proper, injective, surjective) then so is $\bar{f}$. (To prove that if f is proper then $\bar{f}$ is proper, show that the inverse image of a point of X′/G under $\bar{f}$ is quasi-compact.)

b) Suppose that G acts properly and freely on both X and X′. Show that if $\bar{f}$ is proper, then so is f [observe that the restriction of f to an orbit $G.x$ is a homeomorphism onto $G.f(x)$].

¶ 10) Let G be a topological group and let X, Y be two topological spaces on which G operates continuously. Then G operates continuously on $X \times Y$ according to the law $(s, (x,y)) \to (s.x, s.y)$. Let $(X \times Y)/G$ be denoted by $X \times {}^{G}Y$.

a) Show that $X \times {}^{G}G$ (G operating by left translations on itself) is canonically homeomorphic to X. (Remark that if U is open in X, then the union of the orbits of the points of $U \times \{e\}$ is open in $X \times G$).

b) Let Y′ be a third topological space on which G operates continuously, and let $f : Y \to Y'$ be a continuous mapping compatible with the identity mapping of G. The mapping $1_X \times f$ induces a continuous mapping $\bar{f} : X \times {}^G Y \to X \times {}^G Y'$. If f is open (resp. proper, injective, surjective) then so is $\bar{f}$ (Exercise 9). Give an example in which f is closed, G operates properly on X, but $\bar{f}$ is not closed [take $X = G$, $X' = P$ (a space consisting of a single point) and use Exercise 4].

c) Show that if G operates properly on X, then G operates properly on $X \times Y$.

d) Show that, if Y is compact, the canonical mapping $X \times {}^G Y \to X/G$ is proper [use *b*)].

e) Suppose that G is locally compact, and let G′ be the one-point compactification of G. If ω is the point at infinity in G′, then G operates continuously on G′ according to the external law given by $s.t = st$ if $t \in G$, and $s.\omega = \omega$; X is then homeomorphic to an open subspace of $X \times {}^G G'$ (use *b*)). Deduce that if G operates properly on X and if in addition X/G is locally compact, then X is locally compact [use *c*) to show that $X \times {}^G G'$ is Hausdorff].

**f*) Give an example of a locally compact group G operating continuously on a space X which is not locally compact, such that X/G consists of a single point (cf. § 2, Exercise 29). *

¶ 11) Let G be a topological group and let H, K be two closed subgroups of G.

a) Show that the following three conditions are equivalent: (i) $H \times K$ operates properly on G according to the external law $((h, k), s) \to hsk^{-1}$; (ii) H operates properly on the homogeneous space G/K; (iii) K operates properly on the homogeneous space G/H.

b) If G is locally compact, then the (equivalent) conditions of *a*) are also equivalent to the following: for every pair of compact subsets A, B of G, the intersection $HA \cap BK$ is compact. In particular these conditions are satisfied if one of the subgroups H, K is compact.

c) Suppose that G is locally compact, and show that the following conditions are equivalent: (i) for each $x \in G$ the mapping $h \to h.xK$ of H into G/K is proper; (ii) for each $x \in G$ the mapping $k \to k.xH$ of K into G/H is proper; (iii) for each $x \in G$ and each compact subset A of G, $Hx \cap AK$ is compact; (iv) for each $x \in G$ and each compact subset A of G, $Kx \cap AH$ is compact. If one of the subgroups H, K is normal or if the left and right uniformities on G coincide (§ 3, Exercise 3), show that these conditions imply those of *a*).

12) *a*) Let X be a compact space and G a topological group operating continuously on X. Suppose that each orbit A with respect to G has a non-empty interior relative to its closure $\bar{A}$. Show that there is at least one compact orbit (consider a minimal element in the set of compact subsets of X which are stable under G).

$_*$ *b*) Give an example of a locally compact group operating continuously on a compact space, such that none of the orbits is compact (§ 2, Exercise 29). $_*$

¶ 13) Let G be a locally compact group and let D be a discrete subgroup of G such that the homogeneous space G/D is compact. Show that, for each $d \in \mathrm{D}$, the set of all sds^{-1}, as s runs through G, is closed in G. (Prove first, with the help of Proposition 10, that there is a compact subset K of G such that $\mathrm{G} = \mathrm{K}.\mathrm{D}$; then use Corollary 1 to Proposition 1).

14) *a*) Let G be a topological group operating continuously on a topological space X, and let x_0 be a point of X such that $s.x_0 = x_0$ for all $s \in \mathrm{G}$. For each neighbourhood V of x_0 and each quasi-compact subset K of G, show that the set $\bigcap_{s \in \mathrm{K}} s.\mathrm{V}$ is a neighbourhood of x_0.

b) Deduce from *a*) that if G is a locally compact group, V a neighbourhood of e, K a compact subset of G, then the set $\bigcap_{s \in \mathrm{K}} s\mathrm{V}s^{-1}$ is a neighbourhood of e.

¶ 15) Let G be a locally compact group and let G_0 be the identity component in G; suppose that the quotient group G/G_0 is *compact.*

a) Show that G is σ-compact (remark that G_0 is σ-compact and use Proposition 10).

b) Let (U_n) be a decreasing sequence of neighbourhoods of e in G. Show that there is a compact normal subgroup K of G, contained in the intersection of the U_n, and such that the identity element of G/K has a countable fundamental system of neighbourhoods. [There exists a compact symmetric neighbourhood V_0 of e in G such that $\mathrm{G} = \mathrm{V}_0\mathrm{G}_0$. Define recursively a sequence of compact symmetric neighbourhoods V_n of e such that $\mathrm{V}_n^2 \subset \mathrm{V}_{n-1} \cap \mathrm{U}_n$ and $x\mathrm{V}_n x^{-1} \subset \mathrm{V}_{n-1}$ for each $x \in \mathrm{V}_0$ (Exercise 14 *b*); show that the intersection K of the V_n answers the question].

c) Let X be a locally compact space which has a countable base; suppose G operates continuously on X in such a way that there is no $s \neq e$ in G for which $s.x = x$ for all $x \in \mathrm{X}$. Show that the identity element of G has a countable fundamental system of neighbourhoods. [Let (W_n) be a countable base of the topology of X, consisting of relatively compact sets; for each pair of integers (m, n) such that $\overline{\mathrm{W}}_m \subset \mathrm{W}_n$, let U_{mn} be the set of all $s \in \mathrm{G}$ such that $s.\overline{\mathrm{W}}_m \subset \mathrm{W}_n$; show that the U_{mn} are open in G and apply the result of *b*).]

16) Let G be a locally compact connected group, let K be a compact normal subgroup of G and let N be a closed normal subgroup *of* K

such that K/N has no arbitrarily small subgroups (§ 2, Exercise 30). Show that in this case N is a normal subgroup *of* G. (The hypothesis on K/N implies that there is a compact neighbourhood U of e in G such that $xNx^{-1} \subset N$ for each $x \in U$).

¶ 17) Let G be a locally compact group with no arbitrarily small subgroups [§ 2, Exercise 30].

a) Show that every point of G has a countable fundamental system of neighbourhoods (Exercise 15 *b*)).

b) There is a compact neighbourhood V of e such that for any two points x, y of V, the relation $x^2 = y^2$ implies $x = y$. (We may suppose G not commutative. If the result were false, there would exist two sequences (x_n), (y_n) of points of G, tending to e, such that $x_n^2 = y_n^2$ and $x_n y_n^{-1} = a_n \neq e$. Let U be a compact neighbourhood of e which contains no subgroup of G other than $\{e\}$, and let p_n be the smallest integer $p > 0$ such that $a_n^{p+1} \notin U$; show (by passing to a subsequence if necessary) that we may suppose that $a = \lim_{n\to\infty} a_n^{p_n}$ exists, is not equal to e and belongs to U. Show that $a^{-1} = a$ and thus obtain a contradiction).

c) Let U be a compact symmetric neighbourhood of e which does not contain any subgroup of G other than $\{e\}$, and let V be a neighbourhood of e. Show that there exists a number $c(V) > 0$ such that, for each pair of integers $p > 0$, $q > 0$ such that $p \leqslant c(V)q$, and each element $x \in G$ such that $x, x^2, \dots, x^q$ belong to U, we have $x^p \in V$. [Argue by contradiction, by supposing that there exist two sequences of integers (p_n), (q_n) such that $\lim_{n\to\infty} (p_n/q_n) = 0$ and for each n an element $a_n \in G$ such that $a_n^h \in U$ for $1 \leqslant h \leqslant q_n$, but $a_n^{p_n} \notin V$; we may also suppose that the sequence $(a_n^{p_n})$ has a limit $a \neq e$ lying in U. Show that we should then have $a^m \in U$ for every integer $m > 0$ and hence arrive at a contradiction].

18) Let G be a locally compact totally disconnected topological group whose left and right uniformities coincide. Show that every neighbourhood of e in G contains a compact open *normal* subgroup of G (use Corollary 1 to Proposition 14 of no. 6, and Exercise 3 of § 3).

¶ 19) Let G be a locally compact group, let G_0 be the identity component of G, and let H be a closed subgroup of G. Show that if the homogeneous space G/H is locally connected, then G_0H is open in G. Let H_0 be the identity component of H; then there is an open subgroup L of H such that L/H_0 is compact (no. 6, Corollary 1 to Proposition 14); show that on the one hand LG_0 is closed in G, by using Corollary 1 of Proposition 1, and on the other hand that LG_0 is open in G, by considering the canonical image of G_0 in G/L and using Corollary 3 of Proposition 14 and the fact that a quotient space of a locally connected space is locally connected [Chapter I, § 11, no. 6, Proposition 12)].

* 20) Let φ be the canonical homomorphism $\mathbf{R} \to \mathbf{T}$ and let θ be an irrational number. Define a group structure on the topological space $\mathbf{R}^2 \times \mathbf{T}^2$ by the law

$$(x_1, x_2, t_1, t_2)\,(x_1', x_2', t_1', t_2') \\ = (x_1 + x_1', x_2 + x_2', t_1 + t_1' + \varphi(x_2x_1'), t_2 + t_2' + \varphi(\theta x_2 x_1')).$$

G thus becomes a locally compact topological group (and even a Lie group). Show that the commutator subgroup of G is not closed. *

¶ 21) Let M be a compact space endowed with a semigroup structure such that (i) the mapping $(x, y) \to xy$ of $M \times M$ into M is continuous; (ii) for each $a \in M$ the relation $ax = ay$ implies $x = y$.

a) Show that if F is a closed subset of M and $x \in M$ is such that $xF \subset F$, then $xF = F$. [Let y be a cluster point of the sequence $(x^n)_{n \geqslant 1}$ in M; show that $yF = \bigcap_{n \geqslant 1} x^n F$, and deduce that $yxF = yF$].

b) Deduce from *a*) that if, in addition, for each $a \in M$ the relation $xa = ya$ implies $x = y$, then M is a compact topological group. (Start by showing that there is an identity element e; to show that $x \to x^{-1}$ is continuous, argue by *reductio ad absurdum*, considering an ultrafilter on M which converges to e).

¶ 22) *a*) Let G be a Hausdorff topological group. Show that every stable subset S of G which is either compact, or open and relatively compact, is a subgroup of G [use Exercise 21 above, and Exercise 28 *a*) of § 2]. Deduce that a compact commutative group is a radical group [§ 2, Exercise 28 *d*)].

b) Deduce from *a*) that every locally compact stable subset of a compact group K is a subgroup of K.

c) Let G be a Hausdorff topological group and let S be a stable closed subset of G with the property that there is a subset K of G which is precompact with respect to the right uniformity of G, and such that $SK = G$. Show that S is a subgroup of G. (Given $s \in S$, define recursively a sequence of elements $x_i \in K$ and a sequence of elements $s_i \in S$ such that $s^{-1}x_i = s_{i+1}x_{i+1}$; observe that if $i < j$ then $s^{-1}x_i x^{-1}{}_j \in S$, and that for every neighbourhood V of e in G there is a pair of elements x_i, x_j such that $i < j$ and $x_i x^{-1}{}_j \in V$).

23) Let p be a prime number, let (G_n) be an infinite sequence of topological groups each identical with the discrete group $\mathbf{Z}/(p^2)$, and

let H_n be the subgroup $(p)/(p^2)$ of G_n. Let G be the local direct product of the G_n relative to the H_n (§ 2, Exercise 26); then G is a locally compact group but is not compact. Show that the continuous homomorphism $u : x \to px$ of G into itself is not a strict morphism of G onto $u(G)$, and that $u(G)$ is not closed in G.

§ 5

1) Let $(x_\iota)_{\iota \in I}$ be a summable family in a complete Hausdorff commutative group G, and let $\mathfrak{S}$ be a set of subsets of I, directed with respect to the relation $\subset$ and forming a covering of I. Show that $\sum_{\iota \in I} x_\iota = \lim_J \sum_{\iota \in J} x_\iota$, where the limit is taken with respect to the section filter of $\mathfrak{S}$.

2) Let G be a complete Hausdorff commutative group such that every neighbourhood of 0 in G contains an open subgroup of G (this is the case in particular if G is locally compact and totally disconnected; cf. § 4, Exercise 18). A family $(x_\iota)_{\iota \in I}$ of points of G is summable if and only if $\lim x_\iota = 0$ with respect to the filter of complements of finite subsets of I. Hence show that every convergent series in G is commutatively convergent.

¶ 3) Let $(x_\iota)_{\iota \in I}$ be a family of points of a Hausdorff commutative group G. For every finite subset H of I, let s_H denote $\sum_{\iota \in H} x_\iota$, and let A be the set of cluster points of the mapping $H \to s_H$ with respect to the directed set $\mathfrak{F}(I)$. For every finite subset J of I, let $\Phi(J)$ denote the closure of the set of points s_H, where H runs through the finite subsets of I such that $H \cap J = \emptyset$. Show that if $A \neq \emptyset$ then the set $A - A$ (i.e. the set of all $x - x'$ where $x \in A$ and $x' \in A$) is the same as the set $B = \bigcap_{J \in \mathfrak{F}(I)} \Phi(J)$; and that in any case $B + B \subset B$. Deduce that if A is not empty, then B is a closed subgroup of G and that A is a coset of B in G.

4) With the notation of Exercise 3, take G to be the discrete additive group **Z** of rational integers and I to be the set **N**. Give an example of a non-summable sequence (x_n) for which the set A consists of the single point 0 (choose the x_n so that every finite partial sum, all of whose terms have indices $\geqslant m$, is an integral multiple of m).

¶ 5) Let (x_n) be a sequence of points of a Hausdorff commutative group. If, for every infinite subset I of **N**, the series defined by the sequence $(x_n)_{n \in I}$ is convergent, then the sequence (x_n) is summable (argue by *reductio ad absurdum* as in the proof of Proposition 9).

¶ 6) *a*) Let σ be a permutation of $\mathbf{N}$ and let $\varphi(n)$ be the smallest number of intervals of $\mathbf{N}$ whose union is $\sigma([0, n])$. Suppose that $\varphi(n)$ is *bounded* as n runs through $\mathbf{N}$. Then if (u_n) is any convergent series whose terms belong to a Hausdorff commutative group G, the series $(u_{\sigma(n)})$ is convergent and has the same sum as (u_n).
* *b*) Suppose that $\varphi(n)$ is not bounded in $\mathbf{N}$. Construct a series (u_n) whose terms belong to the additive group $\mathbf{R}$ and which converges in $\mathbf{R}$ but is such that the series $(u_{\sigma(n)})$ does not converge. [Consider a strictly increasing sequence (m_k) of integers, defined by induction on k, satisfying the following conditions: (i) if $[0, n_k]$ is the largest interval of $\mathbf{N}$ with left-hand extremity 0 which is contained in $\sigma([0, m_k])$, then

$$\sigma([0, m_k]) \subset [0, n_{k+1}];$$

(ii) $\varphi(m_k) \geqslant k + 1$. Then define u_n appropriately for $n_k < n < n_{k+1}$]. Generalize to the case of a complete Hausdorff commutative group G which is generated by every neighbourhood of 0. *

7) Let (x_{mn}) be a double sequence of points of a Hausdorff commutative group G, satisfying the following conditions:
(a) The series defined by the sequence $(x_{mn})_{n\in\mathbf{N}}$ is convergent for each $m \geqslant 0$; let y_m be its sum.
(b) If we put $r_{mn} = \sum_{p=n}^{\infty} x_{mp}$, then the series defined by the sequence $(r_{mn})_{m\in\mathbf{N}}$ is convergent for each $n \geqslant 0$; let t_n be its sum.

Show that for each $n \geqslant 0$ the series defined by the sequence $(x_{mn})_{m\in\mathbf{N}}$ is convergent, and let z_n be its sum. The series whose general term is y_m has the same sum as the series whose general term is z_n if and only if t_n tends to 0 as n tends to infinity.

§ 6

1) In a Hausdorff topological ring A, the commutant of every subset of A (and in particular the centre of A) is closed, and so is the left (resp. right) annihilator of every subset of A.

2) *a*) Let $\mathfrak{a}$ be a discrete left ideal in a topological ring A. Show that, for every $x \in \mathfrak{a}$, the left annihilator of x in A is open. Hence show that if A is a non-discrete ring with no zero-divisors, it contains no discrete (left or right) ideal other than $\{0\}$.
b) Let A be a locally connected topological ring, and let $\mathfrak{a}$ be a totally disconnected left ideal of A. Show that, for each $x \in \mathfrak{a}$, the left annihilator of x in A is open. Hence show that if A has no zero-divisors it contains no totally disconnected (left or right) ideal other than $\{0\}$.

3) The connected component of 0 in a topological ring A is a two-sided ideal. Deduce that if A is quasi-simple and is not connected (in particular, if A is a non-connected topological division ring) then A is totally disconnected.

4) Let (x_ι) be a summable family in a Hausdorff topological ring A, and let s be its sum. Then for each $a \in A$ the family (ax_ι) [resp. $(x_\iota a)$] is summable and its sum is as (resp. sa). If (x_λ) and (y_μ) are two summable families in A such that the family $(x_\lambda y_\mu)$ is summable, then $\sum_{\lambda,\mu} x_\lambda y_\mu = \left(\sum_\lambda x_\lambda\right)\left(\sum_\mu y_\mu\right)$. If moreover A is complete and if one of the x_λ is a unit, then the summability of the family $(x_\lambda y_\mu)$ implies the summability of (y_μ).

5) A *rectangle* is a subset of $\mathbf{N} \times \mathbf{N}$ which is the product of two intervals of $\mathbf{N}$. Given a finite subset E of $\mathbf{N} \times \mathbf{N}$, let $\varphi(E)$ be the smallest number of mutually disjoint rectangles whose union is E. Let (E_n) be an increasing sequence of finite subsets of $\mathbf{N} \times \mathbf{N}$ which cover $\mathbf{N} \times \mathbf{N}$ and are such that the sequence $(\varphi(E_n))$ is bounded. If (x_n), (y_n) are two convergent series whose terms belong to a Hausdorff topological ring A, then we have

$$\lim_{n\to\infty} \sum_{(h,k)\in E_n} x_h y_k = \left(\mathop{\mathrm{S}}_{n=0}^{\infty} x_n\right)\left(\mathop{\mathrm{S}}_{n=0}^{\infty} y_n\right).$$

Construct an example of a sequence (E_n) satisfying the conditions above and such that $E_{n+1} - E_n$ contains only one element for each n.

6) Let A be a topological ring with an identity element and let $\mathfrak{a}$, $\mathfrak{b}$ be two two-sided ideals of A such that $\mathfrak{a} + \mathfrak{b} = A$ and $\mathfrak{a} \cap \mathfrak{b} = \{0\}$. Show that the topological ring A is isomorphic to the product of the topological rings $\mathfrak{a}$ and $\mathfrak{b}$.

7) Let A be a topological ring with no identity element, and let B be the ring obtained by adjoining an identity element to A show that the topology on B which is the product of the topology of A and the discrete topology on $\mathbf{Z}$ is compatible with the ring structure of B, and induces the given topology on A (a two-sided ideal of B).

8) Let A be a topological ring with an identity element.

a) The set A^* of units of A is open in A if and only if there is a neighbourhood V of 1 all of whose elements are units in A.

b) If A^* is open in A then every maximal (left or right) ideal of A is closed, and the radical of A is closed. If A has no closed left ideals other than $\{0\}$ and A, then A is a division ring.

c) Let A be the subring of the topological field **Q** consisting of all rational numbers $k/2^n$ ($k \in \mathbf{Z}$, $n \in \mathbf{N}$). Show that A contains no closed ideal other than $\{0\}$ and A, but is not a field.

9) *a*) An element x (resp. an ideal $\mathfrak{a}$) of a topological ring A is said to be *topologically nilpotent* if the sequence $(x^n)_{n \geqslant 1}$ converges to 0 [resp. if the filter base formed by the $\mathfrak{a}^n$ $(n \geqslant 1)$ converges to 0). Every element of a topologically nilpotent ideal is topologically nilpotent.

b) Suppose that A has an identity element and that A^* is open in A. Show that if x is any topologically nilpotent element of A, then $1 - x$ is a unit (observe that $1 - x^n$ is a unit for sufficiently large n). Hence show that if all the elements of a (left or right) ideal $\mathfrak{a}$ are topologically nilpotent, then $\mathfrak{a}$ is contained in the radical of A.

c) Suppose that A is complete and has an identity element, and that there exists a fundamental system of neighbourhoods of 0 which are subgroups of the additive group of A. Show that if x is topologically nilpotent, then $1 - x$ is a unit in A.

d) Under the hypotheses of *c*), show that if $y \in A$ is topologically nilpotent, then the equation $x^2 + x = y$ has a root $x \in A$, and that x is topologically nilpotent.

¶ 10) *a*) Let B be a ring with an identity element, let E be a free B-module $B^{(I)}_d$, and let A be the ring $\mathcal{L}(E)$ of endomorphisms of E. For every finite subset F of E, let V_F be the set of all $u \in A$ such that $u(x) = 0$ for all $x \in F$. Show that the V_F from a fundamental system of neighbourhoods of 0 for a Hausdorff topology on A, which is compatible with the ring structure of A, and with respect to which A is totally disconnected.

b) Take $I = \mathbf{N}$ and take B to be a commutative ring which has no zero-divisors but whose radical $\mathfrak{R}$ is not $\{0\}$. Show that the radical of A is not closed. [Let (e_n) be the canonical basis of E. Remark on the one hand that each $u \in A$, such that $u(e_n) = 0$ for all but a finite number of indices and such that $u(E) \subset \mathfrak{R}.E$, belongs to the radical of A; consider on the other hand the element u_0 of A such that $u_0(e_n) = e_n + te_{n+1}$ with $t \neq 0$ in $\mathfrak{R}$, for each n].

¶ 11) *a*) A topological ring with an identity element is said to be a *Gelfand ring* if A^* is open in A and if the topology induced on A^* by the topology of A is compatible with the multiplicative group structure of A^*. Every Hausdorff topological division ring is a Gelfand ring [cf. Exercise 20 *e*)].

b) Show that if A is a Gelfand ring then so is every matrix ring $\mathbf{M}_n(A)$ endowed with the product topology (on A^{n^2}). (Proof by induction on n).

12) A subset M of a topological ring is said to be *right bounded* (resp. *left bounded*) if, for each neighbourhood U of 0 in A, there is a neighbourhood V of 0 such that $VM \subset U$ (resp. $MV \subset U$); M is *bounded* if it is both left and right bounded. The topology of A is said to be *locally bounded* (and A is said to be a *locally bounded* topological ring) if there is a bounded neighbourhood of 0 in A.

a) Every topological ring which has a fundamental system of neighbourhoods of 0 consisting of right ideals is right bounded. Show that if I is infinite and B is a division ring in Exercise 10 *a*), then the ring A is left bounded, but there is no right bounded neighbourhood of 0 in A.

b) If a topological ring A is right bounded (resp. bounded) and has a fundamental system of neighbourhoods of 0 which are subgroups of the additive group of A, then it has a fundamental system of neighbourhoods of 0 which are right (resp. two-sided) ideals.

c) Every finite union of bounded sets is bounded; the closure of a bounded set is bounded; if M and N are bounded, then so are $M + N$ and MN.

d) Every precompact set in a topological ring A is bounded.

e) If M and N are two bounded subsets of A, show that the mapping $(x, y) \rightarrow xy$ elements of $M \times N$ into A is uniformly continuous.

f) Let A be a Hausdorff topological ring which has an identity element. Show that if (x_n) is a sequence of units which tends to 0 in A, then the set of elements x_n^{-1} is neither left nor right bounded.

g) If A is a bounded ring with an identity element, show that the mapping $x \rightarrow x^{-1}$ is uniformly continuous on the set A^* of units of A.

h) Show that if A is a complete Hausdorff bounded ring with an identity element, then the set A^* of units and the radical of A are closed [use *g*)].

i) A subset of a product $\prod_\iota A_\iota$ of topological rings is bounded if and only if all its projections are bounded.

j) Every subring of a locally bounded ring and every product of locally bounded rings are locally bounded rings.

k) The completion of a locally bounded Hausdorff ring is locally bounded.

13) Let A be a bounded topological ring (Exercise 12) which has an identity element and is such that the set A^* of units of A is open in A. Show that the radical of A is open. Hence if A is without radical, it is discrete; in particular, a bounded Hausdorff topological division ring is discrete. A compact ring A without radical, which has an identity element and in which A^* is open, is finite; in particular, a compact topological division ring is finite.

¶ 14) Let A be a compact ring with an identity element, totally disconnected (*) and without radical. Then A is isomorphic to the product of a family of finite simple rings. [Show first that A has a fundamental system of neighbourhoods of 0 which are two-sided ideals, using Exercise 12 *b*) of § 6 and Proposition 14 of § 4. Hence show that there is a maximal set Φ of two-sided ideals of A such that (i) every two-sided ideal $\mathfrak{N} \in \Phi$ is a maximal open two-sided ideal of A; (ii) no ideal $\mathfrak{N} \in \Phi$ contains a finite intersection of ideals of Φ other than $\mathfrak{N}$. Show then that A is isomorphic to the product of the rings $A/\mathfrak{N}$ as $\mathfrak{N}$ runs through Φ; prove first that the intersection of all the $\mathfrak{N} \in \Phi$ consists only of the zero element of A by using Exercise 9 *b*)].

15) Let A be a totally disconnected compact ring with an identity element. Show that the radical $\mathfrak{R}$ of A is topologically nilpotent [Exercise 9 *a*); use Exercise 12 *b*) of § 6 and Proposition 14 of § 4]. Hence show that $\mathfrak{R}$ is the set of all $x \in A$ such that ax is topologically nilpotent for all $a \in A$ [cf. Exercise 9 *c*)].

16) Show that on a ring A (resp. a division ring K), the least upper bound $\mathfrak{T}$ of a family $(\mathfrak{T}_\iota)$ of topologies compatible with the ring structure of A (resp. the division ring structure of K) is compatible with this structure. A set M which is left (resp. right) bounded for each of the topologies $\mathfrak{T}_\iota$ is also left (resp. right) bounded for $\mathfrak{T}$.

17) If K is a Hausdorff topological division ring and is not discrete, then neither of the multiplicative uniformities on K is comparable with the uniformity induced on K^* by the additive uniformity of K (cf. Exercise 13).

18) Let K be a Hausdorff topological division ring, let A be a closed subset of K and let B be a compact subset of K such that $0 \notin B$. Show that AB and BA are closed in K (cf. § 4, no. 1, Proposition 1, Corollary 1). * In the field **R** of real numbers, give an example where A is closed, and B is compact, $0 \in B$ and AB is not closed. *

19) *a*) Let K be a division ring endowed with a Hausdorff topology $\mathfrak{T}$ compatible with the *ring* structure of K. Show that there is a neighbourhood V of 0 in K such that $(\complement V) \cup (\complement V)^{-1} = K^*$ [this is equivalent to $V \cap (V \cap K^*)^{-1} = \varnothing$].

b) Suppose moreover that $\mathfrak{T}$ is not discrete. Show that if U is any neighbourhood of 0 in K and if U^* denotes $U \cap K^*$, then

$$U^*(U^*)^{-1} = K^*.$$

(*) It can be shown that a compact ring with an identity element is necessarily totally disconnected.

¶ 20) *a*) Let K be a division ring endowed with a non-discrete Hausdorff topology $\mathfrak{T}$ compatible with the *ring* structure of K. A subset M of K is right (resp. left) bounded if and only if, for each neighbourhood U of 0 there exists $a \in K^*$ such that $aM \subset U$ (resp. $Ma \subset U$); M is then right (resp. left) bounded for every Hausdorff topology $\mathfrak{T}'$ which is coarser than $\mathfrak{T}$ and is compatible with the ring structure of K.

b) If $\mathfrak{T}$ is locally bounded (Exercise 12) and if U is a bounded neighbourhood of 0 with respect to $\mathfrak{T}$, then the sets xU (resp. Ux) form a fundamental system of neighbourhoods of 0 (with respect to $\mathfrak{T}$) as x runs through K*. Every point of K has a countable fundamental system of neighbourhoods of 0 with respect to $\mathfrak{T}$ if and only if there is a sequence (a_n) of points of K which has 0 as a cluster point.

c) The least upper bound of any finite family of locally bounded topologies on K (compatible with the ring structure of K) is locally bounded. Conversely, if the least upper bound $\mathfrak{T}$ of a family $(\mathfrak{T}_\iota)$ of locally bounded topologies on K is locally bounded, then $\mathfrak{T}$ is the least upper bound of a *finite* subfamily of $(\mathfrak{T}_\iota)$. (Remark that if U is a neighbourhood of 0 which is bounded with respect to $\mathfrak{T}$, then there is a finite system of indices (ι_k) and for each k a neighbourhood V_k of 0, bounded with respect to $\mathfrak{T}_{\iota_k}$, such that $\bigcap V_k \subset U$; on the other hand, there exists $a_k \in K^*$ such that $U \subset a_k V_k$.)

d) A subset F of a division ring K is said to be a *quasi-ring* if: (i) $F \neq K$, $0 \in F$, $1 \in F$, $-F = F$, $FF \subset F$; (ii) there is an element $c \in F^* = F \cap K^*$ such that $c(F + F) \subset F$; and (iii) for each $x \in F$, there exists $y \in F$ such that $yF \subset Fx$ and $Fy \subset xF$. Show that if $\mathfrak{T}$ is a non-discrete Hausdorff topology which is locally bounded and compatible with the ring structure of K, then every symmetric bounded neighbourhood U of 0 (with respect to $\mathfrak{T}$), such that $UU \subset U$ and $1 \in U$, is a quasi-ring, and that $U^*(U^*)^{-1} = K^*$ [Exercise 19 *b*)]. If V is any symmetric bounded neighbourhood of 0 with respect to $\mathfrak{T}$, then the set U of elements $x \in K$ such that $xV \subset V$ (or $Vx \subset V$) is a symmetric bounded neighbourhood of 0 with respect to $\mathfrak{T}$ such that $UU \subset U$ and $1 \in U$, and hence is a quasi-ring.

e) Conversely, let F be a quasi-ring in K such that $F^*(F^*)^{-1} = K^*$. Show that there is a Hausdorff topology $\mathfrak{T}$ which is not discrete and is compatible with the ring structure of K, such that F is a bounded neighbourhood of 0 with respect to $\mathfrak{T}$. In particular we may take F to be any subring A of K such that K is the division ring of right fractions of A. In particular, if $K = \mathbf{Q}$ and $F = \mathbf{Z}$, the corresponding topology $\mathfrak{T}$ on K is not compatible with the field structure of K.

¶ 21) *a*) Let $\mathcal{T}$ be a locally bounded Hausdorff topology on a division ring K, with respect to which K is connected. Show that there are non-zero topologically nilpotent elements [Exercise 9 *a*)] in K. (If U is a symmetric bounded neighbourhood of 0, and t is a non-zero element of K such that $Ut + tU \subset U$, show that K is the union of the sets Ut^{-n}, by using Proposition 6 of § 2, no. 2).

b) Let $\mathcal{T}$ be a locally bounded Hausdorff topology on K, with respect to which K has non-zero topologically nilpotent elements. Show that there is a neighbourhood U of 0 such that, for each neighbourhood V of 0, there is an integer n_0 such that $U^n \subset V$ whenever $n \geqslant n_0$; this implies that all the elements of U are topologically nilpotent and that every point of K has a countable fundamental system of neighbourhoods with respect to $\mathcal{T}$. (Let $t \neq 0$ be topologically nilpotent and let W be a bounded neighbourhood of 0; take U to be a bounded neighbourhood of 0 such that $UW \subset Wt$.) The set T of topologically nilpotent elements of K is therefore a neighbourhood of 0 with respect to $\mathcal{T}$.

¶ 22) In a division ring K endowed with a Hausdorff topology $\mathcal{T}$ compatible with the *ring* structure of K, a set R containing 0 is said to be *retrobounded* if $(\complement R)^{-1}$ is bounded. $\mathcal{T}$ is said to be *locally retrobounded* if there is a fundamental system of retrobounded neighbourhoods of 0 in K with respect to $\mathcal{T}$.

a) A locally retrobounded topology T is locally bounded, and is a minimal element in the set of all Hausdorff topologies compatible with the ring structure of K; also $\mathcal{T}$ is compatible with the division ring structure of K [use Exercise 19 *a*)].

b) Suppose from now on, in this exercise, that $\mathcal{T}$ is a locally retrobounded topology on K. Show that, for every neighbourhood V of 0 in K, $x \to x^{-1}$ is uniformly continuous on $\complement V$. Hence show that the completion $\hat{K}$ of K is a topological division ring whose topology is locally retrobounded.

c) Every neighbourhood of 0 in K (with respect to $\mathcal{T}$) contains a neighbourhood V of 0 such that $E(V) = (\complement V) \cap (\complement V)^{-1}$ is not empty. Show that the three uniformities on E(V), induced by the additive uniformity and the two multiplicative uniformities on K, coincide. Deduce that, if K is complete, then K^* is complete with respect to either multiplicative uniformity.

d) Show that if x is an element of K^* then exactly one of the following three statements is true: (i) x is topologically nilpotent; (ii) x^{-1} is topologically nilpotent; (iii) the sequence $(x^n)_{n \in \mathbf{Z}}$ is bounded. [By considering a bounded neighbourhood U of 0 such that $UU \subset U$

(Exercise 20 d)), show that if 0 is a cluster point of the sequence $(x^n)_{n \geqslant 0}$, then x is topologically nilpotent.] If K contains non-zero topologically nilpotent elements, show that the set B, consisting of 0 and all $x \in K^*$ such that x^{-1} is not topologically nilpotent, is a bounded neighbourhood of 0 in K which is invariant under all inner automorphisms [use Exercise 21 b)]; B then contains all bounded neighbourhoods U of 0 such that $UU \subset U$; the set of such neighbourhoods which are invariant under all inner automorphisms has a greatest element, which coincides with B if the commutator subgroup of K^* is bounded, in particular if K is a field.

e) If U is a bounded symmetric neighbourhood of 0 in K such that $UU \subset U$, then there exists $b \in U^*$ such that $K^* = U^* \cup ((U^*)^{-1} b)$. Let $a \in U^*$ be such that $Ua \subset bU$, and suppose that the sequence $(a^{-n})_{n \geqslant 0}$ is bounded. Show that the set V of all $x \in K$ such that xU is contained in the union of the sets Ua^{-n} is a bounded neighbourhood of 0 such that $VV \subset V$ and $K^* = V^* \cup (V^*)^{-1}$.

f) Suppose that K contains no non-zero topologically nilpotent elements. Show that there is a bounded neighbourhood of 0 in K which is a subring A of K such that $K^* = A^* \cup (A^*)^{-1}$. [Begin with a bounded neighbourhood V of 0 such that $K^* = V^* \cup (V^*)^{-1}$ (see e)]; if $c \in V^*$ is such that $c(V + V) \subset V$, observe that the ring A generated by V is contained in the union of the sets Vc^{-n} $(n \geqslant 0)$.

¶ 23) If p is any prime number and x is any non-zero rational number, let $v_p(x)$ denote the exponent of p in the decomposition of x into prime factors. The mapping v_p of $\mathbf{Q}^*$ into $\mathbf{Z}$ is called the *p-adic valuation* on $\mathbf{Q}$. The set of all $x \in \mathbf{Q}$ such that either $x = 0$ or $v_p(x) \geqslant m$ (for some $m \in \mathbf{Z}$) is the fractional ideal (p^m) of $\mathbf{Q}$.

a) Show that the ideals (p^m) $(m \in \mathbf{Z})$ form a fundamental system of neighbourhoods of 0 in $\mathbf{Q}$ for a locally retrobounded topology (Exercise 22) $\mathcal{T}_p$ on $\mathbf{Q}$, called the *p-adic topology*. The completion $\mathbf{Q}_p$ of $\mathbf{Q}$ in this topology is a field called the *p-adic field*, whose elements are called *p-adic numbers*. The closure $\mathbf{Z}_p$ of $\mathbf{Z}$ in $\mathbf{Q}_p$ is a compact open principal ideal domain in $\mathbf{Q}_p$, in which $\mathfrak{p} = p\mathbf{Z}_p$ is the only non-zero prime ideal; $\mathbf{Z}_p/\mathfrak{p}$ is isomorphic to the prime field $\mathbf{F}_p = \mathbf{Z}/(p)$, and more generally the additive quotient group $\mathfrak{p}^m/\mathfrak{p}^n$ is isomorphic to

$$\mathbf{Z}/(p^{n-m}) \text{ if } m < n.$$

The elements of $\mathbf{Z}_p$ are called *p-adic integers*.

b) Show that every continuous homomorphism of the additive group $\mathbf{Q}_p$ into itself is of the form $x \to ax$, where $a \in \mathbf{Q}_p$ [if f is such a homomorphism, show that $f(rx) = rf(x)$ for all $r \in \mathbf{Q}$, and then pass to the limit in $\mathbf{Q}_p$].

c) Show that every *compact* subgroup $G \neq \{0\}$ of the additive group $\mathbf{Q}_p$ is identical with one of the $\mathfrak{p}^n$ $(n \in \mathbf{Z})$, and that there is no non-compact closed subgroup of the additive group $\mathbf{Q}_p$, except $\mathbf{Q}_p$ itself. [Let m be the largest integer such that $G \subset \mathfrak{p}^m$; by considering the quotient group $(G + \mathfrak{p}^n)/\mathfrak{p}^n$ for $n > m$, show that $G + \mathfrak{p}^n = \mathfrak{p}^m$, and then use formula (1) of § 3, no. 1].

¶ 24) *a*) Show that every subgroup of the multiplicative group $\mathbf{Q}_p^*$ of the p-adic field $\mathbf{Q}_p$ is isomorphic to the product of a subgroup of the multiplicative group U of units of $\mathbf{Z}_p$ and a discrete additive subgroup isomorphic to $\mathbf{Z}$ or $\{0\}$.

b) Show that the compact subgroups of the subgroup $V = 1 + \mathfrak{p}$ of U are the groups $1 + \mathfrak{p}^n$ [same reasoning as in Exercise 23 *c*)].

c) Show that, for each $a \in U$, the sequence $(a^{p^n})_{n \in \mathbf{N}}$ tends to a limit $\alpha \equiv a \pmod{\mathfrak{p}}$ and that $\alpha^p = \alpha$ [show that $a^{p^n} \equiv a^{p^{n-1}} \pmod{\mathfrak{p}^n}$ by induction on n]. Show that all the roots of the polynomial $X^{p-1} - 1$ (in an algebraically closed extension of $\mathbf{Q}_p$) belong to $\mathbf{Q}_p$ and are pairwise incongruent mod $\mathfrak{p}$ [apply what precedes to the roots of the congruence $x^{p-1} - 1 \equiv 0$ [mod p)]. If $p > 2$ and if d is the highest common factor of n and $p - 1$, the polynomial $X^n - 1$ has exactly d roots in $\mathbf{Q}_p$, which are roots of $X^d - 1$ (same method : take a to be a root of $X^n - 1$).
Hence show that, if $p > 2$, every compact subgroup of the multiplicative group $\mathbf{Q}_p^*$ is the direct product of a *finite* subgroup of U [consisting of the $(p - 1)$th roots of unity] and a subgroup of the form $1 + \mathfrak{p}^n$ [use *b*)]. How are these results to be modified for $p = 2$? (Make the group $1 + \mathfrak{p}^2$ play the part previously played by $1 + \mathfrak{p}$).

¶ 25) *a*) Let $a \neq 0$ be a p-adic number. Then the mapping $n \to a^n$ is continuous on $\mathbf{Z}$ (as a subspace of $\mathbf{Q}_p$) if and only if $a \in 1 + \mathfrak{p}$. If this condition is satisfied, show that $n \to a^n$ is uniformly continuous on $\mathbf{Z}$ and extends to a continuous homomorphism of $\mathbf{Z}_p$ into U, denoted by $x \to a^x$, which is injective if $a \neq 1$. If $p > 2$ and if m is the largest integer such that $a \in 1 + \mathfrak{p}^m$, then $x \to a^x$ is an isomorphism of $\mathbf{Z}_p$ onto $1 + \mathfrak{p}^m$. How must this statement be modified when $p = 2$?

b) Show that every continuous homomorphism of $\mathbf{Z}_p$ into U is of the form $x \to a^x$ for some $a \in 1 + \mathfrak{p}$ [if f is such a homomorphism and $f(1) = a$, then $f(n) = a^n$ for $n \in \mathbf{Z}$].

c) Show that the mapping $(x, y) \to x^y$ is continuous on $(1 + \mathfrak{p}) \times \mathbf{Z}_p$. If $p > 2$ and $b \in \mathbf{Z}_p$, and if m is the largest integer such that $b \in \mathfrak{p}^m$, then the mapping $x \to x^b$ is an isomorphism of the multiplicative group $1 + \mathfrak{p}$ onto the multiplicative subgroup $1 + \mathfrak{p}^{m+1}$ [use Exercise 24 *b*)]. How must this statement be modified when $p = 2$?

d) If n is an integer prime to both $p - 1$ and p, show that the mapping $x = x^n$ is an automorphism of the multiplicative group U [use Exercise 24 *c*)].

26) *a*) Let p, q be two distinct prime numbers and let $\mathfrak{T}$ be the topology on $\mathbf{Q}$ which is the least upper bound of the topologies $\mathfrak{T}_p$ and $\mathfrak{T}_q$. Then $\mathfrak{T}$ is locally bounded [Exercise 20 *c*)] and is compatible with the field structure of $\mathbf{Q}$. With respect to this topology neither of the two sequences $((p/q)^n)_{n \geqslant 0}$, $((q/p)^n)_{n \geqslant 0}$ is bounded; the sequence $(1/p^n)_{n \geqslant 0}$ is not bounded, but the sequence $(p^n)_{n \geqslant 0}$, which is bounded, does not tend to 0 [cf. Exercise 22 *d*)]. Show that the completion of $\mathbf{Q}$ with respect to the topology $\mathfrak{T}$ is isomorphic to the product of the topological fields $\mathbf{Q}_p$ and $\mathbf{Q}_q$.
b) If P is the set of all prime numbers then the least upper bound $\mathfrak{T}_0$ of the topologies $\mathfrak{T}_p$ for $p \in \mathrm{P}$ is compatible with the field structure of $\mathbf{Q}$ but is not locally bounded [Exercise 20 *c*)]. What is the completion of $\mathbf{Q}$ with respect to the topology $\mathfrak{T}_0$?

§ 7

1) *a*) Show that the ring $\mathbf{Z}_p$ of p-adic integers [§ 6, Exercise 23 *a*)] is isomorphic to the inverse limit of the sequence of discrete rings $\mathbf{Z}/(p^n)$, $f_{nm} : \mathbf{Z}/(p^m) \to \mathbf{Z}/(p^n)$ for $n \leqslant m$ being the canonical homomorphism when $\mathbf{Z}/(p^n)$ is considered as a quotient ring of $\mathbf{Z}/(p^m)$.
b) For each integer $n > 0$, let G_n be the group $\mathbf{Z}$ of rational integers, and let X_n be its quotient group $\mathbf{Z}/(p^n)$ (p prime), G_n and X_n carrying the discrete topology; consider (G_n) as an inverse system, $\mathrm{G}_m \to \mathrm{G}_n$ being the identity mapping; and consider (X_n) as an inverse system, $f_{nm} : \mathrm{X}_m \to \mathrm{X}_n$ being the canonical homomorphism (see *a*) for $n \leqslant m$. Then the orbit of each point of X_n with respect to G_n is compact, but the orbit of a point $x = (x_n)$ of $\mathrm{X} = \varprojlim \mathrm{X}_n$ with respect to $\mathrm{G} = \varprojlim \mathrm{G}_n$ is not compact, and is not isomorphic to the inverse limit of the orbits of the x_n; also the canonical mapping $\mathrm{X}/\mathrm{G} \to \varprojlim \mathrm{X}_n/\mathrm{G}_n$ is not injective.
c) For each $n > 0$, let G_n be the subgroup $p^n\mathbf{Z}$ of $\mathbf{Z}$ (p prime), and let X_n be the group $\mathbf{Z}$, where G_n and X_n carry the discrete topology and G_n operates by translations on X_n. Consider (G_n) and (X_n) as inverse systems, $\mathrm{X}_m \to \mathrm{X}_n$ being the identity mapping and $\mathrm{G}_m \to \mathrm{G}_n$ the canonical injection ($n \leqslant m$). Then the stabilizer of each point of X_n (with respect to G_n) is compact, the orbit of each point of $\mathrm{X} = \varprojlim \mathrm{X}_n$ with respect to $\mathrm{G} = \varprojlim \mathrm{G}_n$ is compact, but the canonical mapping $\mathrm{X}/\mathrm{G} = \varprojlim \mathrm{X}_n/\mathrm{G}_n$ is not surjective.
d) Construct, using *b*) and *c*), an example of an inverse system of spaces with operators (X_α) with respect to an inverse system of groups (G_α),

such that, if $X = \varprojlim X_\alpha$ and $G = \varprojlim G_\alpha$, the canonical mapping $X/G \to \varprojlim X_\alpha/G_\alpha$ is neither injective nor surjective.

2) Let $(X_\alpha, f_{\alpha\beta})$ be an inverse system of non-empty sets such that $\varprojlim X_\alpha = \varnothing$. Let G_α denote the free **Z**-module of formal linear combinations of elements of X_α, with coefficients in **Z**, and let $g_{\alpha\beta}$ denote the homomorphism $G_\beta \to G_\alpha$ which agrees with $f_{\alpha\beta}$ on X_β. Show that the group $G = \varprojlim G_\alpha$ consists of the identity element alone. [Argue by contradiction: if $z = (z_\alpha)$ is an element of $\varprojlim G_\alpha$, consider, for each α, the finite set F_α of elements of X_α whose coefficient in z_α is not zero, and observe that $f_{\alpha\beta}(F_\beta) = F_\alpha$.] Hence construct an example of an inverse system $(G_\alpha, g_{\alpha\beta})$ of groups such that the $g_{\alpha\beta}$ are surjective, the G_α are infinite and $\varprojlim G_\alpha$ consists only of the identity element.

¶ 3) *a*) Show that every totally disconnected compact group is the inverse limit of a family of finite discrete groups (use Exercise 18 of § 4).

b) Let G be a totally disconnected compact group and let L be a closed subgroup of G. Show that there is a continuous section $G/L \to G$ associated with the canonical mapping $G \to G/L$. [Use *a*) and Proposition 3 of § 3; for each α consider the finite set F_α of sections $G_\alpha/L_\alpha \to G_\alpha$, and remark that these sets from an inverse system with respect to the canonical surjective mappings $h_{\alpha\beta} : F_\beta \to F_\alpha$].

4) Let G be a Hausdorff topological group and let (H_α) be a family of compact normal subgroups of G, directed with respect to the relation $\supset$, and satisfying condition (AP) of no. 3. Let (L_α) be a family of closed subgroups of G such that $H_\alpha \subset L_\alpha$ for each $\alpha \in I$ and $L_\alpha = H_\alpha L_\beta$ for $\alpha \leqslant \beta$, and let $L = \bigcap_\alpha L_\alpha$. Show that $L_\alpha = H_\alpha L$ for each α (use Proposition 3).

5) Let G be a Hausdorff topological group and let X be a Hausdorff topological space on which G operates continuously; let (H_α) be a family of compact normal subgroups of G, directed with respect to the relation $\supset$, and satisfying condition (AP) of no. 3. Let X_α denote X/H_α. Show that the canonical mapping $X \to \varprojlim X_\alpha$ is a homeomorphism.

¶ * 6) Let $(G_n, f_{nm})_{n\in\mathbf{N}}$ be the inverse system of compact groups such that $G_n = \mathbf{T} = \mathbf{R}/\mathbf{Z}$ for each n, and f_{nm} is the continuous homomorphism $x \to p^{m-n}x$ of **T** onto itself, for $n \leqslant m$ (p being a given prime number). The topological group $\mathbf{T}_p = \varprojlim G_n$ is called the *p-adic solenoid*; it is a compact connected commutative topological group.

a) For each n, the continuous homomorphism $f_n : \mathbf{T}_p \to G_n$ is surjective, and its kernel is isomorphic to the group $\mathbf{Z}_p$ of p-adic integers [cf. Exercise 1 *a*)].

b) Let φ be the canonical homomorphism $\mathbf{R} \to \mathbf{R}/\mathbf{Z} = \mathbf{T}$. For each $x \in \mathbf{R}$, put $\theta(x) = (\varphi(x/p^n))_{n \in \mathbf{N}}$; show that θ is an injective continuous homomorphism of $\mathbf{R}$ into $\mathbf{T}_p$, and that $\theta(\mathbf{R})$ is a dense subgroup of $\mathbf{T}_p$.

c) Let I be an open interval in $\mathbf{R}$, with centre 0 and length < 1. Show that the subspace $\overset{-1}{f}_0(\varphi(\mathrm{I}))$ of $\mathbf{T}_p$ is homeomorphic to the product $\mathrm{I} \times \mathbf{Z}_p$. In particular, the group $\mathbf{T}_p$ is not locally connected.

d) Show that every closed subgroup H of $\mathbf{T}_p$, other than $\mathbf{T}_p$ or $\{0\}$, is totally disconnected and isomorphic to a group of the form $\mathbf{Z}/n\mathbf{Z}$ or $(\mathbf{Z}/n\mathbf{Z}) \times \mathbf{Z}_p$, where n is an integer prime to p (use Proposition 3 of no. 3).

e) Show that $\mathbf{T}_p$ is an *indecomposable* compact connected space, i.e. that $\mathbf{T}_p$ cannot be covered by two compact connected sets P, Q neither of which is equal to $\mathbf{T}_p$. (Observe that there is an integer n such that $f_n(\mathrm{P}) \neq \mathrm{G}_n$ and $f_n(\mathrm{Q}) \neq \mathrm{G}_n$, and examine the sets $f_{n+1}(\mathrm{P})$ and $f_{n+1}(\mathrm{Q})$ to get a contradiction). *

HISTORICAL NOTE

(Numbers in brackets refer to the bibliography at the end of this note.)

The general theory of topological groups is one of the most recent branches of analysis. However, particular topological groups were well known a long time ago; and in the second half of the nineteenth century Sophus Lie built up the vast theory of those topological groups which he called "continuous groups" and which are nowadays known as "Lie groups". The reader will find fuller information on the genesis and development of this theory in the Historical Notes attached to the volume on Lie groups in this series.

The study of general topological groups was initiated by O. Schreier in 1926 [1]. Since then it has been the subject of much work, which has, among other things, to a large extent elucidated the structure of locally compact groups. Here our intention has been to give only the most elementary definitions and results of the theory, and we refer the reader who wishes to go deeper into the subject to the monographs of L. Pontrjagin [2], A. Weil [3], and D. Montgomery and L. Zippin [4].

BIBLIOGRAPHY

[1] O. Schreier, Abstrakte kontinuierliche Gruppen, *Hamb. Abh.* **4** (1926), p. 15.
[2] L. Pontrjagin, *Topologische Gruppen*, 2 volumes, Leipzig (Teubner), 1957.
[3] A. Weil, L'intégration dans les groupes topologiques et ses applications, *Act. Sci. et Ind.* no. 869, Paris (Hermann) 1940 [2nd edition, no. 869-1145, Paris (Hermann), 1953].
[4] D. Montgomery and L. Zippin, *Topological Transformation Groups*, New York (Interscience), 1955.

CHAPTER IV

Real Numbers

1. DEFINITION OF REAL NUMBERS

1. THE ORDERED GROUP OF RATIONAL NUMBERS

We have defined the ordering $x \leqslant y$ on the set $\mathbf{Q}$ of rational numbers; we have seen that this ordering makes $\mathbf{Q}$ a *linearly ordered* set, and that it is *compatible* with the *additive group* structure of $\mathbf{Q}$, i.e. for each $z \in \mathbf{Q}$ the relation $x \leqslant y$ is equivalent to $x + z \leqslant y + z$ (that is, *the ordering is invariant under translations*). We recall the notation (which is used in any linearly ordered group)

$$x^+ = \sup(x, 0),$$
$$x^- = \sup(-x, 0) = (-x)^+,$$
$$|x| = \sup(x, -x);$$

$|x|$ is called the *absolute value* of x, and we have

$$x = x^+ - x^-, \qquad |x| = x^+ + x^-$$

and the *triangle inequality*

(1) $$|x + y| \leqslant |x| + |y|,$$

together with the inequality

(2) $$\big|\, |x| - |y| \,\big| \leqslant |x - y|$$

which is an immediate consequence of (1); moreover

(3) $$|x^+ - y^+| \leqslant |x - y|.$$

The relations $x \geqslant 0$, $x = x^+$, $x^- = 0$, $|x| = x$ (resp. $x \leqslant 0$, $x = -x^-$, $x^+ = 0$, $|x| = -x$) are *equivalent*. The relation $|x| = 0$ is equivalent to $x = 0$; if $a \geqslant 0$, the relation $|x| \leqslant a$ is equivalent to $-a \leqslant x \leqslant a$, and the relation $|x| \geqslant a$ is equivalent to "$x \geqslant a$ or $x \leqslant -a$". For all x, y, in $\mathbf{Q}$, we have

$$\sup(x, y) + z = \sup(x + z, y + z), \tag{4}$$
$$\inf(x, y) = -\sup(-x, -y), \tag{5}$$

and, as particular cases,

$$\sup(x, y) = x + (y - x)^{+} = x + (x - y)^{-}, \tag{6}$$
$$\inf(x, y) = x - (y - x)^{-} = x - (x - y)^{+}. \tag{7}$$

Finally, let $\mathbf{Q}_{+}$ denote the set of rational numbers $\geqslant 0$; we have then

$$\mathbf{Q}_{+} + \mathbf{Q}_{+} \subset \mathbf{Q}_{+}, \tag{8}$$
$$\mathbf{Q}_{+} \cap (-\mathbf{Q}_{+}) = \{0\}, \tag{9}$$
$$\mathbf{Q}_{+} \cup (-\mathbf{Q}_{+}) = \mathbf{Q}. \tag{10}$$

The relation $x \leqslant y$ is *equivalent* to $y - x \in \mathbf{Q}_{+}$.

We shall use this ordering to define *a topology on* $\mathbf{Q}$ *compatible with its additive group structure.*

2. THE RATIONAL LINE

Consider the set $\mathfrak{F}$ of *symmetric open intervals* $]-a, +a[$, where a runs through the set of rational numbers > 0; we shall show that $\mathfrak{F}$ is a *fundamental system of neighbourhoods of* 0 in a topology compatible with the additive group structure of $\mathbf{Q}$.

The group $\mathbf{Q}$ is commutative, and axiom $(\mathrm{GV}'_{\mathrm{II}})$ is clearly satisfied; it is therefore enough to show that axiom $(\mathrm{GV}'_{\mathrm{I}})$ is also satisfied, in other words, that for each $a > 0$ there exists $b > 0$ such that the conditions $|x| < b$ and $|y| < b$ together imply $|x + y| < a$. The triangle inequality shows that we may take $b = a/2$.

DEFINITION 1. *The rational line is the topological space consisting of the set* $\mathbf{Q}$ *together with the additive group topology for which the symmetric open intervals* $]-a, +a[$ $(a > 0)$ *form a fundamental system of neighbourhoods of* 0.

The topological group $\mathbf{Q}$ *thus defined is called the additive group of the rational line.*

If a is any rational number > 0, there is an integer $n > 0$ such that $1/n < a$; hence the open intervals $]-\frac{1}{n}, +\frac{1}{n}[$ $(n = 1, 2, \ldots)$ form a fundamental system of neighbourhoods of 0 on the rational line.

We obtain a fundamental system of neighbourhoods of any point $x \in \mathbf{Q}$ by taking the open intervals $]x - a, x + a[$, where a runs through the set of rational numbers > 0 (or the set of numbers $1/n$).

Definition 1 is therefore equivalent to that given in Chapter I, § 1, no. 2.

For each pair of rational numbers (a, b) such that $a < b$, there exists $c \in \mathbf{Q}$ such that $a < c < b$ [for example $c = (a + b)/2$]; it follows that the rational line is a *non-discrete Hausdorff space.*

For each $a > 0$, let U_a be the set of pairs (x, y) in $\mathbf{Q} \times \mathbf{Q}$ such that $|x - y| < a$. As a runs through the set of rational numbers > 0 (or just the set of numbers $1/n$), the sets U_a form a fundamental system of entourages of the uniformity of the additive group $\mathbf{Q}$ of the rational line. Relations (2) and (3) show that $|x|$, x^+ and x^- are *uniformly continuous* on $\mathbf{Q}$. It follows that the functions $\sup(x, y)$ and $\inf(x, y)$ are uniformly continuous on $\mathbf{Q} \times \mathbf{Q}$.

3. THE REAL LINE AND REAL NUMBERS

DEFINITION 2. *Let* $\mathbf{R}$ *denote the topological group which is the completion of the additive group* $\mathbf{Q}$ *of the rational line. The elements of* $\mathbf{R}$ *are called real numbers; as a topological space,* $\mathbf{R}$ *is called the real line; as a topological group,* $\mathbf{R}$ *is called the additive group of the real line.*

We shall always identify $\mathbf{Q}$ with the dense subgroup of $\mathbf{R}$ to which it is canonically isomorphic. With this convention, every rational number is a real number. Every real number which is not rational is said to be *irrational*; we have seen in Chapter II, § 3, no. 3 that such numbers exist (we shall show this in another way in § 3, no. 3 of this chapter; see also Exercise 2 to § 2); hence (Chapter III, § 2, no. 1) the set $\complement\mathbf{Q}$ of irrational numbers is *dense* in $\mathbf{R}$.

We shall show that the *order structure* of $\mathbf{Q}$ can be *extended* to $\mathbf{R}$ in such a way that the extended ordering is still compatible with the additive group structure of $\mathbf{R}$:

PROPOSITION 1. *The relation* $y - x \in \overline{\mathbf{Q}}_+$ *is an ordering on* $\mathbf{R}$ *which makes* $\mathbf{R}$ *into a linearly ordered set; is compatible with the additive group structure on* $\mathbf{R}$, *and induces the ordering* $x \leqslant y$ *on* $\mathbf{Q}$.

We begin by showing that the relations $y - x \in \overline{\mathbf{Q}}_+$ and $z - y \in \overline{\mathbf{Q}}_+$ imply $z - x \in \overline{\mathbf{Q}}_+$. Indeed, the function $x + y$ is continuous on $\mathbf{R} \times \mathbf{R}$, and therefore by (8) we have $\overline{\mathbf{Q}}_+ + \overline{\mathbf{Q}}_+ \subset \overline{\mathbf{Q}}_+$ (Chapter I, § 2, no. 1, Theorem 1). Next, we shall show that the relations $y - x \in \overline{\mathbf{Q}}_+$ and $x - y \in \overline{\mathbf{Q}}_+$ imply $x = y$; this will establish that $y - x \in \overline{\mathbf{Q}}_+$ is an *ordering* on $\mathbf{R}$. It is enough to show that $\overline{\mathbf{Q}}_+ \cap (-\overline{\mathbf{Q}}_+) = \{0\}$. Now the functions $x \to x^+$ and $x \to x^-$ are uniformly continuous on $\mathbf{Q}$ and can therefore be extended by continuity to $\mathbf{R}$ (Chapter II, § 3, no. 6, Theorem 2); let f and g be their respective extensions. By extension we have $x = f(x) - g(x)$ for all $x \in \mathbf{R}$; if $x \in \overline{\mathbf{Q}}_+$ then $g(x) = 0$, and since $-\overline{\mathbf{Q}}_+$ is the closure of $-\mathbf{Q}_+$ by the continuity of $-x$, it follows that if $x \in -\overline{\mathbf{Q}}_+$ then $f(x) = 0$. Hence if $x \in \overline{\mathbf{Q}}_+ \cap (-\overline{\mathbf{Q}}_+)$ we have $f(x) = g(x) = 0$ and therefore $x = 0$.

By (10), we have $\overline{\mathbf{Q}}_+ \cup (-\overline{\mathbf{Q}}_+) = \mathbf{R}$, and hence $\mathbf{R}$ is *linearly ordered* by the ordering $y - x \in \overline{\mathbf{Q}}_+$.

Furthermore, since the relations $y - x \in \overline{\mathbf{Q}}_+$ and $(y+z)-(x+z) \in \overline{\mathbf{Q}}_+$ are equivalent, the ordering $y - x \in \overline{\mathbf{Q}}_+$ is compatible with the additive group structure of $\mathbf{R}$.

Finally, if x and y belong to $\mathbf{Q}$ the relations $y - x \in \overline{\mathbf{Q}}_+$ and $y - x \in \mathbf{Q}_+$ are equivalent, and therefore the relation $y - x \in \overline{\mathbf{Q}}_+$ induces the relation $x \leqslant y$ on $\mathbf{Q}$. This completes the proof.

The relation $y - x \in \overline{\mathbf{Q}}_+$ is again denoted by $x \leqslant y$. The set $\overline{\mathbf{Q}}_+$ is the set of all $x \geqslant 0$ in $\mathbf{R}$ and is denoted by $\mathbf{R}_+$; it is a *closed set*. The set of all $x > 0$ is denoted by $\mathbf{R}_+^*$; it is the complement of $-\mathbf{R}_+$ and is therefore *open* in $\mathbf{R}$.

4. PROPERTIES OF INTERVALS IN R

PROPOSITION 2. *Every closed* (resp. *open*) *interval in* $\mathbf{R}$ *is a closed* (resp. *open*) *set in* $\mathbf{R}$.

The sets $[a, \to[\, = a + \mathbf{R}_+$ and $]\leftarrow, a] = a - \mathbf{R}_+$ are obtained by translation from $\mathbf{R}_+$ and $-\mathbf{R}_+$ respectively and are therefore closed (Chapter III, § 1, no. 1); the sets $]\leftarrow, a[$ and $]a, \to[$, which are their complements, are open; finally, the closed interval $[a, b]$ (resp. the open interval $]a, b[$) is the intersection of $[a, \to[$ and $]\leftarrow, b]$ (resp. of $]a, \to[$ and $]\leftarrow, b[$) and is therefore a closed (resp. open) set.

The closed intervals $[-a, +a]$ $(a > 0)$ in $\mathbf{R}$ are therefore neighbourhoods of 0. Let us show that they form a *fundamental system of neighbourhoods* of 0 as a runs through $\mathbf{R}_+^*$. For this it is enough to establish the following proposition:

PROPOSITION 3. *As r runs through the set of rational numbers > 0, the intervals $s_r = [-r, +r]$ in $\mathbf{R}$ form a fundamental system of neighbourhoods of* 0.

By Proposition 7 of Chapter III, § 3, no. 4 we obtain a fundamental system of neighbourhoods of 0 in $\mathbf{R}$ by taking the *closures* in $\mathbf{R}$ of the intervals

$S_r \cap \mathbf{Q} = [-r, +r]$ *of* $\mathbf{Q}$. The proof will be complete if we show that S_r is the closure of $S_r \cap \mathbf{Q}$. Now S_r is closed in $\mathbf{R}$, and we need therefore only prove that, if x is a real number such that $-r < x < r$, then x is in the closure of $S_r \cap \mathbf{Q}$. The interval $]-r, +r[$ is an open set in $\mathbf{R}$ and therefore for all sufficiently small neighbourhoods V of 0 in $\mathbf{R}$ we have $x + V \subset]-r, +r[$; but $\mathbf{Q}$ being dense in $\mathbf{R}$, there is a rational number $r' \in x + V$, so that $-r < r' < r$ and therefore $r' \in S_r \cap \mathbf{Q}$.

COROLLARY. *Every point of the real line has a countable fundamental system of neighbourhoods.*

PROPOSITION 4. *If* (x, y) *is any pair of real numbers such that* $x < y$, *there is a rational number* r *such that* $x < r < y$.

Since $\mathbf{Q}$ is dense in $\mathbf{R}$, it is enough to show that $]x, y[$ *is not empty*; by translation we may assume $x = 0$ and $y > 0$. Now $\mathbf{R}$ is a Hausdorff space and therefore, by Proposition 3, there is a rational number $r > 0$ such that $y \notin [-r, +r]$, and this implies that $0 < r < y$.

PROPOSITION 5. *Let* I *be any interval in* $\mathbf{R}$. *Then the topology induced on* I *by the topology of* $\mathbf{R}$ *is generated by the open intervals of* I (where I is considered as *linearly ordered by the relation* $x \leqslant y$).

Every open interval of I is the trace on I of an open interval of $\mathbf{R}$. This is clear for a bounded interval, and the unbounded interval $]a, \rightarrow[$ of I is the trace of the unbounded interval $]a, \rightarrow[$ of $\mathbf{R}$. We may therefore restrict ourselves to the case $I = \mathbf{R}$; but in this case the result follows from Proposition 3, since every neighbourhood of a point $x \in \mathbf{R}$ contains an open interval $]x - a, x + a[$.

Remark. If A is a *dense* subset of $\mathbf{R}$, the topology of $\mathbf{R}$ is generated by the open intervals whose end-points belong to A. For if

$$]x - a, \quad x + a[$$

is an open interval containing x, there exist two points y, z of A such that $x - a < y < x$ and $x < z < x + a$; hence $]y, z[$ contains x and is contained in $]x - a, x + a[$. This proof shows, moreover, that the intervals under consideration form a *base* (Chapter I, § 1, no. 3) of the topology of $\mathbf{R}$. In particular, if we take $A = \mathbf{Q}$, we see that the topology of $\mathbf{R}$ has a *countable base*.

5. LENGTH OF AN INTERVAL

DEFINITION 3. *The length of a bounded interval with end-points* a *and* b $(a \leqslant b)$ *is defined to be* $b - a$.

Every bounded interval which contains more than one point therefore has length > 0. If $a \leqslant b$, the four intervals $[a, b]$, $]a, b[$, $[a, b[$ and $]a, b[$ all have the same length. An interval with end-points $a + c$ and $b + c$ has the same length as an interval with end-points a and b; in other words, *the length of an interval is invariant under translation.*
If $a \leqslant c \leqslant d \leqslant b$ we have $d - c \leqslant b - a$. Hence if a bounded interval I is contained in a bounded interval I', the length of I is less than or equal to the length of I'.

If n mutually disjoint open intervals $\mathrm{I}_1, \mathrm{I}_2, \ldots, \mathrm{I}_n$ are contained in the interval $[a, b]$ $(a < b)$ it is easily seen by induction on n that, if $\mathrm{I}_k =]c_k, d_k[$, there is a permutation σ of the indices k $(1 \leqslant k \leqslant n)$ such that $d_{\sigma(k)} \leqslant c_{\sigma(k+1)}$ for $1 \leqslant k \leqslant n - 1$. It follows immediately that the sum of the lengths of the intervals I_k is at most equal to the length of $[a, b]$, and that equality holds if and only if $c_{\sigma(1)} = a$, $d_{\sigma(n)} = b$ and $d_{\sigma(k)} = c_{\sigma(k+1)}$ for $1 \leqslant k \leqslant n - 1$.

6. ADDITIVE UNIFORMITY OF R

Since the group $\mathbf{R}$ is linearly ordered, the functions x^+, x^- and $|x|$ are defined on $\mathbf{R}$ in the same way as on $\mathbf{Q}$ and satisfy all the relations listed above for $\mathbf{Q}$, notably relations (1) to (7). Let a be a real number > 0 and let U_a be the set of all pairs $(x, y) \in \mathbf{R} \times \mathbf{R}$ such that $|x - y| < a$; as a runs through the set of real numbers > 0 (or the set of numbers $1/n$), the sets U_a form a fundamental system of entourages of the uniformity of the additive group $\mathbf{R}$ of the real line (called the *additive uniformity of the real line*).

The functions $|x|$, x^+ and x^- are *uniformly continuous* on $\mathbf{R}$, and the functions $\sup(x, y)$ and $\inf(x, y)$ are *uniformly continuous* on $\mathbf{R} \times \mathbf{R}$; these functions therefore coincide with those obtained by extending by continuity the corresponding functions defined on $\mathbf{Q}$ and $\mathbf{Q} \times \mathbf{Q}$, respectively.

2. FUNDAMENTAL TOPOLOGICAL PROPERTIES OF THE REAL LINE

1. ARCHIMEDES' AXIOM

The topological properties of the real line which are the subject of this section are all consequences of the following theorem:

THEOREM 1. *If x and y are any two real numbers > 0, then there is an integer $n > 0$ such that $y < nx$.*

There exist two rational numbers p/q and r/s such that $0 < p/q < x$ and $y < r/s$, since the open intervals $]0, x[$ and $]y, \rightarrow[$ are not empty (§ 1, no. 4, Proposition 4); take n such that $nps > qr$, and we have $y < nx$.

Remark. An axiomatic construction of the theory of real numbers will be found in Chapter V, § 2, in which Theorem 1 appears as an axiom; for more details about this axiom, see the Historical Note to Chapter IV.

2. COMPACT SUBSETS OF R

THEOREM 2 (Borel-Lebesgue). *For a subset of the real line* **R** *to be compact it is necessary and sufficient that it be closed and bounded.*

1) The condition is *necessary*. Let A be a compact subset of **R** and let a be a real number > 0. The set A is closed (Chapter I, § 9, no. 3, Proposition 4) and there exists a finite number of points x_i $(1 \leqslant i \leqslant n)$ of **R** such that A is contained in the union of the neighbourhoods $[x_i - a, x_i + a]$ (Chapter I, § 9, no. 3). Let b be the maximum of the numbers $|x_i|$; then we have $A \subset [-b - a, b + a]$.

2) The condition is *sufficient*. It is enough to show that every interval $[-a, +a]$ $(a > 0)$ is *compact*, and since this interval is a closed subset of a complete uniform space, it is enough to show that, for each $b > 0$, we can cover $[-a, +a]$ by a *finite* number of intervals of the form $[x - b, x + b]$ (Chapter II, § 4, no. 2, Corollary to Theorem 3). Now, let n be an integer > 0 such that $a < nb$; if $x \in [-a, +a]$ and if m is the largest integer (positive or negative) such that $mb \leqslant x$, then we have $-n \leqslant m \leqslant n$ and $mb \leqslant x \leqslant (m + 1)b$. Hence the $2n + 1$ intervals $[(k - 1)b, (k + 1)b]$ $(-n \leqslant k \leqslant n)$ form a covering of the required type.

COROLLARY 1. *A subset of the real line* **R** *is relatively compact if and only if it is bounded.*

COROLLARY 2. *The real line is a locally compact space and is not compact.*

Remark. Theorem 2 is often referred to as the "Heine-Borel Theorem"; see the Historical Notes to Chapters II and IV.

3. LEAST UPPER BOUND OF A SUBSET OF R

THEOREM 3. *Every non-empty subset of the real line which is bounded above* (resp. *bounded below*) *has a least upper bound* (resp. *greatest lower bound*).

Let A be a non-empty subset of **R**, bounded above, and let b be an upper bound of A, so that $A \subset]\leftarrow, b]$. For each $x \in A$ consider the set A_x of numbers $\geqslant x$ which belong to A; the sets A_x form a *filter base* $\mathfrak{B}$ on **R**, since $A_y \subset A_x$ if $y \geqslant x$. Let a be a point of A. For each $x \geqslant a$ belonging to A, A_x is contained in the *compact* interval $[a, b]$ and thus the filter base $\mathfrak{B}$ has a cluster point c. Since the intervals $[x, \rightarrow[$ are closed, c belongs to their intersection and therefore c is an *upper bound* of A. But, on the other hand, every upper bound z of A is $\geqslant c$, otherwise the neighbourhood $]z, \rightarrow[$ of c would not contain any point of A. Hence c is the *least upper bound* of A.

We can argue similarly for a non-empty set B bounded below, or else simply remark that $-B$ is non-empty and bounded above, and that if c is the least upper bound of $-B$, then $-c$ is the greatest lower bound of B.

The least upper bound c of A can be characterized by the following two properties:

(i) For each $x \in A$, $x \leqslant c_*$.

(ii) For each $a < c$, there exists $x \in A$ such that $a < x \leqslant c$.

The least upper bound of a *closed* set (non-empty and bounded above) belongs to the set and is its *greatest element*; and the least upper bound of any non-empty subset A of **R** which is bounded above may be defined as the *largest real number in the closure of* A.

4. CHARACTERIZATION OF INTERVALS

PROPOSITION 1. *A non-empty subset* A *of* **R** *is an interval if and only if, whenever* a *and* b *are any two points of* A *such that* $a < b$, *the closed interval* $[a, b]$ *is contained in* A.

The condition is clearly necessary. Conversely, suppose that it is satisfied. If A is neither bounded above nor below it must be the whole of **R**, for if x is any point of **R** there are then two points a, b of A such that $a < x < b$. If A is bounded above but not below, let k be its least upper bound; then for any $x < k$ there exist a and b in A such that $a < x < b \leqslant k$, hence $x \in A$, and therefore A can only be one of the two intervals $]\leftarrow, k]$, $]\leftarrow, k[$. The argument is similar in the other cases.

5. CONNECTED SUBSETS OF R

THEOREM 4. *A subset* A *of* **R** *is connected if and only if* A *is an interval.*

1) The condition is *necessary*. Suppose that A is connected: if A consists of a single point, it is an interval. If A has more than one

point, let a and b be two points of A such that $a < b$; by Proposition 1 of no. 4 it is enough to show that every x such that $a < x < b$ belongs to A. Now, if $x \notin A$ we should have $A \subset \complement\{x\}$; but $\complement\{x\}$ is the union of two disjoint open sets $]\leftarrow, x[$ and $]x, \rightarrow[$, each of which meets A, and therefore A would not be connected, contrary to hypothesis.

2) The condition is *sufficient*. Let us show first that every *compact* interval $[a, b]$ is connected. For each integer $n > 0$, let $V_{1/n}$ be the entourage consisting of all pairs (x, y) such that $|x - y| \leqslant 1/n$; by Proposition 6 of Chapter II, § 4, no. 4, it is enough to show that every pair of points x, y of $[a, b]$ can be joined by a $V_{1/n}$-chain. Let p be the greatest integer such that $p/n \leqslant x$ and let q be the greatest integer such that $q/n \leqslant y$ (p and q exist by reason of Theorem 1 of no. 1); then $p \leqslant q$. If $p = q$ then $y - x < 1/n$ and the points x and y form a $V_{1/n}$-chain. If $q > p$, put $x_i = (p + i)\ /n$ $(i = 1, 2, \ldots, q - p)$; we have $x_1 - x \leqslant 1/n$, $y - x_{q-p} \leqslant 1/n$ and $x_{i+1} - x_i = 1/n$, hence the points $x, x_1, x_2, \ldots, x_{q-p}, y$ form a $V_{1/n}$-chain joining x and y.

If now I is any interval not consisting of a single point, and if a and b are any two points of I such that $a < b$, then the interval $[a, b]$ is contained in I and is connected, and hence I is connected.

Corollary 1. *The real line is a connected and locally connected space.*

Corollary 2. *The only compact connected subsets of* **R** *are the bounded closed intervals.*

By Theorem 4, a subset of **R** which does not contain any interval consisting of more than one point is *totally disconnected*; this is true, for example, of the set **Q** of rational numbers, since the set $\complement$**Q** of irrational numbers is dense in **R**.

Proposition 2. *Every non-empty open set in* **R** *is the union of a countable family of mutually disjoint open intervals.*

Let A be a non-empty open set in **R**. Since **R** is locally connected, every *component* of A is a connected open set (Chapter I, § 11, no. 6, Proposition 11) and therefore an *open interval* by Theorem 4. Any two of these open intervals are disjoint; on the other hand, each of them contains a rational number; hence the set of these intervals has a power less than or equal to that of **Q**, i.e. is *countable*.

It follows that every closed set in **R** is the complement of the union of a (finite or infinite) sequence (I_n) of mutually disjoint open intervals. These intervals are said to be *contiguous* to the closed set under consideration. Conversely, given such a sequence of intervals, the complement of their union is a closed set to which these intervals are contiguous.

Example. Let us define by induction a countable family $(I_{n,p})$ of mutually disjoint open intervals as follows:

The integer n takes all values $\geqslant 0$, and for each value of n, p takes the values $1, 2, 3, \ldots, 2^n$. All the intervals $I_{n,p}$ are contained in

$$A = [0, 1],$$

and we take $I_{0,1} =]1/3, 2/3[$ (the "middle third" of $]0, 1[$). Suppose now that the $2^{m+1} - 1$ intervals $I_{n,p}$ have been defined for $0 \leqslant n \leqslant m$ in such a way that, if J_m is their union, the set $A \cap \complement J_m$ is the union of 2^{m+1} mutually disjoint closed intervals $K_{m,p}$ $(1 \leqslant p \leqslant 2^{m+1})$ each of length $\frac{1}{3^{m+1}}$. If $K_{m,p} = [a, b]$ we then take $I_{m+1,p}$ to be the open interval $\left]a + \frac{b-a}{3}, b - \frac{b-a}{3}\right[$ (the "middle third" of the interval $]a, b[$), and it is immediately verified that the induction can continue in this way (Fig. 3).

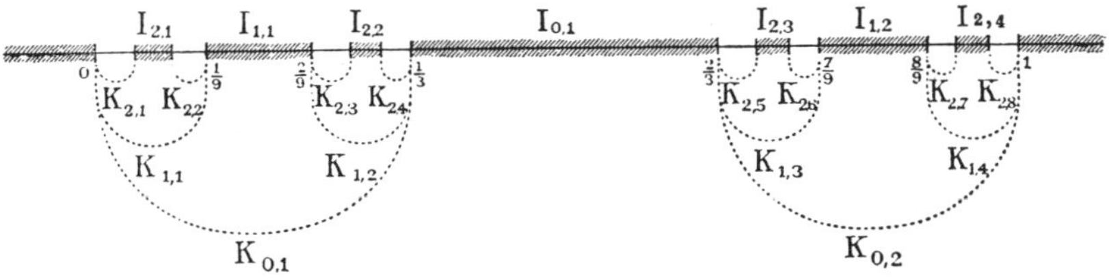

Figure 3.

If K' is the complement of the union of the $I_{n,p}$, the closed set $K = A \cap K'$ is called *Cantor's triadic set*. Clearly K is *compact* (no. 2, Theorem 2); also K is *totally disconnected*. For if K contained an interval I of length > 0, then I would be contained in some interval $K_{m,p}$; hence its length would be $\leqslant 1/3^{m+1}$ for each m, which is absurd.

6. HOMEOMORPHISMS OF AN INTERVAL ONTO AN INTERVAL

THEOREM 5. *Let* I *be an interval in* $\mathbf{R}$. *Then a mapping* f *of* I *into* $\mathbf{R}$ *is a homeomorphism of* I *onto* $f(I)$ *if and only if* f *is strictly monotonic and continuous on* I; *and* $f(I)$ *is then an interval in* $\mathbf{R}$.

1) The condition is *necessary*. Let a and b be two points of I such that $a < b$, and suppose for example that $f(a) < f(b)$. Let us show that f is strictly increasing on I. First, if $a < c < b$ then we must have $f(a) < f(c) < f(b)$; if, for example, we had $f(a) < f(b) < f(c)$

then the image of the interval $[a, c]$ under f would be a connected set (Chapter I, § 11, no. 2, Proposition 4) and would therefore contain the interval $[f(a), f(c)]$; hence there would exist $x \in [a, c]$ such that $f(x) = f(b)$, contrary to the hypothesis that f is injective.

It follows that if x and y are any two points of I such that $x < y$, then $f(x) < f(y)$; for we have $f(a) < f(x) < f(b)$ if $a < x < b$, $f(a) < f(b) < f(x)$ if $b < x$, and $f(x) < f(a) < f(b)$ if $x < a$; repeating the argument with a, x, y in place of a, b, x respectively, we see that $f(x) < f(y)$.

2) The condition is *sufficient*. Suppose that f is continuous and strictly monotonic on I (say, strictly increasing): $f(\mathrm{I})$ is connected and is therefore an interval, and since f is strictly increasing, f is a bijective mapping of I onto $f(\mathrm{I})$. Moreover, the image under f of an open interval *in* I is an open interval *in* $f(\mathrm{I})$, and therefore (§1, no, 4, Proposition 5) f is a homeomorphism of I onto $f(\mathrm{I})$.

Remark. The first part of the preceding proof shows in fact that a *continuous injective* mapping of I into **R** is strictly monotonic; from the second part of the proof it therefore follows that *every continuous injective mapping* f *of an interval* I *into* **R** *is a homeomorphism of* I *onto* $f(\mathrm{I})$.

3. THE FIELD OF REAL NUMBERS

1. MULTIPLICATION IN R

The topology of the rational line **Q** is compatible not only with the *additive group* structure, but also with the *field* structure of **Q**. For the function xy is continuous at $(0, 0) \in \mathbf{Q} \times \mathbf{Q}$, since for each integer $n > 0$ the relations $|x| \leqslant 1/n$ and $|y| \leqslant 1/n$ together imply that $|xy| \leqslant 1/n^2 \leqslant 1/n$; on the other hand, if a is any non-zero rational number, the function ax is continuous at $x = 0$, since for each integer $n > 0$ the relation $|x| \leqslant 1/n|a|$ implies that $|ax| \leqslant 1/n$. This shows that xy is continuous at every point of $\mathbf{Q} \times \mathbf{Q}$ (Chapter III, § 6, no. 3).

To show that $1/x$ is continuous on $\mathbf{Q}^*$ we shall establish more precisely that $1/x$ is *uniformly continuous* (with respect to the additive structure) in the complement of any neighbourhood V of 0. Namely we have $\left|\frac{1}{x} - \frac{1}{y}\right| = \frac{|x-y|}{xy}$; there exists an integer $m > 0$ such that $|x| \geqslant 1/m$ for each $x \in \complement V$; if x and y are any two points of $\complement V$ such that $|x - y| \leqslant 1/m^2 n$, we shall then have $\left|\frac{1}{x} - \frac{1}{y}\right| \leqslant \frac{1}{n}$.

The image, under the function $1/x$, of any Cauchy filter on $\mathbf{Q}^*$ (with respect to the additive uniformity) which does not have 0 as a cluster point, is a Cauchy filter (with respect to the additive uniformity). Hence (Chapter III, § 6, no. 8, Proposition 7):

PROPOSITION 1. *The functions xy and $1/x$, defined respectively on $\mathbf{Q} \times \mathbf{Q}$ and $\mathbf{Q}^*$, can be extended by continuity to $\mathbf{R} \times \mathbf{R}$ and $\mathbf{R}^*$ respectively, and define a field structure on* $\mathbf{R}$. *Endowed with this structure,* $\mathbf{R}$ *is called the field of real numbers.*

All the properties of topological fields established in § 6 of Chapter III are of course applicable; in particular, every *rational function* of n real variables, with real coefficients, is *continuous* at every point of $\mathbf{R}^n$ where its denominator does not vanish.

2. THE MULTIPLICATIVE GROUP $\mathbf{R}^*$

We know from Chapter III, § 6, no. 7, that the topology induced on $\mathbf{R}^*$ by the topology of the real line is *compatible* with the multiplicative group structure of $\mathbf{R}^*$; since $\mathbf{R}^*$ is an *open* subset of the locally compact space $\mathbf{R}$, it follows that $\mathbf{R}^*$ is a *locally compact* topological group (Chapter I, § 9, no. 7, Proposition 13) and is therefore *complete* (Chapter III, § 3, no. 3, Corollary 1 to Proposition 4; this follows also from Chapter III, § 6, no. 8, Proposition 8); of course, this latter property relates to the *multiplicative* uniformity on $\mathbf{R}^*$ and not to the uniformity induced on $\mathbf{R}^*$ by the additive uniformity of $\mathbf{R}$.

The function xy maps the set $\mathbf{Q}_+ \times \mathbf{Q}_+$ into $\mathbf{Q}_+$, and therefore it maps $\mathbf{R}_+ \times \mathbf{R}_+$ into $\mathbf{R}_+$ (Chapter I, § 2, no. 1, Theorem 1); in other words, *the product of two real numbers $\geqslant 0$ is $\geqslant 0$*. The formulae $(-x)y = -xy$ and $(-x)(-y) = xy$ then show that the product of a number $\geqslant 0$ and a number $\leqslant 0$ is $\leqslant 0$, and that the product of two numbers $\leqslant 0$ is $\geqslant 0$; from this it follows that

$$|xy| = |x| \cdot |y| \tag{1}$$

(which might also have been obtained by extension of the corresponding relation on $\mathbf{Q} \times \mathbf{Q}$).

If $x > 0$ and $y > 0$ we have $xy \neq 0$, and therefore $xy > 0$; likewise, if $x < 0$ and $y > 0$, then $xy < 0$; and if $x < 0$ and $y < 0$, then $xy > 0$. In particular, if $x \neq 0$ we have $x^2 > 0$, so that a *sum of squares* of real numbers cannot be zero unless each of the numbers is zero.

If $x > 0$ and $y \leqslant z$ (resp. $y < z$) we have $xy \leqslant xz$ (resp. $xy < xz$) in other words, *a homothety of ratio > 0 preserves order on* $\mathbf{R}$. Since $(-x)y = -xy$, a homothety of ratio < 0 changes the ordering on $\mathbf{R}$ into the opposite ordering.

If $x > 0$ we have $1/x > 0$, since $x.(1/x) = 1 > 0$. If $0 < x < y$ we have $xy > 0$, hence $x.(1/xy) < y.(1/xy)$, that is $1/y < 1/x$. Hence the mapping $x \to 1/x$ of the set $\mathbf{R}_+^*$ of real numbers > 0 onto itself is *strictly decreasing.*

We see in the same way that the function $1/x$ is strictly decreasing in $]\leftarrow, 0[$, and therefore the function $\frac{1}{x - a}$ is strictly decreasing in each of the intervals $]\leftarrow, a[$ and $]a, \to[$.

It follows from what precedes that $\mathbf{R}_+^*$ is a *subgroup* of the multiplicative group $\mathbf{R}^*$; moreover, the order relation $x \leqslant y$ is *compatible* with the multiplicative, group structure of $\mathbf{R}_+^*$; in other words, $\mathbf{R}_+^*$ is a *linearly ordered group.*

The fact that the product of two real numbers $\geqslant 0$ is $\geqslant 0$ can be expressed by saying that $\mathbf{R}$ is an *ordered field*: all the above properties are common to all ordered fields.

PROPOSITION 2. *The multiplicative group* $\mathbf{R}^*$ *of real numbers* $\neq 0$ *is a topological group isomorphic to the product of its subgroups* $\mathbf{R}_+^*$ *and* U_0, *where*

$$U_0 = \{-1, +1\}.$$

For each $x \neq 0$ let $\operatorname{sgn} x$ denote $\frac{x}{|x|}$ (*sign* of x). The function sgn is a homomorphism of $\mathbf{R}^*$ onto U_0. We have $x = |x| \operatorname{sgn} x$, and this decomposition of x as the product of an element of $\mathbf{R}_+^*$ and an element of U_0 is unique; hence the group structure of $\mathbf{R}^*$ is the product of the group structures of $\mathbf{R}_+^*$ and U_0. On the other hand, the mapping $x \to |x|$ is continuous, and so is $x \to \operatorname{sgn} x = \frac{x}{|x|}$, since $x \neq 0$. Hence the result.

We extend the function sgn to the whole of $\mathbf{R}$ by putting $\operatorname{sgn} 0 = 0$.

We shall see in Chapter V (§ 4, no. 1, Theorem 1) that the topological group $\mathbf{R}_+^*$ is *isomorphic* to the *additive* group $\mathbf{R}$; this will complete the determination of the structure of the topological group $\mathbf{R}_*$.

3. nTH ROOTS

Let n be any integer > 0. From the relation $0 < x < y$ we deduce, by induction on n, that $0 < x^n < y^n$. In other words, the function $x \to x^n$ is *strictly increasing* for $x \geqslant 0$; it is clearly continuous at every point and therefore (§ 2, no. 6, Theorem 5) it is a *homeomorphism* of $\mathbf{R}_+$ onto an interval I. On the other hand, since $x \geqslant 1$ implies $x^{n-1} \geqslant 1$ and therefore $x^n \geqslant x$, it follows that I is not bounded and hence $\mathrm{I} = \mathbf{R}_+$. The value, for $x \geqslant 0$, of the *inverse* of the mapping $x \rightarrow x^n$ is denoted by $x^{1/n}$ or $\sqrt[n]{x}$ and is called x *to the power* $1/n$ or *the nth root of* x (for $n = 2, 3$ we say *square* root, *cube* root; for $n = 2$ we write $\sqrt{x}$ in place of $\sqrt[2]{x}$). The positive number $x^{1/n}$ is thus defined as the unique positive solution of the equation

$$y^n = x \qquad (x \geqslant 0). \tag{2}$$

In particular we see that there is a real number x such that $x^2 = 2$, whereas no rational number has this property; thus we recover the fact that the rational line $\mathbf{Q}$ is not a complete space.

The mapping $x \rightarrow x^{1/n}$ of $\mathbf{R}_+$ onto itself is *strictly increasing* and *continuous*. By (2) we have $0^{1/n} = 0$, $1^{1/n} = 1$, also

$$(xy)^{1/n} = x^{1/n} y^{1/n}; \tag{3}$$

hence $x \rightarrow x^{1/n}$ is an *automorphism* of the topological group $\mathbf{R}_+^*$.

In Chapter V, § 4, no. 1, we shall generalize this result by finding *all* the automorphisms of the multiplicative group $\mathbf{R}_+^*$.

4. THE EXTENDED REAL LINE

1. HOMEOMORPHISMS OF OPEN INTERVALS OF R

PROPOSITION 1. *All non-empty open intervals of* $\mathbf{R}$ *are homeomorphic to* $\mathbf{R}$.

Consider first a *bounded* open interval $\mathrm{I} =]a, b[$ $(a < b)$. For each $x \in \mathrm{I}$ put $f(x) = -\left(\frac{1}{x-a} + \frac{1}{x-b}\right)$. This function is continuous and strictly increasing on I, for we have seen that $\frac{1}{x-b}$ is strictly decreasing in $]\leftarrow, b[$ and $\frac{1}{x-a}$ strictly decreasing in $]a, \rightarrow[$. It follows that f is a homeomorphism of I onto an interval $f(\mathrm{I})$ of $\mathbf{R}$ (§ 2, no. 6, Theorem 5). $f(\mathrm{I})$ is neither bounded above nor below; if, for example, we had $f(x) \leqslant c$ for all $x \in \mathrm{I}$, it would follow that $1 - \frac{b-x}{x-a} \leqslant c(b-x)$, since $b - x > 0$; and this leads to a contradiction when x is sufficiently near b (by virtue of the continuity of the two sides of the inequality, which are rational functions, at the point b). Hence $f(\mathrm{I}) = \mathbf{R}$, and therefore every bounded open interval is homeomorphic to $\mathbf{R}$. Let g be the inverse of f: it maps every unbounded open interval of $\mathbf{R}$ onto an interval J contained in I and open in I. Since I is open in $\mathbf{R}$, J is also open in $\mathbf{R}$; since it is bounded, it is homeomorphic to $\mathbf{R}$; and thus we have proved that every unbounded open interval is homeomorphic to $\mathbf{R}$.

Remark. To show that all *bounded* open intervals are homeomorphic to each other, it would be enough to remark that, if $a \neq b$ and $a' \neq b'$, there is a homeomorphism of $\mathbf{R}$ onto itself, of the form $x \to \alpha x + \beta$ (and only one) which maps a to a' and b to b', and hence maps the open (resp. half-open, closed) interval with end-points a, b onto the open (resp. half-open, closed) interval with end-points a', b'; the reader may easily verify this by calculating α and β.

2. THE EXTENDED LINE

We shall now define, by *adjoining* two new elements to $\mathbf{R}$, a topological space $\overline{\mathbf{R}}$ such that every homeomorphism of $\mathbf{R}$ onto a bounded open interval I of $\mathbf{R}$ can be extended to a homeomorphism of $\overline{\mathbf{R}}$ onto the closed interval having the same end-points as I.

Let $\overline{\mathbf{R}}$ be the set obtained by adjoining (*Set Theory*, R, § 4, no. 5) two new elements to $\mathbf{R}$, denoted by $-\infty$ and $+\infty$. We extend the ordering on $\mathbf{R}$ to $\overline{\mathbf{R}}$ by putting $-\infty < a$ and $a < +\infty$ for all $a \in \mathbf{R}$, and $-\infty < +\infty$; it is clear that we thus obtain a linearly ordered set, whose ordering induces the ordering of the real line on $\mathbf{R}$. Next, consider the topology on $\overline{\mathbf{R}}$ generated by the set of open intervals of $\overline{\mathbf{R}}$. Since the trace on $\mathbf{R}$ of an open interval of $\overline{\mathbf{R}}$ is an open interval of $\mathbf{R}$, this topology induces on $\mathbf{R}$ the topology of the real line.

DEFINITION 1. *The set* $\overline{\mathbf{R}}$ *endowed with the order structure and the topology defined above is called the extended real line.*

When using the extended line $\overline{\mathbf{R}}$ it is often convenient to refer to its points as *real numbers*, by abuse of language; the points of $\mathbf{R}$ are then called *finite real numbers*. We shall adopt this convention *in this section and the three following sections of this chapter*; whenever we adopt this convention in future we shall indicate explicitly to which part of the text it extends.

If a is a finite real number, the intervals $[a, +\infty[$ and $]-\infty, a]$ (resp. $]a, +\infty[$ and $]-\infty, a[$) of $\overline{\mathbf{R}}$ are contained in $\mathbf{R}$ and coincide with the intervals of $\mathbf{R}$ hitherto denoted by $[a, \rightarrow[$ and $]\leftarrow, a]$ (resp. $]a, \rightarrow[$ and $]\leftarrow, a[$); this new notation is much more often used. Again, $\mathbf{R}$ coincides with the interval $]-\infty, +\infty[$ of $\overline{\mathbf{R}}$, and is sometimes so denoted.

Proposition 2. *Every homeomorphism f of $\mathbf{R}$ onto an interval $]a, b[$ can be extended to a homeomorphism $\bar{f}$ of $\overline{\mathbf{R}}$ onto $[a, b]$. If f is an increasing function, then $\bar{f}$ is an order isomorphism of $\overline{\mathbf{R}}$ onto $[a, b]$.*

Let f be an increasing homeomorphism. If we extend f to $\overline{\mathbf{R}}$ by putting $\bar{f}(-\infty) = a$ and $\bar{f}(+\infty) = b$, it is obvious that $\bar{f}$ is a strictly increasing mapping (and therefore a bijection) of $\overline{\mathbf{R}}$ onto $[a, b]$. Hence $\bar{f}$ maps every open interval of $\overline{\mathbf{R}}$ onto an interval which is open with respect to $[a, b]$, and is therefore a homeomorphism of $\overline{\mathbf{R}}$ onto $[a, b]$ by reason of Definition 1 and Proposition 5 of § 1, no. 4.

If f is decreasing, we apply what has been proved to the increasing homeomorphism $x \rightarrow -f(x)$ of $\mathbf{R}$ onto $]-b, -a[$.

All the properties of the interval $[a, b]$ obtained in § 2, which involve only the order structure and the topology of the interval, can therefore be transported to $\overline{\mathbf{R}}$; hence the following propositions:

Proposition 3. *The extended real line is compact.*

Hence (Chapter II, § 4, no. 1, Theorem 1) there is a unique uniformity on $\overline{\mathbf{R}}$ compatible with its topology; this uniformity is isomorphic with the uniformity induced on $[a, b]$ by the additive uniformity of $\mathbf{R}$. But it should be remarked that the uniformity *induced* on $\mathbf{R}$ by that of $\overline{\mathbf{R}}$ *is not the additive uniformity of* $\mathbf{R}$ (although it is compatible with the topology of the real line); for $\mathbf{R}$ is a *complete* space with respect to its additive uniformity, but is not a complete subspace of $\overline{\mathbf{R}}$, since it is not closed in $\overline{\mathbf{R}}$.

Proposition 4. *Every non-empty subset of $\overline{\mathbf{R}}$ has a least upper bound and a greatest lower bound.*

The least upper bound (resp. greatest lower bound) of a non-empty subset A of $\overline{\mathbf{R}}$ is denoted by sup A (resp. inf A). Clearly

$$\inf A \leqslant \sup A. \tag{1}$$

If $A \subset B$, then $\sup A \leqslant \sup B$ and $\inf A \geqslant \inf B$ (*Set Theory*, Chapter III, § 1, no. 9, Proposition 4).

PROPOSITION 5. *A subset* A *of* $\overline{\mathbf{R}}$ *is connected if and only if* A *is an interval.*

COROLLARY. *The extended real line is a connected, locally connected space.*

PROPOSITION 6. *A mapping* f *of an interval* I *of* $\overline{\mathbf{R}}$ *into* $\overline{\mathbf{R}}$ *is a homeomorphism of* I *onto* $f(\mathrm{I})$ *if and only if* f *is strictly monotonic and continuous on* I; $f(\mathrm{I})$ *is then an interval of* $\overline{\mathbf{R}}$.

Finally, the functions $\sup(x, y)$ and $\inf(x, y)$ are *continuous* on $\overline{\mathbf{R}} \times \overline{\mathbf{R}}$.

3. ADDITION AND MULTIPLICATION IN $\overline{\mathbf{R}}$

Note first that the function $-x$ can be extended by continuity to $\overline{\mathbf{R}}$, according to the formulae $-(-\infty) = +\infty$ and $-(+\infty) = -\infty$; the function thus extended is a homeomorphism of $\overline{\mathbf{R}}$ onto itself.

Next, consider, the functions $x + y$ and xy, defined on $\mathbf{R} \times \mathbf{R}$, with values in $\mathbf{R}$; by considering that they take their values *in the topological space* $\overline{\mathbf{R}}$, we shall see that they too can be extended by continuity at certain points of $\overline{\mathbf{R}} \times \overline{\mathbf{R}}$.

So far as $x + y$ is concerned, let $\mathrm{A}' =]-\infty, +\infty]$ and $\mathrm{A}'' = [-\infty, +\infty[$. Then we have the following proposition:

PROPOSITION 7. *The function* $x + y$ *can be extended by continuity to each of the sets* $\mathrm{A}' \times \mathrm{A}'$ *and* $\mathrm{A}'' \times \mathrm{A}''$, *according to the formulae*

$$
(2) \qquad \begin{cases} x + (+\infty) = (+\infty) + x = +\infty & (x \neq -\infty), \\ x + (-\infty) = (-\infty) + x = -\infty & (x \neq +\infty). \end{cases}
$$

Let us show, for example, that as (x, y) tends to the point $(a, +\infty)$ $(a \neq -\infty)$ while remaining in $\mathbf{R} \times \mathbf{R}$, $x + y$ tends to $+\infty$. There exists a finite number $b < a$, and the interval $]b, +\infty[$ is a neighbourhood of a in $\overline{\mathbf{R}}$; given any finite c, the relations $x > b$ and $y > c - b$ imply $x + y > c$, and this shows that $x + y$ is as near as we please to $+\infty$ whenever (x, y) is near enough to $(a, +\infty)$. The argument is similar in the other cases.

On the contrary, $x + y$ *has no limit* at the points $(-\infty, +\infty)$ and $(+\infty, -\infty)$ of $\overline{\mathbf{R}} \times \overline{\mathbf{R}}$. For if $x + y$ had limit k (finite or infinite) as (x, y) tends to $(+\infty, -\infty)$ while remaining in $\mathbf{R} \times \mathbf{R}$, it would follow that, for each finite a, the function $(x + a) - x$ would tend to k as x tends to $+\infty$ while remaining in $\mathbf{R}$; and this is absurd, since $(x + a) - x = a$ and a is arbitrary.

The function $x + y$ maps $A' \times A'$ (resp. $A'' \times A''$) into A' (resp. A''). It is therefore a *law of composition* on A' (resp. A'') which extends the law of addition on $\mathbf{R}$. By the principle of extension of identities (Chapter I, § 8, no. 1, Corollary 1 to Proposition 2) this law is *commutative* and *associative*; 0 is the *identity element* for this law; and the only *non-regular* element (*Algebra*, Chapter I, § 2, no. 2) of A' is $+\infty$, by formulae (2).

If x, y, z, t are points of $\overline{\mathbf{R}}$ such that $x \leqslant y$ and $z \leqslant t$, then $x + z \leqslant y + t$ whenever both sides of this inequality are defined.

Note that in $\overline{\mathbf{R}}$ the relation $x < y$ implies $x + z < y + z$ only if z is *finite*, by formulae (2); it is easily verified that the relations $x < y$ and $z < t$ imply $x + z < y + t$ whenever both sides of this inequality are defined.

In $\overline{\mathbf{R}}$ we put $x^+ = \sup(x, 0)$, $x^- = \sup(-x, 0)$, $|x| = \sup(x, -x)$; thus $(+\infty)^+ = (-\infty)^- = +\infty$ and $(+\infty)^- = (-\infty)^+ = 0$, and $|+\infty| = |-\infty| = +\infty$. The sums $x^+ - x^-$ and $x^+ + x^-$ are defined for all $x \in \overline{\mathbf{R}}$ and are therefore equal to x and $|x|$ respectively by the principle of extension of identities. Also, whenever the sum $x + y$ is defined, we have $|x + y| \leqslant |x| + |y|$.

Note on the contrary that formulae (6) and (7) of § 1, no. 1 may no longer make sense for certain values of x and y in $\overline{\mathbf{R}}$; for example, if $x = -\infty$ and $y = 0$ we have $\sup(x, y) = 0$, but the sum $x + (y - x)^+$ is not defined, because $(y - x)^+ = +\infty$.

Let $\overline{\mathbf{R}}^*$ denote the complement of 0 in $\overline{\mathbf{R}}$. Then the analogue of Proposition 7 for multiplication runs as follows:

PROPOSITION 8. *The function xy can be extended by continuity to the set $\overline{\mathbf{R}}^* \times \overline{\mathbf{R}}^*$ according to the formulae*

$$(3)\quad \begin{cases} x.(+\infty) = (+\infty).x = \begin{cases} +\infty & \text{if } x > 0 \\ -\infty & \text{if } x < 0 \end{cases} \\ x.(-\infty) = (-\infty).x = \begin{cases} -\infty & \text{if } x > 0 \\ +\infty & \text{if } x < 0. \end{cases} \end{cases}$$

We leave the proof to the reader; it is analogous to the proof of Proposition 7.

Likewise, we see that xy *has no limit* at the points $(0, +\infty)$, $(+\infty, 0)$, $(0, -\infty)$, $(-\infty, 0)$ of $\overline{\mathbf{R}} \times \overline{\mathbf{R}}$.

The function xy is a *law of composition* on $\overline{\mathbf{R}}^*$ which extends the law of multiplication on $\mathbf{R}$; this law is *associative* and *commutative* (principle of extension of identities); it has 1 as identity element; and the non-regular elements in $\overline{\mathbf{R}}^*$ are $+\infty$ and $-\infty$.

If $x \leqslant y$ and $z > 0$ we have $xz \leqslant yz$ whenever both sides of this inequality are defined. If the product xy is defined, then so is $|x| . |y|$ and we have $|xy| = |x| . |y|$.

Finally, the distributivity formula

$$x(y + z) = xy + xz \tag{4}$$

is still valid, by virtue of the principle of extension of identities, whenever *all* the operations which figure on either side are defined.

Note that it can happen that the left-hand side of (4) is defined while the right-hand side is not: for example, consider the case where $x = +\infty$, $y = 2$ and $z = -1$. The distributivity formula should therefore be used with caution in $\overline{\mathbf{R}}$.

5. REAL-VALUED FUNCTIONS

1. REAL-VALUED FUNCTIONS

DEFINITION 1. *A mapping of a set* X *into the real line is called a real-valued function (or real function) defined on* X.

By an abuse of language analogous to that mentioned in § 4, no. 2, mappings of X into $\overline{\mathbf{R}}$ will also be called *real-valued functions* defined on X *in this and the following section.* Mappings of X into $\mathbf{R}$ will be called *finite real-valued functions.*

If f and g are two real-valued functions defined on X, the relation $f \leqslant g$ is by definition equivalent to "$f(x) \leqslant g(x)$ for all $x \in X$;" this relation is an *ordering* on the set $\overline{\mathbf{R}}^X$ of all real-valued functions on X. The set $\overline{\mathbf{R}}$,^{X} ordered by this relation, is a *lattice*; for if f and g are any two real-valued functions, the function h defined by $h(x) = \sup (f(x), g(x))$ for all $x \in X$ is the smallest of the real-valued functions on X which are both $\geqslant f$ and $\geqslant g$; in accordance with general notation, we denote this function (which is the least upper bound of f and g in $\overline{\mathbf{R}}^X$) by $\sup(f, g)$. Similarly, the function whose value at each $x \in X$ is $\inf (f(x), g(x))$ is denoted by $\inf(f, g)$.

Note that $\sup(f, g)$ is the composition of the mapping

$$(u, v) \rightarrow \sup(u, v)$$

of $\overline{\mathbf{R}} \times \overline{\mathbf{R}}$ into $\overline{\mathbf{R}}$ and the mapping $x \rightarrow (f(x), g(x))$ of X into $\overline{\mathbf{R}} \times \overline{\mathbf{R}}$. Similarly for $\inf(f, g)$.

A real-valued function f defined on a set X is said to be *bounded above* (resp. *bounded below*) in X, if $f(X)$ is a subset of $A'' = [-\infty, +\infty]$ and is bounded above (resp. if $f(X)$ is a subset of $A' =]-\infty, +\infty]$ and is bounded below). f is said to be *bounded* in X if it is bounded both above and below, that is if $f(X)$ is a bounded subset *of* $\mathbf{R}$.

Every bounded function is therefore *finite*. The converse is false, as is shown by the function $1/x$ on $R_+^* =]0, +\infty[$.

2. REAL-VALUED FUNCTIONS DEFINED ON A FILTERED SET

PROPOSITION 1. *Let f and g be two real-valued functions defined on a set X filtered by a filter $\mathfrak{F}$. If $\lim_{\mathfrak{F}} f$ and $\lim_{\mathfrak{F}} g$ exist, and if, for each subset $A \in \mathfrak{F}$, there exists $x \in A$ such that $f(x) \leqslant g(x)$, then we have $\lim_{\mathfrak{F}} f \leqslant \lim_{\mathfrak{F}} g$.*

To prove this result we shall prove the following equivalent statement:

PROPOSITION 2. *Let f and g be two real-valued functions defined on a set X filtered by a filter $\mathfrak{F}$. If $\lim_{\mathfrak{F}} f$ and $\lim_{\mathfrak{F}} g$ exist, and if $\lim_{\mathfrak{F}} f > \lim_{\mathfrak{F}} g$, then there is a set $A \in \mathfrak{F}$ such that $f(x) > g(x)$ for all $x \in A$.*

Let $a = \lim_{\mathfrak{F}} f$, let $b = \lim_{\mathfrak{F}} g$ and let c be such that $b < c < a$. The interval $]c, +\infty]$ of $\overline{\mathbf{R}}$ (resp. $[-\infty, c[$) is a neighbourhood of a (resp. b); hence there is a set $M \in \mathfrak{F}$ (resp. a set $N \in \mathfrak{F}$) such that $f(x) > c$ for all $x \in M$ [resp. $g(x) < c$ for all $x \in N$]. The set $A = M \cap N$ belongs to $\mathfrak{F}$, and we have $f(x) > c > g(x)$ for all $x \in A$.

As a particular case of Proposition 1 we have the following theorem:

THEOREM 1 (Principle of extension of inequalities). *Let f and g be two real-valued functions, defined on a set X filtered by a filter $\mathfrak{F}$. If $\lim_{\mathfrak{F}} f$ and $\lim_{\mathfrak{F}} g$ exist, and if $f \leqslant g$, then $\lim_{\mathfrak{F}} f \leqslant \lim_{\mathfrak{F}} g$.*

Remark. If in particular we have $f(x) < g(x)$ for all $x \in X$ (or only for all points of a set of the filter $\overline{\mathbf{R}}$) we can infer, by Theorem 1, that $\lim_{\mathfrak{F}} f \leqslant \lim_{\mathfrak{F}} g$; *but we cannot infer the strong inequality* $\lim_{\mathfrak{F}} f < \lim_{\mathfrak{F}} g$. For example, if we take X to be the set $\mathbf{N}$ of natural numbers, filtered by the Fréchet filter, and if $f(n) = 0$ and $g(n) = 1/n$, then $f(n) < g(n)$ for all n, but $\lim_{n \to \infty} f(n) = \lim_{n \to \infty} f(n) = 0$. Thus we *lose strictness* when we pass to the limit in a *strict* inequality.

THEOREM 2 (Theorem of the monotone limit). *Let* X *be an ordered set and let* A *be a directed subset of* X (*). *Every monotonic real-valued function* f *defined on* A *has a limit with respect to* A (Chapter I, § 7, no. 3); *if* f *is increasing* (resp. *decreasing*), *this limit is equal to the least upper bound* (resp. *greatest lower bound*) *of the set* $f(\mathrm{A}) \subset \overline{\mathbf{R}}$.

Suppose for example that f is increasing, and let $a = \sup f(\mathrm{A})$. If $a = -\infty$, the theorem is trivial. If $a > -\infty$, then for each $b < a$ there exists $x \in \mathrm{A}$ such that $b < f(x) \leqslant a$; hence, if S_x is the *section* of A relative to x (i.e. the set of all $y \geqslant x$, cf. Chapter I, § 6, no. 3), $f(\mathrm{S}_x)$ is contained in the neighbourhood $]b, +\infty]$ of a, and the theorem follows. The proof is analogous if f is decreasing.

COROLLARY. *An increasing* (resp. *decreasing*) *real-valued function, defined on a directed subset* A *of an ordered set* X, *has a finite limit with respect to* A *if and only if it is bounded above* (resp. *bounded below*) *in* A.

If we apply Theorem 2 to the case where $\mathrm{A} = \mathrm{X} = \mathbf{N}$ (ordered by the relation $\leqslant$), we have the following proposition:

PROPOSITION 3. *Every monotonic sequence of real numbers has a limit in* $\overline{\mathbf{R}}$. In particular, every increasing (resp. decreasing) sequence of *finite* numbers converges to a finite real number if it is bounded above (resp. bounded below) and to $+\infty$ (resp. $-\infty$) otherwise. For example, the sequence of positive integers converges to $+\infty$.

This fact is the origin of the notation $\lim\limits_{n \to \infty} u_n$ to denote the limit of a sequence (Chapter I, § 7, no. 3).

Likewise, every *strictly increasing* sequence of integers (p_n) converges to $+\infty$; for we see by induction that $p_n \geqslant p_0 + n$ for all n.

3. LIMITS ON THE RIGHT AND ON THE LEFT OF A FUNCTION OF A REAL VARIABLE

Let A be a non-empty subset of $\overline{\mathbf{R}}$ and let $a \neq -\infty$ be a point of $\overline{\mathbf{R}}$ lying in the closure of the set $\mathrm{B} = \mathrm{A} \cap [-\infty, a[$. The set B is directed with respect to the relation $\leqslant$, and its section filter $\mathfrak{F}$ is the same as the *trace* on B of the neighbourhood filter of a in $\overline{\mathbf{R}}$.

(*) This statement assumes implicitly that the ordering in X is written $x \leqslant y$. If this relation is written $x(\sigma)\,y$, where (σ) is a certain sign or group of signs characteristic of the relation envisaged, then the word " directed " in the statement must be replaced by "directed with respect to (σ)".

DEFINITION 2. *Let f be a function defined on a non-empty subset* A *of* $\overline{\mathbf{R}}$, *with values in a topological space* X. *A limit of f with respect to the filter* $\mathfrak{F}$, *if it exists, is called a limit of f on the left at the point a, relative to* A, *and is denoted by* $\lim_{x \to a,\, x < a,\, x \in A} f(x)$, *or* $f(a-)$, *if* X *is Hausdorff.*

Likewise, if a is in the closure of the set $A \cap]a, +\infty]$, we define a *limit on the right* (if it exists) of f at the point a, and denote it by $\lim_{x \to a,\, x > a,\, x \in A} f(x)$, or $f(a+)$, if X is Hausdorff.

The following proposition is an immediate consequence of Theorem 2:

PROPOSITION 4. *Let* A *be a subset of* $\overline{\mathbf{R}}$ *and let* $a \neq -\infty$ *be a point in the closure of the intersection* $A \cap [-\infty, a[$. *If f is a monotonic real-valued function defined on* A, *then f has a limit $f(a-)$ on the left at a, relative to* A.

4. BOUNDS OF A REAL-VALUED FUNCTION

DEFINITION 3. *Let f be a real-valued function defined on a set* X, *and let* A *be a non-empty subset of* X. *Then the* least upper bound (resp. *greatest lower bound*) *of the set $f(A)$ in* $\overline{\mathbf{R}}$ *is called the* least upper bound (resp. *greatest lower bound*) *of f in* A, *and is denoted by* $\sup_{x \in A} f(x)$ $\left[\text{resp. } \inf_{x \in A} f(x)\right]$.

In particular, if A is a non-empty subset of $\overline{\mathbf{R}}$, then

$$\sup A = \sup_{x \in A} x. \tag{1}$$

It is often more convenient to use the notation on the right-hand side to denote the least upper bound of A.

The number $a = \sup_{x \in A} f(x)$ is characterized by the following two properties:

(i) For all $x \in A$, $f(x) \leqslant a$.

(ii) For every $b < a$, there exists $x \in A$ such that $b < f(x) \leqslant a$.

The numbers $\sup_{x \in A} f(x)$ and $\inf_{x \in A} f(x)$ belong to the *closure* of $f(A)$ in $\overline{\mathbf{R}}$. We have $\inf_{x \in A} f(x) \leqslant \sup_{x \in A} f(x)$; and these two numbers are *equal* if and only if f is *constant* on A.

A real-valued function, defined on a set X, is *bounded above* (resp. *bounded below*) in a non-empty subset A of X if and only if $\sup_{x \in A} f(x) < +\infty$ [resp. $\inf_{x \in A} f(x) > -\infty$]. f is *bounded* in A if and only if $|f|$ is bounded above in A, hence if and only if $\sup_{x \in A} |f(x)| < +\infty$.

We have

$$(2) \qquad \inf_{x\in A} f(x) = -\sup_{x\in A} (-f(x)).$$

This relation reduces all properties of the greatest lower bound to those of the least upper bound; hence in general we shall speak only of the latter.

PROPOSITION 5. *Let f be a real-valued function defined on a set X. On the set $\mathfrak{F}(X)$ of all finite subsets of X, directed with respect to the relation $\subset$, the real-valued function $H \to \sup_{x\in H} f(x)$ is increasing, the real-valued function $H \to \inf_{x\in H} f(x)$ is decreasing, and we have*

$$(3) \qquad \begin{cases} \sup_{x\in A} f(x) = \lim_{H\in\mathfrak{F}(X)} (\sup_{x\in H} f(x)), \\ \inf_{x\in A} f(x) = \lim_{H\in\mathfrak{F}(X)} (\inf_{x\in H} f(x)). \end{cases}$$

Let $\varphi(H) = \sup_{x\in H} f(x)$. Clearly φ is increasing, and therefore has a limit a (no. 2, Theorem 2); and since $\varphi(H) \leqslant \sup_{x\in A} f(x)$ for all H, we have $a \leqslant \sup_{x\in A} f(x)$ (no. 2, Theorem 1). If we had $a < \sup_{x\in A} f(x)$, then there would exist $x_0 \in X$ such that $a < f(x_0)$; but this is a contradiction since $\varphi(H) \geqslant f(x_0)$ whenever $x_0 \in H$.

In particular, by (1), if A is any non-empty subset of $\overline{\mathbf{R}}$, we have

$$(4) \qquad \sup A = \lim_{H\in\mathfrak{F}(A)} (\sup_{x\in H} x).$$

PROPOSITION 6. *Let f and g be two real-valued functions defined on X. If $f(x) \leqslant g(x)$ at every point x of a non-empty subset A of X, then we have*

$$(5) \qquad \begin{cases} \sup_{x\in A} f(x) \leqslant \sup_{x\in A} g(x), \\ \inf_{x\in A} f(x) \leqslant \inf_{x\in A} g(x). \end{cases}$$

PROPOSITION 7. *Let f be a real-valued function defined on X. If A and B are two non-empty subsets of X such that $A \subset B$, then*

$$(6) \qquad \sup_{x\in A} f(x) \leqslant \sup_{x\in B} f(x).$$

PROPOSITION 8. *Let f be a real-valued function defined on X, and let $(A_\iota)_{\iota\in I}$ be a non-empty family of non-empty subsets of X; then*

$$(7) \qquad \sup_{x\in \bigcup_{\iota\in I} A_\iota} f(x) = \sup_{\iota\in I} (\sup_{x\in A_\iota} f(x)).$$

Let f be a real-valued function defined on a product set $X_1 \times X_2$. If A_2 is a non-empty subset of X_2 we shall denote by $\sup_{x_2 \in A_2} f(x_1, x_2)$ the least upper bound in A_2 of the real-valued function $x_2 \to f(x_1, x_2)$ defined on X_2. From Proposition 8 we deduce in particular:

Proposition 9. *Let f be a real-valued function defined on a product set $X_1 \times X_2$. If A_1, A_2 are any non-empty subsets of X_1, X_2 respectively, then*

$$\sup_{(x_1, x_2) \in A_1 \times A_2} f(x_1, x_2) = \sup_{x_1 \in A_1} (\sup_{x_2 \in A_2} f(x_1, x_2)) = \sup_{x_2 \in A_2} (\sup_{x_1 \in A_1} f(x_1, x_2)). \tag{8}$$

5. ENVELOPES OF A FAMILY OF REAL-VALUED FUNCTIONS

Definition 4. *Let $(f_\iota)_{\iota \in I}$ be a family of real-valued functions defined on a set X. The real-valued function on X whose value at each point $x \in X$ is* $\sup_{\iota \in I} (f_\iota(x))$ [resp. $\inf_{\iota \in I} (f_\iota(x))$] *is called the upper* (resp. *lower*) *envelope of the family* (f_ι), *and is denoted by* $\sup_{\iota \in I} f_\iota$ *or* $\sup_\iota f_\iota$ $\left(\text{resp. } \inf_{\iota \in I} f_\iota \text{ or } \inf_\iota f_\iota\right)$.

The upper envelope of the family (f_ι) is thus the *least upper bound* of this family in the lattice $\overline{\mathbf{R}}$ of real-valued functions defined on X, and this justifies the notation $\sup_\iota f_\iota$.

Furthermore, if we endow $\overline{\mathbf{R}}^X$ with the topology which is the product of the topologies of its factors (all identical with $\overline{\mathbf{R}}$), we have the following proposition:

Proposition 10. *In the product space $\overline{\mathbf{R}}^X$ the upper envelope $\sup_\iota f_\iota$ of a family of real-valued functions $(f_\iota)_{\iota \in I}$ is the limit, with respect to the directed set $\mathfrak{F}(I)$ of finite subsets of I, of the mapping* $H \to \sup_{\iota \in H} f_\iota$ [which maps each finite subset H of I to the upper envelope of the finite subfamily $(f_\iota)_{\iota \in H}$].

This follows immediately from Proposition 5 of no. 4 and from Chapter I, § 7, no. 6, Corollary 1 to Proposition 10.

We may therefore write

$$\sup_{\iota \in I} f_\iota = \lim_{H \in \mathfrak{F}(I)} \left(\sup_{\iota \in H} f_\iota\right). \tag{9}$$

Definition 5. *A family $(f_\iota)_{\iota \in I}$ of real-valued functions defined on a set X is said to be uniformly bounded above* (resp. *uniformly bounded below*) *in* X, *if there exists a finite number* a *such that* $f_\iota(x) \leqslant a$ [resp. $f_\iota(x) \geqslant a$] *for all* $x \in X$ *and all* $\iota \in I$. *The family* (f_ι) *is said to be uniformly bounded in* X *if it is uniformly bounded above and below in* X.

Thus (f_ι) is uniformly bounded above in X if and only if the *upper* envelope of this family is *bounded above* in X. (f_ι) is uniformly bounded in X if and only if the upper envelope of the family $(|f_\iota|)$ is bounded above in X [i.e. if and only if there is a finite real number $a \geqslant 0$ such that $|f_\iota(x)| \leqslant a$ for all $x \in X$ and all $\iota \in I$].

6. UPPER LIMIT AND LOWER LIMIT OF A REAL-VALUED FUNCTION WITH RESPECT TO A FILTER

Let f be a real-valued function defined on a set X filtered by a filter $\mathfrak{G}$. $\mathfrak{G}$ is a *directed* set with respect to the relation $\supset$ (Chapter I, § 6). For each $M \in \mathfrak{G}$ consider the real number $\sup_{x \in M} f(x)$: we have a function $M \to \sup_{x \in M} f(x)$ of $\mathfrak{G}$ into $\overline{\mathbf{R}}$, which is a *decreasing* function on $\mathfrak{G}$, by Proposition 7 of no. 4. Hence, by Theorem 2 of no. 2, it has a limit with respect to the directed set $\mathfrak{G}$.

Definition 6. *The limit of the real-valued function* $M \to \sup_{x \in M} f(x)$ *with respect to the directed set* $\mathfrak{G}$ *is called the upper limit of* f *with respect to the filter* $\mathfrak{G}$, *and is denoted by* $\limsup_{\mathfrak{G}} f$, *or by* $\limsup_{x, \mathfrak{G}} f(x)$.

The *lower limit* of f with respect to the filter $\mathfrak{G}$ is defined similarly, and is denoted by $\liminf_{\mathfrak{G}} f$ or $\liminf_{x, \mathfrak{G}} f(x)$. Thus we have

$$(10)\qquad \begin{cases} \limsup_{\mathfrak{G}} f = \lim_{M \in \mathfrak{G}} (\sup_{x \in M} f(x)), \\ \liminf_{\mathfrak{G}} f = \lim_{M \in \mathfrak{G}} (\inf_{x \in M} f(x)). \end{cases}$$

Often the filter $\mathfrak{G}$ is suppressed from the notation, and we write simply $\limsup f$ or $\limsup_x f(x)$, or $\limsup f(x)$ when there is no risk of confusion.

From formulae (10) and Theorem 1 we have

$$(11)\qquad \inf_{x \in X} f(x) \leqslant \liminf_{\mathfrak{G}} f \leqslant \limsup_{\mathfrak{G}} f \leqslant \sup_{x \in X} f(x).$$

By Theorem 2 of no. 2 we may also write

$$(12)\qquad \begin{cases} \limsup_{\mathfrak{G}} f = \inf_{M \in \mathfrak{G}} (\sup_{x \in M} f(x)), \\ \liminf_{\mathfrak{G}} f = \sup_{M \in \mathfrak{G}} (\inf_{x \in M} f(x)). \end{cases}$$

Also we may replace the filter $\mathfrak{G}$, on the right-hand sides of formulae (10) and (12), by any *base* $\mathfrak{B}$ of $\mathfrak{G}$.

From (2) and (10),

$$(13) \qquad \liminf_{\mathfrak{G}} f = -\limsup_{\mathfrak{G}} (-f)$$

and therefore we need consider only the upper limit.

THEOREM 3. *The upper limit of a real-valued function f with respect to a filter $\mathfrak{G}$ is equal to the largest cluster value of f with respect to $\mathfrak{G}$.*

Let b be a cluster point of f with respect to $\mathfrak{G}$. For each $M \in \mathfrak{G}$, b lies in the closure of $f(M)$, hence $b \leqslant \sup_{x \in M} f(x)$, and therefore, by (12), $b \leqslant \limsup_{\mathfrak{G}} f = a$.

On the other hand, let V be any open neighbourhood of a in $\overline{\mathbf{R}}$. Then there exists a set M_0 in $\mathfrak{G}$ such that, for each $M \in \mathfrak{G}$ contained in M_0, we have $\sup_{x \in M} f(x) \in V$; since V is open it follows that $f(M)$ meets V, and therefore a is a *cluster point* of f with respect to $\mathfrak{G}$, and the proof is complete.

COROLLARY 1. *In order that $\limsup_{\mathfrak{G}} f = \liminf_{\mathfrak{G}} f$, it is necessary and sufficient that f has a limit with respect to the filter $\mathfrak{G}$, and then*

$$\lim_{\mathfrak{G}} f = \limsup_{\mathfrak{G}} f = \liminf_{\mathfrak{G}} f.$$

For since $\overline{\mathbf{R}}$ is compact, the filter base $f(\mathfrak{G})$ has a limit point if and only if it has only one cluster point (Chapter I, § 9, no. 1, Corollary to Theorem 1).

COROLLARY 2. *If $\mathfrak{H}$ is a filter finer than $\mathfrak{G}$, we have*

$$\liminf_{\mathfrak{G}} f \leqslant \liminf_{\mathfrak{H}} f \leqslant \limsup_{\mathfrak{H}} f \leqslant \limsup_{\mathfrak{G}} f.$$

For every cluster point of f with respect to $\mathfrak{H}$ is also a cluster point of f with respect to $\mathfrak{G}$ (Chapter I, § 7, no. 3).

In particular, if $\lim_{\mathfrak{H}} f$ exists, then

$$\liminf_{\mathfrak{G}} f \leqslant \lim_{\mathfrak{H}} f \leqslant \limsup_{\mathfrak{G}} f.$$

COROLLARY 3. *Let A be a set of the filter $\mathfrak{G}$, let $\mathfrak{G}_A$ be the filter induced on A by $\mathfrak{G}$, and let f_A be the restriction of f to A; then*

$$\limsup_{\mathfrak{G}_A} f_A = \limsup_{\mathfrak{G}} f.$$

For every cluster point of the filter base $f(\mathfrak{G})$ is a cluster point of the filter base $f_A(\mathfrak{G}_A)$, and conversely.

For this reason, if f is defined only on a subset A of X belonging to $\mathfrak{G}$, we shall often write $\limsup_{\mathfrak{G}} f$ instead of $\limsup_{\mathfrak{G}_A} f_A$, by abuse of language.

PROPOSITION 11. *Let f and g be two real-valued functions defined on a filtered set* X. *Then the relation $f \leqslant g$ implies*

$$(14) \qquad \begin{cases} \limsup f \leqslant \limsup g, \\ \liminf f \leqslant \liminf g. \end{cases}$$

This is an immediate consequence of the relations (12).

When X is a *topological space* and $\mathfrak{G}$ is the *neighbourhood filter* of a point a of X, we write $\limsup_{x \to a} f(x)$ [resp. $\liminf_{x \to a} f(x)$] in place of $\limsup_{\mathfrak{G}} f$ [resp. $\liminf_{\mathfrak{G}} f$]; clearly we have

$$(15) \qquad \liminf_{x \to a} f(x) \leqslant f(a) \leqslant \limsup_{x \to a} f(x).$$

More generally, if X is a *subspace* of a topological space Y, and if $\mathfrak{G}$ is the trace on X of the neighbourhood filter of a point $a \in \overline{X}$, we write $\limsup_{x \to a,\, x \in X} f(x)$ [resp. $\liminf_{x \to a,\, x \in X} f(x)$] instead of $\limsup_{\mathfrak{G}} f$ [resp. $\liminf_{\mathfrak{G}} f$]; $\limsup_{x \to a,\, x \in X}$ is called the *upper limit* of $f(x)$ *as x tends to a while remaining in* X. If X is the complement of $\{a\}$ we write "$x \neq a$" in place of "$x \in X$" in these notations.

If A is a subset of X such that $a \in \overline{A}$, then (Corollary 2 to Theorem 3)

$$\liminf_{x \to a,\, x \in X} f(x) \leqslant \liminf_{x \to a,\, x \in A} f(x) \leqslant \limsup_{x \to a,\, x \in A} f(x) \leqslant \limsup_{x \to a,\, x \in X} f(x).$$

If V is a neighbourhood of a in Y, we have (Corollary 3 to Theorem 3)

$$\limsup_{x \to a,\, x \in V \cap X} f(x) = \limsup_{x \to a,\, x \in X} f(x).$$

Hence the notions of upper and lower limits at a point of a topological space are, like the notion of limit, of *local* character.

Finally, if $\mathfrak{G}$ is the Fréchet filter on **N**, the upper (resp. lower) limit, with respect to $\mathfrak{G}$, of the mapping $n \to u_n$ of **N** into $\overline{\mathbf{R}}$ is denoted by $\limsup_{n \to \infty} u_n$ (resp. $\liminf_{n \to \infty} u_n$) and is called the *upper* (resp. *lower*) *limit of the sequence of real numbers* u_n.

The relation $\limsup_{n \to \infty} u_n = a \in \mathbf{R}$ is therefore equivalent to the following: given any $\varepsilon > 0$ there exists an integer n_0 such that, for each $n \geqslant n_0$ we have $u_n \leqslant a + \varepsilon$, and for an infinity of values of n we have $u_n \geqslant a - \varepsilon$. The definition of the upper limit of a sequence may be translated similarly when its value is $+\infty$ or $-\infty$.

Given a sequence (f_n) of real-valued functions defined on a set X, we denote by $\limsup_{n\to\infty} f_n$ (resp. $\liminf_{n\to\infty} f_n$) the real-valued function whose value at any point $x \in X$ is $\limsup_{n\to\infty} f_n(x)$ [resp. $\liminf_{n\to\infty} f_n(x)$]. From (10) and (12) we deduce

$$\text{(16)} \qquad \begin{cases} \limsup\limits_{n\to\infty} f_n = \inf\limits_{n\in\mathbf{N}} (\sup\limits_{m\geqslant n} f_m) = \lim\limits_{n\to\infty} (\sup\limits_{m\geqslant n} f_m), \\ \liminf\limits_{n\to\infty} f_n = \sup\limits_{n\in\mathbf{N}} (\inf\limits_{m\geqslant n} f_m) = \lim\limits_{n\to\infty} (\inf\limits_{m\geqslant n} f_m), \end{cases}$$

the limits being taken in the *product space* $\overline{\mathbf{R}}^X$. The sequence (f_n) has a *limit* in $\overline{\mathbf{R}}^X$ if and only if $\limsup_{n\to\infty} f_n = \liminf_{n\to\infty} f_n$ (Corollary 1 to Theorem 3, and Chapter I, § 7, no. 6, Corollary 1 to Proposition 10).

7. ALGEBRAIC OPERATIONS ON REAL-VALUED FUNCTIONS

Let f and g be two real-valued functions defined on a set X; if the sum $f(x) + g(x)$ [resp. the product $f(x)g(x)$] is defined for all $x \in X$, then we denote by $f + g$ (resp. fg) the real-valued function

$$x \to f(x) + g(x) \qquad [\text{resp. } x \to f(x)g(x)].$$

Again, if $1/f(x)$ is dcfined for all $x \in X$, then $1/f$ denotes the function $x \to 1/f(x)$.

This last function is therefore defined provided f does not take the value 0; when f takes its values in the interval $[0, +\infty]$ (resp. in $[-\infty, 0]$) $1/f(x)$ as everywhere defined by putting $1/0 = +\infty$ (resp. $1/0 = -\infty$); in this case the function $1/f$ is defined.

Suppose that X is filtered by a filter $\mathfrak{F}$, and that $\lim_{\mathfrak{F}} f$ and $\lim_{\mathfrak{F}} g$ exist. If on the one hand the function $f + g$ (resp. fg, $1/g$) is defined, and if on the other hand the expression $\lim_{\mathfrak{F}} f + \lim_{\mathfrak{F}} g$ (resp. $\lim_{\mathfrak{F}} f . \lim_{\mathfrak{F}} g$, $1/\lim_{\mathfrak{F}} f$) has a sense, then $\lim_{\mathfrak{F}} (f + g)$ [resp. $\lim_{\mathfrak{F}} fg$, $\lim_{\mathfrak{F}} (1/f)$] exists and is equal to this expression by reason of the continuity of the function $x + y$ (resp. xy, $1/x$) at points where it is defined.

PROPOSITION 12. *Let f and g be two real-valued functions defined on a set* X, *and let* A *be a non-empty subset of* X.

(i) *We have*

$$\text{(17)} \qquad \sup_{x\in A} (f(x) + g(x)) \leqslant \sup_{x\in A} f(x) + \sup_{x\in A} g(x),$$

$$\text{(18)} \qquad \sup_{x\in A} f(x) + \inf_{x\in A} g(x) \leqslant \sup_{x\in A} (f(x) + g(x)),$$

whenever both sides of these inequalities are defined.

(ii) *If* $f(x)$ *and* $g(x)$ *are* $\geqslant 0$ *for each* $x \in \mathrm{A}$, *then*

$$\sup_{x\in A} (f(x)g(x)) \leqslant \sup_{x\in A} f(x) \sup_{x\in A} g(x), \tag{19}$$

$$\sup_{x\in A} f(x) \inf_{x\in A} g(x) \leqslant \sup_{x\in A} (f(x)g(x)) \tag{20}$$

whenever both sides of these inequalities are defined.

(iii) *If* $f(x) \geqslant 0$ *for all* $x \in \mathrm{A}$, *then*

$$\sup_{x\in A} (1/f(x)) = 1/\inf_{x\in A} f(x) \tag{21}$$

(putting $1/0 = +\infty$).

Let H be any *finite* subset of A. If x_0 is one of the points of H where $f + g$ takes its greatest value, then we have

$$f(x_0) + g(x_0) \leqslant \sup_{x\in H} f(x) + \sup_{x\in H} g(x);$$

on the other hand, if x_1 is one of the points of H where f takes its greatest value, then

$$f(x_1) + g(x_1) \geqslant \sup_{x\in H} f(x) + \inf_{x\in H} g(x);$$

therefore

$$\sup_{x\in H} f(x) + \inf_{x\in H} g(x) \leqslant \sup_{x\in H} (f(x) + g(x)) \leqslant \sup_{x\in H} f(x) + \sup_{x\in H} g(x).$$

The inequalities (17) and (18) follow from this by applying Proposition 5 of no. 4 and Theorem 1 of no. 2. The proofs of the other inequalities are analogous.

COROLLARY 1. *Let* f *be a real-valued function defined on* X, *and let* k *be a real number. Then*

$$\sup_{x\in A} (f(x) + k) = k + \sup_{x\in A} f(x) \tag{22}$$

whenever both sides are defined, and, if $k \geqslant 0$,

$$\sup_{x\in A} (kf(x)) = k.\sup_{x\in A} f(x) \tag{23}$$

whenever both sides are defined.

COROLLARY 2. *Let* f_1 *and* f_2 *be two real-valued functions defined on sets* X_1, X_2 *respectively; then if* A_1, A_2 *are any non-empty subsets of* X_1, X_2 *respectively, we have*

$$\sup_{(x_1,x_2)\in A_1\times A_2} (f_1(x_1) + f_2(x_2)) = \sup_{x_1\in A_1} f_1(x_1) + \sup_{x_2\in A_2} f_2(x_2) \tag{24}$$

whenever both sides are defined; and if f_1, f_2 *are* $\geqslant 0$ *in* A_1, A_2 *respectively we have*

$$(25)\qquad \sup_{(x_1,x_2)\in A_1\times A_2} (f_1(x_1)f_2(x_2)) = \sup_{x_1\in A_1} f_1(x_1) \sup_{x_2\in A_2} f_2(x_2),$$

whenever both sides are defined.

This is an immediate consequence of the preceding corollary and Proposition 9 of no. 4.

In particular, if A and B are two subsets of $\overline{\mathbf{R}}$ such that the set $A + B$ of sums $x + y$ $(x \in A, y \in B)$ is defined, we have

$$(26)\qquad \sup (A + B) = \sup A + \sup B$$

if the right-hand side is defined. Again, if A and B are two subsets of $[0, +\infty]$, we have

$$(27)\qquad \sup AB = \sup A . \sup B$$

whenever both sides are defined.

PROPOSITION 13. *Let* f *and* g *be two real-valued functions defined on a filtered set* X.

(i) *We have*

$$(28)\qquad \limsup (f + g) \leqslant \limsup f + \limsup g,$$

$$(29)\qquad \limsup f + \liminf g \leqslant \limsup (f + g)$$

whenever both sides of these inequalities are defined.

(ii) *If* f *and* g *are* $\geqslant 0$ *on* X, *we have*

$$(30)\qquad \limsup fg \leqslant (\limsup f)(\limsup g),$$

$$(31)\qquad (\limsup f)(\liminf g) \leqslant \limsup fg$$

whenever both sides of these inequalities are defined.

(iii) *If* $f \geqslant 0$ *on* X, *then*

$$(32)\qquad \limsup (1/f) = 1/(\liminf f)$$

(putting $1/0 = +\infty$).

These relations are consequences of Proposition 12 and relations (10)

COROLLARY 1. *Let* f *and* g *be two real-valued functions defined on a filtered set* X. *If* $\lim g$ *exists, then*

$$(33)\qquad \limsup (f + g) = \limsup f + \lim g$$

whenever both sides are defined, and

$$\text{(34)} \qquad \limsup fg = (\limsup f)(\lim g)$$

whenever both sides are defined and f *and* g *are* $\geqslant 0$.

COROLLARY 2. *Let* f *and* g *be two real-valued functions defined on a filtered set* X. *If* $\lim f = +\infty$ *and* $\liminf g > -\infty$ *and* $f+g$ *is defined, then* $\lim (f+g) = +\infty$. *If* $\lim f = +\infty$ *and* $\liminf g > 0$ *and* fg *is defined, then* $\lim fg = +\infty$.

6. CONTINUOUS AND SEMI-CONTINUOUS REAL-VALUED FUNCTIONS

1. CONTINUOUS REAL-VALUED FUNCTIONS

Besides the general properties of continuous functions with values in an arbitrary topological space (Chapter I, § 2), continuous real-valued functions have the following two fundamental properties:

THEOREM 1 (Weierstrass). *Let* f *be a continuous real-valued function defined on a non-empty quasi-compact space* X. *Then there is at least one point* $a \in \mathrm{X}$ *such that* $f(a) = \sup_{x \in \mathrm{X}} (f(x))$, *and at least one point* $b \in \mathrm{X}$ *such that* $f(b) = \inf_{x \in \mathrm{X}} f(x)$. For $f(\mathrm{X})$ is compact (Chapter I, § 9, no. 4, Theorem 2) and therefore closed in $\overline{\mathbf{R}}$; hence $f(\mathrm{X})$ contains its bounds.

This theorem is often stated in the form that *a continuous real-valued function on a non-empty quasi-compact space attains its bounds.*

COROLLARY. *If a real-valued function defined on a non-empty quasi-compact space* X *is continuous and finite on* X, *then it is bounded in* X.

THEOREM 2 (Bolzano). *Let* f *be a continuous real-valued function defined on a connected space* X. *If* a *and* b *are any two points of* X, *and if* α *is a real number belonging to the closed interval whose end-points are* $f(a)$ *and* $f(b)$, *then there is at least one point* $x \in \mathrm{X}$ *such that* $f(x) = \alpha$.

For $f(\mathrm{X})$ is connected (Chapter I, § 11, no. 2, Proposition 4) and is therefore an interval of $\overline{\mathbf{R}}$ (§ 4, no. 2, Proposition 5); hence it contains the closed interval with end-points $f(a)$ and $f(b)$.

This theorem is often expressed in the form that *a continuous real-valued function on a connected space cannot pass from one value to another without passing through all intermediate values.*

This property is by no means characteristic of continuous functions; there are examples of functions defined on a connected space and *discontinuous at every point* which have this property (Exercise 2).

2. SEMI-CONTINUOUS FUNCTIONS

Let f be a real-valued function defined on a topological space X. For f to be continuous at a point $a \in X$ it is necessary and sufficient that: (i) given any real number $h < f(a)$, there exists a neighbourhood V of a such that at each point $x \in V$ we have $h < f(x)$; (ii) given any real number $k > f(a)$, there exists a neighbourhood W of a such that at each point $x \in W$ we have $k > f(x)$.

Functions which satisfy *only one* of these conditions play an important part in analysis. To be precise, we make the following definition:

DEFINITION 1. *A real-valued function f, defined on a topological space* X, *is said to be lower semi-continuous* (resp. *upper semi-continuous*) *at a point* $a \in X$, *if for each* $h < f(a)$ [resp. *each* $k > f(a)$] *there is a neighbourhood* V *of a such that* $h < f(x)$ [resp. $k > f(x)$] *for each* $x \in V$.

A real-valued function is said to be lower semi-continuous (resp. *upper semi-continuous*) *on* X *if it is lower semi-continuous* (resp. *upper semi-continuous*) *at every point of* X.

A real-valued function f is therefore *continuous* at a point a if and only if it is *both upper and lower semi-continuous* at a.

If f is lower semi-continuous at a point, then $-f$ is upper semi-continuous at this point, and conversely; hence we may restrict ourselves in what follows, to considering properties of *lower* semi-continuous functions.

It is clear that a function which is lower semi-continuous on X is also lower semi-continuous on every *subspace* of X.

Examples. 1) If f has a *relative minimum* at a point a, that is to say if there is a neighbourhood V of a such that, for each $x \in V$, we have $f(a) \leqslant f(x)$, then f is lower semi-continuous at a. In particular, if $f(a) = -\infty$, f is lower semi-continuous at a.

2) Define a real-valued function f on **R** by putting $f(x) = 0$ if x is irrational, and $f(x) = 1/q$ if x is rational and equal to the irreducible fraction p/q $(q > 0)$. For each integer $n > 0$, the set of rational numbers p/q such that $q < n$ is closed, and its points are isolated; hence for every irrational x there is a neighbourhood V of x such that $f(y) \leqslant 1/n$ for all $y \in V$, which shows that f is continuous at x; and on the other hand, f has a *relative maximum* at every rational point x. Hence f is upper semi-continuous on **R**.

The condition for f to be lower semi-continuous at a may be expressed by saying that, for each $h < f(a)$, the set $\overset{-1}{f}(]h, +\infty])$ must be a *neighbourhood* of a.

It is enough that this condition should be satisfied for an *increasing sequence* (h_n) of real numbers $< f(a)$ which tend to $f(a)$.

Endow $\overline{\mathbf{R}}$ with the topology in which the open sets are $\emptyset$ and all *open intervals* of $\overline{\mathbf{R}}$ *unbounded on the right* (that is, all intervals $]a, +\infty[$ for finite a, and the interval $[-\infty, +\infty] =]\leftarrow, \rightarrow[$). Then the real-valued function f is lower semi-continuous at a if and only if it is *continuous* at a when considered as a mapping into $\overline{\mathbf{R}}$ endowed with this topology.

PROPOSITION 1. *A real-valued function f on a topological space X is lower semi-continuous if and only if, for each finite real number k, $\overset{-1}{f}(]k, +\infty])$* [the set of all $x \in X$ such that $f(x) > k$] *is an open set in* X [or, equivalently, $\overset{-1}{f}([-\infty, k])$ *is a closed set in* X].

For this condition shows that $\overset{-1}{f}(]k, +\infty])$ is a neighbourhood of each of its points.

For f to be lower semi-continuous on X it is sufficient that $f^{-1}(]k, +\infty])$ is open in X for all real numbers k belonging to a dense subset of $\mathbf{R}$.

COROLLARY. *A subset* A *of a topological space* X *is open* (resp. *closed*) *in* X *if and only if its characteristic function* (*) φ_A *is lower* (resp. *upper*) *semi-continuous on* X.

For $\overset{-1}{\varphi}_A(]k, +\infty])$ is empty for $k \geqslant 1$, is equal to A for $0 \leqslant k < 1$ and is equal to X for $k < 0$.

THEOREM 3. *Let f be a lower semi-continuous function on a non-empty quasi-compact space* X. *Then there is at least one point $a \in E$ such that* $f(a) = \inf_{x \in X} f(x)$ (in other words, f *attains its greatest lower bound in* X).

For each $k \in f(X)$, consider the set $A_k = \overset{-1}{f}([-\infty, k])$. These sets are non-empty and form a *filter base* on X; since they are *closed* by Proposition 1, they have at least one common point a [axiom (C″) for quasi-compact spaces]. For each $x \in X$ we have therefore $f(a) \leqslant f(x)$, and the theorem follows.

(*) We recall (*Set Theory*, Chapter III, § 5, no. 5) that the *characteristic function* φ_A of a subset A of a set X is the function defined on X such that $\varphi_A(x) = 1$ for all $x \in A$, and $\varphi_A(x) = 0$ for all $x \in \complement A$.

COROLLARY. *Let f be a lower semi-continuous function on a non-empty quasi-compact space* X. *If $f(x) > -\infty$ for all $x \in X$, then f is bounded below in* X.

Note that this theorem and the corresponding theorem for upper semi-continuous functions include Weierstrass's theorem as a particular case (no. 1, Theorem 1).

PROPOSITION 2. *Let f and g be two real-valued functions, lower semi-continuous at a point $a \in X$. Then the functions* $\inf(f, g)$ *and* $\sup(f, g)$ *are lower semi-continuous at a; so is $f + g$ whenever it is defined, and so is fg if f and g are $\geqslant 0$ and the product fg is defined.*

We give the proof for $f + g$; the argument is analogous in the other cases. The result is clear if either $f(a)$ or $g(a)$ is equal to $-\infty$; if not, then $f(a) + g(a) > -\infty$. Every finite number $h < f(a) + g(a)$ can be written as $h = r + s$, where $r < f(a)$ and $s < g(a)$ are finite [it is enough to take s so that $h - f(a) < s < g(a)$]; by hypothesis, there is a neighbourhood V of a such that, for each $x \in V$, we have $r < f(x)$, and a neighbourhood W such that for each $x \in W$ we have $s < g(x)$; therefore $h = r + s < f(x) + g(x)$ for all points x of the neighbourhood $V \cap W$.

In the same way we see that, if f is lower semi-continuous at a point a, and if $f \geqslant 0$, then $1/f$ is upper semi-continuous at a.

THEOREM 4. *The upper envelope of a family (f_ι) of functions which are lower semi-continuous at a point $a \in X$ is lower semi-continuous at a.*

Let g be the upper envelope. For each $h < g(a)$ there is an index ι such that $h < f_\iota(a) \leqslant g(a)$, and a neighbourhood V of a such that $h < f_\iota(x)$ for all $x \in V$; hence *a fortiori* $h < g(x)$ for all $x \in V$.

It follows from Proposition 2 that the *lower* envelope of a *finite* number of lower semi-continuous functions is again lower semi-continuous; but this is not in general true for the lower envelope of an *infinite* family of lower semi-continuous functions. For example, if r is any rational number, let f_r denote the function which is equal to 0 at r and equal to 1 for all real numbers $x \neq r$; the lower envelope of the f_r is the function g which is equal to 0 for every rational number and 1 for every irrational number ("Dirichlet's function"), and this function is not lower semi-continuous at irrational points.

COROLLARY. *The upper envelope of a family of continuous real-valued functions on a space* X *is lower semi-continuous on* X.

In Chapter IX, § 1, no. 6, Proposition 5, we shall show that the *converse* of this proposition is true if X is *uniformizable* (and only in this case): every lower semi-continuous function on a uniformizable space is the upper envelope of a family of continuous functions.

PROPOSITION 3. *A real-valued function f, defined on a topological space X, is lower semi-continuous at a point $a \in X$ if and only if* $\liminf_{x \to a} f(x) = f(a)$ [or, equivalently, if and only if $\liminf_{x \to a} f(x) \geqslant f(a)$].

The condition is *necessary.* For, given any $h < f(a)$, there is a neighbourhood V of a such that $h < f(x)$ for all $x \in V$; therefore

$$h \leqslant \inf_{x \in V} f(x) \leqslant \liminf_{x \to a} f(x)$$

(§ 5, no. 6, formulae (12)), and so $f(a) \leqslant \liminf_{x \to a} f(x)$. The condition is *sufficient*; for if it is satisfied, then for each $h < f(a)$ there is a neighbourhood V of a such that $h \leqslant \inf_{x \in V} f(x)$, and therefore f is lower semi-continuous at a.

PROPOSITION 4. *Let f be any real-valued function defined on a dense subset A of a topological space X. If, for each $x \in X$, we put* $g(x) = \liminf_{y \to x, y \in A} f(y)$, *then g is lower semi-continuous on X.*

For, given any $h < g(x)$, there is an *open* neighbourhood V of x such that, for all $z \in V \cap A$, we have $h < f(z)$; now V is a neighbourhood of each of its points y; thus we have $\liminf_{z \to y, z \in A} f(z) = g(y) \geqslant h$ for all $y \in V$, and the result follows.

The function g is called the *lower semi-continuous regularization* of f. We define the *upper semi-continuous regularization* of f similarly.

We may also define g as the *greatest* of the lower semi-continuous functions φ on X which are such that $\varphi(x) \leqslant f(x)$ for all $x \in A$. If f is *lower semi-continuous* on A, then g is an *extension* of f to X, by Proposition 3.

7. INFINITE SUMS AND PRODUCTS OF REAL NUMBERS

Since every point of $\mathbf{R}$ has a *countable* fundamental system of neighbourhoods (§ 1, no. 4, Corollary to Proposition 3), it follows that a family (x_ι) of *finite* real numbers is summable in $\mathbf{R}$ only if the set of indices ι such that $x_\iota \neq 0$ is *countable* (Chapter III, § 5, no. 2, Corollary to Proposition 1). The study of summable families in $\mathbf{R}$ thus reduces essentially to the study of summable *sequences.* However, we shall later have to consider uncountable families (x_ι) of finite real numbers, whose terms are functions of a parameter t; it can happen that this family is summable for all t, but that the (countable) set of indices ι such that $x_\iota \neq 0$ depends on t. For this reason we shall not impose any hypothesis on the power of the index set in what follows.

1. FAMILIES OF POSITIVE FINITE NUMBERS SUMMABLE IN R

THEOREM 1. *A family* (x_ι) *of finite real numbers* $\geqslant 0$ *is summable in* **R** *if and only if the set of partial finite sums of the family is bounded above in* **R**. *If so, the least upper bound of this set is the sum of the family* (x_ι).

For each finite subset H of the index set I, let $s_{\mathrm{H}} = \sum_{\iota \in \mathrm{H}} x_\iota$; since the x_ι are $\geqslant 0$, the relation $\mathrm{H} \subset \mathrm{H}'$ implies $s_{\mathrm{H}} \leqslant s_{\mathrm{H}'}$. In other words, the mapping $\mathrm{H} \to s_{\mathrm{H}}$ is an *increasing* function on the directed set $\mathfrak{F}(\mathrm{I})$ of finite subsets of I; therefore (§ 5, no. 2, Corollary to Theorem 2) it has a finite limit if and only if it is *bounded above*.

Remark. Let (H_λ) be a family of finite subsets of I such that, for each finite subset H of I, there is an index λ such that $\mathrm{H} \subset \mathrm{H}_\lambda$; then (x_ι) is summable if and only if the family of the s_{H_λ} is *bounded above* in **R**. In particular, let (x_n) be a sequence of finite real numbers $\geqslant 0$, and for each integer n let $s_n = \sum_{p=0}^{n} x_p$; then the sequence (x_n) is summable in **R** if and only if, for *one sequence* of strictly increasing integers (n_k), the partial sequence (s_{n_k}) is *bounded above* in **R**.

Examples. 1) For each number q such that $0 \leqslant q < 1$, the sequence (q^n) ("geometric progression of ratio q") is summable in **R**, since $s_n = \dfrac{1 - q^{n+1}}{1 - q} \leqslant \dfrac{1}{1-q}$; the sum of this sequence is $\lim_{n \to \infty} s_n = \dfrac{1}{1-q}$.

2) Let a and b be two numbers such that $0 \leqslant a < 1$ and $0 \leqslant b < 1$; then the family $(a^m b^n)_{(m,n) \in \mathbf{N} \times \mathbf{N}}$ is summable in **R**. For each finite subset of $\mathbf{N} \times \mathbf{N}$ is contained in a subset of the form $[0, p] \times [0, p]$, and we have

$$\sum_{m=0}^{p} \sum_{n=0}^{p} a^m b^n = \left(\sum_{m=0}^{p} a^m \right) \left(\sum_{n=0}^{p} b^n \right) = \frac{1 - a^{p+1}}{1 - a} \cdot \frac{1 - b^{p+1}}{1 - b} \leqslant \frac{1}{(1-a)(1-b)}.$$

3) For each integer $p > 1$, the sequence (n^{-p}) $(n > 0)$ is summable, since

$$s_{2^{n+1}} - s_{2^n} = \sum_{k=1}^{2^n} (2^n + k)^{-p} < 2^n . (2^n)^{-p}$$

and therefore, adding these inequalities together,

$$s_{2^n} < \frac{1}{1 - 2^{1-p}}.$$

4) The sequence $(1/n)$ $(n > 0)$ is not summable in $\mathbf{R}$, since

$$s_{2^{n+1}} - s_{2^n} = \sum_{k=1}^{2^n} \frac{1}{2^n + k} > \frac{2^n}{2^{n+1}} = \frac{1}{2}$$

and therefore, adding these inequalities together,

$$s_{2^n} > n/2$$

so that the criterion of Theorem 1 is not satisfied.

5) Let (I_n) be a sequence of mutually disjoint non-empty open intervals, all contained in an interval of finite length l. The sum of the lengths of a finite number of intervals of this family is $\leqslant l$ (§ 1, no. 5) and therefore the family of lengths of the I_n is summable in $\mathbf{R}$, and its sum is $\leqslant l$.

THEOREM 2 (Principle of comparison). *Let $(x_\iota)_{\iota \in \mathrm{I}}$ and $(y_\iota)_{\iota \in \mathrm{I}}$ be two families of finite real numbers $\geqslant 0$, such that $x_\iota \leqslant y_\iota$ for all ι. If (y_ι) is summable in $\mathbf{R}$ then so is (x_ι), and we have $\sum_\iota x_\iota \leqslant \sum_\iota y_\iota$; if in addition there is an index $\varkappa$ such that $x_\varkappa < y_\varkappa$, then $\sum_\iota x_\iota < \sum_\iota y_\iota$.*

The hypothesis implies that,

$$\sum_{\iota \in \mathrm{H}} x_\iota \leqslant \sum_{\iota \in \mathrm{H}} y_\iota,$$

for every finite subset H of I and the first part of the theorem follows from this; the inequality relating the sums follows from the principle of extension of inequalities (§ 5, no. 2, Theorem 1). If $x_\varkappa < y_\varkappa$, then

$$\sum_\iota x_\iota = x_\varkappa + \sum_{\iota \neq \varkappa} x_\iota < y_\varkappa + \sum_{\iota \neq \varkappa} y_\iota = \sum_\iota y_\iota.$$

This theorem provides the most commonly used criterion for deciding whether or not a sequence (x_n) of real numbers $\geqslant 0$ is summable in $\mathbf{R}$: we try to *compare* the given sequence with a simpler sequence (y_n) of which we already know whether it is summable or not. If there exists a finite number $a > 0$ such that $x_n \leqslant a y_n$ for all n from a certain point onwards, and if (y_n) is summable, then (x_n) is summable; if on the other hand there is a finite number $b > 0$ such that $x_n \geqslant b y_n$ for all n from a certain point onwards, and if (y_n) is not summable in $\mathbf{R}$, then (x_n) is not summable in $\mathbf{R}$. We shall see in a later volume how such comparison sequences may be obtained in the cases which arise most frequently.

Examples. 1) Let a be a finite real number > 0, and consider the sequence $\left(\frac{a^n}{n!}\right)$; let n_0 be the smallest integer such that $a < n_0$. Then for each $n \geqslant n_0$ we have

$$\frac{a^n}{n!} \leqslant \frac{a^{n_0}}{n_0!} \cdot \left(\frac{a}{n_0}\right)^{n-n_0};$$

since $q = \frac{a}{n_0} < 1$, the sequence (q^{n-n_0}) is summable, and therefore so is the sequence $\left(\frac{a^n}{n!}\right)$.

2) Let (a_n) be a summable sequence of positive numbers. Since

$$\lim_{n \to \infty} a_n = 0,$$

there exists an integer n_0 such that $a_n \leqslant 1$ whenever $n \geqslant n_0$. Consequently, for each $n \geqslant n_0$ we have $a_n^2 \leqslant a_n$, and therefore the sequence (a_n^2) is summable in **R**. So is the sequence (a_n^p) for each integer $p > 1$.

3) Let a and b be two numbers < 1; then

$$\frac{1}{a^m + b^n} \leqslant \frac{1}{2(\sqrt{a})^m(\sqrt{b})^n}$$

and hence the family $\left(\frac{1}{a^m + b^n}\right)$ is summable in **R**.

COROLLARY. *Let $(x_\iota)_{\iota \in I}$ be a summable family of finite numbers $\geqslant 0$ in* **R**. *If H is any subset of* I, *we have*

$$\sum_{\iota \in H} x_\iota \leqslant \sum_{\iota \in I} x_\iota,$$

and equality holds only if $x_\iota = 0$ for all $\iota \in \complement H$.

2. FAMILIES OF FINITE NUMBERS OF ARBITRARY SIGN SUMMABLE IN R

THEOREM 3. *Let $(x_\iota)_{\iota \in I}$ be a family of finite real numbers; then the following statements are equivalent:*

a) *The family (x_ι) is summable in* **R**.

b) *The family $(|x_\iota|)$ is summable in* **R**.

c) *The set of finite partial sums of the family (x_ι) is bounded in* **R**.

Let I_1 be the set of all $\iota \in I$ such that $x_\iota \geqslant 0$, and I_2 the set of all $\iota \in I$ such that $x_\iota < 0$. The family $(x_\iota)_{\iota \in I}$ [resp. $(|x_\iota|)_{\iota \in I}$] is summable if and only if each of the families $(x_\iota)_{\iota \in I_1}$ and $(x_\iota)_{\iota \in I_2}$ [resp. $(|x_\iota|)_{\iota \in I_1}$ and $(|x_\iota|)_{\iota \in I_2}$] is summable (Chapter III, § 5, no. 3, Propositions 2 and 3). Now it comes to the same thing to say that $(x_\iota)_{\iota \in I_1}$ is summable, or that $(|x_\iota|)_{\iota \in I_1}$ is summable, or that the set of finite partial sums of the family $(x_\iota)_{\iota \in I_1}$ is bounded (no. 1, Theorem 1); and the same is true with I_1 replaced by I_2. The theorem follows immediately.

Theorem 3 shows that the summability in $\mathbf{R}$ of a family of finite real numbers depends only on the summability of the family of their absolute values.

We recall (Chapter III, § 5, no. 5, Proposition 6) that if (x_ι) and (y_ι) are two summable families of finite real numbers, then the family $(x_\iota + y_\iota)$ is summable and $\sum_\iota (x_\iota + y_\iota) = \sum_\iota x_\iota + \sum_\iota y_\iota$. Moreover, if (x_ι) is a summable family of finite real numbers and a is any finite real number, then the family (ax_ι) is summable in $\mathbf{R}$, and we have $\sum_\iota ax_\iota = a . \sum_\iota x_\iota$.

3. PRODUCT OF TWO INFINITE SUME

PROPOSITION 1. *If the families $(x_\lambda)_{\lambda \in L}$ and $(y_\mu)_{\mu \in M}$ of finite real numbers are summable in $\mathbf{R}$, then so is the family $(x_\lambda y_\mu)_{(\lambda, \mu) \in L \times M}$ and we have*

$$(1) \qquad \sum_{(\lambda, \mu) \in L \times M} x_\lambda y_\mu = \Big(\sum_{\lambda \in L} x_\lambda\Big)\Big(\sum_{\mu \in M} y_\mu\Big).$$

Every finite subset of $L \times M$ is contained in a finite subset of the form $H \times K$, where H is a finite subset of L and K is a finite subset of M. By hypothesis, there exists a number $a > 0$ such that $\sum_{\lambda \in H} |x_\lambda| \leqslant a$ and $\sum_{\mu \in K} |y_\mu| \leqslant a$ for all finite subsets H and K of L and M respectively; therefore

$$\sum_{(\lambda, \mu) \in H \times K} |x_\lambda y_\mu| = \Big(\sum_{\lambda \in H} |x_\lambda|\Big)\Big(\sum_{\mu \in K} |y_\mu|\Big) \leqslant a^2,$$

and this shows that the family $(x_\lambda y_\mu)$ is summable in $\mathbf{R}$, by Theorems 1 and 3. By associativity we can write [Chapter III, § 5, no. 3, formula (2)]

$$\sum_{(\lambda, \mu) \in L \times M} x_\lambda y_\mu = \sum_{\lambda \in L}\Big(\sum_{\mu \in M} x_\lambda y_\mu\Big) = \sum_{\lambda \in L} x_\lambda\Big(\sum_{\mu \in M} y_\mu\Big) = \Big(\sum_{\lambda \in L} x_\lambda\Big)\Big(\sum_{\mu \in M} y_\mu\Big);$$

hence the result.

4. FAMILIES MULTIPLIABLE IN R*

In the multiplicative group $\mathbf{R}^*$ of finite non-zero real numbers, a family $(x_\iota)_{\iota \in I}$ can be multipliable only if $\lim x_\iota = 1$ with respect to the filter of complements of finite subsets of I (Chapter III, § 5, no. 2, Proposition 1). In particular there can be only a *finite* number of indices ι such that $x_\iota < 0$. We may therefore limit ourselves to considering only families x_ι

all of whose terms are *strictly positive*; it is then convenient to put $x_\iota = 1 + u_\iota$, where the u_ι are subject to the inequalities $-1 < u_\iota < +\infty$ for all ι. Since each point of $\mathbf{R}^*$ has a countable fundamental system of neighbourhoods, the set of indices ι such that $u_\iota \neq 0$ is *countable*, if the family $(1 + u_\iota)$ is multipliable in $\mathbf{R}^*$.

THEOREM 4. *The family* $(1 + u_\iota)$ *is multipliable in* $\mathbf{R}^*$ *if and only if the family* (u_ι) *is summable in* $\mathbf{R}$.

Lemma. (i) *If* $(a_i)_{1 \leqslant i \leqslant p}$ *is a finite sequence of numbers* > 0, *then*

$$\prod_{i=1}^{p}(1 + a_i) \geqslant 1 + \sum_{i=1}^{p} a_i. \tag{2}$$

(ii) *If also* $a_i < 1$ *for each* i, *then*

$$\prod_{i=1}^{p}(1 - a_i) \geqslant 1 - \sum_{i=1}^{p} a_i. \tag{3}$$

These inequalities are clear if $p = 1$, and are proved by induction on p. If

$$\prod_{i=1}^{p-1}(1 + a_i) \geqslant 1 + \sum_{i=1}^{p-1} a_i,$$

we have

$$\prod_{i=1}^{p}(1 + a_\iota) \geqslant (1 + a_p)\left(1 + \sum_{i=1}^{p-1} a_i\right) = 1 + \sum_{i=1}^{p} a_i + a_p \cdot \sum_{i=1}^{p-1} a_i \geqslant 1 + \sum_{i=1}^{p} a_i.$$

Again, if

$$\prod_{i=1}^{p-1}(1 - a_i) \geqslant 1 - \sum_{i=1}^{p-1} a_i,$$

we have

$$\prod_{i=1}^{p}(1 - a_i) \geqslant (1 - a_p)\left(1 - \sum_{i=1}^{p-1} a_i\right) = 1 - \sum_{i=1}^{p} a_i + a_p \cdot \sum_{i=1}^{p-1} a_i \geqslant 1 - \sum_{i=1}^{p} a_i.$$

Having proved the lemma, we note that if the family $(1 + u_\iota)$ is multipliable then so are the families $(1 + u_\iota^+)$ and $(1 - u_\iota^-)$, since $\mathbf{R}^*$ is a complete group (Chapter III, § 5, no. 3, Proposition 2); and conversely, if the families $(1 + u_\iota^+)$ and $(1 - u_\iota^-)$ are multipliable, then so is $(1 + u_\iota)$ (Chapter III, § 5, no. 3, Proposition 3). We need therefore consider only the cases in which all the u_ι are $\geqslant 0$ and in which they are all $\geqslant 0$.

Suppose first that $u_\iota \geqslant 0$ for all ι. If the family $(1 + u_\iota)$ is multipliable, then for each $\varepsilon > 0$ there is a finite subset J of the index set in I such that, for each finite subset H of I disjoint from J, we have

$1 \leqslant \prod_{\iota \in H} (1 + u_\iota) \leqslant 1 + \varepsilon$; by (2) it follows that $\sum_{\iota \in H} u_\iota \leqslant \varepsilon$, which shows that (u_ι) is summable in **R** by virtue of Cauchy's criterion (Chapter III, § 5, no. 2, Theorem 1).

Conversely, suppose that (u_ι) is summable in **R**. For each ε such that $0 < \varepsilon < 1$, there is a finite subset J of I such that, for each finite subset H of I disjoint from J, we have $0 \leqslant \sum_{\iota \in H} u_\iota \leqslant \varepsilon$. By (3), we have therefore $\prod_{\iota \in H} (1 - u_\iota) \geqslant 1 - \varepsilon$; but since $1 + u \leqslant \frac{1}{1 - u}$ for every number u such that $0 \leqslant u < 1$, it follows that

$$1 \leqslant \prod_{\iota \in H} (1 + u_\iota) \leqslant \frac{1}{1 - \varepsilon},$$

and this shows that $(1 + u_\iota)$ is multipliable (Cauchy's criterion).

The proof is similar when all the u_ι are $\leqslant 0$. To show that (u_ι) is summable if $(1 + u_\iota)$ is multipliable, use formula (2) and the inequality $1 - u \leqslant \frac{1}{1 + u}$ $(0 \leqslant u < 1)$; to show that $(1 + u_\iota)$ is multipliable if (u_ι) is summable, use formula (3).

In Chapter V (§ 4), the topological study of the group $\mathbf{R}^*$ will enable us to give another criterion for multipliability of a family in $\mathbf{R}^*$ with the help of the logarithmic function; in a later volume we shall establish the equivalence of this criterion with the one above, using the differential properties of the logarithm.

5. SUMMABLE FAMILIES AND MULTIPLIABLE FAMILIES IN $\overline{\mathbf{R}}$

On the interval $[0, +\infty]$ of $\overline{\mathbf{R}}$, addition is an associative and commutative law of composition (§ 4, no. 3); hence the notion of a *summable* family of numbers in this interval is defined (Chapter III, § 5, no. 1, Remark 3).

PROPOSITION 2. *Every family* (x_ι) *of real numbers* $\geqslant 0$ *is summable in* $\overline{\mathbf{R}}$.

For the mapping $H \to s_H$ of the directed set $\mathfrak{F}(I)$ into $\overline{\mathbf{R}}$ is *increasing*; hence (§ 5, no. 2, Theorem 2) has a limit.

The same reasoning shows that every family of real numbers $\leqslant 0$ is summable in $\overline{\mathbf{R}}$.

Similarly, multiplication is an associative and commutative law of composition on each of the intervals $[0, 1]$ and $[1, +\infty]$ of $\overline{\mathbf{R}}$, and therefore the notion of a multipliable family is defined on each of these intervals.

Proposition 3. *Every family* $(1 + u_\iota)$ [resp. $(1 - u_\iota)$] *of numbers* $\geqslant 1$ (resp. $\geqslant 0$ *and* $\leqslant 1$) *is multipliable in* $\overline{\mathbf{R}}$.

Same proof as for Proposition 2.

Corollary. *The product* $\prod_\iota (1 + u_\iota)$ [resp. $\prod_\iota (1 - u_\iota)$) *of numbers* $\geqslant 1$ (resp. > 0 *and* $\leqslant 1$) *is equal to* $+\infty$ (resp. 0) *if and only if* $\sum_\iota u_\iota = +\infty$.

For if $\sum_\iota u_\iota$ is finite then $\prod_\iota (1 + u_\iota)$ and $\prod_\iota (1 - u_\iota)$ are in $\mathbf{R}^*$, and conversely (no. 4, Theorem 4).

Remark. The theorem of associativity (Chapter III, § 5, no. 3, Theorem 2) remains valid when G is replaced by $\overline{\mathbf{R}}$ and the x_ι are assumed to be $\geqslant 0$. This is clear if $\sum_{\iota \in I} x_\iota$ is finite. Suppose, on the contrary, that $\sum_{\iota \in I} x_\iota = +\infty$. Then for each finite $a > 0$ there is a finite subset H of I such that $\sum_{\iota \in H} x_\iota \geqslant a$. Let K be a finite subset of L such that $H \subset \bigcup_{\lambda \in K} I_\lambda$; then since $s_\lambda \geqslant \sum_{\iota \in I_\lambda \cap H} x_\iota$ for all $\lambda \in K$, we have

$$\sum_{\lambda \in K} s_\lambda \geqslant \sum_{\iota \in H} x_\iota \geqslant a,$$

which shows that $\sum_{\lambda \in L} s_\lambda = +\infty$. We leave to the reader the task of formulating the analogous proposition for multipliable families of numbers in $[0, 1]$ or in $[1, +\infty]$.

6. INFINITE SERIES AND INFINITE PRODUCTS OF REAL NUMBERS

A series of finite real numbers is simply said to be *convergent* if it is *convergent in* **R**.

Definition 1. *A series of finite real numbers is said to be absolutely convergent if the series of absolute values of its terms is convergent.*

Proposition 4. *A series of finite real numbers is commutatively convergent if and only if it is absolutely convergent.*

This follows from Chapter III, § 5, no. 7, Proposition 9 and from Theorem 3 of no. 2.

In other words, if (u_n) is a sequence of finite real numbers, it comes to the same thing to say that the *series* whose general term is u_n is *commutatively convergent*, or that it is *absolutely convergent*, or that the *sequence* (u_n)

is *summable in* **R**. All the properties of summable families proved in Chapter III, § 5 therefore apply to absolutely convergent series. In particular, if the series whose general term is u_n is absolutely convergent, then the sum $\sum_{n \in H} u_n$ exists for all subsets H of **N**; and, if (H_p) is a partition of **N**, we have

$$\sum_{n=0}^{\infty} u_n = \sum_{p} \left(\sum_{n \in H_p} u_n \right)$$

(*associativity* of absolutely convergent series).

As we have already remarked in Chapter III, § 5, a series of real numbers can be convergent without being commutatively convergent, that is, without being absolutely convergent.

Example. Alternating series. A series defined by a sequence (u_n) of finite real numbers is called *alternating* if $u_n = (-1)^n v_n$, where $v_n \geqslant 0$ for all n. Let us show that a *sufficient* condition for the convergence of such a series is that the *sequence* (v_n) *decreases and has* 0 *as its limit.* For if s_n denotes

$$\sum_{p=0}^{n} u_p,$$

the hypothesis that (v_n) is decreasing implies that

$$s_{2n+1} \leqslant s_{2n+3} \leqslant s_{2n+2} \leqslant s_{2n}$$

for all $n \geqslant 0$. The sequence (s_{2n}) [resp. (s_{2n+1})] is decreasing and bounded below (resp. increasing and bounded above) and therefore has a finite limit a (resp. b), and $b \leqslant a$; since

$$a - b = \lim_{n \to \infty} (s_{2n} - s_{2n+1}) = \lim_{n \to \infty} v_{2n+1} = 0,$$

the assertion is proved.

If we take for example $v_n = 1/n$, the conditions above are satisfied, and therefore the series whose general term is $(-1)^n/n$ (the "alternating harmonic series") is convergent. We have seen in no. 1 that the series whose general term is $1/n$ (the "harmonic series") is not convergent, and thus the alternating harmonic series is not absolutely convergent.

We recall (Chapter III, § 5, no. 6, Proposition 7) that, if (u_n) and (v_n) are two convergent series of finite real numbers, then the series $(u_n + v_n)$ is convergent, and

$$\mathop{\mathrm{S}}_{n=0}^{\infty} (u_n + v_n) = \mathop{\mathrm{S}}_{n=0}^{\infty} u_n + \mathop{\mathrm{S}}_{n=0}^{\infty} v_n;$$

also, if the series (u_n) is convergent then the series (au_n) is convergent for all finite real numbers a, and $\mathop{\mathrm{S}}_{n=0}^{\infty} au_n = a \,.\, \mathop{\mathrm{S}}_{n=0}^{\infty} u_n.$

Finally, if the series (u_n) and (v_n) are convergent, and if $u_n \leqslant v_n$ for all n, we have $\sum_{n=0}^{\infty} u_n \leqslant \sum_{n=0}^{\infty} v_n$ by the principle of extension of inequalities (§ 5, no. 2, Theorem 1).

It should be noted that, if we suppose the series (v_n) to be convergent but not absolutely convergent, and if $|u_n| \leqslant |v_n|$ for each n, we can by no means infer that the series (u_n) is convergent, as is seen by taking $u_n = |v_n|$.

An infinite product of finite non-zero real numbers is said simply to be *convergent* if it is *convergent in* $\mathbf{R}^*$; its value is then a *finite non-zero* real number.

Definition 2. *An infinite product whose general factor is* $1 + u_n$ *is said to be absolutely convergent if the product whose general factor is* $1 + |u_n|$ *is convergent.*

Proposition 5. *An infinite product of finite real numbers is commutatively convergent if and only if it is absolutely convergent.*

This follows from Chapter III, § 5, no. 7, Proposition 9, and from Theorem 4 above.

Moreover, the product whose general factor is $1 + u_n$ is *absolutely convergent* if and only if the series whose general term is u_n is *absolutely convergent.*

A product of non-zero real numbers can be convergent without being commutatively convergent, i.e. without being absolutely convergent.

Example. If we take $u_{2n-1} = -1/n$ and $u_{2n} = 1/n$ for $n \geqslant 2$, the product $(1 + u_n)$ is not absolutely convergent, since the series (u_n) is not absolutely convergent; but since

$$\prod_{p=3}^{n} (1 + u_p) = \prod_{p=2}^{n} \left(1 - \frac{1}{p^2}\right),$$

and

$$\prod_{p=3}^{2n+1} (1 + u_p) = \left(1 - \frac{1}{n+1}\right) \prod_{p=2}^{n} \left(1 - \frac{1}{p^2}\right),$$

it follows from Theorem 4 that the product is convergent and that its value is

$$\prod_{n=2}^{\infty} \left(1 - \frac{1}{n^2}\right).$$

Moreover, it should be observed that the *convergence* of the series whose general term is u_n *is neither necessary nor sufficient* for the *convergence* of the product whose general factor is $1 + u_n$ (see Exercises 21 and 22).

8. USUAL EXPANSIONS OF REAL NUMBERS; THE POWER OF R

1. APPROXIMATIONS TO A REAL NUMBER

DEFINITION 1. *Given a number* $\varepsilon > 0$, *a real number* r *is said to be an approximation to within* ε *to a real number* x, *if* $|x - r| \leqslant \varepsilon$; r *is said to be an approximation by defect if* $r \leqslant x$, *by excess if* $r \geqslant x$.

Let A be a *dense* subset of **R**. For each $x \in \mathbf{R}$ and each $\varepsilon > 0$ there is an approximation by defect (resp. excess) to x to within ε belonging to A, since the interval $]x - \varepsilon, x[$ (resp. $]x, x + \varepsilon[$) contains at least one point of A. If we now consider a *given strictly decreasing* sequence (ε_n) of numbers > 0, *tending to* 0, and if r_n is an approximation to x to within ε_n, then the sequence (r_n) has x as *limit* as n tends to infinity.

In the case where A is a *subgroup* of the additive group **R**, and we restrict the ε_n to *belong* to A, we can define canonically for each $x \in \mathbf{R}$ a sequence (r_n) of approximations to x by defect, belonging to A.

For by Archimedes' axiom (§ 2, no. 1, Theorem 1), the set of integers p such that $p\varepsilon_n \leqslant x$ has a *greatest element* p_n; in other words, there is a unique integer p_n such that

$$(1) \qquad p_n\varepsilon_n \leqslant x < (p_n + 1)\varepsilon_n.$$

Since $|x - p_n\varepsilon_n| \leqslant \varepsilon_n$, it follows that $p_n\varepsilon_n$ is an approximation to x to within ε_n by defect, and belongs to A by hypothesis; similarly $(p_n + 1)\varepsilon_n$ is an approximation to x to within ε_n by excess, belonging to A, and the two sequences $(p_n\varepsilon_n)$ and $((p_n + 1)\varepsilon_n)$ have x as their limit.

2. EXPANSIONS OF REAL NUMBERS RELATIVE TO A BASE SEQUENCE

We shall limit ourselves to studying the case where $\varepsilon_n = 1/d_n$, where (d_n) is a strictly increasing sequence of integers such that $d_0 = 1$ and d_n is a *multiple* of d_{n-1} for $n \geqslant 1$. Let $a_n = d_n/d_{n-1}$ $(n \geqslant 1)$: a_n is an integer > 1. In this case, the sequence of approximations by defect $r_n = p_n/d_n$ is *increasing*, for p_n is the largest integer such that $p_n/d_n \leqslant x$;

but we have $$\frac{p_{n-1}}{d_{n-1}} = \frac{p_{n-1}a_n}{d_n} \leqslant x < \frac{p_{n-1} + 1}{d_{n-1}} = \frac{p_{n-1}a_n + a_n}{d_n}$$

so that $a_n p_{n-1} \leqslant p_n < a_n p_{n-1} + a_n$, and therefore $r_{n-1} \leqslant r_n \leqslant x$. Let

(2) $$p_n = a_n p_{n-1} + u_n;$$

then $0 \leqslant u_n - a_n$, which is equivalent to $0 \leqslant u_n \leqslant a_n - 1$, since u_n is an integer. Hence

(3) $$r_n = r_{n-1} + \frac{u_n}{d_n} = p_0 + \sum_{k=1}^{n} \frac{u_k}{d_k},$$

and, since $x = \lim_{n \to \infty} r_n$,

(4) $$x = p_0 + \sum_{n=1}^{\infty} \frac{u_n}{d_n}.$$

The series on the right-hand side of (4), whose sum is x, is called the *expansion of* x *relative to the base sequence* (d_n). All the coefficients u_n are $\geqslant 0$; p_0 is, by definition, the largest integer p such that $p \leqslant x$; it is called the *integral part of* x, and is often denoted by $[x]$.

3. DEFINITION OF A REAL NUMBER BY MEANS OF ITS EXPANSION

Conversely, suppose we are given an integer q_0 and a sequence (v_n) $(n \geqslant 1)$ of integers such that $0 \leqslant v_n \leqslant a_n - 1$; we ask whether there is a number x whose expansion (4) is such that $p_0 = q_0$ and $u_n = v_n$ for all n. If such a number exists it is *unique*, because it is equal to $q_0 + \sum_{n=1}^{\infty} \frac{v_n}{d_n}$.

For each integer $m > 0$ we have (by the principle of comparison)

$$\sum_{n=m+1}^{\infty} \frac{v_n}{d_n} \leqslant \sum_{n=m+1}^{\infty} \frac{a_n - 1}{d_n} = \sum_{n=m+1}^{\infty} \left(\frac{1}{d_{n-1}} - \frac{1}{d_n} \right) = \frac{1}{d_m}$$

and the extreme left-hand and right-hand terms are equal only if $v_n = a_n - 1$ for each $n > m$ (§ 7, no. 1, Theorem 2). Hence the series whose general term is $\frac{v_n}{d_n}$ is convergent; moreover, if $x = q_0 + \sum_{n=1}^{\infty} \frac{v_n}{d_n}$, we have

$$s_m = q_0 + \sum_{n=1}^{m} \frac{v_n}{d_n} \leqslant x \leqslant s_m + \frac{1}{d_m}$$

and $x = s_m + \frac{1}{d_m}$ only if $v_n = a_n - 1$ for each $n > m$. Since s_m is a fraction with denominator d_m, the approximation r_m to x to within $1/d_m$ by defect is equal to s_m or $s_m + \frac{1}{d_m}$; and the latter alternative can occur only if $v_n = a_n - 1$ for all $n > m$. Thus:

(i) There is an *infinity* of values of n such that $v_n < a_n - 1$: then the series $q_0 + \sum_{n=1}^{\infty} \frac{v_n}{d_n}$ is identical with the expansion of its sum x.

(ii) There is an integer $m \geqslant 0$ such that $v_n = a_n - 1$ whenever $n > m$, and $v_m < a_m - 1$ (if $m > 0$); then the sum x of the series $q_0 + \sum_{n=1}^{\infty} \frac{v_n}{d_n}$ is equal to the rational number

$$q_0 + \sum_{n=1}^{m} \frac{v_n}{d_n} + \frac{1}{d_m} \tag{5}$$

which is of the form k/d_m (k an integer); the *expansion* of x is identical with the series (5), in which all the terms with indices $> m$ are zero; such an expansion is said to be *terminating*, or to *terminate*. The series

$$q_0 + \sum_{n=1}^{\infty} \frac{v_n}{d_n} = q_0 + \sum_{n=1}^{m} \frac{v_n}{d_n} + \sum_{n=m+1}^{\infty} \frac{a_n - 1}{d_n} \tag{6}$$

is called the *improper expansion* of the number x.

Conversely, let x be a rational number which can be written in the form of a fraction with denominator d_n for some value of n. Let m be the smallest integer such that x is of the form k/d_m (k an integer); we have $r_n < x$ for $n < m$, and $r_m = x$, and therefore the expansion of x is of the form (5), and x has an improper expansion given by (6); moreover this improper expansion is *unique*.

A rational number, written in its irreducible form p/q, is equal to a fraction with denominator d_n if and only if q *divides* d_n (the number m will therefore be the smallest integer n such that q divides d_n). It can happen that *every rational number* has this property (for a suitably chosen n) : this will be the case if and only if every integer > 0 divides some d_n, e.g., if $d_n = n!$ If the d_n have this property, then a number is rational if and only if its expansion, relative to the sequence (d_n), terminates.

To summarize: to every sequence s whose initial term q_0 is an arbitrary integer, and whose term v_n $(n \geqslant 1)$ is such that $0 \leqslant v_n \leqslant a_n - 1$, corresponds a real number equal to $q_0 + \sum_{n=1}^{\infty} \frac{v_n}{d_n}$; if I_n denotes the interval $[0, a_n - 1]$ of $\mathbf{N}$, we thus define a mapping φ of $X = \mathbf{Z} \times \prod_{n=1}^{\infty} I_n$ *onto* the real line $\mathbf{R}$; moreover the equation $\varphi(s) = x$, where $x \in \mathbf{R}$ is given, has *one* solution if x is not a fraction with denominator d_n (for some n), and *two* solutions otherwise.

4. COMPARISON OF EXPANSIONS

If we know the expansions of two distinct real numbers x and y, we can determine whether $x < y$ or $x > y$.

Let $x = p_0 + \sum_{n=1}^{\infty} u_n/d_n$, $y = q_0 + \sum_{n=1}^{\infty} v_n/d_n$ be the expansions of x and y. If $p_0 < q_0$ then $x < y$, since

$$p_0 \leqslant x < p_0 + 1 \leqslant q_0 \leqslant y.$$

More generally, suppose that $p_0 = q_0$ and $u_n = v_n$ for $1 \leqslant n \leqslant m$, but that $u_m < v_m$; if

$$r_n = p_0 + \sum_{k=1}^{n} u_k/d_k, \qquad s_n = q_0 + \sum_{k=1}^{n} v_k/d_k,$$

then $r_n = s_n$ for $n < m$, and since $u_m + 1 \leqslant v_m$, $r_m + \frac{1}{d_m} \leqslant s_m$; but $r_m \leqslant x < r_m + \frac{1}{d_m} \leqslant s_m \leqslant y$, hence again we have $x < y$. In other words, *the order of x and y is the same as the order of the first two distinct terms of their respective expansions.*

It follows that, if $p_0 = q_0$ and $u_n = v_n$ for $n < m$, then the first m terms of the expansion of every number z belonging to the closed interval with end-points x and y are the *same* as those of the expansions of x and y.

We remark also that, in this case, we have $|y - x| \leqslant \frac{1}{d_{m-1}}$. If we endow $\mathbf{Z}$ and the intervals I_n with the *discrete* topology, it follows that the mapping φ defined above is *continuous* on the *product space* X.

5. EXPANSIONS TO BASE *A*

The most important base sequences are those for which $d_n = a^n$, where a is an integer > 1; a is then said to be the *base number* (or simply the *base*) of the corresponding expansions. For numerical calculations, expansions to base 10, which are called *decimal expansions*, are used; also expansions to base 2 (*dyadic* expansions) and base 3 (*triadic* expansions) are often used.

To represent the approximations by defect r_n to a number $x \geqslant 0$, in its expansion to base a, the following symbolism is employed: each integer u such that $0 \leqslant u \leqslant a - 1$ is denoted by a particular sign; if

$$r_n = p_0 + \sum_{k=1}^{n} \frac{u_k}{d_k},$$

we first write down, with the aid of these signs, the representation to base a of the integer $p_0 = [x] \geqslant 0$ (*Set Theory*, Chapter III, § 5, no. 7), then we put a point ("decimal point") and we write successively the signs representing the numbers $u_1, u_2, \ldots, u_n$. If S is the symbol thus obtained, it is customary to write $x = \mathrm{S}\ldots$, by abuse of language. It should be understood once and for all that such a relation is only an abbreviated method of indicating that the right-hand side is the approximation to x *to within* $1/a^n$ *by defect.*

For negative numbers the established usage is different: we write an approximation to $x' = -x > 0$ in the symbolism described above, and precede it by the sign "—"; it is thus an approximation to x to within $1/a^n$ *by excess* that is so denoted.

This usage has its inconveniences for numerical calculations. In the notation for negative logarithms the same symbolism is adopted as for positive numbers, by putting a bar over the integral part, to indicate that it is equal to the negative of the number written.

6. THE POWER OF R

We have $\mathbf{R} = \bigcup_{n\in\mathbf{Z}} [n, n+1[$, and all the intervals $[n, n+1[$ are equipotent to $[0, 1[$. Since $[0, 1[$ is an infinite set it follows that $\mathbf{R}$ is *equipotent to the interval* $[0, 1[$. By considering the *dyadic* expansion of the numbers of the interval $[0, 1[$ we shall show that this interval is equipotent to the set S of all sequences (u_n) in which each term is equal to 0 or 1.

First, it is equipotent to the subset S′ of S consisting of sequences (u_n) such that $u_n = 0$ for an infinity of values of n. On the other hand, the complement S″ of S′ in S is equipotent to the set of all *improper* expansions of rational numbers which are equal to a fraction with denominator 2^n; these numbers form a subset of $\mathbf{Q}$, hence the set of them is *countable* and therefore so is S″. Since S′ is infinite, it is *equipotent to* S, hence the result.

Now S is *equipotent* to $\mathfrak{P}(\mathbf{N})$; for we can define a bijection of $\mathfrak{P}(\mathbf{N})$ onto S by mapping each subset A of $\mathbf{N}$ to the sequence (u_n) such that $u_n = 0$ if $n \in \mathrm{A}$ and $u_n = 1$ if $n \in \complement\mathrm{A}$. We have thus proved:

THEOREM 1 (Cantor). *The set of real numbers is equipotent to the set of all subsets of a countably infinite set.*

COROLLARY. *The set of real numbers has a cardinal strictly greater than the cardinal of a countable set.*

A set equipotent to $\mathbf{R}$ is said to *have the power of the continuum.* By Proposition 1 of § 4, no. 1, every *interval* which contains more than one point has the power of the continuum. Again, the complement of a countable subset of $\mathbf{R}$ has the power of the continuum in particular, *the set of all irrational numbers has the power of the continuum.*

EXERCISES

§ 1

1) A topology $\mathfrak{T}$ on an ordered commutative group G is said to be *compatible* with the ordered group structure of G if it is compatible with the group structure of G and if in addition the set G_+ of elements $x \geqslant 0$ in G is *closed* with respect to $\mathfrak{T}$.

a) Let $\mathfrak{T}$ be a Hausdorff topology compatible with the ordered group structure of G, and let $\hat{G}$ be the completion of G (with respect to $\mathfrak{T}$). In the group $\hat{G}$, the closure P of G_+ is such that $y - x \in P$ is a pre-ordering compatible with the group structure of $\hat{G}$.

b) Let θ be an irrational number, and consider the ordering on $\mathbf{Q}^2$ for which the set of positive elements consists of $(0, 0)$ and all pairs (x, y) such that $y - \theta x \geqslant 0$. With respect to this ordering, $\mathbf{Q}^2$ is a linearly ordered group, and the topology on $\mathbf{Q}^2$ which is the product of the topology of the rational line by itself is compatible with this ordered group structure. But on $\hat{G} = \mathbf{R}^2$, the relation $y - x \in P = \overline{G}_+$ is not an ordering.

c) On a linearly ordered group G, the topology $\mathfrak{T}_0(G)$ (Chapter I, § 2, Exercise 5) is the coarsest of all topologies compatible with the ordered group structure (distinguish two cases, according as the set of all $x > 0$ has or has not a smallest element). Every topology on G which is compatible with the group structure of G and is finer than $\mathfrak{T}_0(G)$ is compatible with the ordered group structure of G, and the mapping $x \to x^+$ is continuous with respect to such a topology. This mapping is uniformly continuous with respect to $\mathfrak{T}_0(G)$, but not necessarily with respect to a group topology $\mathfrak{T}$ strictly finer than $\mathfrak{T}_0(G)$ [see *b*)].

2) *a*) Let G be a lattice-ordered commutative group. For each $x \geqslant 0$ in G, let $I(x)$ denote the interval $[-x, x]$ in G. A non-empty family (c_α) of elements $\geqslant 0$ in G is such that the intervals $I(c_\alpha)$

form a fundamental system of neighbourhoods of 0 for a topology $\mathcal{T}$ on G compatible with the group structure of G if and only if the set of c_α is directed (with respect to the relation $\geqslant$) and for each α there exists β such that $2c_\beta \leqslant c_\alpha$. If these conditions are satisfied, the mapping $x \to x^+$ of G into G is uniformly continuous. $\mathcal{T}$ is Hausdorff if and only if $\inf c_\alpha = 0$; and in this case $\mathcal{T}$ is compatible with the ordered group structure of G.

b) On the group $G = \mathbf{Z}^N$, the product of a countable infinity of linearly ordered groups $\mathbf{Z}$, there is no non-discrete topology defined by the procedure of *a*), but the product of the discrete topologies on the factors $\mathbf{Z}$ is a topology on G, compatible with the ordered group structure of G, which is Hausdorff and not discrete.

c) On a linearly ordered group G the only Hausdorff topology defined by the procedure of *a*) is the topology $\mathcal{T}_0(G)$ (Chapter I, § 2, Exercise 5).

¶ 3) Let G be a lattice-ordered commutative group, and let M be the semigroup of *major* subsets of G. Every subset of M which is bounded above (resp. below) has a least upper bound (resp. greatest lower bound), and G can be canonically identified with a subset of M. Let G′ denote the largest subgroup of M (the set of all symmetrizable elements of M).

a) Consider a topology $\mathcal{T}$ on G defined by the process of Exercise 2 *a*); let V_α be the set of all pairs (z, z') in $M \times M$ such that

$$z - c_\alpha \leqslant z' \leqslant z + c_\alpha;$$

show that the V_α form a fundamental system of entourages of a uniformity $\mathcal{U}$ on M, and that the topology $\mathcal{T}'$ induced by $\mathcal{U}$ induces $\mathcal{T}$ on G. If $\mathcal{T}$ is Hausdorff, then $\mathcal{U}$ is a Hausdorff uniformity (remark that the elements of M, considered as subsets of G, are then *closed*), and the uniform space M thus defined is *complete*. Show that the mapping $(z, z') \to z + z'$ of $M \times M$ into M is uniformly continuous. Supposing that $\mathcal{T}$ is Hausdorff, show that in M the set of upper (resp. lower) bounds of any subset of M is closed with respect to $\mathcal{T}'$ (prove this first for the set of upper or lower bounds of an element of M, by considering the elements of M as subsets of D), and that the mapping

$$(z, z') \to \inf (z, z')$$

of $M \times M$ into M is uniformly continuous (same method).

b) Suppose from now on in this exercise that $\mathcal{T}$ is Hausdorff. Then the closure $\overline{G}$ of G in M is a lattice-ordered group which is complete in the topology induced by $\mathcal{T}'$. The group G′ is closed in M (remark that its closure is a subgroup of M) and the topology induced by $\mathcal{T}'$ on G′ is compatible with the ordered group structure of G′. For G′ to be equal to $\overline{G}$ it is sufficient that every non-empty open interval $]a, b[$ in G′ should contain an element of G, and this is always the case if G is linearly ordered. In this case the topology induced on G′ by $\mathcal{T}'$ is $\mathcal{T}_0(G')$.

c) Take G to be the ordered group $\mathbf{Q}^2$, the product of the linearly ordered group $\mathbf{Q}$ by itself; then $G' = M = \mathbf{R}^2$ with the product ordering. Show that there are only three distinct non-discrete Hausdorff topologies on G defined by the procedure of Exercise 2 *a*), and that they are obtained by taking the set of elements c_α to be either the set of pairs (x, y) such that $x > 0$ and $y > 0$, or the set of pairs $(x, 0)$ such that $x > 0$, or the set of pairs $(0, y)$ such that $y > 0$. For the first of these three topologies we have $\overline{G} = G'$, although there are non-empty open intervals of G′ which contain no element of G; for the other two topologies, we have $\overline{G} \neq G'$. For each of these three topologies, the set of elements $z > 0$ in G is not open.

d) Take G to be the group $\mathbf{Q}^2$ endowed with the lexicographic order (*Set Theory*, Chapter III, § 2, no. 6); G is a non-Archimedean linearly ordered group. Then $\overline{G} = G' \neq M$; M is linearly ordered but the topology $\mathcal{T}'$ on M is distinct from the topology $\mathcal{T}_0(M)$; G′ is isomorphic to the group $\mathbf{R} \times \mathbf{Q}$ with the lexicographic order; in G′ the subgroup $H = \mathbf{R} \times \{0\}$ is open and isolated, but on the quotient group G′/H the quotient topology is distinct from $\mathcal{T}_0(G'/H)$.

¶ 4) *a*) Let E be a linearly ordered set. Let F be the $\mathbf{Z}$-module of formal finite linear combinations of elements of E with coefficients in $\mathbf{Z}$, and define a linearly ordered group structure on F by taking the set F_+ of elements $\geqslant 0$ to be the set consisting of 0 and all linear combinations $\sum_{\xi \in E} n(\xi)\xi \neq 0$ such that $n(\xi) > 0$ for the largest element $\xi \in E$ such that $n(\xi) \neq 0$. Show that E can be identified with a cofinal subset of F_+.

b) Suppose that E is well-ordered, and consider the group G of bounded mappings of E into F. Define a linearly ordered group structure on G by taking the set G_+ of elements $\geqslant 0$ to be the set consisting of 0 and all bounded mappings x of E into F such that $x(\xi) > 0$ for the smallest $\xi \in E$ such that $x(\xi) \neq 0$ (lexicographic ordering). Show that G is complete with respect to the topology $\mathcal{T}_0(G)$; if moreover E is uncountable and is such that every segment $]\leftarrow, \xi]$ ($\xi \in E$) is countable, then every countable intersection of open sets in G is open, and every compact subset of G is finite (cf. Chapter I, § 9, Exercise 4).

c) Suppose henceforth that E is well-ordered and uncountable but that every segment $]\leftarrow, \xi]$ is countable. For each $\xi \in E$, let c_ξ be the constant mapping of E into E whose value at every point is ξ. The c_ξ form a cofinal subset in G_+. Let E_0 be a set obtained by adjoining an element ω to E, and consider the topology $\mathcal{T}$, on the set $X = G \times E_0$, generated by the following sets: (i) $V_{a,b,\xi} =]a, b[\times \{\xi\}$ for $a < b$ in G and $\xi \in E$; (ii) $W_{a,b,\xi}$ — for $a < b$ in G and $\xi \in E$ such that $c_\xi > \sup(|a|, |b|)$ — consisting of the pairs (x, ω) such that $a < x < b$, and the pairs (x, ζ) such that $\zeta \in E$, $\zeta \geqslant \xi$ and either $a < x < b$ or $4c_\zeta + a < x < 4c_\zeta + b$. Show that $\mathcal{T}$ is Hausdorff and that every compact subset of X (with respect to $\mathcal{T}$) is finite. Moreover G operates continuously on X according to the law $(x, (y, \zeta)) \to (x + y, \zeta)$; but G does not operate properly on X, although conditions a), b), c) and d) of Chapter III, § 4, no. 2, Proposition 4 are satisfied and although, for each pair of compact subsets K, L of X, the set P(K, L) (Chapter III, § 4, no. 5, Theorem 1) is compact.

§ 2

1) A point $a \in \mathbf{R}$ is said to be *left adherent* to a subset A of $\mathbf{R}$ if it lies in the closure of A and if there is an interval $]a, b[$ $(a < b)$ not containing any point of A. Show that the set of left adherent points of a subset A of $\mathbf{R}$ is countable (set up a one-to-one correspondence between the set of left adherent points and a set of pairwise disjoint open intervals). Hence show that every well-ordered subset of $\mathbf{R}$ is countable.

¶) 2) A countable dense subset A of $\mathbf{R}$ is not closed. [If (a_n) is a sequence obtained by arranging the points of A in some order, define a sequence of intervals $[b_n, c_n]$ such that $b_{n-1} < b_n < c_n < c_{n-1}$ for all n, and such that $[b_n, c_n]$ contains no point a_k with index $k \leqslant n$; then use Theorem 2.] Deduce that $\mathbf{R}$ is uncountable (cf. § 8, no. 6).

3) Let (I_n) be an infinite sequence of non-empty open intervals in $\mathbf{R}$ such that $\bar{I}_n \cap \bar{I}_m = \emptyset$ for $m \neq n$. Show that the complement of the union of the I_n is a perfect set (Chapter I, § 2, no. 6); in particular, Cantor's triadic set is perfect.

4) The sum of the lengths of the intervals contiguous to Cantor's triadic set is equal to 1. Define a totally disconnected perfect subset of $[0, 1]$ such that the sum of the lengths of the intervals contiguous to A and contained in $[0, 1]$ is a given number m such that $0 < m \leqslant 1$.

¶ 5) Let l be a number > 0, and suppose that to each point $x \in \mathbf{R}$ corresponds an open interval $I(x)$ with centre x and length $\leqslant l$. Show that every compact interval $[a, b]$ can be covered by a finite number of intervals $I(x_i)$, the sum of whose lengths is $\leqslant l + 2(b - a)$. (Prove that if the proposition is true for each interval $[a, x]$ such that $a \leqslant x < c$, then there exists $d > c$ such that it is true for each interval $[a, y]$ such that $a \leqslant y < d$.) If $l = (b - a)/n$, where n is an integer $\geqslant 1$, show that the result cannot be improved.

¶ 6) *a*) Let X be a non-empty linearly ordered set, endowed with the topology $\mathfrak{T}_0(X)$ (Chapter I, § 2, Exercise 5). Show that X is compact if and only if each subset of X has a least upper bound, in other words if and only if X is a *complete lattice* (*Set Theory*, Chapter III, § 1, Exercise 11). (To show that the condition is necessary, argue as for Theorem 3; for sufficiency, consider a filter $\mathfrak{F}$ on X and show that, if A is the set of greatest lower bounds of sets of $\mathfrak{F}$, then the least upper bound of A is a cluster point of $\mathfrak{F}$).

b) Give an example of a complete lattice X which is not compact in the topology $\mathfrak{T}_0(X)$.

¶ 7) Let X be a non-empty linearly ordered set, endowed with the topology $\mathfrak{T}_0(X)$.

a) If X is connected, show that it has the following two properties:

α) Every non-empty subset of X which is bounded above has a least upper bound (if A is non-empty and bounded above, and if B is the set of upper bounds of A, and C the set of lower bounds of B, show that $B \cup C = X$ and that B and C are closed).

β) The set X is *without gaps*, in other words (*Set Theory*, Chapter III, § 1, Exercise 18) every open interval $]a, b[$ $(a < b)$ of X is non-empty (use the argument of Theorem 4).

b) Show conversely that if X has properties α) and β), then X is connected. (Show first that every closed interval $[a, b]$ is connected: assuming that this interval is the disjoint union of two non-empty closed sets A and B, and that $a \in A$, consider the greatest lower bound of B and hence obtain a contradiction).

c) Show that if X is connected it is locally compact and locally connected, and that the only connected subsets of X are the intervals (bounded or unbounded).

8) Let X and Y be two linearly ordered sets, endowed with the topologies $\mathfrak{T}_0(X)$ and $\mathfrak{T}_0(Y)$ respectively.

a) Suppose X connected and let $f: X \to Y$ be a continuous mapping. Show that, for all x, y in X such that $x < y$, every $z \in Y$ belonging to the closed interval with end-points $f(x)$ and $f(y)$ belongs to $f(X)$

(use Exercise 7). Deduce that f is a homeomorphism of X onto $f(X)$ if and only if f is continuous and strictly monotone.

b) Give an example of a homeomorphism of the rational line **Q** onto itself which is not strictly monotone.

¶ 9) *a*) Let X be a countable linearly ordered set without gaps (Exercise 7). Show that there is a strictly increasing bijective mapping φ of X onto one of the four intervals of **Q** with end-points 0 and 1. [Arrange the elements of X and the points of the appropriate interval of **Q** in sequences (a_n), (b_n), and define φ by induction.] If X is endowed with the topology $\mathfrak{T}_0(X)$, then φ is a homeomorphism of X onto $\varphi(X)$.

b) Deduce from *a*) that every countable dense subset of an open interval $]a, b[$ of **R** is homeomorphic to **Q**.

c) Deduce from *a*) that if X is any countable ordered set endowed with the topology $\mathfrak{T}_0(X)$, then there exists a strictly increasing homeomorphism of X onto a subspace of **Q** [embed X in a countable linearly ordered space X′ without gaps, such that $\mathfrak{T}_0(X)$ is induced by $\mathfrak{T}_0(X')$].

10) Show that every countable subset of Cantor's triadic set K, which is dense in K and contains no end-point of an interval contiguous to K, is homeomorphic to the rational line **Q** (use Exercise 9).

¶ 11) *a*) Let X be a linearly ordered set with the topology $\mathfrak{T}_0(X)$. If X is connected and has a countable dense subset A, show that there exists a homeomorphism (strictly monotonic by virtue of Exercise 8) of X onto one of the intervals of **R** with end-points 0, 1, which maps A onto the intersection of **Q** with this interval (use Exercises 9 and 7).

b) Deduce from *a*) that if B is a subset of **R** whose complement is dense in **R**, then there exists a homeomorphism of **R** onto itself which maps B onto a subset of the set $\complement \mathbf{Q}$ of irrational numbers (consider a dense countable subset A of **R** contained in $\complement B$).

c) In the set $\mathbf{R} \times \{0, 1\}$, linearly ordered by the lexicographic ordering, let **R**′ denote the complement of $\mathbf{Q} \times \{1\}$. Show that **R**′, endowed with the topology $\mathfrak{T}_0(\mathbf{R}')$, is locally compact and totally disconnected, and contains a countable dense subset **Q**′ homeomorphic to **Q**, but that the topology induced on a non-empty open interval of **R**′ does not have a countable base; in particular such an interval cannot be homeomorphic to any subset of **R**.

¶ 12) *a*) Let A be a well-ordered set and let I denote the interval $[0, 1[$ of **R**. Show that the set $X = A \times I$, linearly ordered by the lexicographic ordering and endowed with the topology $\mathfrak{T}_0(X)$, is connected (Exercise 7); there are therefore linearly ordered sets X which are connected in the topology $\mathfrak{T}_0(X)$ and have an arbitrary cardinal [cf. § 4, Exercise 7 *b*)].

b) Take A to be an uncountable well-ordered set in which every segment $]\leftarrow, t]$ is countable : the corresponding topological space X is called the *Alexandroff half-line*. Show that for every interval $[a, b]$ of X $(a < b)$ there exists a strictly increasing homeomorphism of $[a, b]$ onto the interval $[0, 1]$ of **R**. (Prove by transfinite induction that there exists a strictly increasing homeomorphism of $A \cap [a, b]$, A being identified with the set of points $(t, 0)$ of X onto a subspace of $[0, 1]$.)

¶ 13) Let a be an irrational number > 0. For each rational number x, let $f_a(x)$ be the real number in the interval $[0, a[$ such that $x - f_a(x)$ is an integral multiple of a. Show that f_a is a continuous injective mapping of **Q** into $[0, a[$. Deduce with the help of Exercise 9 that there exist continuous bijective mappings of **Q** onto itself, whose inverse mapping is not continuous at any point.

¶ 14) Let X be a connected Hausdorff topological space not consisting of a single point, and let Δ denote the diagonal of $X \times X$.

a) Show that there is no partition of $\complement\Delta$ into two open sets D_1, D_2 such that $\overset{-1}{D_1} = D_1$ and $\overset{-1}{D_2} = D_2$. [Begin by remarking that $\overline{D_1(x)}$ and $\overline{D_2(x)}$ are connected for each $x \in X$ (use Exercise 4 *a*) of Chapter I, § 11); deduce that, if $y \in D_1(x)$, we have $\overline{D_2(x)} \times \{y\} \subset D_1$, and show that this implies that, for each $x \in X$, one or the other of the sets $D_1(x)$, $D_2(x)$ is empty; hence show that one or the other of the sets D_1, D_2 is empty, which is contrary to hypothesis.] Deduce that either $\complement\Delta$ is connected or else $\complement\Delta$ has exactly two components A, B such that $B = \overset{-1}{A}$.

b) A linear ordering on X is said to be *compatible* with the topology $\mathfrak{T}$ of X if $\mathfrak{T}_0(X)$ is coarser than $\mathfrak{T}$. Deduce from *a*) that there exists such a linear ordering on X if and only if $\complement\Delta$ is not connected, and that there are then exactly two such linear orderings. [With the notation of *a*), show that the order relation must be either $(x, y) \in A$ or $(x, y) \in B = \overset{-1}{A}$.] Show that the intervals, for these orderings, are connected in the topology $\mathfrak{T}$, and that they are the only connected subsets of X with respect to $\mathfrak{T}$.

c) Show that if X is either locally connected or locally compact in the topology $\mathfrak{T}$, and if there exists a linear ordering on X compatible with $\mathfrak{T}$, then we must have $\mathfrak{T} = \mathfrak{T}_0(X)$. [If X is locally connected, use *b*). If X is locally compact, show that every compact neighbourhood of a point $x \in X$ is also a neighbourhood of x with respect to $\mathfrak{T}_0(X)$, by considering the frontier of the neighbourhood].

* *d*) Let X be the subspace of $\mathbf{R}^2$ consisting of (0, 0) and all pairs (x, y) such that $x > 0$ and $y = \sin(1/x)$. Show that there exists a linear ordering on X which is compatible with the topology induced on X by the topology of $\mathbf{R}^2$, but that this topology is strictly finer than $\mathcal{T}_0(X)$. *

Give an example of a connected Hausdorff space X, no point of which has a fundamental system of connected neighbourhoods, and on which there is a linear ordering compatible with the topology of X [cf. Chapter I, § 11, Exercise 2 *b*)].

¶ 15) *a*) Let X be a connected Hausdorff space such that, for each $x \in X$, $\complement\{x\}$ has exactly two components. Show that, if K is either of these components, $\overline{K} = K \cup \{x\}$ [Chapter I, § 11, Exercise 4 *a*)]. Let x, y be two distincts points of X, let A and B be the components of $\complement\{x\}$ and let A′, B′ be the components of $\complement\{y\}$. Show that one of the two sets A, B is contained in A′ or in B′.

b) Let x_0 be a point of X and let $A(x_0)$, $B(x_0)$ be the components of $\complement\{x_0\}$. For each $x \neq x_0$ there is exactly one of the two components of $\complement\{x\}$ which is either contained in $A(x_0)$ or contains $A(x_0)$; denote this component by $A(x)$. Show that the relation $A(x) \subset A(y)$ is a linear order relation compatible (Exercise 14) with the topology of X.

c) Extend the conclusion of *b*) to the case where $\complement\{x\}$ has two components for all $x \in X$ except for one point $a \in X$, for which $\complement\{a\}$ is connected.

* *d*) Let X be the subspace of $\mathbf{R}^2$ consisting of the points $a = (0, 1)$, $b = (0, -1)$ and all pairs (x, y) such that $x \neq 0$ and $y = \sin(1/x)$. Show that X is connected in the topology $\mathcal{T}$ induced on X by the topology of $\mathbf{R}^2$, that $\complement\{a\}$ and $\complement\{b\}$ are connected and that, for each point x of X other than a and b, $\complement\{x\}$ has exactly two components; but that there exists no linear ordering on X compatible with the topology $\mathcal{T}$ *.

¶ 16) Let X be a connected Hausdorff space which is *completely irreducible* between two distinct points a and b, that is to say such that there is no connected subset of X other than X itself which contains both a and b. Show that $X' = \complement\{a, b\}$ is connected and that, for each $x \in X'$, the complement of x in X has exactly two components $A(x)$ and $B(x)$ such that $a \in A(x)$, $b \in B(x)$, $a \notin B(x)$, $b \notin A(x)$. [Show that $\complement\{x\}$ cannot be partitioned into three non-empty open sets, by using Exercise 4 *a*) of Chapter I, § 11.] Deduce that there exists a linear ordering on X which is compatible with the topology of X (use Exercise 15).

¶ 17) Let X be a connected Hausdorff space such that whenever X is covered by three non-empty connected sets distinct from X, there are two of these sets whose union is distinct from X.

a) Show that, for every point $x \in X$, $\complement\{x\}$ has at most two components (same method as in Exercise 16).

b) Show that there are at most two points of X whose complements are connected. Moreover, if there are two distinct points a, b with this property, show that, for each x other than a or b, a and b belong to different components of $\complement\{x\}$.

c) Deduce from *a*) and *b*) that there is a linear ordering on X compatible with the topology of X (use Exercise 15).

¶ 18) *a*) Let X be a compact, connected, locally connected space which is irreducible between two of its points a, b (Chapter II, § 4, Exercise 19). Let x be a point of X other than a or b. Show that $\complement\{x\}$ has exactly two components and that a and b belong to different components of $\complement\{x\}$. [Arguing as in Exercise 16, show first that $\complement\{x\}$ cannot have more than two components. Next, if V is any connected closed neighbourhood of x which contains neither a nor b, let A_V, B_V be the components of a, b respectively in $\complement V$; show that $A_V \neq B_V$ by using Exercise 17 of Chapter II, § 4; finally consider the union of the sets A_V (resp. B_V) as V runs through the set of all connected closed neighbourhoods of x in X.] Deduce that there exists a linear ordering on X such that $\mathcal{T}_0(X)$ is the given topology on X [Exercises 15 and 14 *c*)].

b) Let X be a compact connected space which is irreducible between two of its points a, b. Show that if X has more than one prime constituent (Chapter II, § 4, Exercise 20) then the space X′ of prime constituents of X (*loc. cit.*) is irreducible between the prime constituents of a and b, and that there is a linear ordering on X′ such that the topology of X′ is identical with $\mathcal{T}_0(X')$. * Give an example in which X′ is distinct from X [cf. Exercise 15 *d*)]. *

¶ 19) *a*) Let X be a connected Hausdorff space and let a, b be two distinct points of X such that there is no connected *closed* subset of X, other than X itself, which contains both a and b. If x is a point of X other than a or b, then $\complement\{x\}$ is not connected if and only if, for each $y \in \complement\{x\}$, there is in $\complement\{y\}$ a connected subset which is closed in X and contains x and one of the points a, b. [To show that the condition is necessary, use Exercise 4 *a*) of Chapter I, § 11; to show that it is sufficient, let A (resp. B) be the set of all $y \in X$ such that there is a connected subset of $\complement\{y\}$ which is closed in X and contains a and

x (resp. b and x); show that A and B are non-empty and open and disjoint from each other, and that $\complement\{x\} = A \cup B$].

b) Let X be a compact connected space which is irreducible between two of its points a, b (Chapter II, § 4, Exercise 19). Show that if for each pair of distinct points x, y of X there is a compact connected set in X which contains y and one of a, b but does not contain x, then there is a linear ordering on X such that $\mathcal{T}_0(X)$ is the given topology on X (Exercises 15 and 16).

20) In the construction of Exercise 23 of Chapter I, § 9, take X_0 to be the interval [0, 1] of **R**. Show that the homeomorphisms of X onto itself are the same as the homeomorphisms of X_0 onto itself, although the topology of X is strictly finer than the topology of X_0 [use Exercise 20 *b*) of Chapter I, § 8].

21) Show that if $r > 2$ there is no r-fold transitive group of homeomorphisms of **R** onto itself (consider the subgroup leaving two points of **R** fixed).

§ 3

1) Let x be a real number $\geqslant 0$ and let p and q be two integers > 0. Prove the relations

$$(x^{1/p})^{1/q} = x^{1/pq}, \qquad (x^p)^{1/q} = (x^{1/q})^p, \qquad x^{1/p}x^{1/q} = (x^{p+q})^{1/pq}.$$

2) *a*) On the polynomial ring $A = \mathbf{R}[X]$, consider the linearly ordered ring structure for which the elements $\geqslant 0$ are 0 and all polynomials $\neq 0$ whose leading coefficient is > 0. Show that the topology $\mathcal{T}_0(A)$ is not compatible with the ring structure of A.

b) If K is an ordered field the topology $\mathcal{T}_0(K)$ is compatible with the ring structure of K, and the bounded sets for this ring topology (Chapter III, § 6, Exercise 12) are the bounded sets for the order structure. The topology $\mathcal{T}_0(K)$ is locally retrobounded (Chapter III, § 6, Exercise 22) and in particular is compatible with the field structure of K; moreover the completion $\hat{K}$ of K with respect to this topology, which is a field (*loc. cit.*) is canonically endowed with the structure of an ordered field [§ 1, Exercise 3 *b*)].

c) On the field $K = \mathbf{Q}(\sqrt{2})$, considered as an ordered subfield of **R**, consider the topology $\mathcal{T}$ which is such that $(x, y) \to x + y\sqrt{2}$ is a homeomorphism of $\mathbf{Q}^2$ onto K, $\mathbf{Q}^2$ carrying the product topology. This topology $\mathcal{T}$ is finer than $\mathcal{T}_0(K)$ and is compatible with the field structure of K; but the completion of this topological field is not a field.

d) The field $K = \mathbf{Q}(\sqrt[3]{2})$, considered as a vector space over $\mathbf{Q}$, is the direct sum of the vector subspace G' generated by 1 and $\sqrt[3]{2}$, and the vector subspace G'' generated by $\sqrt[3]{4}$. Consider the topologies on G' and G'' induced by the topology of $\mathbf{R}$; let $\mathcal{T}$ be the topology on K which is the product of the topologies of G' and G''. Show that $\mathcal{T}$ is compatible with the additive group structure of K and is finer than $\mathcal{T}_0(K)$ (K being considered as an ordered subfield of $\mathbf{R}$) but is not compatible with the ring structure of K.

¶ 3) *a*) If f is an isomorphism of the field $\mathbf{R}$ onto a subfield of $\mathbf{R}$ show that f must be the identity mapping of $\mathbf{R}$ onto itself. [Remark that all rational numbers are invariant under f and that

$$f(x^2) = (f(x))^2 \geqslant 0,$$

so that f is increasing].

b) Let K_0 be the field $\mathbf{R}(X)$ of rational fractions in one indeterminate over $\mathbf{R}$, ordered by taking the polynomials > 0 to be those whose leading coefficient is > 0. Let K be the maximal ordered algebraic extension of K_0. Show that there is an order-preserving automorphism f of K such that $f(\xi) = \xi$ for all $\xi \in \mathbf{R}$ and $f(X) = X^2$(*).

§ 4

1) Show that every bounded half-open interval of $\mathbf{R}$ is homeomorphic to every unbounded closed interval of $\mathbf{R}$. If $a < b$, then no two of the three intervals $]a, b[$, $[a, b[$, $[a, b]$ are homeomorphic.

2) Show that the mapping $x \to x/(1 + |x|)$ is a homeomorphism of $\mathbf{R}$ onto the open interval $]-1, 1[$.

3) Let I denote the interval $[0, 1]$ of $\mathbf{R}$, and let f be a homeomorphism of I onto itself such that $f(0) = 0$ and $f(1) = 1$. Show that there is a continuous mapping g of $I \times I$ into I such that: (i) for each $x \in I$, $g(x, 0) = x$ and $g(x, 1) = f(x)$; (ii) for each $y \in I$ the partial mapping $x \to g(x, y)$ is a homeomorphism of I onto itself such that $g(0, y) = 0$ and $g(1, y) = 1$.

4) Show that the uniformity induced on $\mathbf{R}$ by the unique uniformity of $\overline{\mathbf{R}}$ is strictly coarser than the additive uniformity of $\mathbf{R}$.

5) As the point (x, y) tends to $(+\infty, -\infty)$ while remaining in $\mathbf{R} \times \mathbf{R}$, shows that the set of cluster points of the function $x + y$ is $\overline{\mathbf{R}}$.

(*) Use the property of uniqueness, up to isomorphism, of the maximal ordered algebraic extension of an ordered field [B. L. van der Waerden, *Moderne Algebra*, vol. 1, lst edition (Berlin, Springer, 1930) pp. 232-234].

Likewise, as (x, y) tends to $(0, +\infty)$ while remaining in $\mathbf{R} \times \mathbf{R}$, the set of cluster points of xy is $\overline{\mathbf{R}}$.

6) Show that every rational function $P(x)/Q(x)$ with coefficients in $\mathbf{R}$, defined at all points x of $\mathbf{R}$ such that $Q(x) \neq 0$, can be extended by continuity to the points $+\infty$ and $-\infty$ (taking its values in $\overline{\mathbf{R}}$).

¶ 7) Let X be a linearly ordered set.

a) Let X_1 be the *completion* of X (*Set Theory*, Chapter III, § 1, Exercise 15). X_1 is a linearly ordered set whose elements are the subsets A of X such that; (i) if $x \in A$ and $y \leqslant x$, then $y \in A$; (ii) if A has a least upper bound in X, this bound belongs to A [which is equivalent to saying that A is closed in $\mathfrak{T}_0(X)$]; the elements of X_1 are ordered by inclusion. X_1 is compact with respect to the topology $\mathfrak{T}_0(X_1)$ [§ 2, Exercise 6 *a*)]. The mapping $x \to]\leftarrow, x]$ of X into X_1 is strictly order-preserving, and if we identify X with its image under this mapping, then X is dense in X_1 and $\mathfrak{T}_0(X)$ is induced by $\mathfrak{T}_0(X_1)$. X_1 is connected if and only if X is without gaps (§ 2, Exercise 7).

b) For each cardinal $\mathfrak{c}$, give an example of a linearly ordered set X which is connected in the topology $\mathfrak{T}_0(X)$ and is such that every fundamental system of neighbourhoods of any point of X has cardinal $\geqslant \mathfrak{c}$ [use *a*) and the method of § 1, Exercise 4 *b*)].

c) With the notation of *a*), let X_2 be the subset of the lexicographic product $X_1 \times \{-1, 0, 1\}$ which is the complement of the set consisting of: (i) the points $(x, 0)$ where $x \notin X$; (ii) the points $(x, 1)$ where $x \in X$ and the set of all $y > x$ in X has a smallest element; (iii) the points $(x, -1)$ where $x \in X$ and the set of all $y < x$ in X has a greatest element. Show that X_2, endowed with the topology $\mathfrak{T}_0(X_2)$, is compact and totally disconnected, that X can be identified, by means of the strictly order-preserving mapping $x \to (x, 0)$, with a dense subset of X_2, and that the topology induced by $\mathfrak{T}_0(X_2)$ on X is discrete.

d) Let X′ be a linearly ordered set containing X which induces the given order structure on X, is compact in the topology $\mathfrak{T}_0(X')$ and is such that X is dense in X′ with respect to this topology. Show that there is a continuous order-preserving surjective mapping $f : X' \to X_1$ and a continuous order-preserving surjective mapping $g : X_2 \to X'$, both of which induce the identity mapping on X.

§ 5

1) Let X_1, X_2 be two directed sets and let f be a real-valued function defined on $X_1 \times X_2$ such that, for each $x_1 \in X_1$, the mapping $x_2 \to f(x_1, x_2)$ is increasing on X_2, and for each $x_2 \in X_2$, the mapping

$x_1 \to f(x_1, x_2)$ is increasing on X_1. Show that f has a limit in $\overline{\mathbf{R}}$ with respect to the product of the section filters of X_1 and X_2, and that this limit is the least upper bound of f.

2) *a*) Let f be a real-valued function defined on a set X, let A be a non-empty subset of X, and let φ be an increasing real-valued function defined on $\overline{f(A)}$. If φ is continuous at the point $a = \sup_{x \in A} f(x)$, then $\sup_{x \in A} \varphi(f(x)) = \varphi(\sup_{x \in A} f(x))$.

b) Let f be a real-valued function defined on a set X filtered by a filter $\mathfrak{F}$, and let φ be a real-valued function defined on an open neighbourhood V of the set of cluster points of f with respect to $\mathfrak{F}$. Show that if φ is increasing and continuous on V, then

$$\lim \sup_{\mathfrak{F}} (\varphi \circ f) = \varphi(\lim \sup_{\mathfrak{F}} f).$$

3) Let f be a real-valued function defined on an infinite set X and let $\mathfrak{G}$ be the filter of complements of finite subsets of X. Show that $\lim \sup_{\mathfrak{G}} f$ is the least upper bound of the set of real numbers x such that the set $\overset{-1}{f}([x, +\infty])$ is infinite.

4) Let f, g be two real-valued functions defined on a set X filtered by a filter $\mathfrak{F}$. Show that

$$\lim \sup_{\mathfrak{F}} (\sup (f, g)) = \sup (\lim \sup_{\mathfrak{F}} f, \lim \sup_{\mathfrak{F}} g).$$

Give an example of a set X filtered by a filter $\mathfrak{F}$, and an infinite family (f_ι) of real-valued functions defined on X, such that

$$\sup_\iota (\lim \sup_{\mathfrak{F}} f_\iota) < \lim \sup_{\mathfrak{F}} (\sup_\iota f_\iota).$$

5) Let f, g be two real-valued functions defined on a filtered set. Show that if $\lim g$ exists and is $\geqslant 0$, then

$$\lim \sup fg = (\lim \sup f)(\lim g)$$

whenever both sides are defined.

6) Let X_1 (resp. X_2) be a set filtered by a filter $\mathfrak{F}_1$ (resp. $\mathfrak{F}_2$) and let f be a real-valued function defined on $X_1 \times X_2$. Show by an example that the three numbers $\lim \sup_{\mathfrak{F}_1 \times \mathfrak{F}_2} f(x_1, x_2)$, $\lim \sup_{\mathfrak{F}_1} (\lim \sup_{\mathfrak{F}_2} f(x_1, x_2))$, $\lim \sup_{\mathfrak{F}_2} (\lim \sup_{\mathfrak{F}_1} f(x_1, x_2))$ are in general distinct.

¶ 7) Let f be a real-valued function defined on a closed subset A of $\overline{\mathbf{R}}$. Let $\lim \sup_{x \to a,\, x \geqslant a} f(x)$ [resp. $\lim \sup_{x \to a,\, x \leqslant a} f(x)$] denote the upper limit of $f(x)$ as x tends to a point $a \in A$ while remaining in $A \cap [a, +\infty]$

(resp. $A \cap [-\infty, a]$). Show that the set of points $a \in A$ such that $\limsup_{x \to a, x \geqslant a} f(x) \neq \limsup_{x \to a, x \leqslant a} f(x)$ is countable. [Prove that, for each pair of rational numbers p, q such that $p < q$, the set $G_{p,q}$ of points $a \in A$ such that

$$\limsup_{x \to a, x \geqslant a} f(x) \leqslant p < q \leqslant \limsup_{x \to a, x \leqslant a} f(x)$$

is countable, by using Exercise 1 of § 2].

8) Deduce from Exercise 7 that a real-valued function f defined on a closed subset A of $\overline{\mathbf{R}}$, and *monotone* on A, is continuous on A except at the points of a countable subset of A.

¶ 9) Let X be a topological space with a countable base and let f be a real-valued function defined on X. Show that the set of points $x \in X$, such that $\lim_{x \to a, x \neq a} f(x)$ exists and differs from $f(a)$, is countable (same method as in Exercise 7).

¶ 10) Let X be a topological space which has a countable base (U_n) and let f be a real-valued function defined on X; f is said to attain a *strict relative maximum* at a point $a \in X$ if there is a neighbourhood V of a such that $f(x) < f(a)$ for each $x \in V$ other than $x = a$. Show that the set M of points of X, at which f attains a strict relative maximum, is countable. (Consider the set of U_n such that f attains, at some point of U_n, a strict relative maximum equal to its least upper bound in U_n, and show that there is a mapping of this set *onto* M).

11) Let (u_n) be a sequence of real numbers > 0 such that $\lim_{n \to \infty} u_n = 0$. Show that there is an infinite number of indices n such that $u_n \geqslant u_m$ for all $m \geqslant n$.

12) Let (u_n) be a sequence of real numbers > 0 such that

$$\liminf_{n \to \infty} u_n = 0.$$

Show that there is an infinite number of indices n such that $u_n \leqslant u_m$ for all $m \leqslant n$.

13) Let (u_n) be a sequence of finite real numbers and let (ε_n) be a sequence of numbers $\geqslant 0$, such that $\lim_{n \to \infty} \varepsilon_n = 0$ and $u_{n+1} \geqslant u_n - \varepsilon_n$ for all integers $n \geqslant 0$. Let $a = \liminf_{n \to \infty} u_n$ and let $b = \limsup_{n \to \infty} u_n$; show that the set of cluster points of the sequence (u_n) is the interval $[a, b]$.

14) Let (r_n) be an increasing sequence of finite real numbers > 0, such that $\lim r_n = +\infty$. For each finite real number $r > 0$, let $N(r)$ denote the largest index n such that $r_n \leqslant r$. Show that

$$\limsup_{r\to\infty} \frac{N(r)}{r} = \limsup_{n\to\infty} \frac{n}{r_n}, \qquad \liminf_{r\to\infty} \frac{N(r)}{r} = \liminf_{n\to\infty} \frac{n}{r_n}.$$

¶ 15) Let (x_n) be a sequence of finite real numbers and let (p_n) be a sequence of finite real numbers $\geqslant 0$ such that $\lim_{n\to\infty}\left(\sum_{i=0}^{n} p_i\right) = +\infty$. Let $y_n = \left(\sum_{i=0}^{n} p_i x_i\right)\Big/\left(\sum_{i=0}^{n} p_i\right)$ for those values of n such that $\sum_{i=1}^{n} p_i \neq 0$. Show that $\liminf_{n\to\infty} x_n \leqslant \liminf_{n\to\infty} y_n \leqslant \limsup_{n\to\infty} y_n \leqslant \limsup_{n\to\infty} x_n$.

Let H be any non-empty subset of the set of cluster points of the sequence (x_n) in $\overline{\mathbf{R}}$. Show that the sequence (p_n) of finite real numbers $\geqslant 0$ can be determined so that $\lim_{n\to\infty}\left(\sum_{i=0}^{n} p_i\right) = +\infty$ and so that the set of cluster points of the corresponding sequence (y_n) contains H. [Reduce to the case where H is countable, then define (p_n) by induction, taking its terms to be 0 or 1].

Deduce that the sequence (x_n) converges in $\overline{\mathbf{R}}$ if and only if the sequence (y_n) converges in $\overline{\mathbf{R}}$ for *every* sequence (p_n) of numbers $\geqslant 0$ such that $\lim_{n\to\infty}\left(\sum_{i=0}^{n} p_i\right) = +\infty$.

16) Let x_0, y_0 be two real numbers such that $0 < y_0 \leqslant x_0$. For $n > 0$ we define two sequences (x_n), (y_n) inductively by the relations

$$x_{n+1} = (x_n + y_n)/2, \qquad y_{n+1} = \sqrt{x_n y_n}.$$

Show that the two sequences (x_n), (y_n) tend to the same limit α (the "arithmetico-geometric mean" of x_0, y_0); also there exist numbers $a > 0$ and γ such that $0 < \gamma < 1$ and $x_n - y_n \leqslant a\gamma^{2^n}$ for each n [observe that $x_{n+1} - y_{n+1} = (x_n - y_n)^2/4(x_{n+1} + y_{n+1})$].

17) Let g be a mapping of $]0, 1]$ into $[-1, 1]$ such that

$$\lim_{x\to 0,\, x>0} g(x) = 0.$$

Show that there is a continuous increasing mapping g_2 and a continuous decreasing mapping g_1 of $[0, 1]$ into $[-1, 1]$ such that $g_1(0) = g_2(0) = 0$

and $g_1(x) \leqslant g(x) \leqslant g_2(x)$ for $0 < x \leqslant 1$. [For each integer $n > 0$ consider the greatest lower bound x_n of the numbers x such that $g(x) \geqslant 1/n$].

18) Extend the definitions and results of no. 1 to no. 6 to functions which take their values in a linearly ordered set X which is compact in the topology $\mathfrak{T}_0(X)$ (cf. § 2, Exercise 6).

§ 6

1) Let f be a continuous mapping of an open interval $I \subset \mathbf{R}$ into $\mathbf{R}$. Show that if $f(I)$ is open and if, for each $y \in \mathbf{R}$, the set $\overset{-1}{f}(y)$ has at most two distinct points, then f is monotone.

¶ 2) Let B be a base of $\mathbf{R}$ considered as a vector space over the field $\mathbf{Q}$ (" Hamel base "). B is uncountable (§ 2, Exercise 2). Let φ be a bijection of a subset $C \neq B$ of B onto the set B. Define a mapping f of $\mathbf{R}$ into $\mathbf{R}$ as follows: if $x = \sum_{\xi \in B} \lambda(\xi)\xi$, where $\lambda(\xi) \in \mathbf{Q}$ and $\lambda(\xi) = 0$ for all but a finite number of elements $\xi \in B$, then $f(x) = \sum_{\xi \in C} \lambda(\xi)\varphi(\xi)$. Show that $f(x+y) = f(x) + f(y)$, but that, for each $z \in \mathbf{R}$, $\overset{-1}{f}(z)$ is a dense subset of $\mathbf{R}$, which implies that f is not bounded above nor below in any interval of $\mathbf{R}$.

¶ 3) *a*) For each interval $I \subset \mathbf{R}$, let $G(I)$ denote the group of all homeomorphisms of I onto itself. Show that if I and J are any two intervals of $\mathbf{R}$, each containing more than one point, then $G(I)$ and $G(J)$ are isomorphic. In $G(I)$, the set $F(I)$ of increasing homeomorphisms is a normal subgroup of index 2.

b) Let $G = G(\mathbf{R})$ and $F = F(\mathbf{R})$. For each $f \in G$ and each $x \in \mathbf{R}$, let $\sigma(x; f)$ denote $\operatorname{sgn}(f(x) - x)$. Let T^+ (resp. T^-) be the set of all $f \in F$ such that $\sigma(x; f)$ is constant and equal to 1 (resp. -1) on $\mathbf{R}$. Show that every element of T^+ (resp. T^-) is conjugate in F to the translation $x \to x + 1$ (resp. $x \to x - 1$). [If $f \in T^+$, consider the sequence of points $f^n(0)$, $n \in \mathbf{Z}$].

c) Let $T = T^+ \cup T^-$. Show that every $f \in F$ which does not belong to T is the product in F of two elements of T_* [Write $f = f_1 f_2$ in F, where $f_1(x) = \sup(x, f(x))$ and $f_2(x) = \inf(x, f(x))$; if g is the translation $x \to x + 1$, then $f_1 g$ and $g^{-1} f_2$ belong to T].

d) Two elements f, g are conjugate in F if and only if there exists $s \in F$ such that $\sigma(x; f) = \sigma(s(x); g)$ for all $x \in \mathbf{R}$. [Consider the components I_n of the open set of all x such that $\sigma(x; f) \neq 0$; observe that $F(I_n)$ and F are isomorphic (by *a*)) and use *b*)]. In particular, if $[a, b]$ is an interval of $\mathbf{R}$ such that $\sigma(a; f) = \sigma(b; f) = 0$ for some

$f \in F$, then the element f' of F which coincides with f for $x \leqslant a$ or $x \geqslant b$ and is such that $f'(x) = x + \varepsilon(f(x) - x)$ for $a < x < b$, where $0 < \varepsilon < 1$, is conjugate to f in F.

e) Let $f \in F$ and let a, b, c, d be four real numbers such that $a < c < b < d$ and $f(x) - x = 0$ for $x = a$, $x = b$, $x = c$, $x = d$, $\sigma(x; f) = 1$ for $a < x < c$ and $b < x < d$. Let a', b' be two numbers such that $a \leqslant a' < c$ and $b < b' \leqslant d$; let $J = [a, b]$, $J' = [a', b']$, and let f_1 denote the restriction of f to J; thus $f_1 \in F(J)$. Show that there is an increasing homeomorphism s of J onto J′ such that the element $f_2 = sf_1s^{-1}$ of $F(J')$ is such that $f_2(x) > f^{-1}(x)$ whenever $a' < x < b'$ [use *d*)].

f) Let $f \in F$ and let $[a, b]$ be an interval of $\mathbf{R}$ such that $f(x) - x = 0$ for $x = a$ and $x = b$, $\sigma(x; f) = -1$ for $x < a$ and $\sigma(x; f) = +1$ for $x > b$. Show that there exists $g \in F$, conjugate to f in F, such that $g(x) > f(x)$ for all $x \in \mathbf{R}$ (same method).

g) Let H^+ (resp. H^-) denote the normal subgroup of F consisting of all $f \in F$ such that $f(x) = x$ in a neighbourhood of $+\infty$ (resp. $-\infty$). Show that H^+, H^- and $H = H^+ \cap H^-$ are the only normal subgroups of F other than F and $\{e\}$. [If N is a normal subgroup of F, other than F, and if $f \in N$ does not belong to H^+, show that there is an element $g \in N$ such that the set of all $x \in \mathbf{R}$ for which $\sigma(x; g) = +1$ is an interval $]a, +\infty]$, and that we have $g(x) = x$ for $x \leqslant a$. To do this use the constructions of *e*) and *f*), as well as *b*) and *c*). To show that N contains no subgroup, other than H and $\{e\}$, which is normal in F, consider an element $f \in H$ distinct from e; let a and b be the greatest lower bound and least upper bound (both finite by hypothesis) of the set of $x \in \mathbf{R}$ such that $\sigma(x; f) \neq 0$. Observe then that $F([a, b])$ is isomorphic to F and use the preceding result].

h) Show that the group H is simple. (Same method as for proving that H contains no non-trivial subgroups normal in F).

4) Let f be a lower semi-continuous function on a topological space X, and let φ be an increasing lower semi-continuous function on $f(E) \subset \overline{\mathbf{R}}$. Show that $\varphi \circ f$ is lower semi-continuous on X.

5) Let f be a lower semi-continuous function on a topological space X. Show that if A is any non-empty subset of X, then $\sup f(\overline{A}) = \sup f(A)$.

¶ 6) Let f be a real-valued function defined on a topological space X. The number (finite or infinite)

$$\omega(a; f) = \limsup_{x \to a} f(x) - \liminf_{x \to a} f(x)$$

is called the *oscillation* of f at $a \in X$, whenever the right-hand side is defined.

a) Show that $x \to \omega(x; f)$ is upper semi-continuous on the subspace A of X where this mapping is defined.

b) If f is *finite* on X, then $A = X$ and for each $a \in X$ we have

$$\omega(a; f) = \limsup_{(x, y) \to (a, a)} (f(x) - f(y)).$$

c) Let f be a lower semi-continuous finite real-valued function on X. Show that if, at a point $a \in X$, $\omega(a; f)$ has a *finite* value, then

$$\liminf_{x \to a} \omega(x; f) = 0.$$

[Assuming the result false, show that there would exist points x arbitrarily near to a such that $f(x)$ is as large as we please].

d) For each rational number $r = p/q$ in irreducible form (with $q > 0$), put $f(r) = q$. Show that f is lower semi-continuous on **Q**, but that at each point $r \in \mathbf{Q}$ we have $\omega(r; f) = +\infty$.

¶ 7) Let X be a locally compact space and f a lower semi-continuous function on X.

a) If $\limsup_{x \to a} f(x) = +\infty$ for every $a \in X$, show that the set $\overset{-1}{f}(+\infty)$ is dense in X [for each $a \in X$ and each neighbourhood V of a, define a sequence (U_n) of relatively compact open sets such that $\overline{U}_{n+1} \subset U_n$ and $f(x) > n$ for all $x \in U_n$].

* *b*) Let $n \to r_n$ be a bijection of **N** onto the set of rational numbers belonging to $[0, 1]$, and let $\varphi(x) = 1/\sqrt{|x|}$ [with $\varphi(0) = +\infty$]; then the function $f(x) = \sum_{n=0}^{\infty} 2^{-n} \varphi(x - r_n)$ is lower semi-continuous on $[0, 1]$, $\overset{-1}{f}(+\infty)$ is dense in this interval, and so is its complement [to prove the last point, observe that the series is convergent whose general term is

$$2^{-n} \int_0^1 \varphi(x - r_n) dx. \; *$$

(*Integration*, Chapter IV, § 3, no. 6, Theorem 5).] *

c) If f is finite, show that the set of points x such that $\omega(x; f)$ is finite is dense [use *a*)].

d) Show that the set of points of X at which f is continuous (f being finite or not) is dense. [Reduce to the case in which f is bounded, by replacing f by $f/(1 + |f|)$ (Exercise 4); observe that $1/\omega(x; f)$ is lower semi-continuous and use *a*) and Exercise 6 *c*)].

8) Let X be a topological space, and let A be a closed subset of $X \times \overline{\mathbf{R}}$. Show that the mapping $x \to \inf(A(x))$ of $\mathrm{pr}_1 A$ into $\overline{\mathbf{R}}$ is lower semi-continuous. Conversely, if $f : X \to \overline{\mathbf{R}}$ is a lower semi-continuous function, then the subset B of $X \times \overline{\mathbf{R}}$ consisting of pairs (x, y) such that $y \geqslant f(x)$ is closed in $X \times \overline{\mathbf{R}}$.

¶ 9) *a)* Let X and Y be two Hausdorff spaces and let $\pi : X \to Y$ be a proper mapping (Chapter I, § 10, no. 1). Let g be a lower semi-continuous real-valued function on X. For each $y \in Y$, let $f(y)$ be the greatest lower bound of g in the set $\overset{-1}{\pi}(y)$ [thus $f(y) = +\infty$ if $\overset{-1}{\pi}(y) = \varnothing$]. Show that f is lower semi-continuous on Y. [Use the fact that $\pi(X)$ is closed, that $\overset{-1}{\pi}(y)$ is compact for all $y \in \pi(X)$, and that every neighbourhood of $\overset{-1}{\pi}(y)$ contains a neighbourhood which is saturated with respect to the equivalence relation $\pi(x) = \pi(x')$.]

b) Let g be the continuous function $|x_1 x_2 - 1|$ defined on $\mathbf{R} \times]0, +\infty[$, and for each $x_1 \in \mathbf{R}$ let $f(x_1) = \inf_{x_2 > 0} g(x_1, x_2)$. Show that f is not lower semi-continuous.

10) Extend Definition 1, Proposition 1 and Theorem 3 to functions defined on a topological space X which take their values in an arbitrary *linearly ordered* set Y. If f and g are two mappings of X into Y, both lower semi-continuous at a point a, then $\inf(f, g)$ and $\sup(f, g)$ are lower semi-continuous at a. So is $f + g$ if Y is a linearly ordered commutative group.

If Y is *compact* in the topology $\mathcal{T}_0(Y)$ (§ 2, Exercise 6), extend Theorem 4 and Propositions 3 and 4 to functions with values in Y.

¶ 11) Let $\overline{\mathbf{R}}$ carry the *right* topology (Chapter I, § 1, Exercise 2). A mapping f of a topological space X into $\overline{\mathbf{R}}$ is continuous in this topology if and only if f has a relative minimum at *every* point of X. If $X = \mathbf{R}$ (with the usual topology), show that the set $f(X)$ is then *countable*. [Suppose that the result is false, and that the interval $[\alpha, \beta] \cap f(X)$ is uncountable; for each $y \in [\alpha, \beta]$, let $U(y) = \overset{-1}{f}([y, +\infty])$; $U(y)$ is an *open* subset in $\mathbf{R}$. Let (I_n) be the sequence (finite or infinite) of components of $U(\beta)$, and for each n let x_n be a point of I_n; for each $y \in [\alpha, \beta]$, let $g_n(y)$ (resp. $h_n(y)$) be the left-hand (resp. right-hand) end-point of the component of $U(y)$ which contains x_n. Show that for at least one value of n one of the functions g_n, h_n must take an uncountable infinity of distinct values in $[\alpha, \beta]$; now use Exercise 8 of § 5].

12) Let f be a continuous non-constant finite real-valued function on a compact interval $[a, b]$ of $\mathbf{R}$, such that $f(a) = f(b) = 0$; let Z be the closed subset of $[a, b]$ consisting of points x for which $f(x) = 0$,

and let c be the largest of the lengths of the intervals contiguous to Z in $[a, b]$. Show that, for each t such that $0 < t < c$, there is a point $x \in [a, b]$ such that $x + t \in [a, b]$ and $f(x + t) = f(x)$.

13) Let f be a continuous finite real-valued function on a compact interval $[a, b]$ of **R**, and suppose that f is not strictly monotone. Show that there exists a point $x_0 \in]a, b[$ such that, for each $\varepsilon > 0$, there are two points y, z in $[a, b]$ such that $x_0 - \varepsilon < y < x_0 < z < x_0 + \varepsilon$ and $f(y) = f(z)$.

14) Let f be a continuous mapping of **R** into itself.

a) Show that if f is uniformly continuous on **R**, then there exist two real numbers $\alpha \geqslant 0$ and $\beta \geqslant 0$ such that $|f(x)| \leqslant \alpha|x| + \beta$ for all $x \in \mathbf{R}$.

b) Show that if f is monotone and bounded in **R**, then f is uniformly continuous on **R**.

¶ 15) Let X be the Alexandroff half-line [§ 2, Exercise 12 *b*)], and let X′ be its compactification by adjunction of a point at infinity ω. Show that, for every continuous mapping f: X → X, either we have $\lim_{x \to \omega,\, x \in X} f(x) = \omega$, or else there exists $z \in X$ such that f is constant in the interval $[z, \to[$. [Show that if, as x tends to ω, f has a cluster point $c \in X$, then we must have $\lim_{x \to \omega,\, x \in X} = c$; prove that f can have no other cluster point in X′, by using the fact that every increasing sequence converges in X. Then use Exercise 12 *b*) of § 2 and the fact that in X′ every countable intersection of neighbourhoods of ω is a neighbourhood of ω].

§ 7

1) Let (x_ι) be a family of real numbers, all of which belong to the interval $A' =]-\infty, +\infty]$ (resp. $A'' = [-\infty, +\infty[$). The family (x_ι) is summable in $\overline{\mathbf{R}}$ if and only if *one* of the following conditions is satisfied:

a) At least one of the numbers x_ι is equal to $+\infty$ (resp. $-\infty$).

b) At least one of the two sums $\sum_\iota x_\iota^+$, $\sum_\iota x_\iota^-$ is finite.

In the first case we have $\sum_\iota x_\iota = +\infty$ (resp. $-\infty$); in the second case, $\sum_\iota x_\iota = \sum_\iota x_\iota^+ - \sum_\iota x_\iota^-$.

2) Let (x_ι) be a family of real numbers all of which belong to the interval A′ (resp. A″) and which satisfies condition *b*) of Exercise 1. Show that every subfamily of (x_ι) is summable in $\overline{\mathbf{R}}$, and that the sum of the x' is associative (cf. Chapter III, § 5, no. 3, Theorem 2).

3) Let $(x_n)_{n\geqslant 0}$ be a *decreasing* sequence of finite real numbers $\geqslant 0$. This sequence is summable in $\mathbf{R}$ if and only if the sequence $(2^n x_{2^n})_{n\geqslant 0}$ is summable in $\mathbf{R}$ ("Cauchy's condensation test"). * Deduce that the series whose general term is $1/n^\alpha$ is convergent if $\alpha > 1$ and not convergent if $0 < \alpha \leqslant 1$. *

4) Let (a_n) be any sequence of finite real numbers $\geqslant 0$. Show that the sequence $\left(\dfrac{a_n}{(1 + a_1)(1 + a_2)\dots(1 + a_n)}\right)$ is summable in $\mathbf{R}$ (express each term of this sequence as a difference).

5) Let (d_n) be a sequence of finite real numbers $\geqslant 0$ such that

$$\sum_{n=0}^{\infty} d_n = +\infty.$$

What can be said about the convergence of the series whose general terms are

$$\frac{d_n}{1 + d_n}, \qquad \frac{d_n}{1 + nd_n}, \qquad \frac{d_n}{1 + n^2 d_n}, \qquad \frac{d_n}{1 + d_n}?$$

6) Prove that $s = \sum_{n=1}^{\infty} (1/n)$ is equal to $+\infty$ by showing that $s \geqslant s + \dfrac{1}{2}$ (find lower bounds, as functions of s, for the sum of the terms with even indices and the sum of the terms with odd indices).

7) Show that, for each integer $p > 1$,

$$\sum_{n=1}^{\infty} \frac{(-1)^{n-1}}{n^p} = (1 - 2^{1-p}) \sum_{n=1}^{\infty} \frac{1}{n^p}.$$

8) Show that, if m is an integer > 0, then $\displaystyle\sum_{n\geqslant 1,\, n\neq m} \frac{1}{m^2 - n^2} = -\frac{3}{4m^2}$ (express the rational fraction $\dfrac{1}{m^2 - x^2}$ as a sum of partial fractions).

9) If the series whose general term is x_n is convergent in $\mathbf{R}$, show that $\liminf_{n\to\infty} nx_n \leqslant 0 \leqslant \limsup_{n\to\infty} nx_n$.

¶ 10) A series (u_n) is convergent in $\mathbf{R}$ if and only if, for *each* increasing sequence (p_n) of numbers > 0 such that $\lim_{n\to\infty} p_n = +\infty$, we have $\lim_{n\to\infty}\left(\left(\sum_{k=0}^{n} p_k u_k\right)\Big/ p_n\right) = 0$ (use Exercise 15 of § 5).

11) Consider a sequence (a_n) every term of which is the sum of a finite sequence of finite real numbers

$$a_n = b_{n,1} + b_{n,2} + \dots + b_{n,k_n}.$$

For each pair (n, p) of positive integers such that $p \leqslant k_n$, if $m = \sum_{i=0}^{n-1} k_i + p$, put $c_m = b_{n,p}$. Show that if the series whose general term is a_n is convergent in $\mathbf{R}$, and if

$$d_n = |b_{n,1}| + |b_{n,2}| + \cdots + |b_{n,k_n}|$$

tends to 0 as n tends to infinity, then the series whose general term is c_m is convergent and has the same sum as the series whose general term is a_n.

¶ 12) Let (u_n) be a sequence of finite real numbers. For each permutation σ of the set $\mathbf{N}$, put

$$r(n) = |\sigma(n) - n| \cdot \sup_{m \geqslant n} |u_m|.$$

a) If the series with general term u_n is convergent, and if $\lim_{n \to \infty} r(n) = 0$, show that the series with general term $v_n = u_{\sigma(n)}$ is convergent to the same sum. [Consider the difference $\sum_{k=0}^{n} v_k - \sum_{k=0}^{n} u_k$ for large values of n, and if h is the smallest integer $\leqslant n$ such that $\sigma(h) > n$, note that in this difference there are at most $\sigma(h) - h$ terms in each of the two sums which do not cancel out].

b) Suppose that the series whose general term is u_n is convergent. Give an example of a permutation σ such that $\lim_{n \to \infty} r(n) = 0$ but such that the criterion of Chapter III, § 5, Exercise 6 *a*) is not satisfied. [Define a family of mutually disjoint intervals $\mathrm{I}_k = [n_k - 2k, n_k + 2k]$ in $\mathbf{N}$, and take σ such that $\sigma(n) = n$ whenever n does not belong to any of the I_k, and such that $\sigma(n_k - 2j) = n_k + 2j$, $\sigma(n_k + 2j) = n_k - 2j$ for all k and $0 \leqslant j \leqslant k$.]

13) Let (u_n) be a sequence of real numbers $\geqslant 0$ such that $\lim_{n \to \infty} u_n = 0$ and $\sum_{n=0}^{\infty} u_n = +\infty$. Let (p_n) be a strictly increasing sequence of numbers > 0 such that $\lim_{n \to \infty} p_n = +\infty$. Show that there exists a permutation σ of $\mathbf{N}$ such that for each $n \in \mathbf{N}$, $\sum_{k=0}^{n} u_{\sigma(n)} \leqslant p_n$

¶ 14) Let (u_n) be a sequence of finite real numbers such that the series whose general term is u_n is convergent but not absolutely convergent; let $s = \mathop{\mathsf{S}}_{n=0}^{\infty} u_n$. Show that, for each number $s' \geqslant s$, there exists a permutation σ of $\mathbf{N}$ such that $\sigma(n) = n$ for those indices n such that $u_n \geqslant 0$, and such that $\sum_{n=0}^{\infty} u_{\sigma(n)} = s'$. [Show by induction on m that there is a permutation σ_m of $\mathbf{N}$ such that $\sigma_m(k) = k$ for all k for which $u_k \geqslant 0$ and such that, if $u_k^{(m)}$ denotes $u_{\sigma_m(k)}$, there is an integer p_m with the property that $\left| s' - \sum_{i=0}^{k} u_i^{(m)} \right| \leqslant 1/m$ for all $k \geqslant p_m$. Moreover, σ_{m+1} is such that $\sigma_{m+1}(k) = \sigma_m(k)$ for all k such that $\sigma_m(k) < p_m$ and all k such that $u_k \leqslant -1/m$.)

15) Let (u_n) be a sequence of finite real numbers such that $\lim_{n\to\infty} u_n = 0$ and $\sum_n u_n^+ = \sum_n u_n^- = +\infty$. If a and b are any two real numbers (finite or not) such that $a \leqslant b$, show that there is a permutation σ of $\mathbf{N}$ such that, if we put $s_n = \sum_{k=0}^{n} u_{\sigma(k)}$, we have $\liminf_{n\to\infty} s_n = a$ and $\limsup_{n\to\infty} s_n = b$. Show that the set of cluster points of the sequence (s_n) is then the interval $[a, b]$.

¶ 16) Let (u_n) be a sequence of finite real numbers such that the series whose general term is u_n is convergent but not absolutely convergent. Show that there exists a permutation σ of $\mathbf{N}$ satisfying the condition of Chapter III, § 5, Exercise 6 *a*) but such that the series whose general term is $u_{\sigma^{-1}(n)}$ is not convergent. [Let h, k, m be three integers with the following properties : if $p_1, \ldots, p_r$ are the integers n such that $h \leqslant n < h+m$ and $u_n \geqslant 0$, then $u_{p_1} + \cdots + u_{p_r} > 2$ and $h + m < k - r$; moreover if μ, ν are any integers such that $h + m \leqslant \mu \leqslant \nu$, then $\left| \sum_{n=\mu}^{\nu} u_n \right| \leqslant 1$. Put $s = m - r$ and let $q_1, \ldots, q_s$ denote the integers n such that $h \leqslant n < h + m$ and $u_n < 0$. Consider the permutation π of the interval $[h, k + s]$ of $\mathbf{N}$ such that $\pi(p_i) = k - i + 1$ for $1 \leqslant i \leqslant r$, $\pi(q_j) = k + j$ for $1 \leqslant j \leqslant s$, $\pi(k + j) = h + s - j$ for $1 \leqslant j \leqslant s$, $\pi(k - i + 1) = h + s + i - 1$ for $1 \leqslant i \leqslant r$ and finally $\pi(n) = n$ for $h + m \leqslant n \leqslant k - r$; show that we have

$$\left| \sum_{i=h}^{k} u_i - \sum_{i=h}^{k} u_{\pi^{-1}(i)} \right| \geqslant 1.$$

17) If $u_{mn} = 1/(m^2 - n^2)$ for $m \neq n$ and $u_{mm} = 0$, show that

$$\sum_{m=1}^{\infty} \left(\sum_{n=1}^{\infty} u_{mn} \right) = - \sum_{n=1}^{\infty} \left(\sum_{m=1}^{\infty} u_{mn} \right) \neq 0.$$

(Use Exercise 8.)

18) Let $(1 + u_\iota)$ be a family of real numbers all belonging to the interval $[0, +\infty[$ (resp. $]0, +\infty]$). Then the family $(1 + u_\iota)$ is multipliable in $\mathbf{R}$ if and only if *one* of the following conditions is satisfied:

a) At least one of the numbers $1 + u_\iota$ is equal to 0 (resp. $+\infty$).

b) The family u_ι is summable in $\overline{\mathbf{R}}$.

19) Let $(1 + u_\iota)_{\iota \in I}$ be a multipliable family in $\mathbf{R}$. For each non-empty subset H of I, let v_H denote $\prod_{\iota \in H} u_\iota$. Show that the family $(v_H)_{H \in \mathfrak{F}(I)}$ is summable in $\mathbf{R}$ and that

$$1 + \sum_H v_H = \prod_{\iota \in I} (1 + u_\iota).$$

Is the converse true?

Deduce that, if $-1 < x < 1$, then $\prod_{n=0}^{\infty} (1 + x^{2^n}) = 1/(1 - x)$.

20) Let (u_n) be a sequence of finite real numbers $\geqslant 0$ such that $u_0 > 0$, and let $s_n = \sum_{k=0}^{n} u_k$ for each $n \geqslant 0$. Then the sequence (u_n) is summable in $\mathbf{R}$ if and only if the sequence (u_n/s_n) is summable in $\mathbf{R}$ [apply Theorem 4 to the sequence $(1 - u_n/s_n)$]. The same holds if the sequence (u_n/s_n) is replaced by the sequence (u_n/s_{n-1}).

21) Let $u_n = (-1)^n/\sqrt{n}$ for $n \geqslant 2$. The product whose general factor is $1 + u_n$ is not convergent, but the series whose general term is u_n is convergent.

22) For $n \geqslant 2$, let $u_{2n-1} = -1/\sqrt{n}$ and $u_{2n} = (1 + \sqrt{n})/n$. The product whose general factor is $1 + u_n$ is convergent, but the series whose general term is u_n is not convergent.

§ 8

1) If x and y are real, we have

$$[x + y] = [x] + [y] + \varepsilon \qquad \text{where } \varepsilon = 0 \quad \text{or} \quad \varepsilon = 1,$$

$$[x - y] = [x] - [y] - \varepsilon \qquad \text{where } \varepsilon = 0 \quad \text{or} \quad \varepsilon = 1,$$

$$[x] + [x + y] + [y] \leqslant [2x] + [2y],$$

$$[x] + \left[x + \frac{1}{n}\right] + \left[x + \frac{2}{n}\right] + \cdots + \left[x + \frac{n-1}{n}\right] = [nx],$$

$$\left[\frac{[nx]}{n}\right] = [x]$$

for any integer $n > 0$.

2) Let (ε_n) be a strictly decreasing sequence of finite numbers > 0 which tend to 0. For each $x \in \mathbf{R}$, let $r_n(x)$ be the multiple of ε_n which is the approximation by defect to x to within ε_n. Then the sequence $(r_n(x))$ is increasing for all $x \in \mathbf{R}$ if and only if ε_n is an integer multiple of ε_{n+1} for each n.

3) Let a be an integer > 1. A real number x is rational if and only if its expansion to base a, say $x = p_0 + \sum_{n=0}^{\infty} u_n a^{-n}$, is *periodic*, that is to say if and only if there exist two integers n_0 and $r > 0$ such that $u_{n+r} = u_n$ for all $n \geqslant n_0$.

¶ 4) Let (u_n) be a summable sequence of numbers > 0 in $\mathbf{R}$ which satisfies the following conditions: $u_{n+1} \leqslant u_n$ and $u_n \leqslant \sum_{k=1}^{\infty} u_{n+k}$ for all $n \geqslant 0$. Show that for every number a such that $0 < a \leqslant s = \sum_{n=0}^{\infty} u_n$, there exists a subset I of $\mathbf{N}$ such that $a = \sum_{n \in \mathrm{I}} u_n$. These conditions are satisfied in particular if $u_{n+1} \leqslant u_n \leqslant 2u_{n+1}$ for all n. Consider the case of dyadic expansions.

¶ 5) For each real number $x \in]0, 1]$, there is a unique infinite increasing sequence (q_n) of integers > 0 such that $x = \sum_{n=1}^{\infty} 1/(q_1 q_2 \ldots q_n)$. The number x is rational if and only if $q_{n+1} = q_n$ from a certain value of n onwards.

¶ 6) For each real number $x > 1$, there is a unique infinite sequence (t_n) of integers $\geqslant 1$ such that $t_{n+1} \geqslant t_n^2$ for all $n \geqslant 1$ and such that $x = \prod_{n=1}^{\infty} \left(1 + \frac{1}{t_n}\right)$ (use Exercise 19 of § 7). The number x is rational if and only if $t_{n+1} = t_n^2$ form a certain value of n onwards. [To show that the condition is necessary, show that if we put $x_n = \prod_{k=n+1}^{\infty} \left(1 + \frac{1}{t_k}\right)$ then the denominators of the fractions $\frac{x_n}{x_n - 1}$ (in their irreducible form) form a decreasing sequence].

¶ * 7) Let $(\varepsilon_n)_{n \geqslant 0}$ be a sequence of numbers each of which is 1 or -1. Show that the real number

$$x_n = \varepsilon_0 \sqrt{2 + \varepsilon_1 \sqrt{2 + \varepsilon_2 \sqrt{2 + \cdots + \varepsilon_n \sqrt{2}}}}$$

is defined and is equal to

$$2 \sin\left(\frac{\pi}{4} \sum_{k=0}^{n} \frac{\varepsilon_0\varepsilon_1 \ldots \varepsilon_k}{2^k}\right).$$

Deduce that $\lim_{n\to\infty} x_n$ exists and that for each real number x, such that $-2 \leqslant x \leqslant 2$, there is a sequence (ε_n) such that x is equal to the limit of the corresponding sequence (x_n). For what values of x is the sequence (ε_n) unique? For what values of x is it periodic? (see Exercise 3). *

8) Show that there is no non-constant mapping ψ of an interval $I \subset \mathbf{R}$ into the set $\mathbf{N}^{\mathbf{N}}$ of all sequences of natural numbers which has the following property for each $x \in I$: for each integer $n \geqslant 0$ there is a neighbourhood V of x in I such that, for each $y \in V$, the first n terms of the sequence $\psi(y)$ are the same as the first n terms of the sequence $\psi(x)$ (remark that this would imply the continuity of ψ, if each factor $\mathbf{N}$ carries the discrete topology and $\mathbf{N}^{\mathbf{N}}$ the product topology).

9) Show that Cantor's triadic set K (§ 2, no. 5) is the set of all real numbers $x \in [0, 1]$ whose triadic expansion (resp. improper triadic expansion if x is the origin of an interval contiguous to K)

$$x = \sum_{n=1}^{\infty} u_n 3^{-n}$$

is such that $u_n \neq 1$ for each n (so that $u_n = 0$ or $u_n = 2$). If $v_n = \frac{1}{2} u_n$ for each n, and if $f(x) = \sum_{n=1}^{\infty} v_n 2^{-n}$, show that f is a *surjective* continuous mapping of K onto $[0, 1]$; deduce that K has the power of the continuum.

¶ 10) Let (d_n) be a base sequence, let $a_n = d_n/d_{n-1}$ $(n \geqslant 1)$, let (n_i) be a strictly increasing sequence of integers, and let (b_i) be a sequence of integers such that $0 < b_i < a_{n_i} - 1$ for each i. Let G be the set of all $x \in [0, 1]$ such that, if $\sum_{n=1}^{\infty} u_n/d_n$ is the expansion of x, or the improper expansion if it exists, then $u_{n_i} \neq b_i$ for all $i \in \mathbf{N}$. Show that G is a totally disconnected perfect set, and that if l is the sum of the lengths of the intervals contiguous to G and contained in $[0, 1]$, then

$$1 - l = \prod_{i=0}^{\infty}\left(1 - \frac{1}{a_{n_i}}\right).$$

¶ 11) Let X be a Hausdorff topological space. Suppose that, for each finite sequence s whose terms are equal to 0 or 1, there is a non-

empty subset $A(s)$ of X which satisfies the following conditions:
(i) If s is a sequence of n terms (each of which is 0 or 1) and s', s'' are the sequences of $n + 1$ terms (each of which is 0 or 1) whose first n terms are the same as those of s, then $A(s) = A(s') \cup A(s'')$; and if s_0 is the empty sequence, $A(s_0) = X$.
(ii) For each infinite sequence $(u_n)_{n \geqslant 0}$ whose terms are equal to 0 or 1, if s_n denotes the finite sequence $(u_k)_{0 \leqslant k \leqslant n}$, then the filter base formed by the $A(s_n)$ converges to a point of X.

Show that under these conditions there is a *surjective* continuous mapping of Cantor's triadic set K onto X.

¶ 12) Deduce from Exercise 11 that:
a) If A is a compact subset of **R**, there is a continuous mapping of Cantor's triadic set K onto A [take the $A(s)$ to be the intersections of A with suitably chosen intervals].
b) If also A is perfect and totally disconnected, it is homeomorphic to K [same method, by arranging that if s_1 and s_2 are two distinct sequences with the same number of terms (equal to 0 or 1) then

$$A(s_1) \cap A(s_2) = \varnothing_1].$$

¶ 13) Let X be a *countable* Hausdorff space. Suppose that, for each finite sequence s all of whose terms are 0 or 1, there is a non-empty subset $B(s)$ of X which satisfies the following conditions:
(i) If s is a finite sequence of n terms (equal to 0 or 1) and s', s'' are the sequences of $n + 1$ terms (equal to 0 or 1) whose first n terms are the same as those of s, then $B(s') \cup B(s'') = B(s)$ and $B(s') \cap B(s'') = \varnothing$; and if s_0 is the empty sequence, $B(s_0) = X$.
(ii) For each $x \in X$, if s_n is the (unique) sequence of n terms equal to 0 or 1 such that $x \in B(s_n)$, then the filter base formed by the $B(s_n)$ converges to x in X.

Show that, under these conditions, E is *homeomorphic to the rational line* **Q**. [Show first that X is homeomorphic to a countable subset of Cantor's triadic set K which is dense in K and contains no end-point of an interval contiguous to K. In order to do this, remark that the hypotheses associate with each $x \in X$ an infinite sequence $(u_n(x))$ whose terms are equal to 0 or 1; using the fact that X is countable, show that the bijection $s \to B(s)$ can be modified in such a way that for no $x \in E$ do the $u_n(x)$ form a stationary sequence; finally use Exercise 10 of § 2].

Deduce that every countable subspace of **R** which has no isolated points is homeomorphic to **Q**.

14) Show that every closed subset of **R** is either countable or has the power of the continuum [use Exercise 12 *b*) above and Exercise 16 of Chapter I, § 9].

15) Show that the set of open subsets of **R** and the set of compact, totally disconnected, perfect subsets of **R** have the power of the continuum.

16) *a*) Let A be a countable closed subset of **R** and let f be a continuous real-valued function defined on **R**. Show that if f is constant in each of the intervals contiguous to A, then f is constant (use Bolzano's theorem).

b) If f is the continuous mapping of Cantor's triadic set K onto $[0, 1]$ defined in Exercise 9, show that f can be extended to a continuous function on **R** which is constant on each of the intervals contiguous to K.

17) Let X be a topological space which has a countable dense subset. Show that the set of all real-valued continuous functions on X has the power of the continuum.

HISTORICAL NOTE

(Numbers in brackets refer to the bibliography at the end of this note.)

Every measurement of quantities implies a vague notion of real numbers (we shall see the exact reasons for this in Chapter V, § 2). From the mathematical point of view, the origins of the theory of real numbers can be traced back to the progressive formation by the Babylonians of a system of numeration which was (in principle) capable of representing arbitrarily close approximations to any real number [1]. The possession of such a system, and the confidence in numerical calculation which naturally resulted from it, inevitably led to a " naïve " notion of real number which differs hardly at all from that which is current today (linked with the decimal system of numeration) in elementary education and among physicists and engineers. This notion cannot be precisely defined, but can be expressed by saying that a number is regarded as defined by the possibility of finding approximations to it and using these approximations in calculation; this necessarily implies a certain amount of confusion between measures of physical quantities, which of course are not susceptible to an infinite series of successively closer and closer approximation, and " numbers " such as $\sqrt{2}$ (assuming that one is in possession of an algorithm which would make possible an infinite series of successively closer and closer approximation of such numbers).

A similar " pragmatic " attitude appears in all mathematical schools in which expertise in calculation is more important than rigour and theory. The latter, however, were predominant in Greek mathematics; and it is to the Greeks that we owe the first rigorous and coherent theory of ratios of magnitudes, that is, essentially, of real numbers. This theory was the culmination of a series of discoveries about proportions and, in particular, incommensurable ratios, whose importance in the history of Greek thought can hardly be exaggerated, but which in the absence of accurate texts can be discerned only in outline. Greek mathematics in its early stages was inextricably bound up with speculations, part scientific and part philosophical and mystical, about proportion, similitude and ratio, especially

" simple ratios " (expressible by fractions with small numerators and denominators); and one of the characteristic tendencies of the Pythagorean school was to attempt to explain all in terms of integers and ratios of integers. But it was the Pythagorean school, in fact, which discovered that the diagonal of a square is incommensurable with its side (in other words, that $\sqrt{2}$ is irrational). This is without doubt the first example of a proof of impossibility in mathematics, and the mere fact of posing such a question implies a clear distinction between a ratio and approximations to it, and indicates the immense gap which separates the Greek mathematicians from their predecessors (*).

We know little about the movement of ideas which accompanied and followed this important discovery (**). We shall give only a brief summary of the main ideas which lie at the base of the theory of ratios of magnitudes, which was constructed by the great mathematician Eudoxus (a contemporary and friend of Plato), definitively adopted by classical Greek mathematics, and is known to us through Euclid's *Elements* [2] where it is described in a masterly fashion (in Book V of the *Elements*) :

1) The word and the idea of *number* are strictly reserved to natural integers > 1 (1 is the monad and not, strictly speaking, a number), to the exclusion not only of our irrational numbers but also of what we call rational numbers : to the Greek mathematicians of the classical period the latter are ratios of numbers. There is much more here than a simple question of terminology : the word " number " was for the Greeks (and for the moderns up to a recent time) linked with the idea of a *system with*

(*) The discovery that $\sqrt{2}$ is irrational is attributed by some to Pythagoras himself but, it seems, without sufficient authority; by others, to some Pythagorean of the fifth century B.C. On the authority of Plato in his *Theaetetus*, Theodorus of Cyrene is credited with the proof of the irrationality of $\sqrt{3}$, $\sqrt{5}$, " *and so on up to* $\sqrt{17}$ ", following which Theaetetus appears either to have obtained a general proof for $\sqrt{N}$ (N being any integer which is not a perfect square) or at least (if, as may have been the case, Theodorus' proof was general in principle) to have gone on to a classification of certain types of irrational number. We do not know whether these first proofs of irrationality were arithmetic or geometric ; on this point see G. H. Hardy and E. M. Wright, *An Introduction to the Theory of Numbers*, Oxford, 1938, Chapter IV; cf. also Sir Thomas Heath, *A History of Greek Mathematics*, 2 volumes, Oxford, 1921; H. Vogt, Entdeckung des Irrationalen..., *Bibliotheca Mathematica* (3) **10** (1909), p. 97, and H. Hasse and H. Scholz, *Die Grundlagenkrise der griechischen Mathematik*, (Pan-Verlag), 1928.

(**) On this subject consult in particular the articles by O. Becker and those by O. Toeplitz in *Quellen und Studien zur Geschichte der Mathematik* (Abt. B, Studien), vols. 1, 2, 3, Berlin (Springer), 1931-36; also the works cited in the preceding note, and B. L. van der Waerden, Zenon und die Grundlagenkrise..., *Math. Ann.*, **117** (1940), p. 141.

two laws of composition (addition and multiplication); ratios of integers were regarded by the classical Greek mathematicians as operators, defined on the set of integers or on some subset of this set [the ratio of p to q is the operator which, applied to N, *if* N *is a multiple of* q, gives the integer $p.(N/q)$], and forming a multiplicative group but not a system with two laws of composition. In this the Greek mathematicians separated themselves voluntarily from the "logisticians" or professional calculators who, like their Egyptian and Babylonian predecessors, had no scruples about treating fractions as if they were numbers, or adding a fraction to an integer. It seems moreover that this self-imposed restriction on the concept of number came from philosophical rather than mathematical motives, and followed the reflections of the first Greek thinkers on the unit and the multiple; the unit (in this system of thought) being incapable of subdivision without thereby losing its character of unit (*).

2) The theory of magnitudes is based on axioms, which applied simultaneously to all types of magnitudes (there are allusions to earlier theories which apparently treated lengths, areas, volumes, times, etc., all separately). Magnitudes of the same type are characterized by the facts that they can be compared (that is to say, it is assumed that equality, which is an equivalence relation, and the relations $>$ and $<$ are defined), that they can be added and subtracted ($A + B$ is defined, and so is $A - B$ if $A > B$) and that they satisfy "Archimedes' axiom" (Theorem 1 of § 2). It is clearly realized from the beginning that this latter fact is the keystone of the whole edifice (it is in fact indispensable in any axiomatic characterization of real numbers; cf. Chapter V, § 2). Its attribution to Archimedes is purely accidental: in the introduction to his "Quadrature of the Parabola" [3], Archimedes emphasizes that this axiom was used by his predecessors, that it played an essential part in the work of Eudoxus, and that its consequences were no less certainly established than determinations of areas and volumes performed without its help (**).

We shall see in Chapter V, § 2, how the theory of real numbers is a necessary consequence of this axiomatic foundation. Note that, for Eudoxus, the magnitudes of a given type form a system with *one* internal

(*) Plato (*Republic*, Book VII, 525 *e*) mocks calculators "who divide up the unit into small change" and tells us that, where calculators divide, philosophers multiply; in mathematical language this means, for example, that the equality of two ratios a/b and c/d is proved, not by dividing a by b and c by d, which in general leads to calculations of fractions (this is how the Egyptians or the Babylonians would have gone about it), but by verifying that $a.d = b.c$; and other similar facts.

(**) This is clearly an allusion to polemics which have not come down to us; it reminds one of a modern mathematician speaking of Zermelo's axiom.

law of composition (addition), but that this system has an external law of composition whose operators are *ratios of magnitudes*, conceived of as forming an abelian *multiplicative group*. If A and A′ are magnitudes of the same type, and if B and B′ are magnitudes of the same type, then the *ratios* of A to A′ and B to B′ are *defined* to be equal if, for all integers m and m', the relation $mA < m'A'$ implies $mB < m'B'$, and $mA > m'A'$ implies $mB > m'B'$. Inequalities between ratios are defined by similar methods. That these ratios form a *domain of operators* for every type of magnitude is equivalent to the axiom (not explicitly stated but frequently used in Euclid's exposition) of the existence of the fourth proportional: if a ratio A/A' is given, and if B is given, then there exists a B′, of the same type as B, such that $B/B' = A/A'$. Thus Eudoxus' brilliant idea allowed him to identify all the domains of operators defined by different types of magnitude (*); in a similar way, the set of ratios of integers (see above) can be identified with a *subset* of the set of ratios of magnitudes, namely the set of rational ratios (or ratios of commensurable magnitudes); nevertheless, since these ratios, regarded as operators of the integers, are (in general) defined only on a subset of the set of integers, it was necessary to develop their theory separately (Book VII of Euclid).

The universal domain of operators thus constructed was the equivalent, for the Greek mathematicians, of what the set of real numbers is for us; moreover it is clear that, with *addition* of magnitudes and *multiplication* of ratios of magnitudes, they possessed the equivalent of what the *field* of real numbers is for us, although in a much less manageable form (**). On the other hand, one may ask whether they regarded these sets (the set of magnitudes of a given type, or the set of ratios of magnitudes) as *complete*

(*) It thus imparts complete rigour to the work of the early Greek mathematicians, who considered a theorem about proportions as established once they were able to prove it for all rational ratios. Apparently before the time of Eudoxus there had been attempts to construct a theory which would have attained the same object by defining the ratio A/A' of two magnitudes by means of what in modern language would be called the terms of the continued fraction which expresses the ratio; these attempts were natural outgrowths of the algorithm (named after Euclid) for finding a common measure of A and A′ if one exists (or for the determination of the highest common factor). Cf. the articles of O. Becker cited above [note (**) on p. 407).

(**) So unwieldy that in order to translate the algebraic science of the Babylonians into their language, the Greek mathematicians were obliged to use systematically a means of quite a different order, namely the correspondence between two *lengths* and the *area* of the rectangle constructed on these two lengths as sides; this is not, strictly speaking, a law of composition, and does not allow convenient expression of algebraic relations of degree higher than two.

It should be observed that throughout this Note we disregard the question of negative numbers.

in our sense. It is not clear, otherwise, why they should have assumed the existence of the fourth proportional (without even perceiving the need to make it an axiom); moreover, some texts seem to refer to ideas of this nature; and they certainly assumed as self-evident the fact that a curve which can be described by continuous motion cannot pass from one side of a line to the other without cutting the line, a principle which they used, for example, in their investigations on the duplication of the cube (construction of $\sqrt[3]{2}$ by intersections of curves) and which is essentially equivalent to the property in question; nevertheless, the texts which have come down to us do not enable us to discern their ideas on this point with complete accuracy.

Such was the state of the theory of real numbers in the classical period of Greek mathematics. Admirable though Eudoxus' construction was, and leaving nothing to be desired in rigour or coherence, nevertheless it must be admitted that it lacked flexibility and did not encourage the development of numerical calculation, still less the development of algebraic calculation. Moreover its logical necessity could not be apparent except to those in love with rigour and familiar with abstraction; thus it is natural that, with the decline of Greek mathematics, the "naïve" point of view, which had been preserved through the tradition of the logisticians, should gradually re-emerge. This point of view is dominant for example in Diophantus [4], who in truth was an upholder of this tradition rather than of official Greek science; he perfunctorily reproduces the Euclidean definition of number, but in reality he uses the word "number" to mean the unknown in algebraic problems whose solution may be either an integer, or a fraction, or an irrational number (*). Although this change of attitude on the subject of number is connected with one of the most important advances in the history of mathematics, namely the development of algebra, it does not of course constitute an advance in itself, but rather a retreat.

We cannot trace here the vicissitudes of the concept of number through Hindu, Arab and western mathematics up to the end of the Middle Ages. The "naïve" notion of number predominated, and although the *Elements* of Euclid served as a basis for the teaching of mathematics during this period, it is most likely that the doctrine of Eudoxus remained generally uncomprehended because the need for it was no longer appreciated. The "ratios" of Euclid were customarily described as "numbers", and the rules for calculating with integers were applied to them without any attempt to analyse the reasons for the success of these methods.

(*) "*The 'number' is therefore irrational*", Diophantus, Book IV, Problem IX. Concerning this return to the naïve notion of number, cf. also Eutocius, in his Commentary on Archimedes ([3], Vol. 3, p. 120-126 of 2nd edition = pp. 140-148 of 1st edition).

Nevertheless we see R. Bombelli, as early as the middle of the 16th century, expounding a point of view on this subject, in his *Algebra* [5] (*), which is essentially correct (provided that the results of Book V of Euclid are assumed to be known); having realized that once the unit of length has been chosen there is a one-to-one correspondence between lengths and ratios of magnitudes, he defines the various algebraic operations *on lengths* (assuming of course that the unit has been fixed) and, representing numbers by lengths, obtains the geometrical definition of the field of real numbers (a point of view which is usually credited to Descartes) and thus gives his algebra a solid geometrical foundation (**).

But Bombelli's *Algebra*, though singularly advanced for its time, did not go beyond the extraction of radicals and the solution by radicals of equations of the second, third and fourth degrees; of course the possibility of extraction of radicals is assumed without any discussion. Simon Stevin [6] adopts an analogous viewpoint on the subject of numbers; for him a number denotes a measure of a magnitude and is regarded as being essentially "continuous" (without giving a precise meaning to this word). If he distinguishes between "geometric numbers" and "arithmetic numbers", it is only because of the accident of their mode of definition and not because of a difference in nature. His last words on this subject run as follows: "*We conclude therefore that there are no absurd, irrational, irregular, inexplicable or surd numbers, but that there is in them such excellence and concordance that we have matter for meditation night and day on their admirable perfection*" ([6], p. 10). On the other hand he was the first to use decimal fractions as a method of calculation and proposed a notation for them which is very close to ours; and he saw clearly that these fractions provide an algorithm for indefinitely close approximation to any real number, as is shown by his *Appendice algebraique* of 1594, "*containing a general rule for all equations*" (the only known copy of this pamphlet was burnt at Louvain in 1914; but see [6], vol. 1, p. 88). Such an equation being written in the form $P(x) = Q(x)$ [where the degree of the polynomial P is greater than the degree of the polynomial Q, and $P(0) < Q(0)$], he substitutes for x the numbers 10, 100, 1 000, ... until he finds $P(x) > Q(x)$ which, he says, determines the number of digits in the root; then (if for example the root has two digits) he substitutes 10, 20, ... which determines the

(*) We are concerned here with Book IV of his *Algebra*, which remained unpublished until modern times; for our purposes it matters little whether or not the ideas of Bombelli on this subject were known to his contemporaries.

(**) We do not enter here into the history of the use of negative numbers, which is within the framework of algebra. Let us nevertheless note that Bombelli, in the same context, gives with perfect clarity the purely formal definition (such as one would find in a modern algebra) not only of negative numbers, but also of complex numbers.

number of tens; then similarly for the number of units, then for the successive decimal digits: "*And proceeding indefinitely in this way*", he says, "*we approach infinitely near to the number required*" ([6], p. 88). As we see, Stevin had exactly the idea of Theorem 2 of § 6 (and was without doubt the first to have it) and recognized that this theorem was the essential tool for the systematic solution of numerical equations; at the same time we can see that he had so clear an intuitive conception of the numerical continuum that little remained to be done to make it definitively precise.

Nevertheless, in the following two centuries the definitive establishment of correct methods was twice retarded by the development of two theories whose history does not come within the framework of this note: the infinitesimal calculus and the theory of series. Throughout the discussions they gave rise to, we perceive, as in all periods of the history of mathematics, the perpetual balance between those who sought to push forward, at the cost of some insecurity, convinced that there would always be time later to consolidate the conquered terrain, and those of a critical mind who (without necessarily being in any way inferior in point of intuitive gifts and inventive genius) believed that their energies were not wasted when they devoted their effort to the precise formulation and rigorous justification of their conceptions. In the seventeenth century the main subject of debate was the notion of "infinitely" small which, though justified *a posteriori* by the results which were obtained with its help, seemed to be in open opposition to the axiom of Archimedes; and we see the most enlightened minds of this period finally adopting a point of view which differed little from that of Bombelli, and which is distinguished above all by the greater attention it paid to the rigorous methods of the ancients. Isaac Barrow (Newton's teacher, who himself played an important part in the creation of the infinitesimal calculus) gave a brilliant exposition of this viewpoint in his Mathematical Lectures given at Cambridge in 1664-5-6 [7]; recognizing the need to return to the theory of Eudoxus in order to regain the proverbial "geometrical certainty" in the subject of number, he presents at length an extremely judicious defence of Eudoxus' theory (which, on Barrow's evidence, had remained unintelligible to many of his contemporaries) against those who charged it with obscurity or even absurdity. On the other hand, defining numbers to be symbols which denote ratios of magnitudes and to be capable of being combined by the operations of arithmetic, Barrow obtains the field of real numbers in terms which Newton took up again in his *Arithmetic*, and which his successors up to Dedekind and Cantor did not change.

But it was in this period that the method of expansion in series was introduced; this rapidly took on an exclusively formal character in the hands of impenitent algebraists and deflected the attention of mathematicians from the questions of convergence which are essential to any sound use of series in the domain of real numbers. Newton, the principal creator of the method, was certainly aware of the necessity of considering these questions; and if he did not elucidate them sufficiently, he at least realized that the power series which he introduced converged "usually" at least as well as a geometric series (whose convergence was already known to the ancients) for small values of the variable [8]. At about the same time, Leibniz had observed that an alternating series whose terms decrease in absolute value and tend to 0 is convergent. In the following century, d'Alembert in 1768 raised doubts about the use of non-convergent series; but the authority of the Bernoullis and above all Euler was such as to make doubts of this nature exceptional at this period.

It is clear that mathematicians who were in the habit of using series for numerical calculations would never have so neglected the notion of convergence; and it is no accident that the first to lead the return to correct methods, in this domain as in many others, was a mathematician who in his early youth was in love with numerical calculation. C.F. Gauss, who, when still almost a child, had practised the algorithm of the arithmetico-geometric mean (*), could scarcely fail to form a clear notion of the limit; and in a fragment dated 1800 (but first published in our times) ([9], Vol. X¹, p. 390) he gives precise definitions of the least upper bound and the greatest lower bound, and the upper and lower limits of a sequence of real numbers; the existence of the bounds (for a bounded sequence) he seems to have assumed as obvious, and the upper and lower limits he defines correctly as the limits, as n tends to $+\infty$, of $\sup_{p\geqslant 0} u_{n+p}$ and $\inf_{p\geqslant 0} u_{n+p}$ respectively. On the other hand, in his memoir of 1812 on the hypergeometric series ([9], vol. III, p. 139), Gauss gives the first model of a discussion of convergence conducted, as he says, "*in full rigour, and made to satisfy those whose preferences are for the rigorous methods of the ancient geometers*". It is true that this discussion, which occupies a secondary place in the memoir, does not go back to the first principles of the theory of series; Cauchy, in his *Cours d'Analyse* of 1821 [10], was the first to establish these in a manner correct in every point, starting from Cauchy's criterion (clearly stated, and assumed as self-evident). Since he takes the point of view of Barrow and Newton on the definition of number, it can be said that for Cauchy the real numbers are defined by the axioms of magnitudes and Cauchy's criterion; and this is in fact enough to define them (see Chapter V, § 2).

(*) If x_0, y_0 are given and > 0, let $x_{n+1} = (x_n + y_n)/2$ and

$$y_{n+1} = \sqrt{x_n y_n};$$

as n tends to $+\infty$, x_n and y_n tend (very rapidly) to a common limit, called the arithmetico-geometric mean of x_0 and y_0 (§ 5, Exercise 16); this function is closely related to the elliptic functions and was the starting-point of Gauss's important work on this subject.

At the same moment another important aspect of the theory of real numbers was definitively cleared up. As we have said, it had always been assumed as geometrically obvious that two continuous curves cannot cross each other without intersecting — a principle which (suitably made precise) is again equivalent to the completeness of the real line (as a uniform space). This principle is at the base of the "rigorous" proof, given by Gauss in 1799, of d'Alembert's theorem, according to which every polynomial with real coefficients has a real or complex root ([9], vol. III, p. 1); the proof of the same theorem given by Gauss in 1815 ([9], vol. III, p. 31) depends, as does an earlier attempt by Lagrange, on the analogous but simpler principle that a polynomial cannot change sign without vanishing; this is a particular case of Theorem 2 of § 6, which we have already seen used by Stevin. In 1817 Bolzano gave a complete proof, founded on Cauchy's criterion, of this latter principle, which he obtains as a particular case of the analogous theorem for continuous real-valued functions of a real variable [11]. He enunciates "Cauchy's criterion" clearly (and before Cauchy did), and seeks to justify it by an argument which, in the absence of any arithmetic definition of real numbers, was only and could only be a vicious circle; but, once this point has been accepted, his work is entirely correct and most remarkable, since it contains not only the modern definition of a continuous function (given here for the first time), with the proof of the continuity of polynomials, but also the proof of the existence of the greatest lower bound of an *arbitrary* bounded set of real numbers (he speaks not of sets, but of properties of real numbers, which comes to the same thing). Cauchy in his *Cours d'Analyse* [10] also defined continuous functions of one or more real variables and proved by, the same reasoning as Simon Stevin used, that a continuous function of one variable cannot change sign without vanishing. This reasoning becomes correct, once continuity has been defined, if Cauchy's criterion is invoked (or if one assumes, as Cauchy does at this point, the equivalent principle of "nested intervals"; the convergence of decimal fractions is of course only a particular case of this principle).

Once arrived at this point, there remained for the mathematicians only the task of making precise and extending the results obtained, by correcting various errors and filling various gaps. For example, Cauchy had believed at one time that a convergent series, whose terms are continuous functions of one variable, has a continuous function as its sum. Abel's rectification of this point, in the course of his important work on series ([12], vol. 1, p. 219; cf. also vol. 2, p. 257 and *passim*) led finally to the elucidation by Weierstrass of the concept of uniform convergence (in his lectures, which remained unpublished but which had a considerable influence; see the Historical Note to Chapter X). Again, Cauchy had assumed, without sufficient justification, the existence of the minimum of a continuous function in one of his proofs of the existence of roots of a polynomial; and again it was Weierstrass who threw light on questions of this nature by proving (in his lectures) Theorem 1 of § 6 for functions of real variables, defined on bounded closed intervals. Following his criticism of unjustified applications of this theorem to sets of functions ("Dirichlet's principle " is the best-known example) there began the movement of ideas which led, as we have seen in the Historical Note to Chapter I, to the general definition of compact spaces and the modern statement of the theorem as we have given it.

At the same time Weierstrass, in his lectures, had perceived the logical importance in making the idea of real number entirely independent of the theory of magnitudes; the latter is effectively equivalent to an axiomatic definition of the points of the line (and thus of the set of real numbers) and the assumption of the existence of such a set; although this method is essentially correct, it is evidently preferable to start only from the rational numbers, and to construct the real numbers from them by completion (*). This was achieved, by diverse methods and independently of each other, by Weierstrass, Dedekind, Méray and Cantor; while the method of "cuts", proposed by Dedekind [13], came very near to the definitions of Eudoxus, the other methods proposed are close to that which is expounded in this series. Simultaneously Cantor began to develop the theory of sets of real numbers, the idea of which was first conceived by Dedekind (see the Bibliography to Chapter I), and thus obtained the principal elementary results on the topology of the real line, the structure of its open and closed sets, the notion of derived set and of totally disconnected perfect set, etc.; he also obtained Theorem 1 of § 8 on the power of the continuum, and deduced from it that the continuum is uncountable, that the set of transcendental numbers has the power of the continuum, and also (a paradoxical result for its time) that the set of points of a plane (or of space) has the same power as the set of points of a line.

With Cantor, the questions studied in this chapter assumed practically their definitive form. We refer to the Historical Note on Chapter I for the immediate impact of Cantor's work; let us indicate briefly the directions in which it has extended. Apart from leading to work on

(*) The question of the existence, i.e., in modern language, the non-contradiction of the theory of real numbers, is thus brought back to the analogous question for the rational numbers, *provided always that abstract set theory is assumed known* (because completion involves the notion of an arbitrary subset of an infinite set); in other words, everything is reduced to abstract set theory, since the rational numbers can be constructed from this theory (See *Set Theory*, Chapter III, § 4). On the other hand, if we do not assume abstract set theory, it is impossible to reduce the non-contradiction of the theory of real numbers to that of arithmetic, and it becomes necessary to provide an independent axiomatic characterization of the theory of real numbers.

general topology (see Chapter I) and applications to integration, which will be dealt with thoroughly elsewhere in this series, Cantor's work has led to investigations of the structure and classification of sets of points on a line, and of real-valued functions of a real variable. These have their origin in the work of Borel [14] which was directed mainly towards measure theory but which led, among other things, to the definition of "Borel sets", i.e., sets belonging to the smallest family of subsets of **R** which contains all intervals and is closed with respect to union and *countable* intersection and with respect to the operation $\complement$ (cf. Chapter IX, § 6, no. 3). These sets are closely related to the so-called "Borel functions" or "Baire functions", that is to say functions which can be obtained from continuous functions by the operation of taking the limit of a sequence, repeated "transfinitely"; they were defined by Baire in the course of an important work in which he entirely abandoned the measure viewpoint for a systematic investigation of the qualitative and "topological" aspect of these questions [15]; in this context he defined and studied semi-continuous functions (and was the first to do so), and in order to characterize functions which are limits of continuous functions he introduced the important notion of a meagre set (set "of the first category" in Baire's terminology) which we shall study in Chapter IX. As to the many works which have followed Baire, and which are mainly products of the Russian and especially the Polish schools, we can do no more here than draw attention to their existence (see for example [16] and the journal *Fundamentae Mathematicae*).

BIBLIOGRAPHY

[1] O. Neugebauer, *Vorlesungen über Geschichte der antiken Mathematik*, Bd. I : Vorgriechische Mathematik, Berlin (Springer), 1934.

[2] *Euclidis Elementa*, 5 volumes, ed. J. L. Heiberg, Lipsiae (Teubner), 1883-88.

[2a] T. L. Heath, *The Thirteen Books of Euclid's Elements...*, 3 volumes, Cambridge, 1908.

[3] *Archimedis Opera Omnia*, 3 volumes, ed. J. L. Heiberg, 2nd edition, 1913-15.

[3a] *Les Œuvres complètes d'Archimède*, translated by P. Ver Eecke, Paris-Bruxelles (Desclée-de Brouwer), 1921.

[4] *Diophanti Alexandrini Opera Omnia...*, 2 volumes, ed. P. Tannery, Lipsiae (Teubner), 1893-95.

[4a] *Diophante d'Alexandrie*, translated by P. Ver Eecke, Bruges (Desclée de Broower), 1926.

[5] R. Bombelli, *L'Algebra*, ed. E. Bortolotti, Bologna (Zanichelli), 1929.

[6] Les Œuvres Mathématiques de Simon Stevin de Bruges, Ou sont insérées les Mémoires Mathématiques, Esquelles s'est exercé le Très-haut et Très-illustre Prince Maurice de Nassau, Prince d'Aurenge, Gouverneur des Provinces des Païs-Bas unis, General par Mer et par Terre, etc., *Le tout reveu, corrigé et augmenté* par Albert Girard Samielois, Mathématicien, A Leyde, Chez Bonaventure et Abraham Elsevier, Imprimeurs ordinaires de l'Université, Anno MDCXXXIV (= 1634), vol. I.

[7] I. Barrow, *Mathematical Works*, Cambridge (University Press), 1860.

[8] I. Newton, *De Analysi per ecquatione numero terminorum infinitas*, in *Commercium Epistolicum D. Johannis Collins et aliorium de Analysi promota*, Londini, 1712.

[9] C. F. Gauss, *Werke*, vols. III (Göttingen, 1816) and X_1 (*ibid.*, 1917).

[10] A. Cauchy, *Cours d'Analyse de l'École Royale Polytechnique*, 1re partie, 1821 = Œuvres (II) vol. III, Paris (Gauthier-Villars), 1897.

[11] B. Bolzano, *Rein Analytischer Beweis des Lehrsatzes, dass zwischen je zwei Werthen, die ein entgegengesetztes Resultat gewähren, wenigstens eine reelle Wurzel liegt*, Ostwald's Klassiker, no. 153, Leipzig, 1905.

[12] N. H. Abel, *Œuvres*, 2 volumes, ed. Sylow and Lie, Christiania, 1881.

[13] R. Dedekind, *Gesammelte mathematische Werke*, vol. II, Braunschweig (Vieweg), 1932, p. 315.

[14] E. Borel, *Leçons sur la théorie des fonctions*, 2nd edition, Paris (Gauthier-Villars), 1914.

[15] R. Baire, *Leçons sur les fonctions discontinues*, Paris (Gauthier-Villars), 1905.

[16] N. Lusin, *Leçons sur les ensembles analytiques et leurs applications*, Paris (Gauthier-Villars), 1930.

INDEX OF NOTATION

The reference numbers indicate the chapter, section and sub-section (or, exercise) in that order.

P(K, L) (K, L subsets of a space with operators) : III, 4, 5.

$\sum_{\iota \in I} x_\iota$, $\sum_\iota x_\iota$, $\sum x_\iota$: III, 5, 1.

$\prod_{\iota \in I} x_\iota$, $\prod_\iota x_\iota$, $\prod x_\iota$: III, 5, 1.

(x_n) (series, by abuse of language) : III, 5, 6.

$\mathop{\mathsf{S}}\limits_{n=0}^{\infty} x_n$, $x_0 + x_1 + \cdots + x_n + \ldots$, $\sum_{n=0}^{\infty} x_n$: III, 5, 6.

$\mathop{\mathsf{P}}\limits_{n=0}^{\infty} x_n$, $\prod_{n=0}^{\infty} x_n$: III, 5, 7.

K^* (K a division ring) : III, 6, 7.

$\mathbf{R}$, $\mathbf{R}^*$, $\mathbf{R}_+$, $\mathbf{R}_*^+$: IV, 1, 3.

$\sqrt[m]{x}$, $\sqrt{x}$, $x^{1/m}$ (x real and $\geqslant 0$) : IV, 3, 3.

$\overline{\mathbf{R}}$, $+\infty$, $-\infty$: IV, 4, 2.

sup A, inf A (A a subset of $\overline{\mathbf{R}}$) : IV, 4, 2.

$\overline{\mathbf{R}}^*$: IV, 4, 3.

$\lim_{x \to a,\, x > a,\, x \in A} f(x)$, $\lim_{x \to a,\, x < a,\, x \in A} f(x)$: IV, 5, 3.

$f(a-)$, $f(a+)$: IV, 5, 3.

$\sup_{x \in A} f(x)$, $\inf_{x \in A} f(x)$ (f a real-valued function) : IV, 5, 4.

$\sup_{\iota \in I} f_\iota$, $\sup_\iota f_\iota$, $\inf_{\iota \in I} f_\iota$, $\inf_\iota f_\iota$ (f_ι real-valued functions) : IV, 5, 5.

$\limsup_{\mathfrak{G}} f$, $\liminf_{\mathfrak{G}} f$, $\limsup_{x,\mathfrak{G}} f(x)$, $\liminf_{x,\mathfrak{G}} f(x)$: IV, 5, 6.

$\limsup f$, $\liminf f$, $\limsup_x f(x)$, $\liminf_x f(x)$: IV, 5, 6.

$\limsup_{x \to a} f(x)$, $\liminf_{x \to a} f(x)$: IV, 5, 6.

$\limsup_{x \to a,\, x \in X} f(x)$, $\liminf_{x \to a,\, x \in X} f(x)$, $\limsup_{x \to a,\, x \neq a} f(x)$, $\liminf_{x \to a,\, x \neq a} f(x)$: IV, 5, 6.

$\limsup_{n \to \infty} u_n$, $\liminf_{n \to \infty} u_n$ [(u_n) a sequence of numbers] : IV, 5, 6.

$\limsup_{n \to \infty} f_n$, $\liminf_{n \to \infty} f_n$ [(f_n) a sequence of functions] : IV, 5, 6.

$f + g$, fg, $1/f$ (f, g functions with values in $\overline{\mathbf{R}}$) : IV, 5, 7.

$[x]$ (x a real number) : IV, 8, 2.

INDEX OF TERMINOLOGY

The reference numbers indicate the chapter, section and sub-section (or exercise) in that order.